INTRODUCTION TO PLASMA PHYSICS

With Space, Laboratory and Astrophysical Applications

Introducing the basic principles of plasma physics and their applications to space, laboratory, and astrophysical plasmas, this new edition provides updated material throughout. Topics covered include single-particle motions, kinetic theory, magnetohydrodynamics, small-amplitude waves in hot and cold plasmas, and collisional effects. New additions include the ponderomotive force, tearing instabilities in resistive plasmas, the magnetorotational instability in accretion disks, charged particle acceleration by shocks, and a more in-depth look at nonlinear phenomena. A broad range of applications are also explored: planetary magnetospheres and radiation belts, the confinement and stability of plasmas in fusion devices, the propagation of discontinuities and shock waves in the solar wind, and analysis of various types of plasma waves and instabilities that can occur in planetary magnetospheres and laboratory plasma devices. With step-by-step derivations and self-contained introductions to mathematical methods, this book is ideal as an advanced undergraduate to graduate-level textbook, or as a reference for researchers.

DONALD A. GURNETT is a pioneer in the study of waves in space plasmas, and has been active in teaching plasma physics and conducting experimental space physics research for over fifty years. He is currently the James A. Van Allen/R. J. Carver Professor of Physics at the University of Iowa and has received numerous awards for both his teaching and research. In 1994 he received the Iowa Board of Regents Award for Faculty Excellence, and in 1998 was elected a member of the National Academy of Sciences.

AMITAVA BHATTACHARJEE is a leading theoretical plasma physicist and has contributed to a wide range of subjects spanning fusion, space, and astrophysical plasma physics. He is currently a Professor of Astrophysical Sciences at Princeton University, and Head of the Princeton Plasma Physics Laboratory Theory Department. He is a Fellow of the American Physical Society, the American Association for the Advancement of Science, and the American Geophysical Union.

INTRODUCTION TO PLASMA PHYSICS

With Space, Laboratory and Astrophysical Applications

DONALD A. GURNETT
University of Iowa

AMITAVA BHATTACHARJEE
Princeton University, New Jersey

CAMBRIDGE
UNIVERSITY PRESS

CAMBRIDGE
UNIVERSITY PRESS

University Printing House, Cambridge CB2 8BS, United Kingdom
One Liberty Plaza, 20th Floor, New York, NY 10006, USA
477 Williamstown Road, Port Melbourne, VIC 3207, Australia
4843/24, 2nd Floor, Ansari Road, Daryaganj, Delhi – 110002, India
79 Anson Road, #06-04/06, Singapore 079906

Cambridge University Press is part of the University of Cambridge.

It furthers the University's mission by disseminating knowledge in the pursuit of education, learning, and research at the highest international levels of excellence.

www.cambridge.org
Information on this title: www.cambridge.org/9781107027374
10.1017/9781139226059

First edition published 2005
Second edition published 2017

A catalogue record for this publication is available from the British Library.

Library of Congress Cataloging-in-Publication Data
Names: Gurnett, Donald A., author. | Bhattacharjee, A. (Amitava), 1955– author.
Title: Introduction to plasma physics : with space, laboratory and
astrophysical applications / Donald A. Gurnett (University of Iowa) and
Amitava Bhattacharjee (Princeton University, New Jersey).
Description: Second edition. | Cambridge, United Kingdom ; New York, NY :
Cambridge University Press, [2016] | Includes bibliographical references and index.
Identifiers: LCCN 2016028027| ISBN 9781107027374 (hardback ; alk. paper) |
Subjects: LCSH: Plasma (Ionized gases) | Space plasmas.
Classification: LCC QC718.G87 2016 | DDC 530.4/4–dc23 LC record available at
https://lccn.loc.gov/2016028027

ISBN 978-1-107-02737-4 Hardback

Dedicated to Marie, my loving wife for over fifty years, who has enthusiastically supported my various endeavors, especially this book.

Don Gurnett

And, to Melissa, without whose love and support this book and so much else would not be possible.

Amitava

Contents

Preface

This textbook is intended for a full year introductory course in plasma physics at the senior undergraduate or first-year graduate level. It is based on lecture notes from courses taught by the authors for more than three decades at the University of Iowa, Columbia University, University of New Hampshire, and Princeton University. During these years, plasma physics has grown increasingly interdisciplinary, and there is a growing realization that diverse applications in laboratory, space, and astrophysical plasmas can be viewed from a common perspective. Since the students who take a course in plasma physics often have a wide range of interests, typically involving some combination of laboratory, space, and astrophysical plasmas, a special effort has been made to discuss applications from these areas of research. The emphasis of the book is on physical principles, less so on mathematical sophistication. An effort has been made to show all relevant steps in the derivations, and to match the level of presentation to the knowledge of students at the advanced undergraduate and early graduate level. The main requirements for students taking this course are that they have taken an advanced undergraduate course in electricity and magnetism and that they are knowledgeable about using the basic principles of vector calculus, i.e., gradient, divergence, and curl, and the various identities involving these vector operators. Although extensive use is made of complex variables, no special background is required in this subject beyond what is covered in an advanced calculus course. Relatively advanced mathematical concepts that are not typically covered in an undergraduate sequence, such as Fourier transforms, Laplace transforms, the Cauchy integral theorem, and the residue theorem, are discussed in sufficient detail that no additional preparation is required. Although this approach has undoubtedly added to the length of the book, we believe that the material covered provides an effective and self-contained textbook for teaching plasma physics. MKS units are used throughout. Problem solutions are available to instructors at www.cambridge.org/9781107027374

For the preparation of this text we would especially like to thank Kathy Kurth who did the typing and steadfastly stuck with us through the many revisions and additions that occurred over the years. We would also like to thank Joyce Chrisinger and Ann Persoon for their outstanding work preparing the illustrations and proofreading, Mr. Feng Chu, Dr. Manasvi Lingam, and Dr. Chung-Sang Ng for checking the accuracy of the equations, and Dr. Robert Decker, Dr. Yi-Min Huang, and Dr. Roscoe White for providing key illustrations. We would also like to thank Professors Iver Cairns, Len Fisk, Paul Kellogg, and Ondřej Santolík for their comments on portions of the manuscript. Don Gurnett would like to acknowledge the salary support provided by the University of Iowa and the Carver Foundation during the preparation of this manuscript, and Amitava Bhattacharjee would like to acknowledge the generous support of Princeton University.

1

Introduction

A plasma is an ionized gas consisting of positively and negatively charged particles with approximately equal charge densities. Plasmas can be produced by heating an ordinary gas to such a high temperature that the random kinetic energy of the molecules exceeds the ionization energy. Collisions then strip some of the electrons from the atoms, forming a mixture of electrons and ions. Because the ionization process starts at a fairly well-defined temperature, usually a few thousand K, a plasma is often referred to as the "fourth" state of matter. Plasmas can also be produced by exposing an ordinary gas to energetic photons, such as ultraviolet light or X-rays. The steady-state ionization density depends on a balance between ionization and recombination. In order to maintain a high degree of ionization, either the ionization source must be very strong, or the plasma must be very tenuous so that the recombination rate is low.

The definition of a plasma requires that any deviation from charge neutrality must be very small. For simplicity, unless stated otherwise, we will assume that the ions are singly charged. The charge neutrality condition is then equivalent to requiring that the electron and ion number densities be approximately the same. In the absence of a loss mechanism, the overall charge neutrality assumption is usually satisfied because all ionization processes produce equal amounts of positive and negative charge. However, deviations from *local* charge neutrality can occur. Usually these deviations are small, since as soon as a charge imbalance develops, large electric fields are produced that act to restore charge neutrality. Systems that display large deviations from charge neutrality, such as vacuum tubes and various electronic devices, are not plasmas, even though some aspects of their physics are similar.

In the most common type of plasma, the charged particles are in an unbound gaseous state. This requirement can be made more specific by requiring that the random kinetic energy be much greater than the average electrostatic energy, and is imposed to provide a distinction between a plasma, in which the particles move relatively freely, and condensed matter, such as metals, where electrostatic forces play a dominant role. In such a plasma, long-range electrical forces are much more important than short-range forces. Because many particles "feel" the same long-range forces, a plasma is dominated by "collective" motions involving correlated movements of large numbers of particles rather than uncorrelated interactions between neighboring particles. Long-range forces lead to many complex effects that do not occur in an ordinary gas.

Plasmas can be divided into two broad categories: natural and man-made. It is an interesting fact that most of the material in the visible universe, over 99% according to some estimates, is in the plasma state. This includes the Sun, most stars, and a significant fraction of the interstellar and intergalactic medium. Thus, plasmas play a major role in the universe. Plasma physics is relevant to the formation of planetary radiation belts, the development of sunspots and solar flares, the acceleration of high velocity winds that flow outward from the Sun and other stars, the generation of radio emissions from the Sun and other astrophysical objects, and the acceleration of cosmic rays.

In Earth's atmosphere, the low temperatures and high pressures that are commonly present are not favorable for the formation of plasmas except under unusual conditions. Probably the most common plasma phenomenon encountered in Earth's atmosphere is lightning. In a lightning discharge the atmospheric gas is ionized and heated to a very high temperature by the electrical currents that are present in the discharge. Because of the high recombination rate, the resulting plasma exists for only a small fraction of a second. Less common is ball lightning, which consists of a small ball of hot luminous plasma that lasts for up to a few tens of seconds. Another terrestrial plasma phenomenon, readily observable at high latitudes, is the aurora, which is produced by energetic electrons and ions striking the atmosphere at altitudes of 80 to 100 km. At higher altitudes, from one hundred to several hundred km, Earth is surrounded by a dense plasma called the ionosphere. The ionospheric plasma is produced by ultraviolet radiation from the Sun, and also exists on the nightside of Earth because the recombination rate is very low at high altitudes. The ionosphere plays an important role in radio communication by acting as a reflector for low-frequency radio waves. At even higher altitudes, Earth is surrounded by a region of magnetized plasma called the magnetosphere. Planetary magnetospheres have now been observed at all the magnetized planets and exhibit many of the plasma processes that are believed to occur at magnetized astronomical objects such as accretion disks and neutron stars.

Numerous applications of basic plasma physics can be found in man-made devices. One of the most important of these is the attempt to achieve controlled thermonuclear fusion. Because fusion requires temperatures of 10^7 K or more to overcome the Coulomb repulsion between nuclei, controlled fusion necessarily involves very high temperatures. Since a fusion plasma would be quickly cooled by the walls of any ordinary container, considerable effort has gone into attempts to contain plasmas by magnetic fields, using a so-called "magnetic bottle." Although the principles of such magnetic confinement may appear at first glance to be straightforward, attempts to achieve controlled fusion using magnetic confinement have been complicated by collective effects that develop when large numbers of particles are introduced into the machine. In another approach, known as "inertial confinement," one attempts to use extremely powerful lasers to compress and heat fusion fuel to very high densities for short periods of time so as to enable self-sustaining fusion reactions. The effort to find a technologically and economically attractive configuration for confining a dense, hot plasma remains one of the main challenges of fusion research. Besides fusion, numerous other devices involving plasmas also exist. Fluorescent lights and various other devices involving plasma discharges, such as electric arc welders and plasma etching machines, are in common daily use. More advanced devices include magnetohydrodynamic generators for producing electricity from high-temperature gas jets, ion engines for spacecraft propulsion, various surface treatment processes that involve the injection of ions into metal surfaces, and high-frequency electronic devices such as traveling wave tubes and magnetrons.

The purpose of this book is to provide the basic principles needed to analyze a broad range of plasma phenomena. Since both natural and man-made plasmas are of potential interest, a special effort has been made in this book to provide examples from space, laboratory, and astrophysical applications.

2

Characteristic Parameters of a Plasma

Before starting with a detailed discussion of the processes that occur in a plasma, it is useful to identify certain fundamental parameters that are relevant to the description of essentially all plasma phenomena.

2.1 Number Density and Temperature

In an ordinary material there are usually three parameters, pressure, density, and temperature, that must be specified to determine the state of the material, any two of which can be selected as the independent variables. A plasma almost always involves considerably more parameters. For a plasma consisting of electrons and various types of ions, it is necessary to define a number density for each species, denoted by n_s, where the subscript s stands for the sth species. Since the electrons and ions respond differently to electromagnetic forces, the number densities of the various species must be regarded as independent variables. In general, a plasma cannot be characterized by a single density. Certain types of plasmas, called non-neutral plasmas (Davidson, 2001), in which the plasma does not have overall charge neutrality, are not within the scope of this book.

The temperature of particles of type s is directly proportional to their average random kinetic energy. In thermal equilibrium, the distribution of velocities for particles of type s is given by the Maxwellian distribution

$$f_s(\mathbf{v}) = n_s \left(\frac{m_s}{2\pi\kappa T_s} \right)^{3/2} \mathrm{e}^{-\frac{m_s v^2}{2\kappa T_s}}, \tag{2.1.1}$$

where $f_s(\mathbf{v})$ is the distribution function, $\mathbf{v}$ is the velocity, v is the magnitude of the velocity, $v^2 = v_x^2 + v_y^2 + v_z^2$, m_s is the mass of the particles, κ is Boltzmann's constant, and T_s is the temperature. The distribution function is normalized such that $f_s(\mathbf{v})$ integrated over all velocities gives the number density of particles of

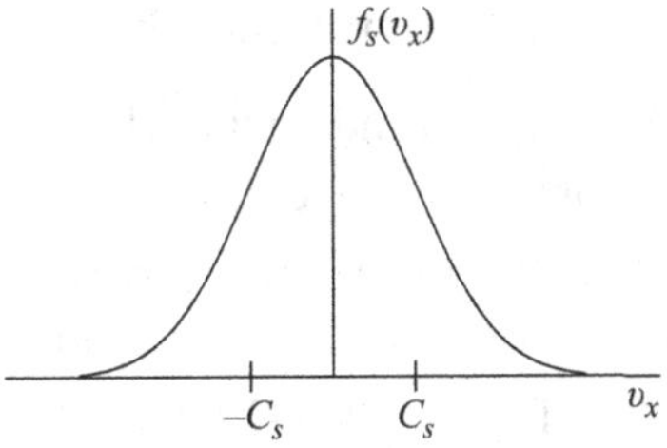

Figure 2.1 The Maxwellian velocity distribution.

type s:

$$\int_{-\infty}^{\infty} f_s(\mathbf{v})\,\mathrm{d}v_x\,\mathrm{d}v_y\,\mathrm{d}v_z = n_s\,. \tag{2.1.2}$$

A plot of the Maxwellian distribution as a function of v_x is shown in Figure 2.1 (for $v_y = v_z = 0$). It is a relatively simple matter to show that the root–mean–square velocity is given by $\sqrt{3}C_s$, where

$$C_s = \sqrt{\frac{\kappa T_s}{m_s}}\,. \tag{2.1.3}$$

Hereafter, C_s will be referred to as the thermal speed. The average kinetic energy is given by

$$\left\langle \frac{1}{2} m_s v^2 \right\rangle = \frac{3}{2}\kappa T_s\,, \tag{2.1.4}$$

where the angle brackets indicate an average. The above equation shows that the temperature is directly proportional to the average kinetic energy of the particles.

According to a general principle of statistical mechanics called the H-theorem, the Maxwellian distribution is the unique distribution function that arises when a gas is in thermal equilibrium (Huang, 1963). For a plasma in thermal equilibrium, not only should the distribution function for each species be a Maxwellian, but the temperature of all species must be equal. However, because collisions occur very infrequently in a tenuous plasma, the approach to thermal equilibrium is often very slow. Therefore, non-equilibrium effects are quite common in plasmas. Since the electron and ion masses are very different, the rate of energy transfer between electrons and ions is much slower than between electrons or between ions. Therefore, when a plasma is heated, substantial temperature differences often develop between the electrons and ions. Non-equilibrium distributions also occur when an electron beam or an ion beam is injected into a plasma. Under these circumstances, the velocity distribution function of the beam usually cannot be represented by a Maxwellian distribution. Such non-thermal distributions produce many interesting effects that will be discussed later.

2.2 Debye Length

All plasmas are characterized by a fundamental length-scale determined by the temperature and number density of the charged particles. To demonstrate the existence of this length-scale, consider what happens when a negative test charge Q is placed in an otherwise homogeneous plasma. Immediately after the charge is introduced, the electrons are repelled and the ions are attracted. Very quickly, the resulting displacement of the electrons and ions produces a polarization charge that acts to shield the plasma from the test charge. This shielding effect is called Debye shielding, after Debye and Hückel (1923) who first studied the effect in dielectric fluids. The characteristic length over which shielding occurs is called the Debye length.

To obtain an expression for the Debye length, it is useful to consider a homogeneous plasma of electrons of number density $n_{\rm e}$ and temperature $T_{\rm e}$, and a fixed background of positive ions of number density n_0. After the negative test charge Q has been inserted and equilibrium has been established, the electrostatic potential Φ is given by Poisson's equation

$$\nabla^2\Phi = -\frac{\rho_{\rm q}}{\epsilon_0} = -\frac{e}{\epsilon_0}(n_0 - n_{\rm e}), \tag{2.2.1}$$

where $\rho_{\rm q}$ is the charge density, ϵ_0 is the permittivity of free space, and e is the electronic charge. To obtain a solution for the electrostatic potential, it is necessary to specify the electron density as a function of the electrostatic potential. We assume that at infinity, where $\Phi = 0$, the electrons have a Maxwellian velocity distribution with a number density n_0. From general principles of kinetic theory it can be shown that the velocity distribution function for the electrons is given by

$$f_{\rm e}(v) = n_0\left(\frac{m_{\rm e}}{2\pi\kappa T_{\rm e}}\right)^{3/2} {\rm e}^{-\frac{(\frac{1}{2}m_{\rm e}v^2 + q\Phi)}{\kappa T_{\rm e}}}, \tag{2.2.2}$$

where $q = -e$. This equation is like the Maxwellian distribution discussed previously, but has an additional factor $\exp[-q\Phi/\kappa T]$. This factor comes from a principle of statistical mechanics that states that the number of particles with velocity $\mathbf{v}$ is proportional to $\exp[-W/\kappa T]$, where W is the total energy (Huang, 1963). The total energy is given by the sum of the kinetic energy and the potential energy, $W = (1/2)m_{\rm e}v^2 + q\Phi$. By integrating the distribution function over velocity space, it is easy to show that the electron density is given by

$$n_{\rm e} = n_0\,{\rm e}^{\frac{e\Phi}{\kappa T_{\rm e}}}. \tag{2.2.3}$$

Substituting the above expression into Poisson's equation (2.2.1), one obtains the nonlinear differential equation

$$\nabla^2\Phi = -\frac{n_0 e}{\epsilon_0}\left(1 - \mathrm{e}^{\frac{e\Phi}{\kappa T_\mathrm{e}}}\right). \tag{2.2.4}$$

This differential equation can be solved analytically if we assume that $e\Phi/\kappa T_\mathrm{e} \ll 1$. Expanding the exponential in a Taylor series and keeping only the first-order term, one obtains the linear differential equation

$$\nabla^2\Phi = \frac{n_0 e^2}{\epsilon_0 \kappa T_\mathrm{e}}\Phi. \tag{2.2.5}$$

Since the plasma is isotropic, the electrostatic potential can be assumed to be spherically symmetric. The above equation then simplifies to

$$\frac{\partial^2}{\partial r^2}(r\Phi) - \frac{n_0 e^2}{\epsilon_0 \kappa T_\mathrm{e}}(r\Phi) = 0, \tag{2.2.6}$$

which has the general solution

$$\Phi = \frac{A}{r}\mathrm{e}^{-r/\lambda_\mathrm{D}}, \tag{2.2.7}$$

where r is the radius and A is a constant. The factor λ_D is the Debye length and is given by

$$\lambda_\mathrm{D}^2 = \frac{\epsilon_0 \kappa T_\mathrm{e}}{n_0 e^2}. \tag{2.2.8}$$

The constant A is determined by requiring that the solution reduce to the Coulomb potential as the radius goes to zero. The complete solution is then given by

$$\Phi = \frac{1}{4\pi\epsilon_0}\frac{Q}{r}\mathrm{e}^{-r/\lambda_\mathrm{D}}, \tag{2.2.9}$$

and is called the Debye–Hückel potential. A plot of the Debye–Hückel potential (for negative Q) is shown in Figure 2.2. As can be seen, the potential decays exponentially, with a length-scale given by the Debye length, λ_D. A simple, practical formula for the Debye length is

$$\lambda_\mathrm{D} = 6.9\sqrt{T_\mathrm{e}/n_0}\,\mathrm{cm}, \tag{2.2.10}$$

where T_e is in K and n_0 is in cm^{-3}.

The derivation of the Debye length given above is deceptively simple and hides some subtleties inherent in the concept, especially in collisionless plasmas for

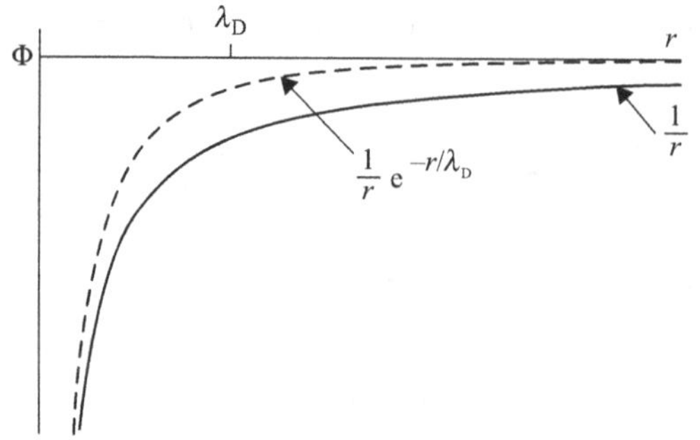

Figure 2.2 A comparison of the Debye–Hückel potential (dashed line) with the Coulomb potential (solid line) for a negative test charge.

which the assumption of a Maxwellian distribution (2.2.2) is open to question. For instance, consider the following interesting paradox involving the role of ions. If the ions were mobile, then by a simple extension of the treatment used for the electrons it would appear that the ion number density should be given by

$$n_{\rm i} = n_0 {\rm e}^{-\frac{e\Phi}{\kappa T_{\rm i}}}. \tag{2.2.11}$$

However, this equation does not provide a correct representation of the ion density. This is because the ions are accelerated by the negative charge Q, and particle flux conservation dictates that as the ion velocity increases, the ion density should decrease. This suggests that the mobility of the ions actually decreases the positive charge density, in contrast with Eq. (2.2.11) which implies that the ion density should increase as r decreases. This effect is called *anti-shielding*, since it decreases the charge density in the shielding region. The resolution of this paradox requires a more sophisticated understanding of distribution functions than we have at the moment, and is postponed until Chapter 11.

2.2.1 Plasma Sheaths

When an object of finite size is placed in a plasma with approximately equal electron and ion temperatures, it acquires a net negative charge because the electron thermal speed, $C_{\rm e} = \sqrt{\kappa T_{\rm e}/m_{\rm e}}$, is much greater than the ion thermal speed, $C_{\rm i} = \sqrt{\kappa T_{\rm i}/m_{\rm i}}$, thereby causing more electrons than ions to hit the object. As the object charges negatively, the electrons start to be repelled, just as when a negative test charge is introduced into the plasma. Equilibrium occurs when the electron current collected by the object balances the incident ion current. An electrically polarized region is thereby formed around the object. This polarized region is called a plasma sheath, or sometimes a positive ion sheath, because the electrons are largely excluded from the sheath. The exact form of the electrostatic potential distribution is a complicated boundary value problem and can only be solved analytically for certain simple geometries such as a sphere, a cylinder, or a planar surface. If the radius of curvature is much larger than the Debye length, so that the surface can be regarded as locally planar, then the potential decays exponentially with a characteristic length-scale given by the Debye length. In these simple cases it is easy to show, by equating the incident electron and ion currents, that the equilibrium potential of the surface is given to a good approximation by

$$V = -\frac{\kappa T_{\mathrm{e}}}{2e}\left[\ln\left(\frac{m_{\mathrm{i}}}{m_{\mathrm{e}}}\right) + \ln\left(\frac{T_{\mathrm{e}}}{T_{\mathrm{i}}}\right)\right]. \tag{2.2.12}$$

Note that for a given ion-to-electron mass ratio the equilibrium potential is controlled dominantly by the electron temperature. Because of the weak logarithmic dependence, this potential is typically only a few times the electron thermal energy. For a proton–electron plasma with equal electron and ion temperatures $V = -3.75\kappa T_{\mathrm{e}}/e$.

If the object is exposed to ultraviolet radiation, as in the case of a spacecraft exposed to sunlight, then the emitted photoelectron current must be added to the equilibrium current balance condition. Under these conditions the object can charge to a positive potential if the emitted photoelectron flux exceeds the incident electron flux. Modifications to the equilibrium potential can also occur if secondary electrons are produced by energetic particles striking the surface.

2.3 Plasma Frequency

If the electrons in a uniform, homogeneous plasma are displaced from their equilibrium position, an electric field arises because of charge separation. This electric field produces a restoring force on the displaced electrons. Since the magnitude of the charge imbalance is directly proportional to the displacement, the restoring force is given by Hooke's law, $F = -k\Delta x$, where Δx is the displacement and k is the effective "spring constant." Since the electrons have inertia, the system behaves as a harmonic oscillator. The resulting oscillations are called electron plasma oscillations or Langmuir oscillations, after Tonks and Langmuir (1929) who first discovered these oscillations.

To compute the oscillation frequency, let us assume that the plasma consists of a uniform slab of electrons of number density n_0 and a fixed background of positive ions of the same density. Suppose we now displace the slab of electrons to the right by a small distance Δx, as shown in Figure 2.3. The slab can be divided into three regions. Region 1 has a net positive charge, region 2 has no net charge, and region 3 has a net negative charge. The electric field in region 2 can be computed

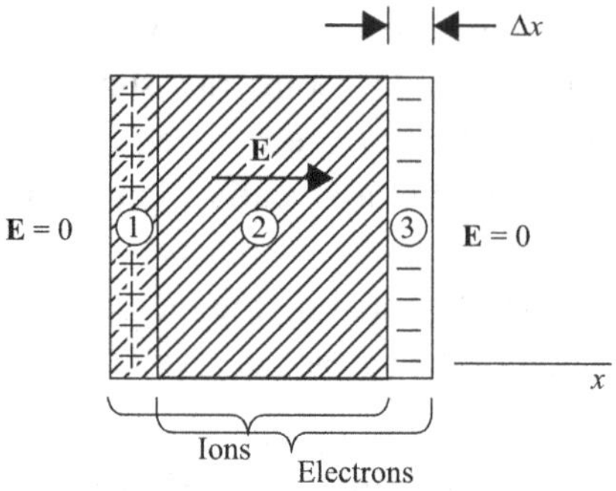

Figure 2.3 A simple slab model that illustrates electron plasma oscillations.

using Gauss' law and is given by

$$E = \frac{n_0 e \Delta x}{\epsilon_0}. \tag{2.3.1}$$

If the slab of electrons is then released, the equation of motion for the electrons is given by

$$m_e \frac{d^2 \Delta x}{dt^2} = (-e)E = -\frac{n_0 e^2}{\epsilon_0} \Delta x, \tag{2.3.2}$$

which simplifies to

$$\frac{d^2 \Delta x}{dt^2} + \left(\frac{n_0 e^2}{\epsilon_0 m_e}\right) \Delta x = 0. \tag{2.3.3}$$

The above equation is just the harmonic oscillator equation. The oscillation frequency ω_{pe} is determined by the term in parentheses via the equation

$$\omega_{pe}^2 = \frac{n_0 e^2}{\epsilon_0 m_e}, \tag{2.3.4}$$

and is called the electron plasma frequency. Note that the electron plasma frequency is proportional to the square root of the electron density and inversely proportional to the square root of the electron mass. A simple formula for the electron plasma frequency (in hertz) is given by

$$f_{pe} = 8980 \sqrt{n_0}\,\text{Hz}, \tag{2.3.5}$$

where the number density n_0 is in electrons cm^{-3}. Note that the electron plasma frequency is determined solely by the number density of the electrons.

If a plasma contains several species, it is customary to define a plasma frequency for each species according to the equation

$$\omega_{ps}^2 = \frac{n_s e_s^2}{\epsilon_0 m_s}, \tag{2.3.6}$$

where e_s, m_s, and n_s are the charge, mass, and number density of the sth species, respectively. It is easily verified that the plasma frequency, the Debye length, and the thermal speed are related by the formula

$$\omega_{ps} \lambda_{Ds} = C_s. \tag{2.3.7}$$

It should *not* be inferred that in a multi-species plasma each species oscillates independently at its own plasma frequency. If the ions are allowed to move in the simple slab model discussed above, it can be shown that the oscillation frequency is given by (Problem 2.2)

$$\omega_p = \sqrt{\omega_{pe}^2 + \omega_{pi}^2}, \tag{2.3.8}$$

which is approximately ω_{pe}, since $\omega_{pi} \ll \omega_{pe}$.

2.4 Cyclotron Frequency

It is well known that if a charged particle of mass m and charge q is injected into a uniform static magnetic field, it moves in a uniform circular motion around the magnetic field at a characteristic frequency called the cyclotron frequency, given by

$$\omega_c = \frac{|q|B}{m}, \tag{2.4.1}$$

where B is the magnetic field strength. Note that the cyclotron frequency is inversely proportional to the particle mass. A simple formula for the electron cyclotron frequency (in Hertz) is given by

$$f_{ce} = 28B \text{ Hz}, \tag{2.4.2}$$

where B is in nT. Here we adopt the standard convention that the cyclotron frequency is a positive quantity, independent of the sign of the charge. However, if a plasma involves several species, it is often convenient to write the cyclotron frequency, which occurs in the equations of motion, in a form that includes the sign of the charge. In this case, to simplify notation, the sign of the charge is included in the cyclotron frequency, with ω_{cs} defined as

$$\omega_{cs} = \frac{e_s B}{m_s}, \tag{2.4.3}$$

where e_s is the charge for that species (including the sign). Since the ions normally encountered in plasmas are positively charged, in this book we always assume that the ion cyclotron frequency, ω_{ci}, is positive. For electrons, the sign of the cyclotron frequency is negative. However, when only electrons are being considered, it will be our convention to use ω_c for the magnitude of the electron cyclotron frequency, as in Eq. (2.4.1), and incorporate the minus sign into the equations involved in the analysis.

2.5 Collision Frequency

To describe the effects of collisions between particles in a plasma, it is convenient to define the collision frequency, ν_{rs}, as the average rate at which particles of type r collide with particles of type s. Two types of collisions can be defined: those between charged particles and neutral particles, and those between charged particles. The physics involved in these two types of collisions is quite different, and is discussed separately.

2.5.1 Collisions between Charged and Neutral Particles

For collisions between charged particles (of type s) and neutral particles (of type n), the interaction force has a very short range and the scattering process is similar to the scattering produced by hard spheres. If we assume that the neutrals are immobile hard spheres, it can be shown that the average collision frequency is given by

$$\nu_{ns} = n_n C_s \sigma_n, \tag{2.5.1}$$

where n_n is the number density of the neutral gas, σ_n is the collision cross section with neutral atoms, and C_s is the thermal speed of charged particles of type s, defined by Eq. (2.1.3). Note that the collision frequency increases as the temperature increases.

The effect of neutral collisions on a plasma is determined by the collision frequency and the time-scale τ of the plasma process of interest. If $\nu\tau$ is much less than one, the effect of neutral collisions on the plasma is small. In this weakly collisional regime, the main effect is a weak damping of disturbances and waves, with a slow rate of energy transfer to the neutrals. To achieve this regime, the number density of neutral atoms (and hence the neutral gas pressure) must be reduced to a low level.

If, on the other hand, $\nu\tau$ is much greater than one, the plasma is said to be collision dominated. In this case, the motion of the ionized component is almost completely controlled by the motion of the neutral gas, and plasma processes are of secondary importance.

2.5.2 Collisions between Charged Particles

Because the Coulomb force is long range, collisions between charged particles are quantitatively different from collisions between charged particles and neutral particles. For such collisions, it is useful to define the differential scattering cross section, $\sigma_C(\chi)$, by

$$\sigma_C(\chi)\mathrm{d}\Omega = \frac{\text{number of particles scattered into d}\Omega}{\text{incident beam intensity}}, \tag{2.5.2}$$

where χ is the scattering angle relative to the incident beam, and $\mathrm{d}\Omega = 2\pi\sin\chi\,\mathrm{d}\chi$ is the differential solid angle. For an electron of mass m_e and charge $-e$ incident on a much heavier ion of charge $+e$ and mass m_i, where $m_i \gg m_e$, the differential scattering cross section is given by the Rutherford formula

$$\sigma_C(\chi) = \frac{1}{4}\left(\frac{e^2}{4\pi\epsilon_0 m_e v^2}\right)^2 \frac{1}{\sin^4(\chi/2)}. \tag{2.5.3}$$

Simple principles of classical mechanics show that small scattering angles are associated with large impact parameters b and vice versa. The scattering geometry is illustrated in Figure 2.4. In contrast to scattering by neutral particles, Coulomb scattering is highly anisotropic and has a strong peak at small angles, as can be readily seen from the inverse $\sin^4(\chi/2)$ dependence. Furthermore, the total scattering cross section, obtained by integrating expression (2.5.3) over a 4π solid angle,

$$\sigma_T = \int_0^{\pi} \sigma_C(\chi) 2\pi \sin\chi\,\mathrm{d}\chi, \tag{2.5.4}$$

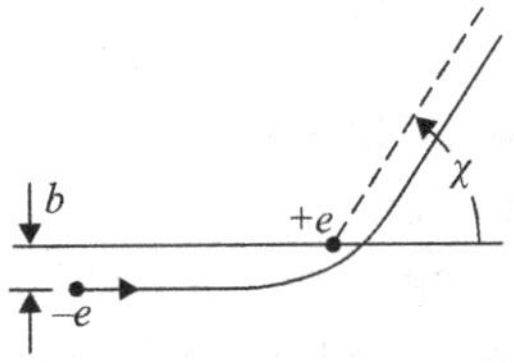

Figure 2.4 A typical trajectory for Coulomb scattering.

is infinite because the integral diverges at the lower limit, which corresponds to large impact parameters. In a plasma, this divergence does not occur because the Coulomb force is strongly reduced by Debye shielding for impact parameters greater than the Debye length. Thus, the Debye length sets a lower limit on the scattering angle (or equivalently, an upper limit on the impact parameter), and eliminates the divergence in Eq. (2.5.4).

It is interesting to note that the Coulomb collision cross section varies as $1/v^4$. If we interpret v as the thermal speed of the incident particles, it follows then that the collision frequency $\nu = nv\sigma_C$ varies as $1/v^3$, or as $1/T^{3/2}$. Thus, the frequency for collisions with charged particles decreases as the temperature increases, which is just the opposite of collisions with neutral particles (see Section 2.5.1). In the absence of neutral collisions, a nearly collisionless regime is attained when the temperature is very large. Later, in Chapter 12, we show that the effective collision frequency for electrons colliding with ions is given by

$$\nu_{ei} = \frac{n_0 e^4}{32\pi^{1/2}\epsilon_0^2 m_e^{1/2}(2\kappa T_e)^{3/2}} \ln(12\pi N_D), \tag{2.5.5}$$

where $N_D = n_0\lambda_D^3$ is the number of electrons per Debye cube. The logarithmic term implies that the numerical value of the collision frequency is weakly dependent on N_D.

2.6 Number of Electrons per Debye Cube

In the previous section, we encountered the dimensionless parameter $N_D = n_0\lambda_D^3$, which is the number of electrons per Debye cube. This parameter plays a fundamental role in the description of a plasma, and is proportional to a quantity called the plasma parameter (see Chapter 12). The importance of the number of electrons per Debye cube can be illustrated from several points of view.

2.6.1 Macroscopic Averages

An implicit assumption in our analysis of a plasma is that macroscopic averages are meaningful. For example, consider the electron density, defined as the average number of electrons per unit volume. In order that the electron density be a smooth function of space, it is necessary that we calculate the average over a volume element that includes many particles.

To consider the effect the number of particles has on this averaging process, let us start with an ensemble of particles of charge e and mass m and divide each particle into M equal parts. Upon division, the charge and mass of each particle become $e' = e/M$ and $m' = m/M$, while the number density becomes $n' = nM$. Since the temperature is proportional to the mass, $(1/2)mv^2 = (3/2)\kappa T$, the new temperature becomes $T' = T/M$. As we take the limit $M \to \infty$, the discrete nature of the particles is suppressed and the system tends to a continuous fluid. It is easy to show that the fundamental length- and time-scales of the plasma are unchanged by this fluidization process, i.e., $\lambda'_D = \lambda_D, \omega'_p = \omega_p$, and $\omega'_c = \omega_c$. On the other hand, in this limiting process, the number of electrons per Debye cube increases without bound because $N'_D = N_D M$. The parameter N_D, therefore, characterizes the extent to which we can treat the plasma as a fluid: a plasma can be considered a continuous fluid only if N_D is very large (i.e., much greater than one).

As another illustration of the role of N_D in macroscopic averages, we return to the discussion of Debye shielding in Section 2.2. In the derivation of the Debye shielding formula (2.2.7), we assumed that $e\Phi/\kappa T_e \ll 1$ in order to obtain a tractable expression for the charge density. At the time, no justification was given for this assumption, which is clearly violated at a sufficiently small radius, since the Coulomb potential varies as $1/r$. Since the charge density involves a macroscopic average, we can ask how small the radius r can possibly be and still have a meaningful average number density. As a rough approximation, we take this minimum radius to correspond to a volume which contains only one electron, i.e., $r_{min} = 1/n_e^{1/3}$. The Coulomb potential at this radius is approximately

$$\Phi = \frac{1}{4\pi\epsilon_0}\frac{e}{r_{min}}. \tag{2.6.1}$$

The $e\Phi/\kappa T_e$ term is then given by

$$\frac{e\Phi}{\kappa T_e} = \frac{1}{4\pi\epsilon_0}\frac{e^2}{\kappa T_e} n_e^{1/3}. \tag{2.6.2}$$

With a slight rearrangement of the terms, it is easy to show that the above equation can be rewritten as

$$\frac{e\Phi}{\kappa T_e} = \frac{1}{4\pi}\frac{1}{N_D^{2/3}}, \tag{2.6.3}$$

where $N_D = n_0\lambda_D^3$. One can see that the assumption $e\Phi/\kappa T_e \ll 1$ is satisfied only if the number of electrons per Debye cube is much greater than one ($N_D \gg 1$).

2.6.2 Ratio of Kinetic Energy to Potential Energy

We next show that the number of electrons in a Debye cube is closely related to the ratio of the average kinetic energy to the average electrostatic potential energy. This ratio can be estimated by noting that the average kinetic energy of a particle is

$(3/2)\kappa T$ and the average electrostatic energy is approximately $e^2/(4\pi\epsilon_0\langle r\rangle)$, where $\langle r\rangle$ is the average separation distance between neighboring particles. The ratio of these two quantities is

$$\frac{\text{kinetic energy}}{\text{potential energy}} = \frac{3\kappa T\, 4\pi\epsilon_0\langle r\rangle}{2e^2}. \tag{2.6.4}$$

Since $\langle r\rangle \simeq 1/n^{1/3}$, the above equation simplifies to

$$\frac{\text{kinetic energy}}{\text{potential energy}} = 6\pi N_{\mathrm{D}}^{2/3}. \tag{2.6.5}$$

If $N_{\mathrm{D}} \gg 1$, one can see that the average kinetic energy greatly exceeds the average electrostatic potential energy. Under these conditions, the plasma is dominated by small-angle collisions. Large-angle collisions caused by the close approach of pairs of particles are extremely rare. As N_{D} decreases, electrostatic interactions become more important and large-angle collisions become more frequent. As N_{D} drops below one, electrostatic interactions become dominant and the motions are strongly controlled by the electric fields in the near vicinity of the individual charges. Electrons become trapped in the potential wells of the ions and the ions organize their average position so as to minimize the average electrostatic energy, leading to lattice-like structures. Systems of this type are said to be strongly coupled, and exhibit properties that are characteristic of solids. An example of such a system is a dusty plasma in which microparticles suspended in a laboratory plasma attract several thousands of electrons and acquire electrostatic potential energies that by far exceed their thermal energies at room temperature. Under such conditions, the dust particles arrange themselves in the form of a crystal. Dusty plasma crystals have been produced under controlled conditions in the laboratory; see Chu and Lin (1994) and Thomas et al. (1994).

2.6.3 Ratio of Collective to Discrete Interactions

Because of the long-range nature of the Coulomb force, a plasma is dominated by large-scale collective motions in which many particles respond in unison to a given electric field. Electron plasma oscillations are an example of such a collective effect. Discrete collisional interactions, on the other hand, tend to inhibit such large-scale collective motions since the collisional forces only act on individual pairs of particles. The importance of collective interactions relative to discrete interactions can be characterized by taking the ratio of the electron plasma frequency to the electron collision frequency. Using Eqs. (2.3.4) and (2.5.5), this ratio can be written entirely as a function of N_D:

$$\frac{\omega_{pe}}{\nu_{ei}} = \sqrt{\frac{\pi}{2} \frac{128 N_D}{\ln(12\pi N_D)}}. \quad (2.6.6)$$

If $N_D \gg 1$, the electron plasma frequency is much greater than the electron–ion collision frequency. This means that collective oscillations, characterized by ω_{pe}, are much more important than discrete-particle effects, which are characterized by ν_{ei}. On the other hand, if $N_D \ll 1$, plasma oscillations are rapidly destroyed by collisions between individual particles. This situation occurs in solids, such as metals, where the oscillations at the electron plasma frequency are strongly damped by collisional interactions.

2.7 The de Broglie Wavelength and Quantum Effects

Quantum effects become important if the typical distance separating particles in a plasma becomes comparable or less than the de Broglie wavelength of the particles. Since the de Broglie wavelength of an electron is much larger than that of an ion, quantum effects occur for electrons before they do for ions. Therefore, to determine the boundary between classical and quantum regimes, it is sufficient to consider only electrons. Heisenberg's uncertainty principle states that

$$\langle \Delta p \rangle \langle \Delta x \rangle \geq \frac{\hbar}{2}, \quad (2.7.1)$$

where $\langle \Delta p \rangle$ and $\langle \Delta x \rangle$ are the uncertainty in the momentum and position, respectively, and $\hbar$ is Planck's constant divided by 2π; see Merzbacher (1998). For a classical description, the largest possible $\langle \Delta p \rangle$ occurs when the uncertainty in the momentum is equal to the root–mean–square momentum, which, using Eq. (2.1.4), is given by

$$\langle \Delta p \rangle = \langle (m_e v)^2 \rangle^{1/2} = (3 m_e \kappa T_e)^{1/2}. \quad (2.7.2)$$

Quantum effects first become important when the uncertainty in the position $\langle \Delta x \rangle$ is comparable to the average distance to the nearest electron, which is approximately

$$\langle \Delta x \rangle = 1/n_e^{1/3}. \quad (2.7.3)$$

Combining conditions (2.7.2) and (2.7.3) with the uncertainty principle (2.7.1) gives the following condition for the validity of the classical description:

$$\frac{(3 m_e \kappa T_e)^{1/2}}{n_e^{1/3}} \gg \hbar. \quad (2.7.4)$$

The above inequality defines a boundary in $n_e - T_e$ space that separates the classical from the quantum regime. The equation defining this boundary is given by

$$\frac{T_e^{3/2}}{n_e} = \frac{(\hbar/2)^3}{(3m_e\kappa)^{3/2}}, \tag{2.7.5}$$

and is shown by the hatched line marked $N_D = 1$ in Figure 2.5.

It is interesting to note that quantum effects become more important as the temperature decreases. This occurs because the de Broglie wavelength increases as the electron velocity (temperature) decreases. The numerical value of the factor on the right-hand side of the above equation is $6.33 \times 10^{-19} K^{3/2} cm^{-3}$, which is very small. So only very moderate temperatures are required for the classical description to be valid. On the other hand, the conditions for the quantum regime are met at very high densities, as in metals and various astrophysical objects such as white dwarfs. The range of temperatures and number densities for various plasmas is summarized in Figure 2.5. For convenience, we also list the temperatures in electron volts, $\kappa T_e/e$, and the number densities in terms of the electron plasma frequency. Lines of constant Debye length and constant N_D are also shown. This book is entirely concerned with plasmas in the classical and weakly coupled regime, with $N_D \gg 1$.

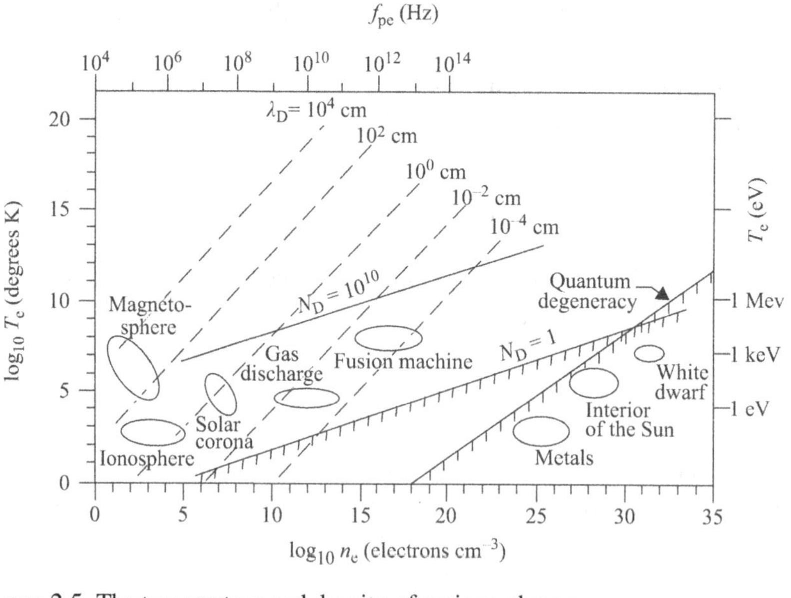

Figure 2.5 The temperature and density of various plasmas.

Problems

2.1. Consider a plasma consisting of electrons and ions of equal number densities, $n = n_e = n_i$, and temperatures T_e and T_i. Assume that the electron density is determined by a term of the form $\exp[e\Phi/\kappa T_e]$ and that the ion density is determined by a term of the form $\exp[-e\Phi/\kappa T_i]$. Show that the Debye length, λ_D, is then given by

$$\left(\frac{1}{\lambda_D}\right)^2 = \left(\frac{1}{\lambda_{De}}\right)^2 + \left(\frac{1}{\lambda_{Di}}\right)^2.$$

How do the electron and ion Debye length terms λ_{De} and λ_{Di} compare if $T_e = T_i$? Can you find any reason to criticize the basic model used in this calculation?

2.2. Consider two superposed slabs of electrons of mass m_e and ions of mass m_i. Using the same type of analysis performed in the text, show that the natural oscillation frequency is

$$\omega_p^2 = \omega_{pe}^2 + \omega_{pi}^2,$$

where ω_{pe} and ω_{pi} are the electron and ion plasma frequencies (assume that the ions have charge e).

2.3. Demonstrate the following relationship between the electron plasma frequency ω_{pe}, the electron thermal speed C_e, and the Debye length λ_{De}:

$$C_e = \omega_{pe}\lambda_{De},$$

or equivalently, defining $k_D = 1/\lambda_D$,

$$k_{De}C_e = \omega_{pe}.$$

2.4. A Maxwellian velocity distribution has the form

$$f(\mathbf{v}) = n_0\left(\frac{m}{2\pi\kappa T}\right)^{3/2} \exp\left[-\frac{mv^2}{2\kappa T}\right],$$

where $v^2 = v_x^2 + v_y^2 + v_z^2$.

(a) Show that

$$\int_{-\infty}^{\infty} f(\mathbf{v})\,dv_x\,dv_y\,dv_z = n_0.$$

(b) Show that

$$\left\langle \frac{1}{2}mv_x^2 \right\rangle = \frac{1}{2}\kappa T.$$

(c) Show that

$$\left\langle \frac{1}{2}mv^2 \right\rangle = \frac{3}{2}\kappa T.$$

2.5. When we compute the equilibrium potential of a metal object in a plasma (assumed to be a flat plate, i.e., one-dimensional) we need to compute the electron and ion currents incident on the plate.

(a) Assuming that the velocity distribution functions of the electrons and ions are Maxwellians, show that the incident currents are

$$J_e = \frac{1}{(2\pi)^{1/2}} e n_e \sqrt{\frac{\kappa T_e}{m_e}}$$

and

$$J_i = \frac{1}{(2\pi)^{1/2}} e n_i \sqrt{\frac{\kappa T_i}{m_i}}.$$

(b) Assuming that the ion density is $n_i = n_0$ and that the electron density varies as $n_e = n_0 \exp[e\Phi/\kappa T_e]$, show that the equilibrium potential V of the plate (i.e., when $J_e + J_i = 0$) is given by

$$V = -\frac{\kappa T_e}{2e}\left[\ln\left(\frac{m_i}{m_e}\right) + \ln\left(\frac{T_e}{T_i}\right)\right].$$

2.6. An infinite plate is immersed in a plasma of number density n_0 and temperature T. The plate is located in the x, y plane and is held at a potential V_0 (negative) relative to the potential at a large distance from the plate. By following a procedure very similar to the Debye sheath problem, show that the potential in the plasma is given by

$$\Phi = V_0 \exp[-z/\lambda_D],$$

where λ_D is the Debye length.

2.7. Show that the number of electrons per Debye cube is given by

$$N_D = 328\frac{T_e^{3/2}}{n_e^{1/2}},$$

where T_e is in K and n_e is in cm^{-3}.

2.8. Show that the boundary between a classical plasma and a quantum plasma is given by

$$T_e^{3/2} = (6.33 \times 10^{-19}) n_e,$$

where T_e is in K and n_e is in cm^{-3}.

References

Chu, J. H., and Lin, I. 1994. Direct observation of Coulomb crystals and liquids in strongly coupled rf dusty plasmas. *Phys. Rev. Lett.* **72**, 4009–4012.

Davidson, R. C. 2001. *Physics of Non-neutral Plasmas*. Singapore: World Scientific, pp. 1–14.

Debye, P., and Hückel, E. 1923. On the Debye length in strong electrolytes. *Physikal. Z.* **24** (9), 185–206.

Huang, K. 1963. *Statistical Mechanics*. New York: Wiley, p. 157.

Merzbacher, E. 1998. *Quantum Mechanics*. New York: Wiley, pp. 217–220.

Thomas, H., Morfill, G. E., Demmel, V., Goree, J., Feuerbacher, B., and Möhlmann, D. 1994. Plasma crystal: Coulomb crystallization in a dusty plasma. *Phys. Rev. Lett.* **73**, 652–655.

Tonks, L., and Langmuir, I. 1929. Oscillations in ionized gases. *Phys. Rev.* **33**, 195–210.

Further Reading

Bellan, P. M. 2008. *Fundamentals of Plasma Physics*. Cambridge: Cambridge University Press, Chapter 1.

Chen, F. F. 1990. *Introduction to Plasma Physics and Controlled Fusion Volume 1: Plasma Physics*. New York: Plenum Press, Chapter 1. Originally published in 1983.

Fitzpatrick, R. 2015. *Plasma Physics: An Introduction*. Boca Raton, FL: CRC Press, pp. 115–119.

Nicholson, D. R. 1992. *Introduction to Plasma Theory*. Malabar, FL: Krieger Publishing, Chapter 1. Originally published in 1983 by Wiley.

3
Single-Particle Motions

A complete mathematical model of a plasma requires three basic elements: first, the motion of all particles must be determined for some assumed electric and magnetic field configuration; second, the current and charge densities must be computed from the particle trajectories; and third, the electric and magnetic fields must be self-consistently determined from the currents and charges, taking into account both internal and external sources. To be self-consistent, the electric and magnetic fields obtained from the last step must correspond to the fields used in the first step. It is this self-consistency requirement that makes the analysis of a plasma difficult.

To develop an understanding of the processes occurring in a plasma, a useful first step is to forget about the self-consistency requirement and concentrate on the motion of a single particle in a specified field configuration. This approach can be useful in a variety of situations. If the external fields are very strong and the plasma is sufficiently tenuous, the internally generated fields are sometimes small and can be safely ignored. This situation arises, for example, in radiation belts at high energies and in various electronic devices such as vacuum tubes and traveling wave amplifiers. In other situations, the self-consistent electric and magnetic fields may be known from direct measurement. In this case, it is often useful to follow the motion of individual tracer particles in the known electric and magnetic fields in order to gain insight into the physical processes involved, such as particle transport and energization. Finally, in some cases it is possible to use the general solution for the particle motion in an assumed field geometry to determine a fully self-consistent solution in which the currents and charges produce the assumed fields.

3.1 Motion in a Static Uniform Magnetic Field

The simplest field configuration of importance in plasma physics is a static uniform magnetic field. The equation of motion for a particle moving at non-relativistic velocities in a static uniform magnetic field is given by

$$m\frac{d\mathbf{v}}{dt} = q(\mathbf{v}\times\mathbf{B}), \tag{3.1.1}$$

where $q(\mathbf{v}\times\mathbf{B})$ is the Lorentz force. Taking the scalar product of the above equation with the velocity vector gives

$$m\mathbf{v}\cdot\frac{d\mathbf{v}}{dt} = \frac{d}{dt}\left(\frac{1}{2}mv^2\right) = q\mathbf{v}\cdot(\mathbf{v}\times\mathbf{B}) = 0, \tag{3.1.2}$$

which shows that the kinetic energy, $w = (1/2)mv^2$, is a constant of the motion. To determine the particle trajectory it is convenient to resolve the velocity into components parallel, $\mathbf{v}_{\parallel}$, and perpendicular, $\mathbf{v}_{\perp}$, to the magnetic field, $\mathbf{v} = \mathbf{v}_{\parallel} + \mathbf{v}_{\perp}$. The kinetic energy can then be written as $w = w_{\parallel} + w_{\perp}$, where $w_{\parallel} = (1/2)mv_{\parallel}^2$ and $w_{\perp} = (1/2)mv_{\perp}^2$. Since the $\mathbf{v}\times\mathbf{B}$ force has no component parallel to the magnetic field, the parallel component of the velocity is constant, so the particle moves at a constant velocity along the magnetic field. Since w and $w_{\parallel}$ are both constants, it follows that $w_{\perp}$, and hence $v_{\perp}$, are also constants of the motion. The radius of curvature, ρ_c, of the motion in a plane perpendicular to the magnetic field can be obtained from the perpendicular component of the equation of motion (3.1.1) which, ignoring signs, is easily shown to be

$$m\frac{v_{\perp}^2}{\rho_c} = |q|v_{\perp}B. \tag{3.1.3}$$

The radius ρ_c is often called the cyclotron radius and from the above equation is given by

$$\rho_c = \frac{mv_{\perp}}{|q|B}. \tag{3.1.4}$$

Since $v_{\perp}$ is constant, the cyclotron radius is constant, which means that the motion in a plane perpendicular to the magnetic field is a circle. The frequency of the circular motion around the magnetic field, known as the cyclotron frequency, is given by

$$\omega_c = \frac{v_{\perp}}{\rho_c} = \frac{|q|B}{m}. \tag{3.1.5}$$

From the above analysis it is evident that the uniform motion along the magnetic field plus the circular motion around the magnetic field generates a helical trajectory, as shown in Figure 3.1. Positively charged particles rotate in the left-hand sense with respect to the magnetic field, and negatively charged particles rotate in the right-hand sense. The instantaneous center of the rotational motion

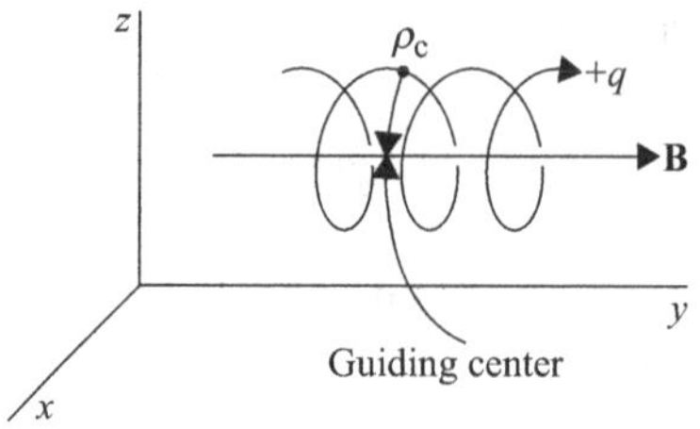

Figure 3.1 The motion of a charged particle in a static magnetic field.

is called the *guiding center*. The concept of an instantaneous guiding center will later be useful for describing motions in non-uniform magnetic fields. For a static uniform magnetic field the guiding center moves with constant velocity along the magnetic field.

3.1.1 Magnetic Moment

Since a moving charge produces a current, it is evident that the uniform circular motion around the magnetic field constitutes a current loop. The magnetic moment μ of this current loop is the product of the current and the area of the loop,

$$\mu = IA. \tag{3.1.6}$$

The magnitude of the current is simply the charge divided by the cyclotron period, and is given by

$$I = \frac{|q|}{T_c} = \frac{q^2 B}{2\pi m}. \tag{3.1.7}$$

The area of the current loop is the area of the circular orbit, which is given by

$$A = \pi \rho_c^2 = \pi \left(\frac{m v_\perp}{qB} \right)^2. \tag{3.1.8}$$

Combining Eqs. (3.1.6), (3.1.7), and (3.1.8), the magnetic moment can be written in the form

$$\mu = \frac{m v_\perp^2}{2B} = \frac{W_\perp}{B}. \tag{3.1.9}$$

From the direction of rotation the right-hand rule shows that the magnetic moment is directed opposite to the magnetic field, as indicated in Figure 3.2.

This relationship holds for both positively and negatively charged particles. (Note that if q is negative the direction of rotation is reversed, but the direction of the current remains the same.) Since the magnetic moment is oriented opposite

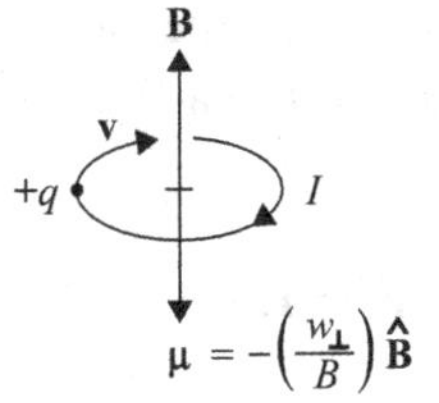

Figure 3.2 The magnetic moment μ is directed opposite the magnetic field **B**.

to the magnetic field, a magnetized plasma is always diamagnetic. In the presence of a large number of particles, diamagnetism appreciably reduces the magnitude of an externally imposed magnetic field. The magnetization, **M**, defined as the magnetic moment per unit volume, is obtained by adding the magnetic moments of all species present in the plasma,

$$\mathbf{M} = -\sum_s n_s \langle \mu_s \rangle \hat{\mathbf{B}}, \tag{3.1.10}$$

where $\langle \mu_s \rangle$ is the average magnetic moment of the sth species. (Note that the average magnetic moment is directly proportional to the average perpendicular kinetic energy.) The corresponding magnetization current is given by $\mathbf{J}_\mathrm{m} = \nabla \times \mathbf{M}$.

3.2 Motion in Static and Uniform Electric and Magnetic Fields

Next we consider the more complicated case of spatially uniform static electric and magnetic fields. For the moment, we restrict our attention to situations in which the electric field is perpendicular to the magnetic field. (If the parallel component of the electric field, $E_{\|}$, is non-zero, it will accelerate the guiding center of the particle parallel to the field line, and we avoid that situation for reasons of simplicity.) The equation of motion for a particle moving in such a configuration can be resolved into components parallel and perpendicular to the magnetic field, i.e.,

$$m\frac{\mathrm{d}v_{\|}}{\mathrm{d}t} = 0 \tag{3.2.1}$$

and

$$m\frac{\mathrm{d}\mathbf{v}_\perp}{\mathrm{d}t} = q[\mathbf{E} + \mathbf{v}_\perp \times \mathbf{B}], \tag{3.2.2}$$

where we assume that the electric and magnetic fields are constant. By transforming to a frame of reference moving perpendicular to the magnetic field, it is possible, under certain conditions, to eliminate the electric field from

Eq. (3.2.2) and reduce the problem to the one solved in the previous section. For non-relativistic velocities, the electric and magnetic fields transform according to the relations

$$\mathbf{E}' = \mathbf{E} + \mathbf{v}_{\mathrm{E}} \times \mathbf{B} \tag{3.2.3}$$

and

$$\mathbf{B}' = \mathbf{B}, \tag{3.2.4}$$

where $\mathbf{v}_{\mathrm{E}}$ is the velocity of a frame of reference (indicated by primes) moving perpendicular to $\mathbf{E}$ and $\mathbf{B}$. The transformed velocities in the moving frame are $v'_{\|} = v_{\|}$ and $\mathbf{v}'_{\perp} = -\mathbf{v}_{\mathrm{E}} + \mathbf{v}_{\perp}$. The equations of motion in the moving frame are then

$$m\frac{\mathrm{d}v'_{\|}}{\mathrm{d}t} = 0 \tag{3.2.5}$$

and

$$m\frac{\mathrm{d}\mathbf{v}'_{\perp}}{\mathrm{d}t} = q[\mathbf{E}' + \mathbf{v}'_{\perp} \times \mathbf{B}]. \tag{3.2.6}$$

As long as the velocity $\mathbf{v}_{\mathrm{E}}$ is less than the speed of light, $\mathbf{v}_{\mathrm{E}}$ can be chosen such that

$$\mathbf{E}' = \mathbf{E} + \mathbf{v}_{\mathrm{E}} \times \mathbf{B} = 0, \tag{3.2.7}$$

which can be inverted, by taking the cross-product with $\mathbf{B}$, to yield

$$\mathbf{v}_{\mathrm{E}} = \frac{\mathbf{E} \times \mathbf{B}}{B^2}. \tag{3.2.8}$$

The velocity $\mathbf{v}_{\mathrm{E}}$ is called the $\mathbf{E} \times \mathbf{B}$ drift velocity. In the moving frame of reference, the equation of motion (3.2.6) then reduces to

$$m\frac{\mathrm{d}\mathbf{v}'}{\mathrm{d}t} = q\mathbf{v}' \times \mathbf{B}, \tag{3.2.9}$$

which simply corresponds to a particle moving in a static uniform magnetic field. Thus, the motion of the particle in the original (unprimed) frame of reference consists of a uniform circular motion around the magnetic field plus a uniform translational motion in the $\mathbf{E} \times \mathbf{B}$ direction with velocity $\mathbf{v}_{\mathrm{E}}$. As shown in Figure 3.3 for the special case $v_{\|} = 0$, the particle trajectories are cycloids in the $\mathbf{E}, \mathbf{v}_{\mathrm{E}}$ plane, with positive charges rotating in the left-hand sense with respect to the magnetic field, and negative charges rotating in the right-hand sense. The guiding center moves at the $\mathbf{E} \times \mathbf{B}$ drift velocity. If the particle initially has a

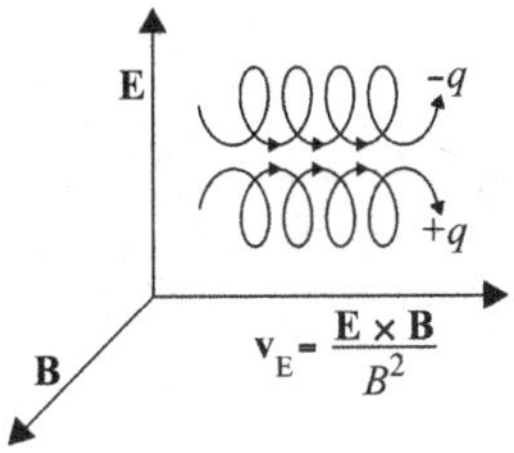

Figure 3.3 Perpendicular electric and magnetic fields produce a drift, $\mathbf{v}_\mathrm{E}$, in the $\mathbf{E}\times\mathbf{B}$ direction.

velocity component along the magnetic field, then the guiding center continues to move with this same velocity parallel to the magnetic field.

If the electric field is so large that the magnitude of $\mathbf{v}_\mathrm{E}$ equals or exceeds the speed of light, Eq. (3.2.8) is no longer valid. Under these conditions, it is not possible to transform to a reference frame in which the particle simply rotates around the magnetic field. This possibility, which rarely occurs in laboratory and space plasma applications, may be realized in some astrophysical applications (near the light cylinder at a pulsar, for example).

The motion of a charge particle in perpendicular electric and magnetic fields has several notable features.

First, all particles drift with the same velocity irrespective of their charge, mass, or energy. Thus, the only effect of adding a constant perpendicular electric field is that all the particles move at the $\mathbf{E}\times\mathbf{B}$ drift velocity. This simple result occurs because the electric field is determined by the frame of reference, and in the frame moving with the $\mathbf{E}\times\mathbf{B}$ drift velocity the electric field vanishes and the particles simply rotate about the magnetic field. Based on this result, the rest frame of the plasma is usually defined as the frame of reference in which $\mathbf{E} = 0$. Note that in a plasma with an inhomogeneous electric field, this definition only applies locally, since in general it is not possible to find a frame of reference for which $\mathbf{E} = 0$ everywhere.

Second, since all particles drift at the same velocity ($\mathbf{v}_{\mathrm{E}s} = \mathbf{v}_\mathrm{E}$) and since the plasma is assumed to be electrically neutral ($\sum_s n_s e_s = 0$), the $\mathbf{E}\times\mathbf{B}$ drift produces no net current density

$$\mathbf{J} = \sum_s e_s n_s \mathbf{v}_{\mathrm{E}s} = \mathbf{v}_\mathrm{E}\left(\sum_s n_s e_s\right) = 0. \tag{3.2.10}$$

Third, since contours of constant electrostatic potential are perpendicular to $\mathbf{E}$, the $\mathbf{E}\times\mathbf{B}$ drift velocity is always along a contour of constant electrostatic potential. This relationship is illustrated in Figure 3.4, which shows a magnetic field $\mathbf{B}$ out of the paper and an inhomogeneous electric field $\mathbf{E}$ in the plane of the paper. Contours

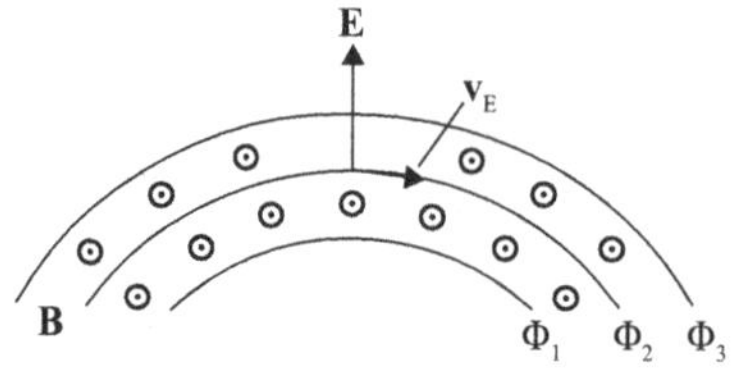

Figure 3.4 In an inhomogeneous electric field, the $\mathbf{E} \times \mathbf{B}$ drift is along a contour of constant electrostatic potential.

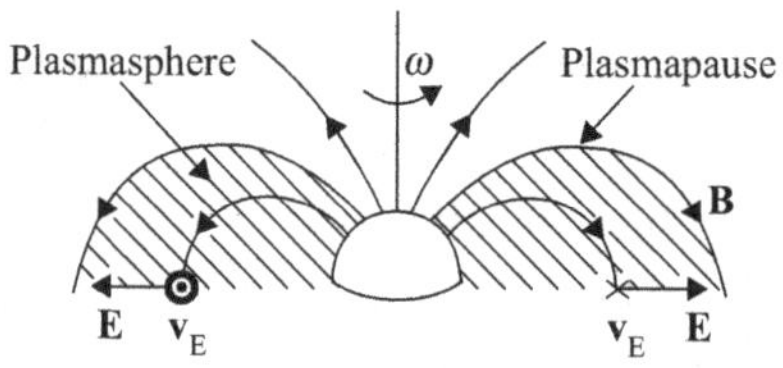

Figure 3.5 The co-rotational electric field forces the plasma to rotate with the planet inside of a boundary called the plasmapause.

of constant electrostatic potential, which must be perpendicular to **E**, are indicated by Φ_1, Φ_2, and Φ_3. Since the $\mathbf{E} \times \mathbf{B}$ drift is along an equipotential contour, the total energy of the particle is not changed by the $\mathbf{E} \times \mathbf{B}$ drift motion. Later, we consider other types of drifts that can cause the guiding center to move away from an equipotential contour, and thus change the kinetic energy of the particle.

3.2.1 Examples of $\mathbf{E} \times \mathbf{B}$ *Drifts*

Numerous examples of $\mathbf{E} \times \mathbf{B}$ drifts occur in space, laboratory, and astrophysical plasmas. In planetary magnetospheres the collisional coupling of the plasma to the neutral atmosphere causes the plasma in the inner regions of the magnetosphere to rotate with the planet. This rotation is associated with the so-called co-rotational electric field, $\mathbf{E} = -\mathbf{v}_E \times \mathbf{B} = -(\omega \times \mathbf{r}) \times \mathbf{B}$, induced by the rotation of the planet, where ω is the angular rotation velocity of the planet. In Earth's magnetosphere the co-rotational electric field is the dominant electric field for magnetic field lines extending out to about three to four Earth radii. The region where co-rotation dominates is called the plasmasphere (see Figure 3.5), and the outer boundary of the plasmasphere is called the plasmapause. In strongly magnetized planets, such as Jupiter, Saturn, Uranus, and Neptune, most of the magnetosphere rotates with the planet.

Co-rotational electric fields are also important in laboratory plasmas. In the laboratory, a rotating column of plasma can be produced by imposing a radial electric field on an axial magnetic field, as shown in Figure 3.6. Such a radial

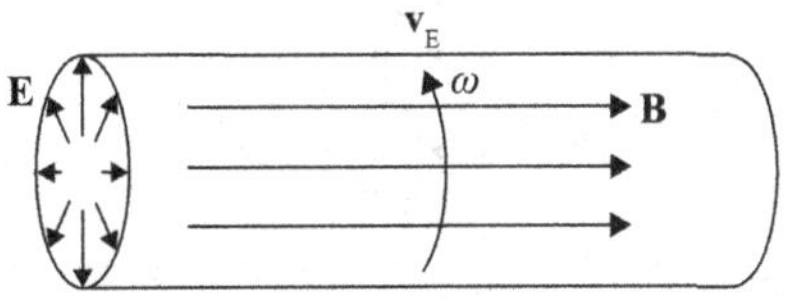

Figure 3.6 A magnetized plasma column can be forced to rotate by imposing a radial electric field at each end of the column.

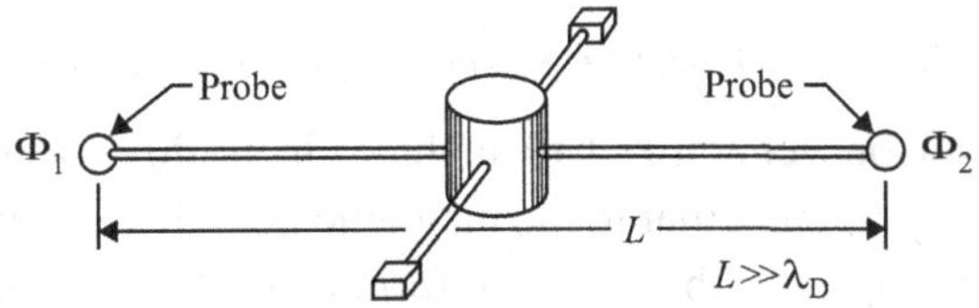

Figure 3.7 Electric fields in space plasmas are often determined by measuring the potential difference between two probes.

electric field can be produced by electrically biasing a system of concentric grids at each end of the column. The plasma then rotates at the $\mathbf{E}\times\mathbf{B}$ velocity. If the electric field increases linearly with radial distance from the central axis, the plasma column can be made to rotate as though it were a rigid body. This type of rotation is similar to magnetospheric co-rotation, except the electric field is imposed by external grids rather than by an interaction with the rotating atmosphere. In some types of magnetic confinement devices, radial electric fields arise due to charge accumulation in the plasma. External bias grids or metal plates at the ends of the plasma column can be used to control the amplitude of these fields.

The relationship between perpendicular electric fields and plasma drifts provides a useful method for measuring plasma flow velocities. In both laboratory and space plasmas the flow velocity perpendicular to the magnetic field can be determined by measuring the perpendicular electric field and then using $\mathbf{v}_E = \mathbf{E}\times\mathbf{B}/B^2$ to compute the flow velocity. In laboratory plasmas the electric field can be determined by using a conducting probe, called a Langmuir probe, to measure the electrostatic potential at various points in the plasma relative to some reference potential. In space plasmas two probes mounted on the ends of booms are used, as shown in Figure 3.7 where $L \gg \lambda_D$. These probes are called double probes. The electric field along the axis of the boom is computed using $E = -(\Phi_2 - \Phi_1)/L$, where Φ_1 and Φ_2 are the potentials of the two probes, L is the separation distance, and it is assumed that the sheath potential of the probes cancel. By making the probe separation large, typically 100 to 200 m, it is possible to measure very small electric fields, sometimes only a fraction of a millivolt per meter.

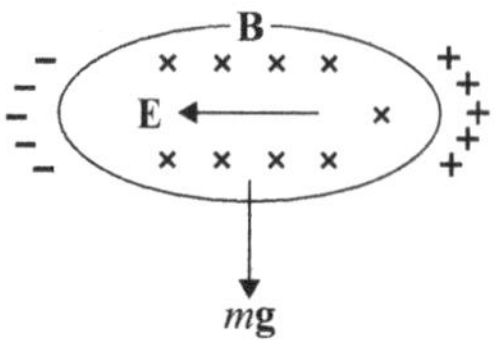

Figure 3.8 A downward gravitational field $m\mathbf{g}$ causes electrons and ions to drift in opposite directions, thereby producing polarization charges at the boundaries of the plasma. The resulting electric field $\mathbf{E}$ then causes a downward $\mathbf{E}\times\mathbf{B}$ drift.

3.2.2 Drift Due to a Force Perpendicular to B

The $\mathbf{E}\times\mathbf{B}$ drift can be generalized to include drifts due to an external force, provided the force is perpendicular to the magnetic field. If we define an equivalent electric field $\mathbf{E}_\perp = \mathbf{F}_\perp/q$ and substitute it in formula (3.2.8), we obtain the formula for the drift due to a perpendicular external force

$$\mathbf{v}_\mathrm{F} = \frac{\mathbf{F}_\perp \times \mathbf{B}}{qB^2}. \tag{3.2.11}$$

A common example is the gravitational force $\mathbf{F} = m\mathbf{g}$. Note that a gravitational force acting perpendicular to the magnetic field does not directly cause the plasma to "fall" because the induced drift velocity is perpendicular to the applied force. The plasma does indeed fall, but the explanation is more complicated. Unlike the $\mathbf{E}\times\mathbf{B}$ drift, the drift velocity produced by a gravitational force causes positive and negative charges to move in opposite directions, thereby causing a current. For a plasma of finite size this current produces a polarization charge at the boundaries of the plasma which, in turn, produces an electric field perpendicular to the gravitational force, as shown in Figure 3.8. This electric field causes an $\mathbf{E}\times\mathbf{B}$ drift in the direction of the gravitational force. Thus, although the drift produced by a gravitational force does not directly cause the plasma to fall, the resulting polarization electric field leads to a downward acceleration. The extent to which a polarization electric field actually develops depends on the boundary conditions. For example, if some mechanism exists to remove the polarization charge, such as grids or metal walls at the sides of the chamber, the polarization electric field can be strongly inhibited. In most laboratory and space plasmas, the effect of the gravitational force turns out to be small compared to the electric and magnetic forces.

3.3 Gradient and Curvature Drifts

If the magnetic field has a transverse gradient or is curved, the guiding center tends to move perpendicular to the magnetic field because of variations in the instantaneous radius of curvature as the particle orbits the magnetic field. Two types of drifts occur, called gradient and curvature drifts. Because they involve spatially varying magnetic fields, their analysis is more difficult than those in the previous sections. The most mathematically elegant approach to analyzing these drifts is based on a guiding-center analysis in which the trajectory of the particle is regarded to be a tightly coiled helical motion around the guiding-center magnetic field line, as in Figure 3.1. However, the magnetic field is no longer assumed to be straight and uniform. In the guiding-center model, the magnetic field at the particle is assumed to be given by $\mathbf{B}(\mathbf{r}) = \mathbf{B}(\mathbf{r}_0) + (\boldsymbol{\rho}_c \cdot \boldsymbol{\nabla})\mathbf{B}|_{\mathbf{r}=\mathbf{r}_0} + \cdots$, where $\mathbf{r} = \mathbf{r}_0$ represents the instantaneous guiding center, and $\boldsymbol{\rho}_c$ is the cyclotron radius vector to the magnetic line at $\mathbf{r}_0$. If the cyclotron radius is sufficiently small relative to the spatial gradients, specifically if $(\rho_c/B)|\boldsymbol{\nabla}B| \ll 1$, a rigorous perturbation expansion can be developed that gives the gradient and curvature drift velocities in terms of spatial gradients of the magnetic field; see Northrop (1963). Because this perturbation analysis is somewhat complicated and tends to obscure the physical reasons for the drifts, we have taken a simpler and more heuristic approach. Specifically, we have chosen to analyze the particle motions for relatively simple magnetic field geometries that are easy to analyze, and from these results deduce the general equations for the drift motions. Our basic strategy is to eliminate the faster cyclotron motion by averaging over the nearly circular orbital motion of the particle, thereby revealing the much slower drift motions caused by the magnetic field gradients. Because the gradient and curvature drifts have somewhat different characteristics, they are discussed separately.

3.3.1 Gradient Drift

To analyze the motion of a particle moving in a magnetic field with a transverse gradient, we consider the idealized case of a magnetic field in the z direction with a magnetic field gradient, $\boldsymbol{\nabla}B$, in the x direction. The motion of a positively charged particle in this magnetic field is sketched in Figure 3.9. As can be seen from Eq. (3.1.4), the instantaneous radius of curvature decreases in the region of stronger magnetic field and increases in the region of weaker magnetic field. The trajectory of the particle then consists of a cycloid-like drift motion along the y axis, perpendicular to the gradient of the magnetic field. A simple description of the orbit can be obtained if the magnetic field variation over the cyclotron orbit is small. The motion is then nearly circular in a frame of reference moving at the drift velocity.

To derive the average drift velocity we make use of the fact that the motion in the x direction is periodic, which implies that the x component of the $\mathbf{v} \times \mathbf{B}$ force,

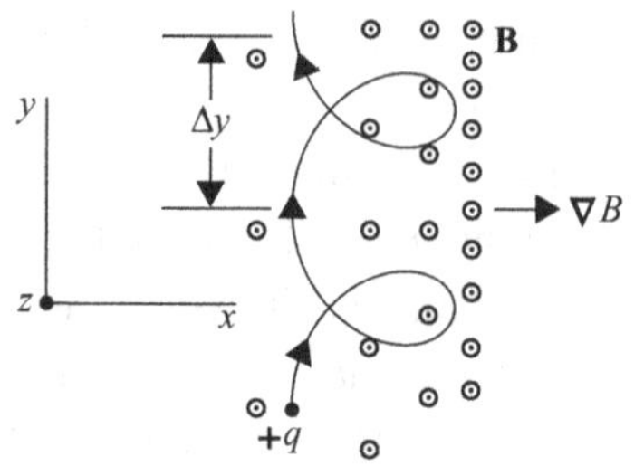

Figure 3.9 A gradient in the magnetic field, ∇B, causes a drift perpendicular to the gradient.

time-averaged over one orbit, is zero; that is,

$$\oint F_x \, dt = q \oint v_y B_z \, dt = 0. \tag{3.3.1}$$

Since the gradient in the magnetic field strength is small, the magnetic field can be expanded about the guiding center using a Taylor series. Keeping only the first-order term, the magnetic field can be written

$$B_z(x) = B_z(x_0) + \frac{\partial B_z}{\partial x}(x - x_0) + \cdots \tag{3.3.2}$$

Substituting $B_z(x)$ into Eq. (3.3.1) and keeping only the first two terms, we obtain

$$B_z(x_0) \oint v_y \, dt + \frac{\partial B_z}{\partial x} \oint v_y (x - x_0) \, dt = 0. \tag{3.3.3}$$

The integral in the first term gives Δy, the distance moved in one orbit. The integral in the second term is the negative of the area, A_z, of the near-circular orbit, so that

$$A_z = \oint v_y (x - x_0) \, dt = \oint (x - x_0) \, dy = -\frac{q}{|q|} \pi \rho_c^2. \tag{3.3.4}$$

In the above equation the $q/|q|$ term occurs because the direction of integration in the x, y plane depends on the sign of the charge. The y component of the gradient drift velocity, v_G, can then be written as

$$v_G = \frac{\Delta y}{\Delta t} = \frac{1}{\Delta t} \frac{1}{B_z} \frac{\partial B_z}{\partial x} \left(\frac{q}{|q|} \pi \rho_c^2 \right), \tag{3.3.5}$$

which, using $\Delta t = 2\pi/\omega_c$ and $w_\perp = (m/2)\omega_c^2 \rho_c^2$, simplifies to

$$v_G = \frac{w_\perp}{q B_z} \left[\frac{1}{B_z} \frac{\partial B_z}{\partial x} \right]. \tag{3.3.6}$$

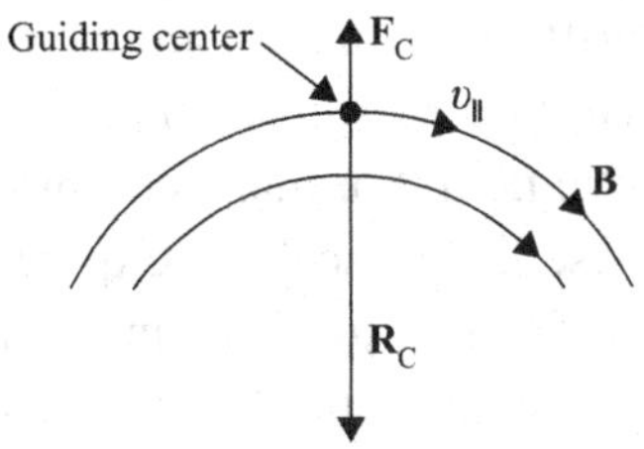

Figure 3.10 The centrifugal force of a particle moving along a curved magnetic field causes a drift motion in the $\mathbf{B} \times \mathbf{R}_C$ direction, where $\mathbf{R}_C$ is the instantaneous radius of curvature.

This expression can be rewritten in a more general vector form as

$$\mathbf{v}_G = \frac{w_\perp}{qB}\left[\frac{\hat{\mathbf{B}} \times \nabla B}{B}\right], \tag{3.3.7}$$

where $\hat{\mathbf{B}}$ is a unit vector in the direction of the magnetic field. Note that the term in brackets has units of inverse length and is a purely geometric factor that characterizes the gradient of the magnetic field.

3.3.2 Curvature Drift

The simplest way to analyze the motion of a charged particle moving in a curved magnetic field is to notice that the particle experiences a centrifugal force $F_C = mv_\parallel^2/R_C$, where R_C is the radius of curvature of the magnetic field. This force is directed outward from the center of curvature, as shown in Figure 3.10.

Since the centrifugal force is perpendicular to the magnetic field, the curvature drift velocity, $\mathbf{v}_C$, can be computed using the general expression (3.2.11) for the drift due to an external force which gives

$$\mathbf{v}_C = \frac{\mathbf{F} \times \mathbf{B}}{qB^2} = -m\frac{v_\parallel^2}{R_C^2}\frac{\mathbf{R}_C \times \mathbf{B}}{qB^2}. \tag{3.3.8}$$

The minus sign on the right-hand side of the equation arises because the centrifugal force is directed opposite to the radius of curvature vector, $\mathbf{R}_C$. By introducing the parallel kinetic energy, $w_\parallel = (1/2)mv_\parallel^2$, the curvature drift can be rewritten in a form that is very similar to the equation for gradient drift,

$$\mathbf{v}_C = \frac{2w_\parallel}{qB}\left[\frac{\hat{\mathbf{B}} \times \hat{\mathbf{R}}_C}{R_C}\right], \tag{3.3.9}$$

where $\hat{\mathbf{R}}_C$ is a unit vector in the direction of the instantaneous radius of curvature. Note that the term in the brackets has units of inverse length, and is a purely geometric factor that characterizes the curvature of the magnetic field.

Gradient and curvature drifts have many similar features. In both cases the drift velocity is proportional to an inverse length-scale that characterizes the inhomogeneity of the magnetic field. For many common field geometries, such as toroidal and dipole fields, these length-scales are quite similar, which means that the gradient and curvature drifts tend to have similar magnitudes. In both cases the drift velocities are proportional to the kinetic energy (perpendicular energy for the gradient drift, and parallel energy for the curvature drift) and inversely proportional to the magnetic field strength. The crucial dimensionless quantity controlling the magnitude of the drift velocity is the ratio of the cyclotron radius to the characteristic length-scale of the magnetic field inhomogeneity. Increasing the kinetic energy or decreasing the magnetic field strength increases the cyclotron radius and, consequently, the drift velocity.

Gradient and curvature drifts differ from the $\mathbf{E} \times \mathbf{B}$ drift in several important respects. First, the gradient and curvature drift velocities are both proportional to the kinetic energy, whereas the $\mathbf{E} \times \mathbf{B}$ drift is independent of the kinetic energy. Thus, the $\mathbf{E} \times \mathbf{B}$ drift tends to control the motion of low-energy ("cold") particles, whereas gradient and curvature drifts tend to control the motion of high-energy ("hot") particles. Second, the gradient and curvature drift velocities both depend on the sign of the charge, whereas the $\mathbf{E} \times \mathbf{B}$ drift is independent of the charge. Gradient and curvature drifts, therefore, give rise to currents, whereas the $\mathbf{E} \times \mathbf{B}$ drifts do not. Third, gradient and curvature drifts do not necessarily occur along contours of constant electrostatic potential. Therefore, these drifts can cause particles to gain or lose energy as the particles drift onto different electrostatic potential contours. The gain or loss of energy is due to the work done by the electric field. In general, only the "cold" plasma particles move with no change in kinetic energy.

3.3.3 Examples of Gradient and Curvature Drifts

In Earth's radiation belt, energetic charged particles experience both gradient and curvature drifts. From the geometry of the dipole magnetic field, it is easy to see that the gradient and curvature drifts are in the same direction, causing electrons to drift eastward around Earth, and causing positively charged ions to drift westward. As shown in Figure 3.11, these drifts produce a current $\mathbf{J}$ that encircles Earth in the westward direction. This current is called the ring current. The magnitude of the ring current depends on the number density and kinetic energy of the particles in the radiation belt. During disturbed periods, called magnetic storms, a dense energetic plasma is injected into the magnetic field, intensifying the ring current and producing measurable magnetic disturbances. Because the ring current

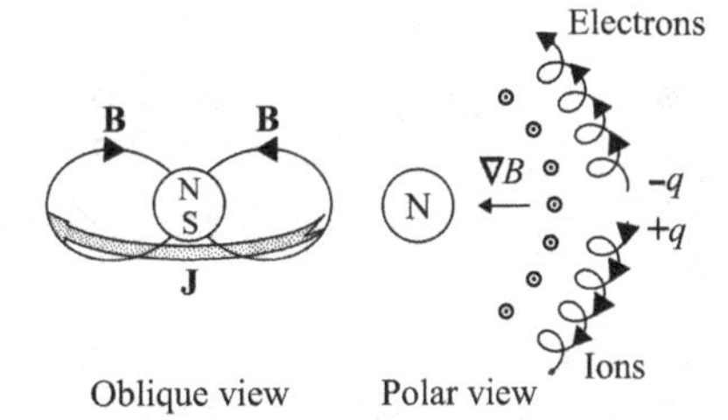

Figure 3.11 In Earth's dipole magnetic field, electrons drift eastward and ions drift westward, producing a current **J** called the ring current.

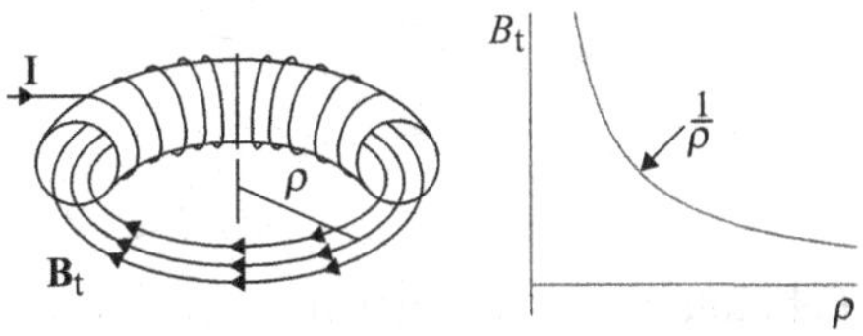

Figure 3.12 A toroidal magnetic field has a gradient directed radially inward, toward the symmetry axis.

is proportional to the kinetic energy of the particles, the magnetic disturbance at Earth gives a rough measure of the total energy of the particles in the ring current.

As an example of gradient and curvature drifts in laboratory plasmas, consider a toroidal magnetic field, $\mathbf{B}_t$. This magnetic field configuration can be produced by wrapping a current-carrying winding around the torus, as shown in Figure 3.12. Since the magnetic field lines close on themselves, it appears at first glance that the guiding-center motion would consist of an azimuthal circulation around the torus on the same closed magnetic field line, thereby ensuring a long confinement time. However, a radial gradient exists in the magnetic field strength, as can be seen from Ampère's law,

$$B_t = \frac{\mu_0 N_\ell}{2\pi} \frac{I}{\rho}, \tag{3.3.10}$$

where μ_0 is the permeability of free space, N_ℓ is the number of turns per unit length, I is the current, and ρ is the distance from the symmetry axis. The gradient is directed inward, toward the center of the torus, and produces a gradient drift along the vertical symmetry axis of the torus. Since the magnetic field lines are curved, a particle spiraling along the magnetic field also experiences curvature drift. It can be verified that the gradient and curvature drifts are both of similar magnitude and in the same direction. Electrons and ions then drift in opposite directions, as illustrated in Figure 3.13.

These gradient and curvature drifts would eventually cause the particles to hit the wall of the machine. However, the actual dynamics is more complicated.

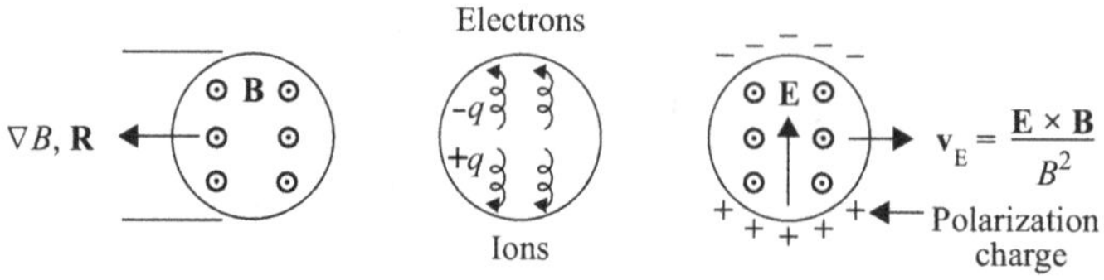

Figure 3.13 Gradient and curvature drifts in a toroidal magnetic field cause electrons and ions to drift in opposite directions, causing polarization charges. The resulting electric field causes an outward $\mathbf{E}\times\mathbf{B}$ drift.

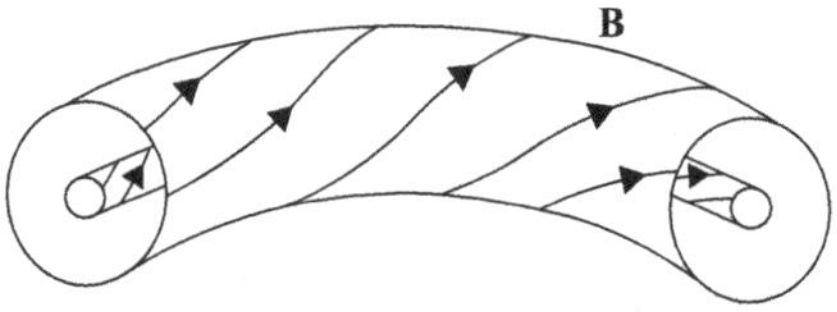

Figure 3.14 Helical field lines in an axisymmetric toroidal plasma produced by a toroidal and a poloidal field.

Since the electrons and ions drift in opposite directions, a polarization charge soon builds up on the upper and lower boundaries of the torus. This polarization charge produces an electric field along the symmetry axis of the torus. This electric field in turn causes an $\mathbf{E}\times\mathbf{B}$ drift that is directed radially outward, causing the plasma to move radially outward until it hits the walls of the machine, where it is lost.

An ingenious solution to this particle confinement problem was found by researchers in the field of controlled thermonuclear fusion. They suggested that the toroidal magnetic field be given a twist, as illustrated in Figure 3.14, either by using an external, helical current-carrying winding, as in a machine called a stellarator, or by passing an axial current through the plasma, as in a machine called a tokamak. The plasma current produces a poloidal magnetic field, B_p, which is much smaller in magnitude than that of the toroidal field, B_t. The magnetic field lines no longer close on themselves as in Figure 3.12, but become helical and spiral around, as shown in Figure 3.14. Upwardly drifting particles now spend about half of the time in the upper part of the torus and half of the time in the lower part of the torus, with the consequence that no polarization charge accumulates on the upper and lower boundaries of the torus.

3.3.4 A Self-consistent static equilibrium

As discussed in the beginning of this chapter, one of our objectives is to obtain a self-consistent plasma configuration. Therefore, it is instructive to analyze a simple self-consistent plasma–magnetic field configuration that involves some of the drift motions described above. To illustrate the procedure, consider an idealized two-dimensional system in which spatially varying magnetic field lines are parallel to the z axis, with $\mathbf{B} = [0, 0, B_z(x, y)]$. If a plasma is placed in such a magnetic field, two types of currents occur, a magnetization current, $\mathbf{J}_{\mathrm{m}}$, and a gradient drift current, $\mathbf{J}_{\mathrm{G}}$. The magnetization current can be computed from the magnetization, $\mathbf{M}$, which is given by

$$\mathbf{M} = -\sum_s n_s \langle \mu_s \rangle \hat{\mathbf{B}} = -\hat{\mathbf{z}} \sum_s n_s \frac{\langle w_{\perp s} \rangle}{B} = -\hat{\mathbf{z}} \frac{W_\perp}{B}, \tag{3.3.11}$$

where $W\perp = \sum_s n_s \langle w_{\perp s} \rangle$ is the average perpendicular kinetic energy density, summed over all species. The magnetization current is given by

$$\mathbf{J}_{\mathrm{m}} = \nabla \times \mathbf{M} = \hat{\mathbf{z}} \times \nabla \left(\frac{W_\perp}{B} \right) = \frac{\hat{\mathbf{z}}}{B} \times \nabla W_\perp - \frac{W_\perp}{B^2} \hat{\mathbf{z}} \times \nabla B, \tag{3.3.12}$$

where we have used the vector identity $\nabla \times (\Phi \mathbf{F}) = \nabla \Phi \times \mathbf{F} + \Phi \nabla \times \mathbf{F}$. The gradient drift current is then given by

$$\mathbf{J}_{\mathrm{G}} = \sum_s n_s e_s \langle \mathbf{v}_{\mathrm{G}s} \rangle = \sum_s n_s e_s \frac{\langle w_{\perp s} \rangle}{e_s B^2} (\hat{\mathbf{z}} \times \nabla B), \tag{3.3.13}$$

which can be simplified to

$$\mathbf{J}_{\mathrm{G}} = \frac{W_\perp}{B^2} \hat{\mathbf{z}} \times \nabla B. \tag{3.3.14}$$

The total current is then

$$\mathbf{J} = \mathbf{J}_{\mathrm{m}} + \mathbf{J}_{\mathrm{G}} = \frac{\hat{\mathbf{z}}}{B} \times \nabla W_\perp. \tag{3.3.15}$$

To provide a self-consistent, static equilibrium the above current must be related to the magnetic field via Ampère's law, so that

$$\nabla \times \mathbf{B} = \mu_0 \mathbf{J} = \frac{\mu_0}{B} \hat{\mathbf{z}} \times \nabla W_\perp. \tag{3.3.16}$$

Using the vector identity $\nabla \times (\hat{\mathbf{z}} B) = -\hat{\mathbf{z}} \times \nabla B$, this equation can be written

$$\hat{\mathbf{z}} \times \left(\nabla B + \frac{\mu_0}{B} \nabla W\perp \right) = 0. \tag{3.3.17}$$

Noting that $\nabla B^2 = 2B \nabla B$, the above equation can be simplified to

$$\hat{\mathbf{z}} \times \nabla \left(\frac{B^2}{2\mu_0} + W\perp \right) = 0, \tag{3.3.18}$$

which is satisfied if

$$\frac{B^2}{2\mu_0} + W_\perp = \text{constant}. \tag{3.3.19}$$

It is readily verified that both $B^2/2\mu_0$ and $W_\perp$ have units of pressure. Later we will demonstrate that the above condition is a special case of a more general magnetostatic equilibrium condition, in which the sum of the magnetic field pressure, $B^2/2\mu_0$, and the particle pressure, $P_\perp$, is a constant. For a more complete treatment of a self-consistent solution that includes both gradient and curvature drifts, see Longmire (1963).

3.4 Motion in a Magnetic Mirror Field

A magnetic mirror field is characterized by a spatial variation in the magnetic field strength along the magnetic field lines. As will be shown, this type of spatial variation leads to a reflection of the parallel component of the guiding-center motion, as in a mirror, which is why the field is called a mirror field. An example of an axially symmetric, magnetic mirror field is shown in Figure 3.15.

For an axially symmetric field, it is convenient to use cylindrical coordinates, with ρ representing the radius from the symmetry axis, ϕ representing the azimuthal angle around the symmetry axis, and z representing the position along the symmetry axis.

If the magnetic field strength increases along the magnetic field line, then Maxwell's equation $\nabla \cdot \mathbf{B} = 0$ shows that the field lines must converge as one moves along the axis of symmetry. The rate of convergence can be determined from Maxwell's equation

$$\nabla \cdot \mathbf{B} = \frac{1}{\rho}\frac{\partial}{\partial \rho}\left(\rho B_\rho\right) + \frac{\partial B_z}{\partial z} = 0, \tag{3.4.1}$$

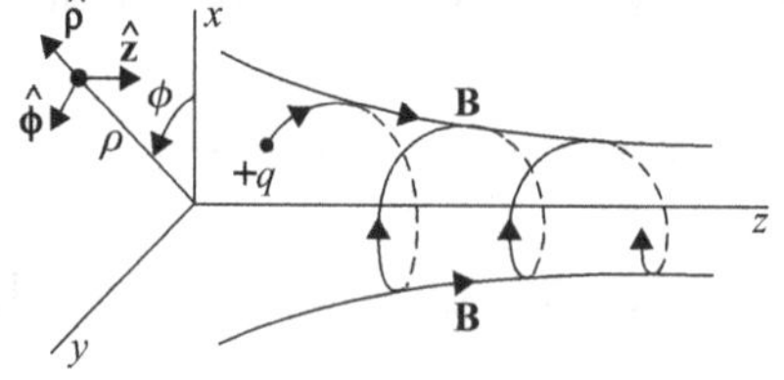

Figure 3.15 The coordinate system used for analyzing the motion of a charged particle in a magnetic mirror field.

which after integrating once, assuming that $\partial B_z/\partial z$ is independent of ρ, gives

$$B_\rho = -\frac{1}{2}\left(\frac{\partial B_z}{\partial z}\right)\rho \tag{3.4.2}$$

or

$$B_x = -\frac{1}{2}\left(\frac{\partial B_z}{\partial z}\right)x \quad \text{and} \quad B_y = -\frac{1}{2}\left(\frac{\partial B_z}{\partial z}\right)y. \tag{3.4.3}$$

If $\partial B_z/\partial z$ is positive, the field lines converge towards the symmetry axis as z increases. These equations are valid only near the axis of symmetry, since in doing the integration we have assumed that $\partial B_z/\partial z$ is independent of ρ.

When analyzing the Lorentz force on a particle moving in a magnetic mirror field, it is convenient to consider the parallel and azimuthal motions separately.

3.4.1 Parallel Motion: The Magnetic Mirror Force

From the Lorentz force equation, it follows that the equation for the parallel motion along the z axis is given by

$$m\frac{\mathrm{d}v_z}{\mathrm{d}t} = F_z = q\,[v_x B_y - v_y B_x]. \tag{3.4.4}$$

Using Eq. (3.4.3), the above equation can be written

$$F_z = -\frac{q}{2}\left(\frac{\partial B_z}{\partial z}\right)(v_x y - v_y x). \tag{3.4.5}$$

If the magnetic field varies sufficiently slowly, the transverse motion is very nearly circular, in which case, to a good approximation, we can write

$$x = \rho_c \sin\omega_c t, \;\; y = \frac{q}{|q|}\rho_c \cos\omega_c t \tag{3.4.6}$$

and

$$v_x = \omega_c\rho_c \cos\omega_c t, \;\; v_y = -\frac{q}{|q|}\omega_c\rho_c \sin\omega_c t. \tag{3.4.7}$$

The z component of the force then becomes

$$F_z = -\frac{\partial B_z}{\partial z}\left(\frac{|q|}{2}\omega_c\rho_c^2\right). \tag{3.4.8}$$

From Section 3.1.1, one can see that the term inside the parentheses, $|q|\omega_c\rho_c^2/2$, is simply the magnetic moment $\mu = w_\perp/B$. Thus, we obtain the following simple

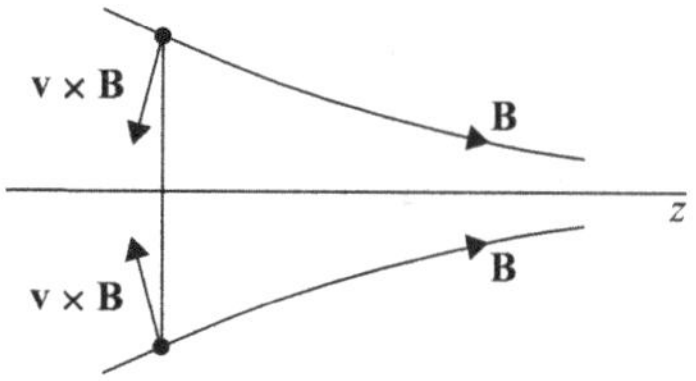

Figure 3.16 In a converging magnetic field, the $\mathbf{v} \times \mathbf{B}$ force has a component directed away from the region of stronger magnetic field.

equation for the z component of the force:

$$F_z = -\mu \frac{\partial B_z}{\partial z}. \tag{3.4.9}$$

If we let $\mathbf{m} = -\mu \hat{\mathbf{B}}$, this equation reduces to the well-known equation for the force on a magnetic moment in an inhomogeneous magnetic field, $\mathbf{F} = (\mathbf{m} \cdot \boldsymbol{\nabla})\mathbf{B}$. Thus, the parallel force arises from the interaction between the magnetic field and the magnetic moment produced by the cyclotron motion of the particle. The direction of the force is always such that the particle tends to be repelled from the region of strong magnetic field. The physical origin of this force is a consequence of the convergence of the magnetic field lines. As shown in Figure 3.16, if the magnetic field lines are converging, the Lorentz ($\mathbf{v} \times \mathbf{B}$) force has a component opposite to the direction of convergence, thereby producing a force away from the region of stronger magnetic field.

3.4.2 Azimuthal Motion: Constancy of the Magnetic Moment

In addition to the parallel component, the Lorentz force also has a component in the azimuthal direction. This force is given by

$$F_\phi = q v_z B_\rho, \tag{3.4.10}$$

and is directed as shown in Figure 3.17. The azimuthal force produces a torque that causes the perpendicular kinetic energy to change as the particle moves along the z

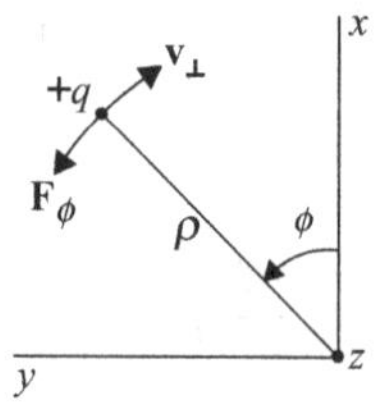

Figure 3.17 The coordinate system used to evaluate the azimuthal torque. Note that for a positive charge $\mathbf{v}_\perp$ is in the minus $\hat{\boldsymbol{\phi}}$ direction, so $v_\phi = -v_\perp$.

axis. The rate of change of the perpendicular kinetic energy can then be computed from the rate at which work is done by the azimuthal force, which is given to a good approximation by

$$\frac{\mathrm{d}w_\perp}{\mathrm{d}t} = v_\phi F_\phi = q v_\phi v_z B_\rho. \tag{3.4.11}$$

Substituting Eq. (3.4.2) for B_ρ and noting that $v_\phi = -(q/|q|)v_\perp$, the above equation becomes

$$\frac{\mathrm{d}w_\perp}{\mathrm{d}t} = |q| v_\perp v_z \left(\frac{\partial B_z}{\partial z}\right)\frac{\rho}{2}. \tag{3.4.12}$$

Note that although the perpendicular kinetic energy changes, the total kinetic energy remains constant because the $\mathbf{v}\times\mathbf{B}$ force is always perpendicular to $\mathbf{v}$. Using the cyclotron radius $\rho_c = mv_\perp/|q|B$ for ρ, the rate of change of the perpendicular kinetic energy can be written

$$\frac{\mathrm{d}w_\perp}{\mathrm{d}t} = \frac{w_\perp v_z}{B}\left(\frac{\partial B_z}{\partial z}\right). \tag{3.4.13}$$

From the above result, we next show that the magnetic moment is a constant of the motion. The rate of change of the magnetic moment is given by

$$\frac{\mathrm{d}\mu}{\mathrm{d}t} = \frac{\mathrm{d}}{\mathrm{d}t}\left(\frac{w_\perp}{B}\right) = \frac{1}{B}\frac{\mathrm{d}w_\perp}{\mathrm{d}t} - \frac{w_\perp}{B^2}\frac{\mathrm{d}B}{\mathrm{d}t}, \tag{3.4.14}$$

which, since $\mathrm{d}B/\mathrm{d}t = v_z(\partial B/\partial z)$, can be written

$$\frac{\mathrm{d}\mu}{\mathrm{d}t} = \frac{1}{B}\frac{\mathrm{d}w_\perp}{\mathrm{d}t} - \frac{w_\perp v_z}{B^2}\left(\frac{\partial B_z}{\partial z}\right). \tag{3.4.15}$$

Substituting expression (3.4.13) for $\mathrm{d}w_\perp/\mathrm{d}t$ into the above equation, we obtain

$$\frac{\mathrm{d}\mu}{\mathrm{d}t} = \frac{w_\perp v_z}{B^2}\left(\frac{\partial B_z}{\partial z}\right) - \frac{w_\perp v_z}{B^2}\left(\frac{\partial B_z}{\partial z}\right) = 0, \tag{3.4.16}$$

or μ = constant. As in the previous section, this result is valid only if the magnetic field gradient is sufficiently small. Later, in Section 3.8, a more specific criterion will be developed to determine the limits of this result.

3.4.3 Turning Points and the Pitch Angle

The parallel equations of motion (3.4.4) together with (3.4.9), which were derived for a small local region, can be easily extended to an arbitrary magnetic field configuration by simply changing z to s, the distance along the field line, v_z to

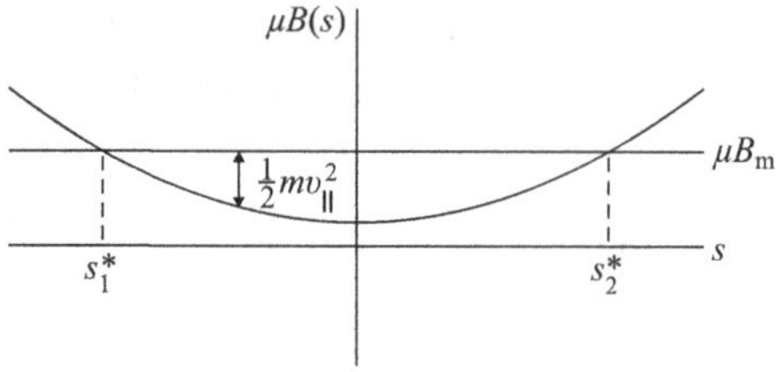

Figure 3.18 The effective potential for motion along the magnetic field.

$v_{\|}$, the parallel component of velocity, and B_z to B, the magnitude of the magnetic field. With these substitutions, the parallel equation of motion simplifies to

$$m\frac{\mathrm{d}v_{\|}}{\mathrm{d}t} = -\mu\frac{\partial B}{\partial s}, \tag{3.4.17}$$

where B is the magnetic field strength at position s. Writing $\mathrm{d}v_{\|}/\mathrm{d}t = v_{\|}\,\mathrm{d}v_{\|}/\mathrm{d}s$ and noting that μ is constant, the above equation can be integrated once to give

$$\frac{\mathrm{d}}{\mathrm{d}s}\left(\frac{1}{2}mv_{\|}^2 + \mu B\right) = 0, \tag{3.4.18}$$

which implies that

$$\frac{1}{2}mv_{\|}^2 + \mu B(s) = \mu B_{\mathrm{m}}, \tag{3.4.19}$$

where μ and B_{m} are constants. The above equation can be interpreted as an energy conservation equation for a particle of mass m moving in a one-dimensional effective potential $\mu B(s)$. The effective potential for a typical magnetic mirror field is shown in Figure 3.18.

The turning points of the motion are at s_1^* and s_2^*, where $B = B_{\mathrm{m}}$. Such a configuration is often called a magnetic bottle, since the particle is trapped between the two turning points. Since the parallel kinetic energy at the turning points is zero, it is easy to see that μB_{m} is the total kinetic energy of the particle, $w = \mu B_{\mathrm{m}}$. The energy conservation equation (3.4.19) can then be solved for the parallel velocity

$$v_{\|} = \frac{\mathrm{d}s}{\mathrm{d}t} = \pm\sqrt{\frac{2\mu}{m}(B_{\mathrm{m}} - B)}, \tag{3.4.20}$$

which in turn can be integrated to completely determine the guiding-center motion of the particle along the field line.

It is often useful to know the angle between the velocity vector and the magnetic field. This angle is called the pitch angle α. The pitch angle at any point can be computed from the magnetic moment by noting that $v_\perp = v \sin\alpha$, which gives

$$\mu = \frac{w_\perp}{B} = \frac{w \sin^2\alpha}{B}. \tag{3.4.21}$$

Since $w = \mu B_{\rm m}$ and $w_\perp = \mu B$, the above equation simplifies to

$$\sin^2\alpha = \frac{B}{B_{\rm m}}. \tag{3.4.22}$$

For a particle of given energy and magnetic moment, this equation can be used to compute the pitch angle at any point along the magnetic field line.

Practical considerations dictate that in any magnetic mirror configuration there is always an upper limit to the magnetic field strength at the ends or "throats" of the magnetic bottle. This upper limit implies that a minimum pitch angle exists, below which particles are not reflected but are lost from the bottle. If the minimum field strength along a field line at the mid-plane is defined to be B_0 as shown in Figure 3.19, one can see that the minimum pitch angle α_0, below which particles are lost, is given by

$$\sin^2\alpha_0 = \frac{B_0}{B_{\rm max}}, \tag{3.4.23}$$

where $B_{\rm max}$ is the maximum magnetic field strength. Any particle with a pitch angle less than α_0 will escape from the system. The cone of directions around the magnetic field at the angle α_0 is called the loss cone. Later we will show that a loss cone in the particle velocity distribution leads to instabilities that have detrimental effects on the confinement of a plasma in a magnetic mirror machine. Charged particles trapped in planetary radiation belts also have a loss cone because any particle moving with a pitch angle less than some minimum angle α_0 will hit the planet and be lost. The geometry involved is illustrated in Figure 3.20. From the illustration one can see that α_0 is also given by Eq. (3.4.23) where $B_{\rm max}$ is now the magnetic field strength at the surface of the planet.

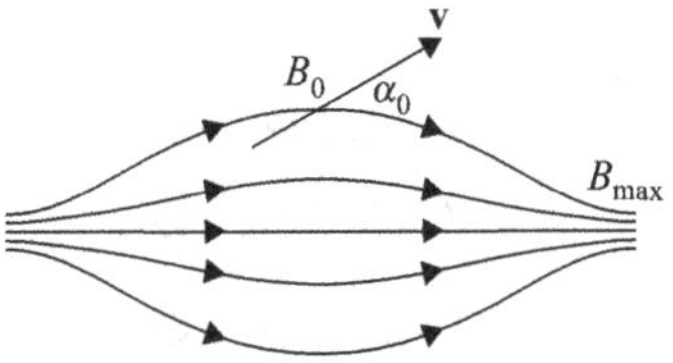

Figure 3.19 The magnetic field in an axially symmetric mirror machine always has an upper limit to the magnetic field, $B_{\rm max}$, at the ends of the mirror. This upper limit causes a lower limit to the pitch angle, α_0, at the middle of the machine. Particles with pitch angles less than α_0 are lost from the machine.

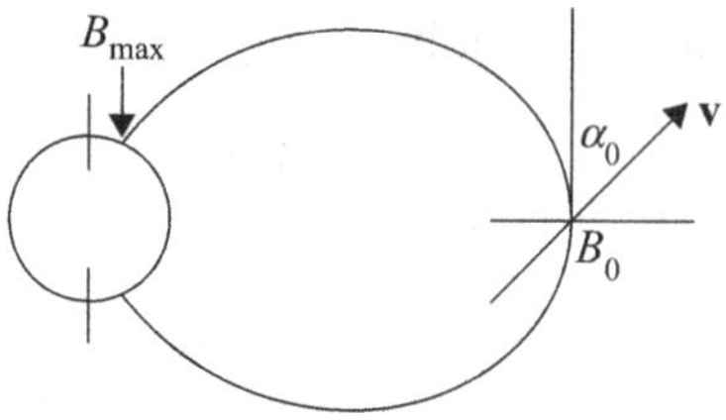

Figure 3.20 Planetary radiation belts also have a lower limit to the pitch angle, α_0. Particles with pitch angles less than α_0 strike the planet and are lost from the radiation belt.

3.5 Motion in a Time Varying Magnetic Field

If the magnetic field varies with time, Faraday's law, $\nabla \times \mathbf{E} = -\partial \mathbf{B}/\partial t$, shows that a non-conservative electric field is present. Therefore, the total energy of the particle is not constant. To illustrate the effect of a non-conservative electric field, consider the idealized case of a uniform magnetic field that is varying on a time-scale which is slow compared with the cyclotron period. If the magnetic field is directed upward, out of the paper, and $\partial \mathbf{B}/\partial t > 0$, the non-conservative electric field $\mathbf{E}$ is in the clockwise sense, as shown in Figure 3.21. A charged particle orbiting in this field gains an amount of kinetic energy per orbit given by

$$\Delta w_\perp = q \int_C \mathbf{E} \cdot \mathrm{d}\ell = -q \int_S \frac{\partial \mathbf{B}}{\partial t} \cdot \mathrm{d}\mathbf{A}, \tag{3.5.1}$$

where Stokes' theorem has been used to change the line integral to an area integral and the positive sense of the curve C is in the direction of the particle motion. To a good approximation the area of the orbit is $\pi \rho_c^2$. If the magnetic field varies slowly with time, the change in the perpendicular energy during one orbit is

$$\Delta w_\perp = |q| \left(\frac{\mathrm{d}B}{\mathrm{d}t} \right) \left(\pi \rho_c^2 \right), \tag{3.5.2}$$

where $|q|$ is used because the energy gain is independent of the sign of the charge. Note that for positive charges the curve C in Eq. (3.5.1) must be clockwise, which

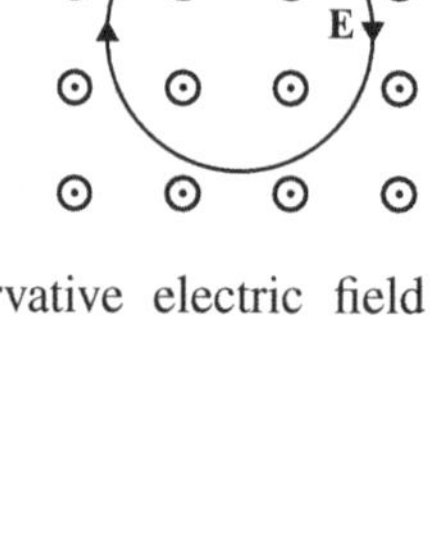

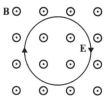

Figure 3.21 The non-conservative electric field induced by a time varying magnetic field.

means that $\mathbf{dA}$ must be directed into the paper. For negative charges the direction of $\mathbf{dA}$ is reversed, which accounts for the $|q|$ in Eq. (3.5.2). Since the cyclotron period is given by $\Delta t = 2\pi/\omega_c$, the rate of change of the perpendicular energy is given by

$$\frac{\mathrm{d}w_\perp}{\mathrm{d}t} = \frac{1}{2\pi}\omega_c|q|\left(\frac{\mathrm{d}B}{\mathrm{d}t}\right)\left(\pi\rho_c^2\right), \tag{3.5.3}$$

which, noting that $\mu = (1/2\pi)\omega_c|q|\pi\rho_c^2$, can be written

$$\frac{\mathrm{d}w_\perp}{\mathrm{d}t} = \mu\left(\frac{\mathrm{d}B}{\mathrm{d}t}\right). \tag{3.5.4}$$

Since $\mu = w_\perp/B$, the equation for the perpendicular energy then becomes

$$\frac{\mathrm{d}w_\perp}{w_\perp} = \frac{\mathrm{d}B}{B}, \tag{3.5.5}$$

which can be integrated directly to give

$$\frac{w_\perp}{B} = \text{constant}. \tag{3.5.6}$$

But this ratio is just the magnetic moment, μ. Thus, for a time-dependent magnetic field we again arrive at the conclusion that the magnetic moment is a constant.

The fact that the magnetic moment is constant in both time-dependent and converging magnetic fields is no accident. By simply changing to a coordinate system moving with the guiding center of the particle, the motion in a converging magnetic field can be converted to a time-dependent magnetic field. Thus, the azimuthal torque encountered in a magnetic mirror field can be thought of as arising from the non-conservative electric field produced by the time varying magnetic field encountered as the guiding center of the particle moves along the magnetic field line.

It is also interesting to note that conservation of the magnetic moment is equivalent to maintaining a constant magnetic flux through the cyclotron orbit. This interpretation follows from the expression for the magnetic moment which can be written

$$\mu = \frac{|q|}{2}\omega_c\rho_c^2 = \frac{1}{2\pi}\frac{q^2}{m}\Phi_B, \tag{3.5.7}$$

where

$$\Phi_B = \pi\rho_c^2 B \tag{3.5.8}$$

is the magnetic flux Φ_B through the orbit. Since μ = constant, the magnetic flux through the orbit remains constant, irrespective of whether the change is due to an

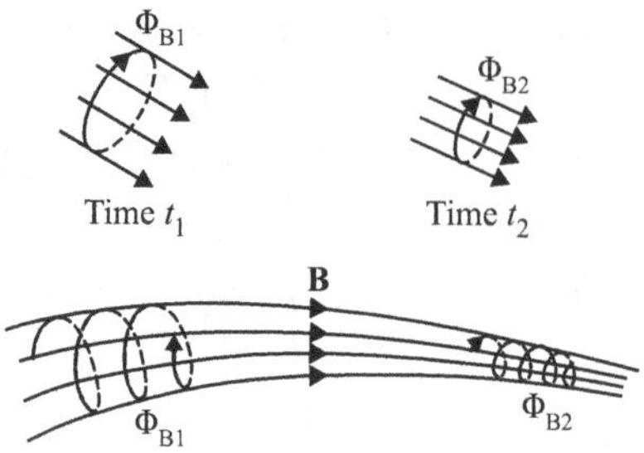

Figure 3.22 The change in the cyclotron orbit due to a time varying magnetic field (top), and a spatially varying magnetic field (bottom). In both cases the magnetic flux through the orbit is conserved, i.e., $\Phi_{B1} = \Phi_{B2}$.

explicit time variation or to the motion in an inhomogeneous magnetic field. This relationship is illustrated in Figure 3.22, where $\Phi_{B1} = \Phi_{B2}$ at two different times, t_1 and t_2, in both a time varying and a spatially varying magnetic field.

3.6 Polarization Drift

If the electric field in a $\mathbf{E} \times \mathbf{B}$ field geometry varies with time, then an additional drift is produced in the direction of $\mathrm{d}\mathbf{E}/\mathrm{d}t$. This drift arises because the acceleration caused by the electric field varies systematically around the cyclotron orbit and does not average to zero as it does when the electric field is constant. The result is a net shift of the guiding center in the direction of $\mathrm{d}\mathbf{E}/\mathrm{d}t$ for positive charges, and in the opposite direction for negative charges. Because positive and negative charges drift in opposite directions, this drift causes a charge polarization in the direction of $\mathrm{d}\mathbf{E}/\mathrm{d}t$, which is the reason the drift is called the polarization drift.

The basic principles of polarization drift are illustrated in Figure 3.23, which shows the motion of a positively charged particle initially orbiting around the origin in a spatially uniform magnetic field $\mathbf{B} = \hat{\mathbf{z}}B_0$ directed out of the paper. At time $t = 0$, a uniform linearly increasing electric field, $\mathbf{E} = \hat{\mathbf{y}}kt$, where k is a constant, is applied upward along the y axis perpendicular to the magnetic field. The guiding center immediately starts to move in the x direction due to the $\mathbf{E} \times \mathbf{B}$ drift. This drift velocity, $\mathbf{v}_E$, increases linearly with time due to the linearly increasing electric field strength. The guiding center is also displaced in the y direction during each orbit due to the polarization drift, $\mathbf{v}_P$. This displacement arises because the y component of the electric field during each orbit is larger during the v_y positive part of the orbit compared to the v_y negative part of the orbit. The guiding center motion, shown by dashed lines, then consists of the electric field drift, $\mathbf{v}_E$, in the $\mathbf{E} \times \mathbf{B}$ direction plus the polarization drift, $\mathbf{v}_P$, in the $\mathrm{d}\mathbf{E}/\mathrm{d}t$ direction.

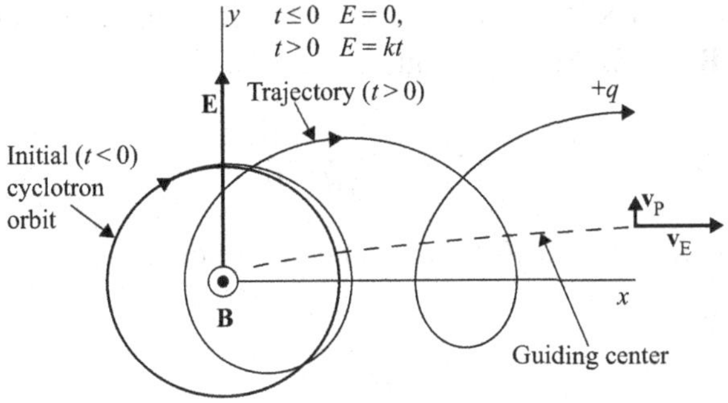

Figure 3.23 The trajectory of a positively charged particle, initially orbiting around a magnetic field out of the page, after a linearly increasing electric field, $\mathbf{E} = \hat{\mathbf{y}}kt$, is applied perpendicular to the magnetic field starting at $t = 0$. The polarization drift, $\mathbf{v}_\mathrm{P}$, is usually relatively small compared to the electric field drift, $\mathbf{v}_\mathrm{E}$.

To derive an equation for the polarization drift, we assume that $\mathrm{d}\mathbf{E}/\mathrm{d}t$ is sufficiently small that the particle motion can be resolved into two very different time-scales: a very rapid time-scale associated with the cyclotron motion and a very slow time-scale associated with the electric field and polarization drifts. It is convenient to start with the Lorentz force equation,

$$m\frac{\mathrm{d}\mathbf{v}}{\mathrm{d}t} = q[\mathbf{E} + \mathbf{v}\times\mathbf{B}], \tag{3.6.1}$$

and compute the drift velocity of the guiding center by averaging the perpendicular velocity, $\langle\mathbf{v}_\perp\rangle$. To carry out this averaging process we start by taking the cross-product of the Lorentz force equation with $\mathbf{B}/qB^2$ to obtain

$$\frac{m}{qB^2}\frac{\mathrm{d}}{\mathrm{d}t}(\mathbf{v}\times\mathbf{B}) = \frac{\mathbf{E}\times\mathbf{B}}{B^2} - \frac{\mathbf{B}\times(\mathbf{v}\times\mathbf{B})}{B^2}. \tag{3.6.2}$$

This step has the immediate benefit of producing the drift velocity, $\mathbf{v}_\mathrm{E} = \mathbf{E}\times\mathbf{B}/B^2$, in the first term on the right. The second term on the right can be expanded using the identity $\mathbf{B}\times(\mathbf{v}\times\mathbf{B}) = \mathbf{v}B^2 - \mathbf{B}(\mathbf{v}\cdot\mathbf{B})$. Recognizing that $\mathbf{v}_\parallel = \mathbf{B}(\mathbf{v}\cdot\mathbf{B})/B^2$ and $\mathbf{v}_\perp = \mathbf{v} - \mathbf{v}_\parallel$, the above equation simplifies to

$$\frac{m}{qB^2}\frac{\mathrm{d}}{\mathrm{d}t}(\mathbf{v}\times\mathbf{B}) = \mathbf{v}_\mathrm{E} - \mathbf{v}_\perp, \tag{3.6.3}$$

which can then be solved for the perpendicular velocity $\mathbf{v}_\perp$. Averaging $\mathbf{v}_\perp$ over one orbit then gives the drift velocity of the guiding center

$$\mathbf{v}_\mathrm{d} = \mathbf{v}_\mathrm{E} - \frac{m}{qB^2}\frac{\mathrm{d}}{\mathrm{d}t}(\langle\mathbf{v}\rangle\times\mathbf{B}), \tag{3.6.4}$$

where in computing $\langle \mathbf{v} \times \mathbf{B} \rangle$ we make use of the fact that $\mathbf{B}$ is constant. Finally, noting that $\langle \mathbf{v} \rangle \times \mathbf{B} = -\mathbf{E}$, the above equation can be written

$$\mathbf{v}_{\mathrm{d}} = \mathbf{v}_{\mathrm{E}} + \frac{m}{qB^2}\frac{\mathrm{d}\mathbf{E}}{\mathrm{d}t}, \tag{3.6.5}$$

from which the polarization drift term can be identified as

$$\mathbf{v}_{\mathrm{P}} = \frac{m}{qB^2}\frac{\mathrm{d}\mathbf{E}}{\mathrm{d}t}. \tag{3.6.6}$$

As expected, the direction of the polarization drift depends on the sign of the charge. Because of the sign dependence the polarization drift produces a current, which for electrons ($q = -e$) and a single species of positively charged ions ($q = +e$) is given by

$$\mathbf{J}_{\mathrm{P}} = \frac{n_0(m_{\mathrm{e}} + m_{\mathrm{i}})}{B^2}\frac{\mathrm{d}\mathbf{E}}{\mathrm{d}t}, \tag{3.6.7}$$

where n_0 is the number density and m_{e} and m_{i} are the electron and ion masses. Note that since $m_{\mathrm{i}} \gg m_{\mathrm{e}}$, the polarization current is dominated by the ion contribution. This is because the ions have a much larger cyclotron radius than the electrons and are therefore displaced by a much larger distance during each orbit.

It is instructive to compare the charge polarization caused by the polarization drift to the charge polarization that develops when an electric field is applied to a solid dielectric material. In a dielectric material the polarization is typically described by a dielectric constant, K, that relates the displacement field to the electric field via the relation $\mathbf{D} = \epsilon_0 K \mathbf{E}$. To compute the dielectric constant one can imagine placing the plasma between the plates of a capacitor, as in Figure 3.24, and then gradually increasing the electric field from zero to some final value by increasing the voltage on the plates of the capacitor. As the electric field increases, a polarization current, $\mathbf{J}_{\mathrm{P}}$, is produced due to the time varying electric field. This current causes a surface charge density $\sigma_{\mathrm{P}} = \int \mathbf{J}_{\mathrm{P}} \cdot \hat{\mathbf{n}} \mathrm{d}t$ to develop on the boundaries of the plasma, where $\hat{\mathbf{n}}$ is the normal to the surface of the plasma. The surface charge density is positive at the top and negative at the bottom. Following the usual development in the analysis of dielectric materials it follows that the polarization, $\mathbf{P}$, which is defined as the dipole moment per unit volume, can be related to the surface charge density by $\mathbf{P} \cdot \hat{\mathbf{n}} = \sigma_{\mathrm{P}}$; see Griffiths (2013). Using Eq. (3.6.7), it follows that the polarization is given by

$$\mathbf{P} = \epsilon_0 \left[\frac{n_0(m_{\mathrm{i}} + m_{\mathrm{e}})}{\epsilon_0 B^2} \right] \mathbf{E}, \tag{3.6.8}$$

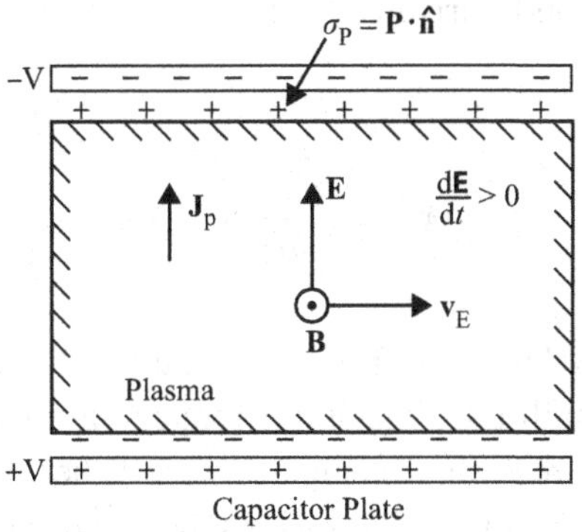

Figure 3.24 The capacitor plate model used to compute the dielectric constant, K. As the electric field increases, a polarization current, $\mathbf{J}_\mathrm{P}$, is produced. This causes a polarization, $\mathbf{P}$, and a surface charge density, σ_P, to develop, which is very similar to the polarization and surface charge that occurs in a solid dielectric.

where we assume that the plasma consists of a single species of ions of charge $q = +e$ and electrons of charge $q = -e$, both with a uniform number density n_0. Again, following the terminology used in the theory of dielectric materials, the quantity in the bracket can be identified as the dielectric susceptibility, χ. Once the susceptibility is known the dielectric constant can then be computed using $K = 1 + \chi$, which gives the following equation for the dielectric constant:

$$K = 1 + \frac{n_0(m_\mathrm{i} + m_\mathrm{e})}{\epsilon_0 B^2}. \qquad (3.6.9)$$

In the next chapter we show that this dielectric constant is exactly the value to account for the propagation of a type of low-frequency electromagnetic plasma wave, called an Alfvén wave, see Sections 4.4 and 6.5.

3.7 Ponderomotive Force

When a charged particle is placed in an inhomogeneous rapidly oscillating electric field, it not only responds at the oscillation frequency, but also experiences a force that acts to repel it from the region of strongest field. This force is called the ponderomotive force. To see how this force arises, consider an electric field $E_0 \cos\omega t$ that produces a force F_x on a particle of mass m and charge q initially located at $x = x_0$. If the electric field amplitude E_0 is uniform, then the equation of motion is simply

$$F_x = m\ddot{x} = qE_0 \cos\omega t, \qquad (3.7.1)$$

where ·· denotes the second derivative with respect to time. The above equation has a solution given by

$$\Delta x = -\frac{qE_0}{m\omega^2}\cos\omega t, \tag{3.7.2}$$

where $\Delta x = x - x_0$ is the deviation from its initial position, x_0. Next, let's assume that the electric field amplitude, E_0, is no longer uniform but instead increases slowly in the positive x direction, varying as $E_0(x)$. When the particle position, Δx, is at its maximum positive value ($\cos\omega t = -1$) the electric field is larger than in the uniform amplitude case, and also in the negative x direction due to the sign change between Eqs. (3.7.1) and (3.7.2). A positive charge then experiences an increased force in the negative x direction compared to the uniform amplitude case. One half-cycle later, when the particle position, Δx, is at its maximum negative value ($\cos\omega t = +1$) the electric field is smaller than in the uniform field case, but now in the positive x direction. Thus, the particle again experiences an increased force in the negative x direction. When averaged over one cycle, the net effect is to produce a force, the ponderomotive force, directed away from the region of highest electric field amplitude. The same effect also occurs for a negative charge. Therefore, when a charged particle is exposed to a spatial modulated high-frequency electric field, as illustrated in Figure 3.25, it always experiences a net force directed away from the regions of largest electric field amplitude.

To derive an equation for the ponderomotive force, we make two simplifying assumptions. First, we assume that the particle motion can be separated into two time-scales, a slow motion of the center of oscillation, x_0, and a rapid motion relative to the center of oscillation. Second, we assume that the amplitude of the electric field varies sufficiently slowly on a spatial scale compared to the rapid oscillation that the electric field amplitude can be represented by the first two terms in a Taylor series expansion, i.e., $E_0(x) = E_0 + (\mathrm{d}E_0/\mathrm{d}x)\Delta x$. Based on these assumptions, and introducing $x = x_0 + \Delta x$, the equation of motion (3.7.1) can be

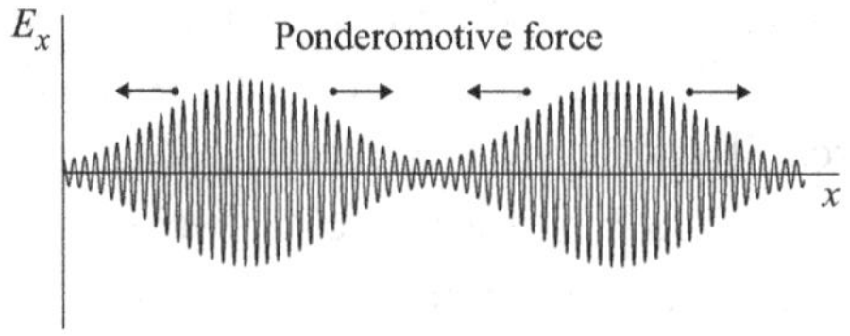

Figure 3.25 When a charged particle is exposed to a spatially modulated high-frequency electric field it experiences a net force, called the ponderomotive force, that is directed away from the regions of largest electric field.

rewritten as

$$m(\ddot{x}_0 + \Delta\ddot{x}) = q\left(E_0 + \frac{\mathrm{d}E_0}{\mathrm{d}x}\bigg|_{x_0} \Delta x\right)\cos\omega t, \tag{3.7.3}$$

where we specifically assume that $\ddot{x}_0 \ll \Delta\ddot{x}$ and $(\mathrm{d}E_0/\mathrm{d}x)\Delta x \ll E_0$. From the two-time-scale assumption one can see that the equation of motion separates nicely into two equations. The first equation, obtained by equating the rapidly oscillating second term on the left-hand side with the first term on the right-hand side, is identical to Eq. (3.7.2). After subtracting these terms one is left with the following equation for the motion of the center of oscillation:

$$m\ddot{x}_0 = q\frac{\mathrm{d}E_0}{\mathrm{d}x}\Delta x \cos\omega t. \tag{3.7.4}$$

The right-hand side of the above equation still has a rapidly oscillating component. This rapidly oscillating component can be eliminated by inserting Eq. (3.7.2) for Δx into Eq. (3.7.4) and averaging over many cycles of the high-frequency oscillations. The result is

$$m\langle\ddot{x}_0\rangle = -\frac{q^2 E_0}{2m\omega^2}\frac{\mathrm{d}E_0}{\mathrm{d}x}, \tag{3.7.5}$$

where $\langle x_0\rangle$ is the average position of the center of oscillation and the factor of 1/2 comes from averaging $\cos^2\omega t$. Noting that $E_0(\mathrm{d}E_0/\mathrm{d}x) = (1/2)(\mathrm{d}/\mathrm{d}x)(E_0^2)$, the above equation can be rewritten as

$$m\langle\ddot{x}_0\rangle = -\frac{q^2}{4m\omega^2}\frac{\mathrm{d}}{\mathrm{d}x}\left(E_0^2\right). \tag{3.7.6}$$

By recognizing that the left-hand side is the effective force acting on the center of oscillation (i.e., the ponderomotive force), and that comparable equations can be derived for the y and z components of the force, the general equation for the ponderomotive force can be written

$$\mathbf{F}_\mathrm{P} = -\frac{q^2}{4m\omega^2}\nabla(E_0^2), \tag{3.7.7}$$

where ∇ is the gradient operator. Note that the ponderomotive force is inversely proportional to the mass of the interacting particle. This means that electrons are much more affected than ions. Note also that the force is proportional to the gradient of the electric field energy density, which is given by $(1/2)\epsilon_0 E_0^2$, where ϵ_0 is the permittivity of free space. Later, in Chapter 11, we show that the pondermotive force can have important nonlinear effects in plasmas.

3.8 Adiabatic Invariants

From Hamiltonian mechanics, it is known that in a nearly periodic system with slowly varying parameters, the action integral

$$J = \oint p \, \mathrm{d}q \tag{3.8.1}$$

is an approximate constant of the motion, often called an adiabatic invariant (Goldstein, 1959). Here p and q are the conjugate momentum and position coordinates, and the integral is performed around a complete cycle while holding the parameters of the system constant. Adiabatic invariants are not exact invariants, but are approximately constant whenever the parameters of the system vary sufficiently slowly.

3.8.1 Adiabatic Invariant for a Harmonic Oscillator

As a simple example it is useful to discuss the action integral for the time-dependent harmonic oscillator, described by the equation

$$\frac{\mathrm{d}^2 x}{\mathrm{d}t^2} + \omega(t)^2 x = 0. \tag{3.8.2}$$

In a spring–mass system, for instance, x represents the displacement of the mass from equilibrium when the spring is unstretched. The frequency $\omega(t)$ is determined by two system parameters, the spring constant and the particle mass. If these parameters are held fixed, then the frequency $\omega(t)$ is constant, and the position $q = x$ and the conjugate momentum $p = m v_x$ are given by

$$q = A \sin \omega t \tag{3.8.3}$$

and

$$p = m\omega A \cos \omega t, \tag{3.8.4}$$

where A is the amplitude of the oscillation. The resulting trajectory in the p, q plane is an ellipse, as shown in Figure 3.26. The action integral is then simply the area

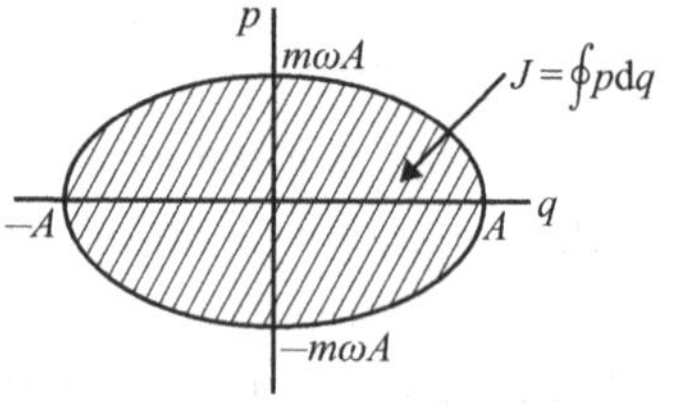

Figure 3.26 For a harmonic oscillator, the motion in the p, q plane is an ellipse. The action integral, J, is the area of the ellipse.

of the ellipse, πab, where a and b are the minor and major radii of the ellipse, so that

$$J = \pi m \omega A^2. \tag{3.8.5}$$

If the oscillator frequency varies sufficiently slowly, the action integral J is an approximate constant of the motion (i.e., an adiabatic invariant). The amplitude of the oscillation must then vary in such a way that $\omega A^2 \simeq$ constant, or $A \propto 1/\sqrt{\omega}$. Since the total energy is given by $W = (1/2)m\omega^2 A^2$, the adiabatic invariance of J is equivalent to the statement that

$$\frac{W}{\omega} = \text{constant}. \tag{3.8.6}$$

This result applies, for example, to a simple pendulum, the length of which is slowly varied. If both W and ω change sufficiently slowly, the ratio of these two quantities remains approximately constant.

Although the action integral is known to be an approximate constant of the motion if the parameters of the system vary sufficiently slowly, it is interesting to ask just how slow these variations must be for the concept to be valid. This question can be answered by investigating the so-called WKB solution of the time-dependent harmonic oscillator equation, which is given by

$$X_{\mathrm{WKB}} = \frac{1}{\sqrt{\omega(t)}} \exp\left[\pm \mathrm{i} \int^t \omega(t')\, dt'\right], \tag{3.8.7}$$

after Wentzel (1926), Kramers (1926), and Brillouin (1926). Note that the amplitude factor in the WKB solution, $A = 1/\sqrt{\omega(t)}$, is such that the action integral $J = \pi m \omega A^2$ is precisely constant. It is tedious but relatively straightforward (Problem 3.11) to show that X_{WKB} is an exact solution to the differential equation

$$\frac{\mathrm{d}^2 X_{\mathrm{WKB}}}{\mathrm{d}t^2} + \left[\omega^2 + \frac{\ddot{\omega}}{2\omega} - \frac{3}{4}\left(\frac{\dot{\omega}}{\omega}\right)^2\right] X_{\mathrm{WKB}} = 0, \tag{3.8.8}$$

where $\cdot$ denotes the time derivative. From the above equation one can see that the WKB solution is a good solution to the time-dependent harmonic oscillator equation whenever

$$\omega^2 \gg \left|\frac{3}{4}\left(\frac{\dot{\omega}}{\omega}\right)^2 - \frac{\ddot{\omega}}{2\omega}\right|. \tag{3.8.9}$$

As a rough rule we can say that the action integral is a "good" constant of the motion if the fractional change in the oscillator frequency is small during one cycle. Note that the WKB solution always fails whenever the frequency goes through zero ($\omega = 0$).

How good an invariant is the adiabatic invariant? This question received considerable attention in the context of the invariance of the magnetic moment in the 1960s; see Kruskal (1962) and Northrop (1963). The question has great practical consequences for the confinement of charged particles in a magnetic configuration. In the context of the present discussion, the invariant J can be written as a perturbative expansion in the small parameter ε; that is,

$$J(\varepsilon) = J_0 + \varepsilon J_1 + \varepsilon^2 J_2 + \cdots \tag{3.8.10}$$

It can be shown that

$$\Delta J(\varepsilon) \equiv J(\varepsilon) - J_0 = c_1 \exp(-c_2 \varepsilon), \tag{3.8.11}$$

where c_1 and c_2 are positive constants of the order unity. Equation (3.8.11) implies that $J(\varepsilon)$ is constant to all orders in ε; that is, $\Delta J(\varepsilon) \to 0$ as $\varepsilon \to 0$ faster than any power of ε.

3.8.2 Adiabatic Invariants for an Axially Symmetric Magnetic Mirror

It is evident from the previous section that an action integral exists for each degree of freedom that exhibits a periodicity. For the case of a charged particle in an axially symmetric magnetic mirror field, three types of periodic motion can be identified: (1) the cyclotron motion of the particle around the magnetic field, (2) the parallel bounce motion along the magnetic field line, and (3) the azimuthal gradient and curvature drift around the symmetry axis. These three types of motion are illustrated in Figure 3.27. The action integrals corresponding to these three motions are called the first, second, and third adiabatic invariants, respectively.

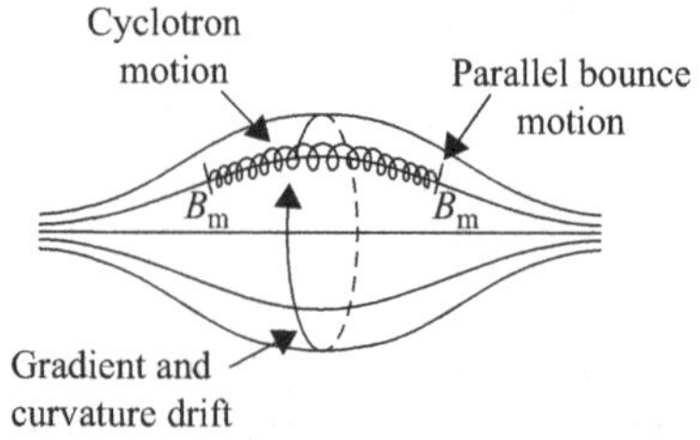

Figure 3.27 The three types of motion for a particle trapped in an axially symmetric magnetic mirror field.

The First Adiabatic Invariant

The first adiabatic invariant can be obtained from the equations of motion for a particle in a uniform magnetic field B_z, which are

$$m\frac{dv_x}{dt} = qv_yB_z \quad \text{and} \quad m\frac{dv_y}{dt} = -qv_xB_z. \tag{3.8.12}$$

By eliminating either v_x or v_y between these two equations, the equations for the x and y motions reduce to two identical harmonic oscillator equations:

$$\frac{d^2x}{dt^2} + \omega_c^2 x = 0 \quad \text{and} \quad \frac{d^2y}{dt^2} + \omega_c^2 y = 0. \tag{3.8.13}$$

Since the amplitude of the x and y motions is just the cyclotron radius ρ_c, the corresponding action integral is

$$J = \pi m \omega_c \rho_c^2. \tag{3.8.14}$$

Except for a multiplicative constant, this action integral is the same as the magnetic moment

$$\mu = \frac{|q|}{2}\omega_c\rho_c^2. \tag{3.8.15}$$

Therefore, the first adiabatic invariant is simply the magnetic moment, which is an approximate constant of the motion whenever the magnetic field varies sufficiently slowly.

The Second Adiabatic Invariant

The action integral for the parallel motion can be written

$$J = m\oint v_\parallel \, ds, \tag{3.8.16}$$

where the integration is along the magnetic field line between the two mirror points. The parallel velocity can be determined from the energy conservation equation (3.4.19),

$$\frac{1}{2}mv_\parallel^2 + \mu B(s) = \mu B_m, \tag{3.8.17}$$

so that

$$J = \sqrt{2\mu m}\oint \sqrt{B_m - B(s)}\, ds. \tag{3.8.18}$$

The quantity J is called the second adiabatic invariant. Since μ is a constant, the second adiabatic invariant is equivalent to requiring that the integral

$$\int_a^b \sqrt{B_{\rm m} - B(s)}\, {\rm d}s \tag{3.8.19}$$

be a constant, where a and b are the two mirror points.

To illustrate the use of the second adiabatic invariant, consider the situation where the magnetic field is slowly changed from some initial configuration $B_{\rm i}(s)$ to some final configuration $B_{\rm f}(s)$. Since the magnetic field changes with time, the total energy, $w = \mu B_{\rm m}$, is no longer a constant of the motion. The locations of the new mirror points can be determined from the invariance of J which requires that

$$\int_{a_{\rm i}}^{b_{\rm i}} \sqrt{B_{\rm mi} - B_{\rm i}(s)}\, {\rm d}s = \int_{a_{\rm f}}^{b_{\rm f}} \sqrt{B_{\rm mf} - B_{\rm f}(s)}\, {\rm d}s. \tag{3.8.20}$$

Since the initial configuration is known, the integral on the left has a known value. The final mirror point field $B_{\rm mf}$ and the corresponding mirror points $a_{\rm f}$ and $b_{\rm f}$ are then adjusted until the integral on the right matches the known value for the integral on the left. This procedure is illustrated schematically in Figure 3.28, where $B_{\rm i}(s)$ is the initial magnetic field and $B_{\rm f}(s)$ is the final magnetic field. Once the final mirror point field is determined, the final total energy is given by $w_{\rm f} = \mu B_{\rm mf}$.

The second adiabatic invariant also applies to geometries where the magnetic field is not azimuthally symmetric. Since gradient and curvature drifts cause an azimuthal drift motion, the problem is usually to determine the field line that the particle is on after it has drifted to a new azimuthal location. If the magnetic field is constant in time, then the total energy is conserved and the mirror point field $B_{\rm m}$ is constant. The second adiabatic invariant can then be written in the form

$$I = \int_a^b \left(1 - \frac{B(s)}{B_{\rm m}}\right)^{1/2} {\rm d}s = \text{constant}, \tag{3.8.21}$$

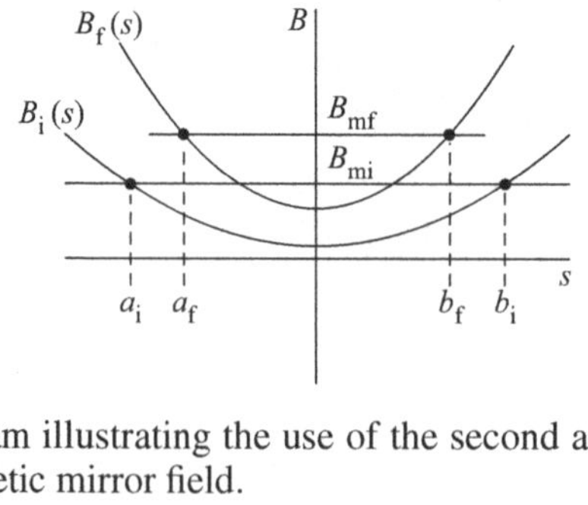

Figure 3.28 A diagram illustrating the use of the second adiabatic invariant in a slowly varying magnetic mirror field.

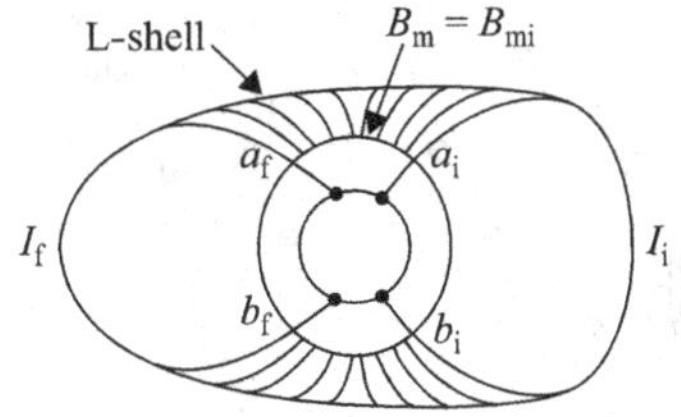

Figure 3.29 As a particle drifts in azimuth in a planetary magnetic field it traces out a surface called an *L-shell.*

where the integral I is called the integral invariant. As the particle drifts in azimuth, the locations of the new mirror points are determined by two conditions: first, that the magnetic field strength at the mirror points remain constant; and second, that the integral I remain constant. This procedure is often used to determine the drift motion of a particle trapped in a planetary magnetic field. Higher-order multipole moments typically cause planetary magnetic fields to have substantial azimuthal distortions relative to an ideal dipole field. Conservation of the first adiabatic invariant assures us that the magnetic field at the mirror points remains constant ($B_m = B_{mi}$). As shown in Figure 3.29, the condition $B_m = B_{mi}$ corresponds to a quasi-spherical surface surrounding the planet on which the mirror points must lie as the particle drifts in azimuth (i.e., local time).

Gradient and curvature drifts cause a particle initially trapped on a field line (a_i, b_i) with integral invariant I_i to drift to a field line (a_f, b_f) that satisfies $I_f = I_i$. This field line can be identified by computing I for a series of field lines on the left side of the planet, starting at B_{mi} in the northern hemisphere and ending at B_{mi} in the southern hemisphere. This process is continued until the magnetic field line is found for which $I_f = I_i$. The locus of all field lines around the planet with the same I for a given B_m generates a surface called an L-shell. As particles drift around the planet they stay on the same L-shell. The L-value for a given shell is defined to be the equatorial radial distance measured in planetary radii of a field line with the same values of I and B_m in a dipole field that has the same magnetic moment as the planet. The L-value and mirror magnetic field B_m provide a very useful set of coordinates for ordering measurements of energetic charged particles trapped in planetary magnetic fields. For a further discussion, see McIlwain (1961).

The Third Adiabatic Invariant

The third adiabatic invariant is associated with azimuthal drift motions and only exists for fields with a well-defined axial symmetry, such that the particles drift around in nearly closed orbits as illustrated in Figure 3.30. For such orbits, the adiabatic invariant is simply the magnetic flux linking the drift orbit. To prove this result, for simplicity we only consider orbits near the axis of symmetry. If

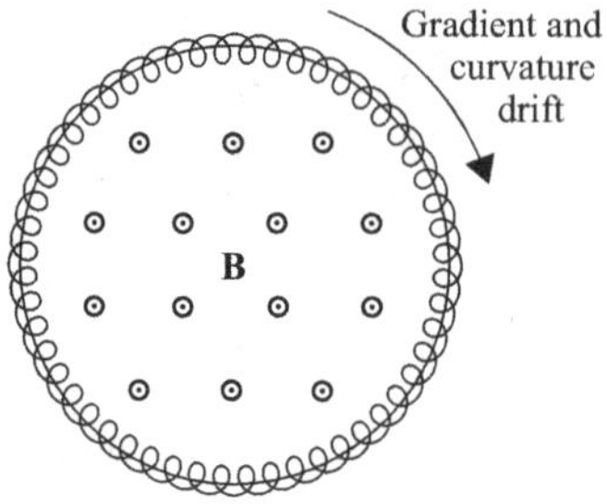

Figure 3.30 In a slowly varying magnetic field, the magnetic flux through the drift orbit is constant.

the magnetic field changes with time, it produces an azimuthal non-conservative electric field given by Faraday's law,

$$\int_C \mathbf{E}\cdot \mathrm{d}\ell = -\int_S \frac{\partial \mathbf{B}}{\partial t}\cdot \mathrm{d}\mathbf{A}, \tag{3.8.22}$$

where the integral is evaluated along the drift contour. Making use of the assumed axial symmetry and the weak radial dependence of B near the axis, the integrals in the above equation can be evaluated to show that

$$2\pi R E = -\frac{\mathrm{d}B}{\mathrm{d}t}(\pi R^2) \tag{3.8.23}$$

or

$$E = -\frac{R}{2}\frac{\mathrm{d}B}{\mathrm{d}t}. \tag{3.8.24}$$

The azimuthal electric field produces a radial $\mathbf{E}\times\mathbf{B}$ drift, given by

$$\upsilon_{\mathrm{E}} = \frac{E}{B} = -\frac{R}{2B}\frac{\mathrm{d}B}{\mathrm{d}t}. \tag{3.8.25}$$

Since $\upsilon_{\mathrm{E}} = \mathrm{d}R/\mathrm{d}t$, it follows that

$$2\frac{\mathrm{d}R}{R} = -\frac{\mathrm{d}B}{B}. \tag{3.8.26}$$

The above equation can be integrated directly to show that the magnetic flux inside the drift orbit is constant; that is,

$$\Phi_{\mathrm{B}} = \pi R^2 B = \text{constant}. \tag{3.8.27}$$

This is the third adiabatic invariant. The third adiabatic invariant shows that the guiding center of a particle drifting azimuthally in a slowly varying axially symmetric magnetic mirror field stays on the surface of a flux tube.

3.8.3 Violations of the Adiabatic Invariants

Exact conservation of the three adiabatic invariants implies that once a particle is trapped in an axially symmetric magnetic mirror field it should remain trapped forever, spiraling around the magnetic field, bouncing back and forth between the mirror points, and drifting azimuthally around the axis of symmetry. Because experiments with magnetic confinement systems often show significant deviations from this idealized description, it is important to consider how the adiabatic invariants can be violated.

Since the adiabatic invariants are all related to action integrals, one can see that there are two ways in which the adiabatic invariants can be violated. First, the parameters of the system may vary so rapidly that the action integral is no longer an approximate constant of the motion. Second, the parameters may change in such a way that the system is no longer periodic, in which case the action integral no longer exists. Examples illustrating the various ways in which the three adiabatic invariants can be violated are discussed separately below.

Violations of the First Invariant

The conditions for violation of the first adiabatic invariant are easy to determine because this invariant is directly related to the WKB solution for the time varying harmonic oscillator. The condition for validity of the WKB solution is given by Eq. (3.8.9) with $\omega = \omega_c$, and is violated whenever either $\dot{\omega}_c$ or $\ddot{\omega}_c$ becomes sufficiently large, or whenever ω_c goes through zero. The variations can be due either to an explicit time dependence, or to the motion of the guiding center through an inhomogeneous magnetic field.

One of the most important examples of an explicit time dependence arises from the presence of waves. Interactions with waves are particularly strong if the wave frequency, ω', in the guiding-center frame of reference is at the cyclotron frequency or is a multiple of the cyclotron frequency

$$\omega' = \omega - k_{\|}v_{\|} = n\omega_c, \tag{3.8.28}$$

where n is a non-zero integer. This type of interaction is called a resonant interaction, and the above condition is called cyclotron resonance. The term $k_{\|}v_{\|}$, where $\mathbf{k}$ is the wave vector, represents the Doppler shift. At cyclotron resonance, the wave field rotates in synchronism with the cyclotron motion of the particle. Even for very small wave intensities, this type of interaction can cause large changes in the perpendicular kinetic energy, thereby violating the first adiabatic invariant.

Abrupt spatial changes in the magnetic field, such as at shock waves and other types of discontinuities, can also violate the first adiabatic invariant. These effects

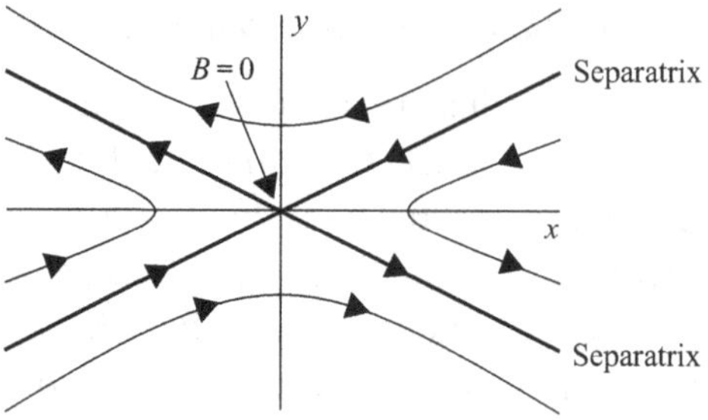

Figure 3.31 The magnetic field geometry around an x-type neutral point.

are particularly important for ions, since ions have much larger cyclotron radii than electrons and are more susceptible to spatial changes.

The first adiabatic invariant is also violated whenever the cyclotron frequency goes through zero. At zero frequency the motion is no longer periodic, so the action integral no longer exists. An important example in which this type of violation occurs is near an x-type neutral line, where the magnetic field, and hence the cyclotron frequency, goes to zero ($B = 0$). It can be shown that the simplest magnetic field geometry near a neutral line consists of an x-shaped separatrix that separates regions of oppositely directed magnetic fields, such as that shown in Figure 3.31. The neutral line in this diagram passes through the origin perpendicular to the plane of the paper. An x-type neutral line of this type is believed to form between Earth's magnetic field and the solar wind (Dungey, 1961), which is a high-speed plasma that flows radially outward from the Sun (Hundhausen, 1972). The geometry is easiest to picture when the magnetic field in the solar wind is directed southward, opposite to Earth's magnetic field near the equator. Under these conditions the neutral line encircles Earth. If a two-dimensional cut is made through the magnetosphere, such as in Figure 3.32, the neutral line appears as two x-type separatrices similar to Figure 3.31, one on the dayside (to the left in the diagram), and the other on the nightside (to the right) in a region called the magnetotail. Since the magnetic field reverses direction across the neutral line, particles moving in the vicinity of the neutral line follow S-shaped trajectories in the y, z plane as shown in Figure 3.33. For trajectories of this type, the first adiabatic invariant is obviously violated since the essential cyclotron motion no longer occurs. Because charges of opposite sign move in opposite directions, the S-type trajectories cause a current to flow along the neutral line. This current must be present to satisfy Ampère's law, $\nabla \times \mathbf{B} = \mu_0 \mathbf{J}$, in the region of oppositely directed field.

A particularly interesting modification to the basic x-type magnetic field configuration arises if a static electric field is present parallel to the neutral line. As shown in Figure 3.34, this electric field causes an $\mathbf{E} \times \mathbf{B}$ drift, $\mathbf{v}_E$, toward the neutral

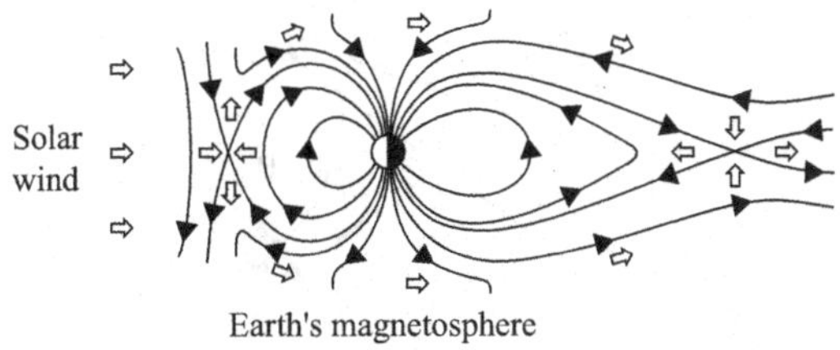

Figure 3.32 The magnetic field configuration near Earth when the solar wind magnetic field is directed southward, opposite to Earth's magnetic field.

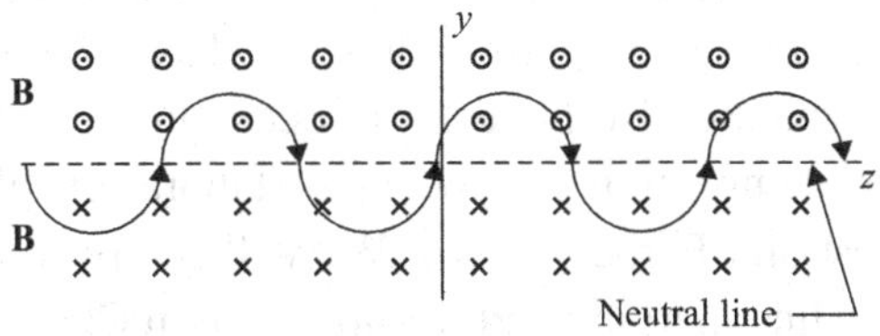

Figure 3.33 Particles moving along the neutral line follow S-shaped trajectories, which violate the first adiabatic invariant.

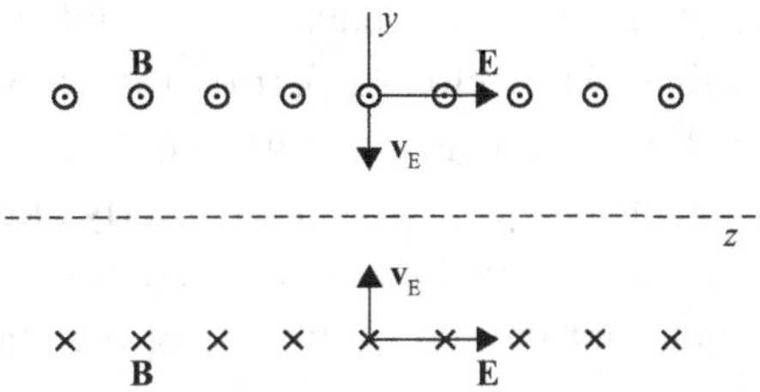

Figure 3.34 If an electric field is imposed along the neutral line, the resulting $\mathbf{E} \times \mathbf{B}$ drift drives particles into the neutral line.

line (downward for $y > 0$, and upward for $y < 0$). This $\mathbf{E} \times \mathbf{B}$ drift continuously transports particles into the region where the first adiabatic invariant is violated. The current produced by the S-type trajectories along the neutral line also results in a positive Joule heating, $\mathbf{E} \cdot \mathbf{J}$, indicating that electrical energy is being dissipated. Viewed in the x, y plane, the $\mathbf{E} \times \mathbf{B}$ drift transports plasma across the x-shaped separatrices, from regions A–A′ to regions B–B′, as shown in Figure 3.35.

This type of plasma transport, across a boundary between regions with topologically different magnetic field configurations, is called magnetic merging or reconnection. The terminology comes from magnetospheric physics. On the dayside of Earth, the solar wind carries magnetic field lines into the nose of the magnetosphere where they are "merged" with Earth's magnetic field lines, as indicated by the open arrows near the dayside x-point in Figure 3.32. These same magnetic field lines are carried over the polar caps by the motion of the solar wind

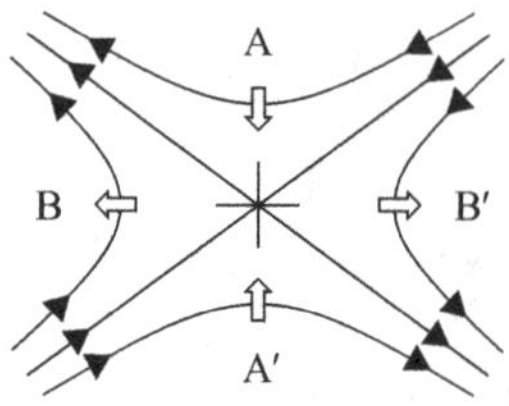

Figure 3.35 The $\mathbf{E} \times \mathbf{B}$ drift velocities (large open arrows) near an x-type neutral line with a superimposed electric field directed upward, out of the paper.

and are then "reconnected" at the nightside x-point. For a further discussion of this field geometry and the associated plasma flow, see Kivelson and Russell (1995). Neutral lines of this type are also believed to occur in the solar corona (Parker, 1963) and have been demonstrated to occur in laboratory plasma experiments (Biskamp, 2000; Priest and Forbes, 2000). We will return to the very important topic of magnetic field line merging and reconnection in Chapter 7.

Violations of the Second Invariant

Electromagnetic perturbations that vary on time-scales comparable to or less than the bounce time can violate the second adiabatic invariant. Since the bounce time is usually much longer than the cyclotron period, the second invariant is more easily violated than the first invariant. As with the first invariant, any type of resonance with the bounce motion provides a particularly effective mechanism for violating the second invariant. A periodic interaction of this type is called a bounce resonance. Because the relevant motion occurs along the magnetic field line, the perturbing force must have a component parallel to the magnetic field. The main effect of a bounce resonance is to change the location of the mirror points, and therefore the total energy of the particle. Magnetohydrodynamic waves, discussed in Chapter 6, provide a common source of such perturbations.

If the magnetic field configuration changes to such an extent that the particle is no longer trapped, then the second adiabatic invariant is violated because the corresponding action integral no longer exists. This can occur, for instance, when the mirror points are brought together by a compression parallel to the magnetic field. Conservation of the second adiabatic invariant,

$$J = m \oint v_{\parallel} \, \mathrm{d}s, \tag{3.8.29}$$

shows that the parallel velocity, and therefore the total energy of the particle, must increase during such a compression. Since the magnetic mirror field strength required to reflect the particle increases in direct proportion to the total energy ($w = \mu B_{\mathrm{m}}$), the particle energy eventually increases to the point where it can no longer remain trapped. The particle then escapes from the system, violating

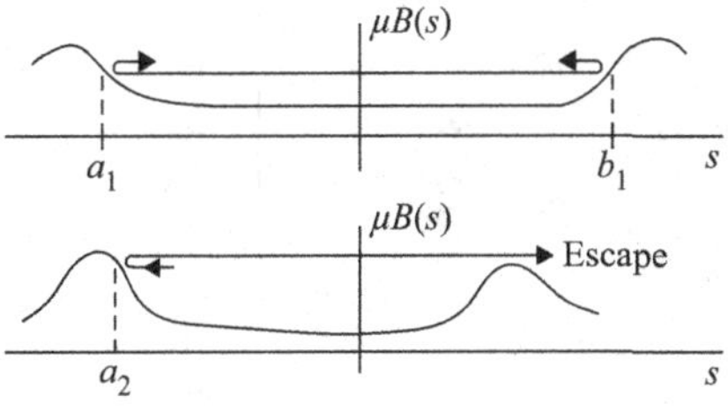

Figure 3.36 Violation of the second adiabatic invariant due to escape of the particle during a parallel compression.

the second invariant. This type of process, illustrated in Figure 3.36, has been suggested by Fermi (1949) as a mechanism for accelerating cosmic rays to high energies. In Fermi's acceleration mechanism, a particle is reflected back and forth between regions of strong magnetic field embedded in two approaching plasma clouds. As the distance between the two approaching clouds decreases, the particle is accelerated by repeated reflection between the two clouds, conserving the second adiabatic invariant. Eventually the particle energy increases to the point where the magnetic field in the clouds cannot reflect the particle, and the particle escapes. Fermi acceleration may also be effective in explaining the acceleration of particles by shock waves associated with solar flares and supernova explosions. As we will show in Chapter 8, a substantial jump in the magnetic field occurs at a shock wave. If a particle is trapped between the shock and a stationary reflection point ahead of the shock, substantial acceleration can occur before the particle escapes from the system.

Violations of the Third Invariant

Since the azimuthal drift has the largest period of the three types of periodic motions that occur in a magnetic mirror, the third invariant is the easiest to violate. Any change in the system that occurs on a time-scale comparable to or shorter than the azimuthal drift period violates the third invariant. In planetary magnetospheres, changes in electric and magnetic fields due to variations in the solar wind pressure can easily cause violations of the third invariant. If the third adiabatic invariant is violated, particles can move from one L-shell to another as they drift azimuthally around the planet. If the perturbing forces are random, the resulting radial motion can be described as a random walk, and is often called radial diffusion. Radial diffusion is believed to be the primary mechanism by which particles are transported radially inward in planetary radiation belts. If a particle diffuses into a dipole magnetic field from a large distance while conserving the first ($\mu = w_\perp / B$) and second ($J = m \int v_\| \, ds$) adiabatic invariants, considerable energization can occur. Since the magnetic field strength of a dipole field varies

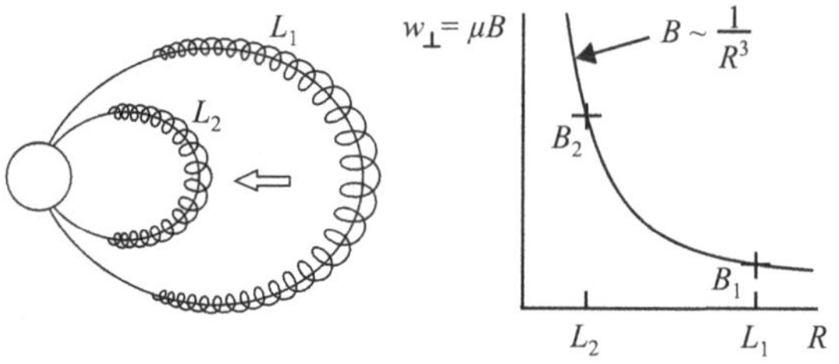

Figure 3.37 Violations of the third adiabatic invariant can cause particles to drift azimuthally onto different L-shells (L_1 to L_2). If the first adiabatic invariant is conserved, the increase in the magnetic field strength (B_1 to B_2) causes a considerable increase in the perpendicular energy, $w_\perp$.

as $1/R^3$, conservation of the first adiabatic invariant implies that the perpendicular energy should vary as $1/R^3$. This process is illustrated in Figure 3.37. As a particle moves from L_1 to L_2, the perpendicular kinetic energy increases by a factor of $(L_2/L_1)^3$. Since the length of the magnetic field decreases substantially in going from L_1 to L_2, conservation of the second invariant shows that the parallel kinetic energy also increases by a substantial factor. The change in the parallel energy is somewhat difficult to analyze since the mirror points change as the particle moves inward.

3.9 The Hamiltonian Method

Up to this point our treatment of charged particle motions in inhomogeneous and time varying magnetic fields has centered on certain adiabatic invariants that are valid in the limit of small spatial gradients and slow temporal variations. As is well known from classical mechanics, if a dynamical system has some geometric symmetry, then there is a corresponding quantity that is an exact constant of the motion. The analysis of such systems is best approached by using Hamilton's equations of motion; see Goldstein (1959). Although the Hamiltonian approach relies on the presence of symmetry, it does not assume adiabaticity, and thus complements the adiabatic invariant approach developed in the previous section.

3.9.1 Hamilton's Equations

In Hamiltonian mechanics, the state of a system is specified by a set of generalized coordinates q_i and their conjugate momenta p_i. The Hamiltonian is defined by

$$H = \sum_i \dot{q}_i p_i - L, \tag{3.9.1}$$

where L is the Lagrangian. For a charged particle moving in an electromagnetic field the Lagrangian is given by

$$L = \frac{1}{2}mv^2 + q\mathbf{A}\cdot\mathbf{v} - q\Phi, \tag{3.9.2}$$

where $\mathbf{A}$ is a vector potential defined by $\mathbf{B} = \nabla \times \mathbf{A}$, and Φ is a scalar potential defined by $\mathbf{E} = -\nabla\Phi - \partial\mathbf{A}/\partial t$. The conjugate momenta p_i are obtained from the Lagrangian by the relation

$$p_i = \frac{\partial L}{\partial \dot{q}_i}. \tag{3.9.3}$$

Once the Hamiltonian for the system is known as a function of the generalized coordinates and momenta, the motion of the system can be determined from Hamiltonian's equations

$$\dot{q}_i = \frac{\partial H}{\partial p_i}, \qquad \dot{p}_i = -\frac{\partial H}{\partial q_i}, \qquad \text{and} \qquad \frac{\partial H}{\partial t} = -\frac{\partial L}{\partial t}. \tag{3.9.4}$$

For a system with f degrees of freedom ($i = 1, \ldots, f$), Hamilton's equations constitute a system of $2f$ first-order differential equations that can be solved for the $2f$ variables q_i and p_i.

To use Hamilton's equations it is first necessary to determine the vector and scalar potentials from the known electric and magnetic fields. This step involves a certain arbitrariness (known as gauge freedom), since $\nabla \cdot \mathbf{A}$ is not specified. Usually the choice of $\mathbf{A}$ is made in such a way as to minimize the number of generalized coordinates by taking advantage of various symmetries. Once $\mathbf{A}$ and Φ are determined, the conjugate momenta are computed using Eq. (3.9.3). The equations for the conjugate momenta are then used to eliminate the $\dot{q}_i$ from Eq. (3.9.1), thus determining the Hamiltonian as a function of the coordinates, conjugate momenta, and time, $H(q_i, p_i, t)$. Hamilton's canonical equations (3.9.4) then yield a system of nonlinear differential equations that describe the evolution of the system.

Although Hamilton's equations can in principle be solved to give the evolution of the system, in most cases it is difficult, if not impossible, to obtain closed-form analytic solutions. Hamilton's equations are often non-integrable and sometimes exhibit "chaotic" solutions; that is, solutions that are so sensitive to initial conditions that extremely small differences in initial conditions produce exponentially large differences in the final solutions. However, if one or more constants of the motion can be identified, it is sometimes possible to obtain a qualitative understanding of the motions involved. If the Hamiltonian does not depend explicitly on a particular coordinate q_j, then from Eq. (3.9.4) one obtains

$$\dot{p}_j = -\frac{\partial H}{\partial q_j} = 0. \tag{3.9.5}$$

The corresponding conjugate momentum p_j is then a constant of the motion. Such a coordinate is called a cyclic coordinate. If, in addition, the Lagrangian is independent of time (i.e., static electric and magnetic fields) then

$$\frac{\partial H}{\partial t} = -\frac{\partial L}{\partial t} = 0, \tag{3.9.6}$$

which implies that the Hamiltonian itself is a constant. The Hamiltonian then corresponds to the total energy, W, of the system.

3.9.2 Hamiltonian for an Axisymmetric Magnetic Field

To illustrate the techniques involved in the Hamiltonian method, we next analyze an axially symmetric magnetic mirror field. The results obtained can then be compared and contrasted with the results from the adiabatic invariant approach.

For an axially symmetric system it is natural to use cylindrical coordinates (ρ, ϕ, z). The Lagrangian in cylindrical coordinates is

$$L = \frac{1}{2} m\left(\dot{\rho}^2 + \rho^2\dot{\phi}^2 + \dot{z}^2\right) + q\left(A_\rho\dot{\rho} + A_\phi\rho\dot{\phi} + A_z\dot{z}\right), \tag{3.9.7}$$

and the conjugate momenta are

$$p_\rho = m\dot{\rho} + qA_\rho, \tag{3.9.8}$$

$$p_\phi = m\rho^2\dot{\phi} + q\rho A_\phi, \quad \text{and} \tag{3.9.9}$$

$$p_z = m\dot{z} + qA_z. \tag{3.9.10}$$

After eliminating $\dot{\rho}, \dot{\phi}$, and $\dot{z}$ from Eq. (3.9.7), the Hamiltonian can be written

$$H = \frac{1}{2m}\left(p_\rho - qA_\rho\right)^2 + \frac{\left(p_\phi - q\rho A_\phi\right)^2}{2m\rho^2} + \frac{1}{2m}(p_z - qA_z)^2. \tag{3.9.11}$$

If the magnetic field has axial symmetry with no B_ϕ component, the field can be derived from a vector potential of the form $\mathbf{A} = [0, A_\phi(\rho, z), 0]$ with

$$B_\rho = -\frac{\partial A_\phi}{\partial z} \quad \text{and} \quad B_z = \frac{1}{\rho}\frac{\partial}{\partial \rho}\left(\rho A_\phi\right). \tag{3.9.12}$$

The conjugate momenta are then

$$p_\rho = m\dot{\rho}, \quad p_\phi = m\rho^2\dot{\phi} + q\rho A_\phi, \quad \text{and} \quad p_z = m\dot{z}. \tag{3.9.13}$$

Since ϕ is a cyclic coordinate ($\partial H/\partial\phi = 0$), from Eq. (3.9.5) one can see that

$$p_\phi = \text{constant}. \tag{3.9.14}$$

If the magnetic field is time-independent, the Hamiltonian is also a constant of the motion, so that

$$H = \frac{p_\rho^2}{2m} + \frac{p_z^2}{2m} + \psi(\rho, z) = W, \tag{3.9.15}$$

where $\psi(\rho, z)$ is an effective potential defined by

$$\psi(\rho, z) = \frac{1}{2m\rho^2}\left(p_\phi - q\rho A_\phi\right)^2, \tag{3.9.16}$$

and W is a constant.

Motion in an Axisymmetric Mirror Field

When the Hamiltonian is written in the form (3.9.15) it is evident that the motion is equivalent to the two-dimensional motion of a particle of mass m and total energy W in a potential $\psi(\rho, z)$. The motion is then bounded by the surface

$$W = \psi(\rho, z). \tag{3.9.17}$$

All that remains is to determine the general form of $\psi(\rho, z)$. Since p_ϕ is determined by the initial conditions, it is obvious that some value for p_ϕ exists such that $p_\phi = q\rho A_\phi$. The effective potential (at a fixed value of z) then has a minimum at a ρ-value determined by the condition $p_\phi = q\rho A_\phi$. A representative plot of $\psi(\rho, z)$ for a constant value of z is shown in Figure 3.38. If the total energy W is sufficiently small, one can see that the motion along the ρ direction is always bounded, with one bound inside and the other bound outside the point where $p_\phi = q\rho A_\phi$.

The location of the minimum at $p_\phi = q\rho A_\phi$ has an interesting interpretation. Let us compute the magnetic flux Φ_B through a surface of radius ρ perpendicular to

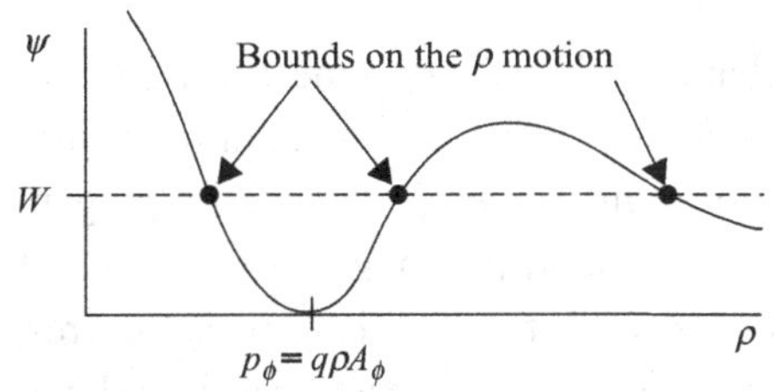

Figure 3.38 The effective potential, ψ, and the bounds on the ρ motion.

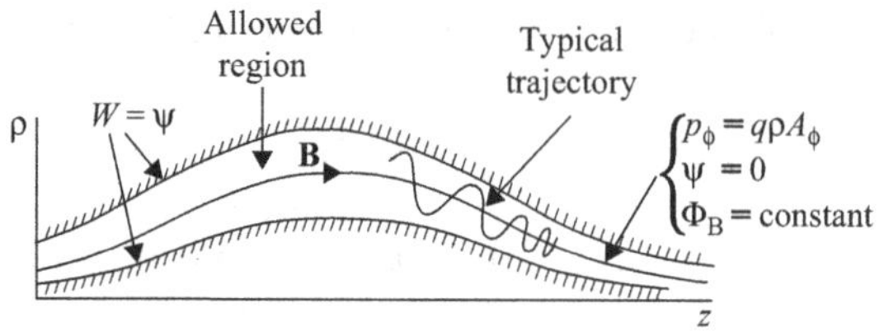

Figure 3.39 The bounds on the motion determined by the condition $W = \psi(\rho, z)$.

the z axis:

$$\Phi_{\mathrm{B}} = \int_0^\rho B_z 2\pi\rho\,\mathrm{d}\rho = \int_0^\rho \left[\frac{1}{\rho}\frac{\partial}{\partial\rho}\left(\rho A_\phi\right)\right] 2\pi\rho\,\mathrm{d}\rho. \tag{3.9.18}$$

The integral can be evaluated directly to give

$$\Phi_{\mathrm{B}} = 2\pi\rho A_\phi. \tag{3.9.19}$$

Thus, the condition

$$p_\phi = q\rho A_\phi(\rho, z) = \text{constant} \tag{3.9.20}$$

corresponds to a surface of a flux tube (i.e., a magnetic field line). Therefore, in the ρ, z plane, the effective potential has a minimum that follows a magnetic field line. As shown in Figure 3.39, the bounds on the motion prescribed by Eq. (3.9.17) represent two surfaces on either side of this magnetic field line. A particle moving in this two-dimensional potential well executes an oscillatory motion in the ρ direction while moving in the z direction, always remaining within the bounds given by $W = \psi(\rho, z)$. A turning point in the z motion occurs as the potential well narrows in the region of strong magnetic field. This turning point corresponds to the mirror point described in Section 3.8 using adiabatic invariants.

The azimuthal motion can be determined from the equation for p_ϕ (see Eq. (3.9.13)), which can be solved for $\dot{\phi}$ to give

$$\dot{\phi} = \frac{p_\phi - q\rho A_\phi}{m\rho^2}. \tag{3.9.21}$$

It is obvious that $\dot{\phi}$ changes sign every time the particle crosses the magnetic field line (where $p_\phi = q\rho A_\phi$). Therefore, the azimuthal motion consists of loops in the ρ, ϕ plane, as shown in Figure 3.40. The looping motion corresponds to the usual cyclotron motion around the magnetic field line. The asymmetry of the loops causes a secular motion in the ϕ direction that corresponds to the gradient drift discussed in Section 3.3.1. If the particle has a total energy sufficiently small to be trapped near the bottom of the potential well (see Figure 3.38) where the

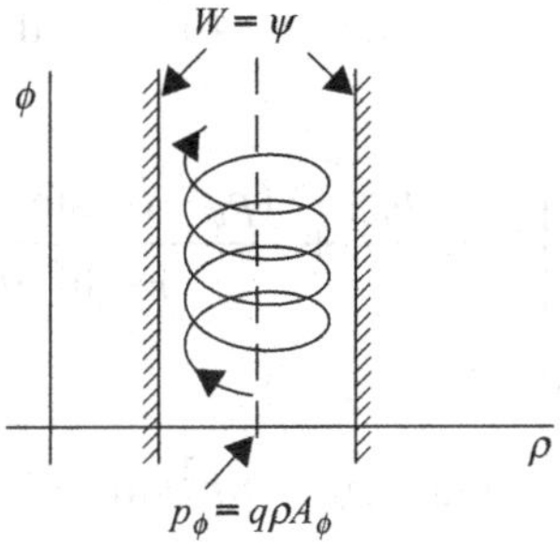

Figure 3.40 The azimuthal cyclotron motion around the magnetic field line.

potential is parabolic, then the ρ and ϕ motions are harmonic, corresponding to the usual nearly circular cyclotron motion.

If the magnetic field varies with time, then the total energy W is no longer constant and the bounds on the motion given by $W = \psi$ change with time. The variations could cause the particle to escape from the potential well. However, if the particle does not gain enough energy to escape from the potential well, it will continue to oscillate around a field line with the same constant magnetic flux Φ_B. This result is essentially equivalent to the conservation of the third adiabatic invariant, but generalized to allow an arbitrary radial dependence for the magnetic field.

It is useful to compare the results of the Hamiltonian and adiabatic invariant methods of analysis. Both methods show that in an axially symmetric static magnetic mirror field the particle performs a cyclotron motion around a magnetic field line while simultaneously bouncing back and forth between regions of strong magnetic field and drifting azimuthally around the axis of symmetry. However, the Hamiltonian method is more robust. The Hamiltonian method requires no assumption about the amplitude of the oscillations or the magnitude of the spatial gradients in the magnetic field. Furthermore, if the magnetic field is time-independent, the Hamiltonian method in some cases can place rigorous bounds on the particle motions.

Motion in a Dipole Magnetic Field

The motion of a charged particle in a dipole magnetic field is a basic problem that has applications to the study of both cosmic rays and planetary radiation belts. The general nature of the motion was first solved in a classic paper by Störmer (1907). Because a dipole field has axial symmetry, the Hamiltonian method is ideally suited to this problem. However, instead of the cylindrical coordinates used in the previous section, it is more convenient to use the spherical coordinates r, θ, ϕ, where r is the radius, θ is the polar angle, and ϕ is the azimuthal angle.

It is straightforward to show (Problem 3.14b) that the Hamiltonian in spherical coordinates is given by

$$H = \frac{1}{2m}\left[p_r^2 + \frac{p_\theta^2}{r^2} + \frac{(p_\phi - qr\sin\theta A_\phi)^2}{r^2\sin^2\theta}\right], \tag{3.9.22}$$

where the conjugate momenta are given by

$$p_r = m\dot{r}, \ p_\theta = mr^2\dot{\theta}, \quad \text{and} \quad p_\phi = mr^2\sin^2\theta\dot{\phi} + qr\sin\theta A_\phi. \tag{3.9.23}$$

An appropriate vector potential for a dipole magnetic field in spherical coordinates is

$$A_r = 0, \ A_\theta = 0, \quad A_\phi = \frac{\mu_0 M}{4\pi}\frac{\sin\theta}{r^2}, \tag{3.9.24}$$

where M is the magnetic moment. The effective potential can be shown to be

$$\psi = \frac{1}{2mr^2\sin^2\theta}\left(p_\phi - \frac{q\mu_0 M\sin^2\theta}{4\pi r}\right)^2. \tag{3.9.25}$$

The minimum in the effective potential, corresponding to $\psi = 0$, is located at

$$r = \frac{q\mu_0 M}{4\pi p_\phi}\sin^2\theta, \tag{3.9.26}$$

which is the equation for a dipole magnetic field line in spherical coordinates (see Problem 3.7a). The bounds on the motion are obtained by setting $\psi = W$, which gives

$$W = \frac{1}{2mr^2\sin^2\theta}\left(p_\phi - \frac{q\mu_0 M\sin^2\theta}{4\pi r}\right)^2. \tag{3.9.27}$$

The above equation has solutions for $1/r$ given by

$$\frac{1}{r} = \frac{2\pi p_\phi}{q\mu_0 M\sin^2\theta}\left[1 \pm \sqrt{1 \pm \frac{q\mu_0 M}{\pi p_\phi^2}\sqrt{2mW}\sin^3\theta}\right]. \tag{3.9.28}$$

If p_ϕ is negative, only one root has a real positive value, whereas if p_ϕ is positive, either one or three roots have real positive values. The number of roots when p_ϕ is positive is controlled by the quantity under the square root, which must be positive to give real values for r. The bounds on the motion can be summarized by the three diagrams in Figure 3.41.

Note that at an infinite distance the quantity p_ϕ is just the angular momentum, which can be written

$$p_\phi(\rho = \infty) = mbv_\infty, \tag{3.9.29}$$

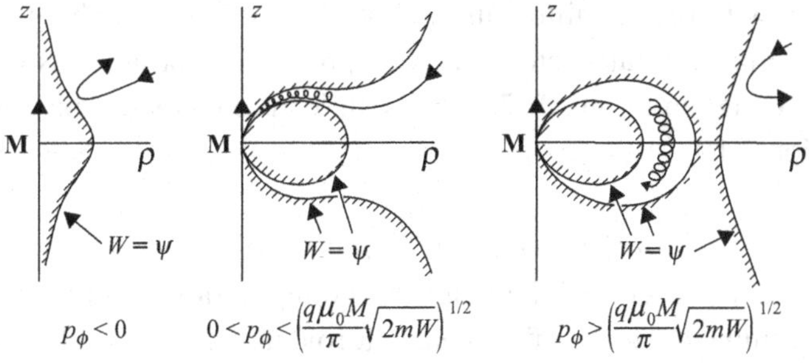

Figure 3.41 The bounds on the motion for a charged particle moving in a dipole magnetic field.

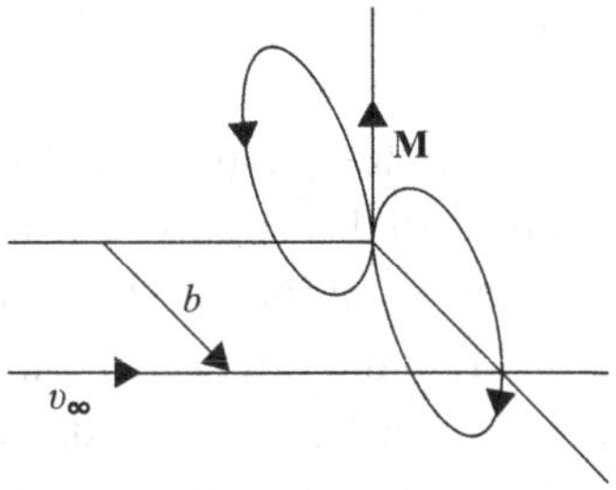

Figure 3.42 The geometry defining the impact parameter, b.

where b is the impact parameter relative to the z axis and v_∞ is the speed at infinity, as shown in Figure 3.42. For negative p_ϕ, all particles are reflected away from the dipole, irrespective of their total energy, as illustrated in the left diagram of Figure 3.41. This occurs because the $\mathbf{v}\times\mathbf{B}$ force has a component directed outward, away from the dipole. For positive p_ϕ, the $\mathbf{v}\times\mathbf{B}$ force has a component of force directed inward, toward the dipole, so the trajectories curve toward the dipole. For certain initial conditions the particles can approach the dipole very closely by going through the "throat" region near the equator and then spiraling down a magnetic field line, as illustrated in the middle diagram of Figure 3.41.

For p_ϕ greater than a critical value, given by

$$p_\phi > \left(\frac{q\mu_0 M}{\pi}\sqrt{2mW}\right)^{1/2}, \tag{3.9.30}$$

the allowed region breaks into two distinct regions, as illustrated in the right diagram of Figure 3.41. One of these regions corresponds to internally trapped particles, and to the radiation belts of Earth and other planets (Van Allen, 1996). It is interesting to note that the Hamiltonian method shows that, in the absence of time variations, this region is rigorously inaccessible to particles arriving from the outside, a result that cannot be claimed from the adiabatic invariant approach.

However, even with the Hamiltonian method, the motion of a charged particle in a dipole magnetic field falls into a class of problems that are "insoluble" in the formal sense (Dragt and Finn, 1976), and can lead to chaotic orbits, which is the topic of Section 3.10.

Motion in an Axisymmetric Tokamak

In Section 3.3.3, we discussed the magnetic configuration of a tokamak in which particles move under the joint influence of a toroidal magnetic field, B_t, and a much smaller poloidal magnetic field, B_p. The toroidal magnetic field can be written in the form

$$B_t = \frac{B_0 \rho_0}{\rho} = \frac{B_0 \rho_0}{\rho_0 + r\cos\theta}, \tag{3.9.31}$$

where (r, θ) are plane polar coordinates parameterizing the poloidal (i.e., meridional) cross section of a tokamak, and $\rho = \rho_0$ is the major and $r = a$ the minor radius of a tokamak of circular cross section, as shown in Figure 3.43. The aspect ratio of such a tokamak is defined by the relation $\varepsilon_t \equiv a/\rho_0$, and is usually smaller than unity. As illustrated in Figure 3.12, the toroidal magnetic field strength is larger on the inner edge ($\theta = \pi$) than on the outer edge ($\theta = 0$) of the torus. This difference creates a mirror-like magnetic geometry between the inner and outer edges, where the magnetic fields are given, respectively, by

$$B_{\max} = B_0 \rho_0 / (R_0 - \rho_0) \tag{3.9.32}$$

and

$$B_{\min} = B_0 \rho_0 / (R_0 + \rho_0). \tag{3.9.33}$$

Consider a particle at the outer edge with total speed v and parallel speed $v_\parallel$. As in Section 3.4.3, we can use the law of conservation of energy and the adiabatic

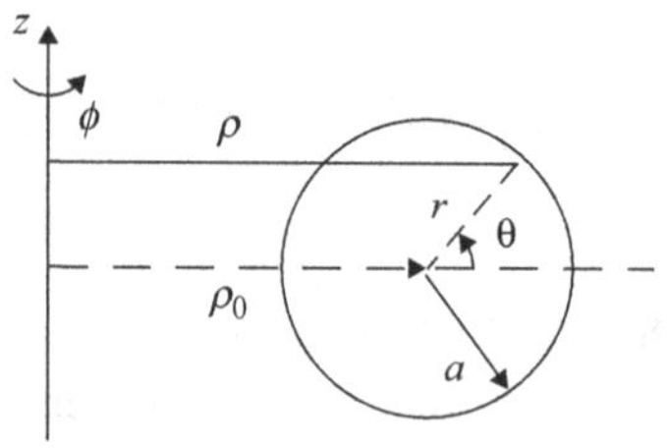

Figure 3.43 The coordinate system used to describe the poloidal cross section of an axisymmetric tokamak. Here $\rho = \rho_0 + r\cos\theta$, where ρ_0 is the major radius of the tokamak.

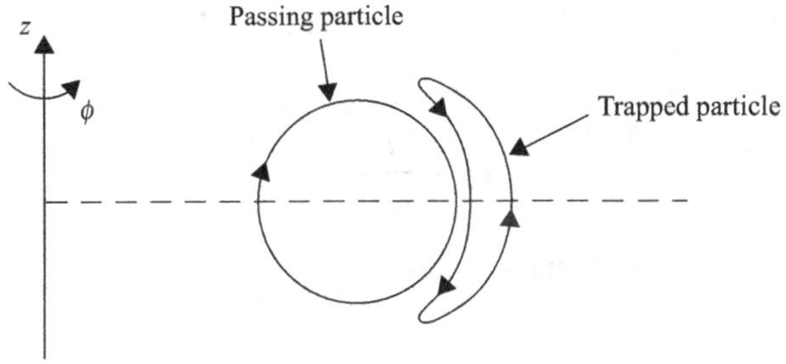

Figure 3.44 The guiding-center orbits of passing and trapped particles in a tokamak. The trapped particles execute banana orbits. The orbits are those of the guiding centers of charged particles but not the particles themselves, which exhibit rapid gyromotion around magnetic field lines (not shown) as their guiding centers drift in the inhomogeneous magnetic field.

invariance of the magnetic moment to obtain the pitch angle

$$\sin^2\alpha_t = \frac{R_0 - \rho_0}{R_0 + \rho_0}. \tag{3.9.34}$$

The above equation implies that all particles that obey the condition

$$\frac{v_\parallel^2}{v^2} < 1 - \frac{R_0 - \rho_0}{R_0 + \rho_0} = 2\frac{\rho_0}{R_0 + \rho_0} \cong 2\varepsilon_t \tag{3.9.35}$$

are trapped between the mirror points. In the last step, we have used the approximation $\varepsilon_t \ll 1$. Particles that do not obey condition (3.9.35) are not trapped, but become passing particles. In contrast with magnetic mirrors, passing particles are not lost from the tokamak. Figure 3.44 shows the projection of the guiding center of a trapped particle on a poloidal plane. Such an orbit is called a "banana orbit." Since the tokamak is axisymmetric, this orbit is independent of the toroidal angle ϕ.

A banana orbit arises due to the combined effects of guiding-center motion parallel to the magnetic field combined with the curvature and gradient drifts caused by the toroidal magnetic field. Using Eqs. (3.3.7) and (3.3.9), the velocity of the guiding center can be written

$$\mathbf{v}_0(t) = v_\parallel\hat{\mathbf{B}} + \mathbf{v}_G + \mathbf{v}_C = v_\parallel\hat{\mathbf{B}} + \frac{w_\perp}{qB}\left[\frac{\hat{\mathbf{B}}\times\nabla B}{B}\right] + \frac{2w_\parallel}{qB}\left[\frac{\hat{\mathbf{B}}\times\hat{\mathbf{R}}_C}{R_C}\right], \tag{3.9.36}$$

where the first term on the right of the above equation represents the rapid motion of a charged particle parallel to a magnetic field line, while the second and third terms represent the gradient and curvature drifts of the guiding center, respectively. We now apply this equation to a discussion of the nature of the orbits of trapped particles in an axisymmetric tokamak. (The detailed steps are left to Problem 3.15.)

To leading order, assuming a dominant toroidal field, $\mathbf{B} \cong B_{\mathrm{t}}\hat{\boldsymbol{\phi}}$, the gradient-B drift becomes

$$\mathbf{v}_{\mathrm{G}} = \frac{v_{\perp}^2}{2\rho\omega_c}\hat{\mathbf{z}}, \tag{3.9.37}$$

where $\omega_c = qB_{\mathrm{t}}/m$. Similarly, the curvature drift can be written

$$\mathbf{v}_{\mathrm{C}} = \frac{v_{\parallel}^2}{\rho\omega_c}\hat{\mathbf{z}}. \tag{3.9.38}$$

Hence, the total drift parallel to the z-axis, which is in opposite directions for electrons and ions, is given by

$$\mathbf{v}_{\mathrm{G}} + \mathbf{v}_{\mathrm{C}} = \frac{v_{\parallel}^2 + (v_{\perp}^2/2)}{\rho\omega_c}\hat{\mathbf{z}}. \tag{3.9.39}$$

We can estimate the width of the banana orbit shown in Figure 3.44 for a barely trapped particle for which the mirror reflection point is located near $\theta = \pi$. For such a trapped particle, by Eq. (3.9.35), we estimate that $v_{\parallel} \approx \sqrt{2\varepsilon_{\mathrm{t}}}C_s$, where C_s is a characteristic thermal speed defined by Eq. (2.1.3). The time taken by this trapped particle to execute half an orbit can be shown to be

$$t_b \approx \frac{2\pi q_0 \rho}{\sqrt{2\varepsilon_{\mathrm{t}}}C_s}, \tag{3.9.40}$$

where $q_0 = r_0 B_t/(\rho B_p)$ is called the safety factor at $r = r_0$ where the orbit is centered. During this time, the particle drifts laterally a characteristic width

$$\Delta_{\mathrm{b}} \approx \frac{v_{\parallel}^2 + (v_{\perp}^2/2)}{\rho\omega_c}t_{\mathrm{b}}, \tag{3.9.41}$$

where t_{b} is the bounce period, which yields

$$\Delta_b \approx \frac{\pi q_0 \rho_c}{\sqrt{2\varepsilon_{\mathrm{t}}}}. \tag{3.9.42}$$

Since tokamaks typically operate with $q_0 > 1$, we note that the banana orbit width of a charged particle is much greater than its Larmor radius.

3.10 Hamiltonian Chaos

In the previous discussion we showed several examples of particle motion where, either for reasons of symmetry or because the particle Hamiltonian depends on parameters that vary slowly with time (or space), the motion of the particle is constrained either by exactly conserved quantities or by adiabatic invariants. For example, in Section 3.8.2, we considered the motion of a charged particle in a time-independent and axially symmetric magnetic field. Time independence implies time translation symmetry ($\partial/\partial t = 0$), and the corresponding conserved quantity is the Hamiltonian given by Eq. (3.9.15) which, in this case, is the total energy, W. Similarly, axial symmetry ($\partial/\partial\phi = 0$) implies that the momentum p_ϕ, given by Eq. (3.9.13), is a constant. The existence of these two conserved quantities, W and p_ϕ, drastically constrains the motion of the particle, as indicated in Figures 3.39 and 3.40. Even in the absence of rigorous symmetries, if the Hamiltonian depends on parameters that vary slowly, adiabatic invariants play the rôle of approximate constants of the motion.

We have discussed in Section 3.8.3 how the three adiabatic invariants of particle motion in a magnetic mirror geometry can be broken if certain parameters vary rapidly. Constants of the motion, such as W and p_ϕ, can also be violated by perturbations that break the symmetries in t and ϕ. When such breakdowns occur, either because existing symmetries are broken or because certain slowly varying parameters vary too rapidly, the particle motion is not constrained by either exact or adiabatic invariants and can become chaotic. Chaotic motions are characterized by an extreme sensitivity to the initial conditions. Typically, the parameter space made up of the initial conditions has regions that lead to bounded integrable orbits, and other regions where the orbits are chaotic. Chaotic motions are a generic feature of Hamiltonian systems, and their occurrence is an important feature that must be considered when analyzing charged particle motions in plasmas.

Sometimes even very simple magnetic field geometries lead to chaotic motions. One of the simplest is a two-dimensional magnetic field containing an x-type neutral point, such as that shown in Figure 3.31. Since the first adiabatic invariant is always violated in the vicinity of a neutral point, the trajectories for such geometries must be determined by numerical computations. An example of such a trajectory, from Chen and Palmadesso (1986), is shown in Figure 3.45. The

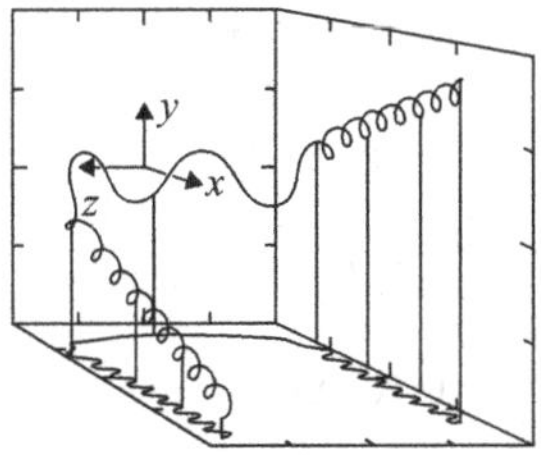

Figure 3.45 A representative particle trajectory near an x-type neutral line using the same coordinate system as in Figure 3.31 (adapted from (Chen and Palmadesso, 1986)).

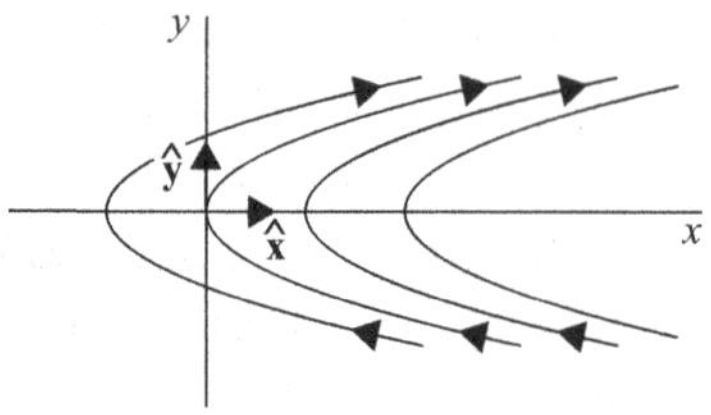

Figure 3.46 A parabolic magnetic field configuration, such as exists along the current sheet near an x-type neutral point (compare with Figure 3.31).

S-shaped trajectory, similar to that expected near an x-type neutral line (see Figure 3.33), is clearly evident near $x = 0$. In their studies of trajectories of this type, Chen and Palmadesso (1986) and also Büchner and Zelenyi (1989) showed that, depending on the initial conditions, certain particles displayed chaotic behavior, i.e., trajectories that diverged exponentially from trajectories with only slightly different initial conditions. The distinction between integrable and chaotic orbits is particularly clear for a parabolic magnetic field configuration of the form $\mathbf{B} = B_0[\hat{\mathbf{y}} + \hat{\mathbf{x}}(y/L)]$, where B_0 and L are constants. Such a field configuration, which is shown in Figure 3.46, is qualitatively similar to the field that exists along the current sheet near an x-type neutral point. It can be shown that the length-scale, L, corresponds to the radius of curvature of the magnetic field line at $y = 0$. Since the magnetic field strength increases with distance both above and below the symmetry plane, $y = 0$, all of the particles are magnetically trapped, irrespective of their pitch angle at $y = 0$. The magnetic field is derivable from the vector potential

$$\mathbf{A} = \hat{\mathbf{z}} B_0(-x + y^2/L). \tag{3.10.1}$$

The equation of motion of charged particles in this magnetic field configuration admits three constants of motion: the energy,

$$W = (1/2)mv^2, \tag{3.10.2}$$

the z component of the conjugate momentum,

$$p_z = mv_z + qA_z, \tag{3.10.3}$$

and the quantity

$$C_x = mv_x + qB_0 z. \tag{3.10.4}$$

The above equation can be derived from the x component of the equation of motion (3.1.1), which can be written

$$m\frac{\mathrm{d}v_x}{\mathrm{d}t} = q(v_y B_z - v_z B_y) = -q\frac{\mathrm{d}z}{\mathrm{d}t} B_0. \tag{3.10.5}$$

From the above equation, it follows that $dC_x/dt = 0$, which implies that C_x is a constant of the motion.

At first glance it might appear that in this problem there are three independent constants of motion, equal to the number of generalized coordinates (or degrees of freedom), and hence the orbits are completely integrable. However, this is not so. According to the Liouville–Arnol'd theorem (Sagdeev et al., 1988), a Hamiltonian system with N degrees of freedom is integrable if and only if there are N first integrals

$$F_i = F_i(p_1, q_1; p_2, q_2; .....p_N, q_N), \tag{3.10.6}$$

which are linearly independent and commute; that is,

$$[F_i, F_j] = 0, \tag{3.10.7}$$

where $[A, B]$ is the Poisson bracket, defined as

$$[A, B] = \sum_{i=1}^{N} \left(\frac{\partial A}{\partial p_i} \frac{\partial B}{\partial q_i} - \frac{\partial A}{\partial q_i} \frac{\partial B}{\partial p_i} \right). \tag{3.10.8}$$

It turns out that the three constants of motion, Eqs. (3.10.2), (3.10.3), and (3.10.4), do not obey Eq. (3.10.7) (Problem 3.16), and hence the orbits are not integrable. For the purposes of plotting the trajectory it is useful to use the normalized variable $X = (x - p_z/qB_0)/L$, where p_z = constant. As the particle moves in the magnetic field, it repeatedly crosses the symmetry plane ($y = 0$), while conserving the total energy, W.

To reveal possible chaotic orbits, a phase-space plot is constructed using the X and $\dot{X}$ coordinates each time a particle crosses the $y = 0$ plane. Such a plot is shown in Figure 3.47. This type of plot is called a Poincaré surface of section, after the French mathematician Henri Poincaré who first used this type of plot to study planetary orbits (Gleick, 1987). In this plot, integrable orbits appear as the concentric closed "curves" in region A, each of which is made up of successive $y = 0$ crossing points. In contrast, a chaotic orbit leaves seemingly "random" crossing points, as in region B. No orbit in region A crosses into region B and vice versa. The outermost solid circle corresponds to an orbit with zero parallel velocity in the $y = 0$ plane, i.e., a trajectory consisting of a simple cyclotron orbit in the $y = 0$ plane. For the example shown, the ratio of the cyclotron radius, ρ_c, of this orbit to the radius of curvature, L, is $\rho_c/L = 31.6$. The transition between the integrable region and the chaotic region is not sharp, and has fractal structures extending into the chaotic region, as often occurs in such plots. Although the trajectories in region B appear random, they are not, since each point is determined by the initial conditions. However, even the smallest random fluctuation in the

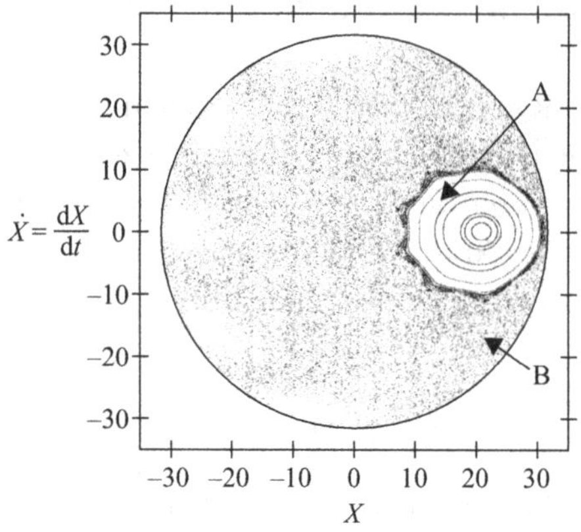

Figure 3.47 The normalized position and velocity coordinates $(X, \dot{X})$ at which particles cross the $y = 0$ plane for a parabolic magnetic field geometry, such as occurs near an x-type neutral line (adapted from Chen and Palmadesso (1986)). The particles in region A are in bounded integrable orbits, and the particles in region B have chaotic orbits.

initial conditions, or the magnetic field (i.e., noise), leads to a random distribution. Such chaotic orbits are now known to be an important feature of particle motions near x-type neutral lines in planetary magnetospheres, and in certain laboratory magnetic confinement machines, such as a tokamak.

Chaotic particle orbits can also arise because the underlying magnetic field structure is itself chaotic. It should be recognized that magnetic field lines, which obey the condition $\boldsymbol{\nabla} \cdot \mathbf{B} = 0$, comprise a Hamiltonian system. The equations governing magnetic field lines can be written as

$$\mathrm{d}\mathbf{r} \times \mathbf{B} = 0, \qquad (3.10.9)$$

where $\mathrm{d}\mathbf{r} = \mathrm{d}x\hat{\mathbf{x}} + \mathrm{d}y\hat{\mathbf{y}} + \mathrm{d}z\hat{\mathbf{z}}$ is a differential line element tangent to a field line. Written out in Cartesian components for a time-independent magnetic field, Eq. (3.10.9) yields

$$\frac{\mathrm{d}x}{\mathrm{d}\ell} = \frac{B_x(x, y, z)}{B(x, y, z)}, \qquad (3.10.10)$$

$$\frac{\mathrm{d}y}{\mathrm{d}\ell} = \frac{B_y(x, y, z)}{B(x, y, z)}, \qquad (3.10.11)$$

$$\frac{\mathrm{d}z}{\mathrm{d}\ell} = \frac{B_z(x, y, z)}{B(x, y, z)}, \qquad (3.10.12)$$

where ℓ is the length along a field line, with $\mathrm{d}\ell = \sqrt{\mathrm{d}x^2 + \mathrm{d}y^2 + \mathrm{d}z^2}$, and $B(x, y, z)$ is the magnitude of $\mathbf{B}$. Equations (3.10.10)–(3.10.12) are the Hamiltonian equations governing the flow of magnetic field lines in three-dimensional space, and as discussed below, these deceptively simple equations generically exhibit chaotic behavior.

To illustrate such a chaotic magnetic field system, consider a straight cylinder parameterized by the coordinate system (r,θ,z), in which the magnetic field depends only on r, with $B_r = 0$, $B_\theta = B_\theta(r)$, $B_z = B_z(r)$. This straight cylinder is topologically equivalent to a torus of major radius ρ_0 if we assume the cylinder to be periodic in the z direction with periodicity length $2\pi\rho_0$. The equations of the magnetic field line flow are

$$\frac{\mathrm{d}r}{\mathrm{d}\ell} = \frac{B_r}{B} = 0, \tag{3.10.13}$$

$$\frac{r\mathrm{d}\theta}{\mathrm{d}\ell} = \frac{B_\theta}{B}, \tag{3.10.14}$$

$$\frac{\mathrm{d}z}{\mathrm{d}\ell} = \frac{B_z}{B}. \tag{3.10.15}$$

It follows from Eq. (3.10.13) that $r =$ constant is an integral of "motion" for field lines. In the present context, this means that magnetic field lines in the periodic cylinder lie on nested surfaces that are concentric circles. In other words, there exists a flux function $\Psi_0 = \Psi_0(r)$ such that

$$\mathbf{B}\cdot\nabla\Psi_0 = 0. \tag{3.10.16}$$

In the language of Hamiltonian mechanics, this is a completely integrable system. In three dimensions, we have two ignorable coordinates θ and z, requiring the orbits to lie on constant Ψ_0 surfaces that depend only on the radial coordinate r.

We now examine the consequences of imposing a helical perturbation on the system, say proportional to $\cos(m\theta - kz)$, where $k = n/\rho_0$, and m and n are integers. Due to this perturbation, the axisymmetric cylindrical configuration will become helically twisted. The twisted configuration will depend now on two coordinates: r and a new helical angle $\theta_h = m\theta - kz$. Transforming partial derivatives in this new helical coordinate system, we write

$$\frac{\partial}{\partial\theta} = \frac{\partial\theta_h}{\partial\theta}\frac{\partial}{\partial\theta_h} = m\frac{\partial}{\partial\theta_h} \quad \text{and} \quad \frac{\partial}{\partial z} = \frac{\partial\theta_h}{\partial z}\frac{\partial}{\partial z} = -k\frac{\partial}{\partial z}. \tag{3.10.17}$$

We then obtain

$$\nabla\cdot\mathbf{B} = \frac{1}{r}\frac{\partial}{\partial r}(rB_r) + \frac{m}{r}\frac{\partial B_\theta}{\partial\theta_h} - k\frac{\partial B_z}{\partial\theta_h} = 0. \tag{3.10.18}$$

The above equation motivates the choice of a helical flux function $\Psi_h(r,\theta_h)$ such that

$$B_r = -\frac{1}{r}\frac{\partial \Psi_h(r,\theta_h)}{\partial \theta_h} \quad \text{and} \quad mB_\theta - krB_z = \frac{\partial \Psi_h(r,\theta_h)}{\partial r}. \tag{3.10.19}$$

With this choice, we satisfy the relation $\mathbf{B}\cdot\nabla\Psi_h = 0$; that is, **B**-lines lie on helical flux surfaces of constant Ψ_h. To be specific, we consider the case of a weak helical perturbation; that is,

$$\Psi_h(r,\theta,z) = \Psi_0(r) + \Psi_1(r)\cos(m\theta - kz), \tag{3.10.20}$$

where $k = n/\rho_0$ and $|\Psi_1| \ll |\Psi_0|$. By Eq. (3.10.19), we obtain $\partial\Psi_h/\partial r = 0$ at those radial locations $r = r_s$ where

$$mB_\theta - krB_z = 0, \quad \text{or} \quad m - nq_s = 0. \tag{3.10.21}$$

Here $q_s = r_s B_z/(\rho_0 B_\theta)$ is the safety factor at the surface $r = r_s$. Such a surface is called a "resonant surface" and has the distinguishing feature that the helical magnetic field line closes on itself after completing m poloidal and n toroidal transits in the periodic cylinder. At the resonant surface, the original circular flux surface breaks up into "magnetic islands." Figure 3.48 shows the meridional projection of an $m = 2$, $n = 1$ island in a tokamak. The width of an island centered at a resonant surface can be calculated by expanding the function Ψ_h in the vicinity of a resonant surface in a Taylor series, as follows:

$$\Psi_h(r,\theta_h) \cong \Psi_h''\frac{(r-r_s)^2}{2} + \Psi_1\cos\theta_h. \tag{3.10.22}$$

Here all the terms on the right side of the above equation are evaluated at $r = r_s$. The leading term in the expansion of the flux function can be set to zero because such a constant has no effect on the magnitude of the magnetic field, and the first derivative vanishes by Eq. (3.10.20). From Eq. (3.10.22), we obtain the island width

$$w = \max(r - r_s) = \left[\frac{4\Psi_1}{\Psi_0''}\right]^{1/2}. \tag{3.10.23}$$

As Figure 3.48 shows, under the influence of a single Fourier perturbation of pitch m/n, magnetic field lines lie on flux surfaces, albeit helically deformed ones with islands. However, if two or more perturbations of different pitches m/n are present, the picture changes drastically. As long as the perturbations are small, they open up small islands in the vicinity of the surfaces where they are resonant. However, if the amplitudes of the perturbations are increased, the islands tend

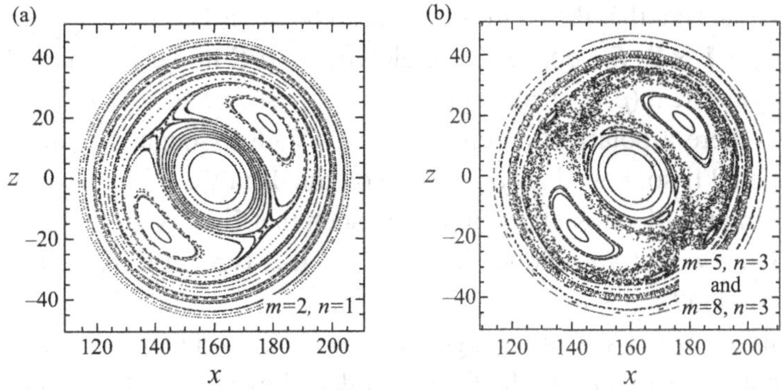

Figure 3.48 (a) Poloidal (i.e., meridional) projection of an $m = 2$, $n = 1$ island in a cylindrical tokamak. The island is obtained by superposing a $m = 2$, $n = 1$ sinusoidal perturbation on a configuration with concentric circular flux surfaces. (b) Poloidal projection of the configuration in (a), but now with $m = 5$, $n = 3$ and $m = 8$, $n = 3$ perturbations added. When the added perturbations have significant amplitudes, magnetic chaos results whereby field lines no longer lie on nice flux surfaces but fill a spatial region. The $m = 5$, $n = 3$ and $m = 8$, $n = 3$ islands are seen embedded in the chaotic regions. (Figure provided by R. B. White.)

to overlap, causing chaos. We note that this chaos is intrinsic to magnetic field lines and does not require the presence of charged particles. However, if charged particles are present and their guiding centers tend to follow the field lines, they will experience chaotic motion by simply following the field lines. Experimental evidence for such magnetic chaos has been observed to occur in toroidal fusion devices, and can cause substantial transport of energy and momentum, and even a loss of equilibrium.

Problems

3.1. A plasma has a constant uniform magnetic field aligned along the z axis, $\mathbf{B} = B_0\hat{\mathbf{z}}$, and has a constant uniform drift along the x axis, $\mathbf{U} = v_E\hat{\mathbf{x}}$. A positively charged particle of charge $+q$ and mass m is released from rest at the origin at $t = 0$.
(a) Compute the electric field in the plasma.
(b) Find the equation of motion of the particle (i.e., $x(t), y(t)$, and $z(t)$).
(c) What is the kinetic energy of the particle in the plasma rest frame?

3.2. A cylindrical column of plasma rotates around its central axis (as though it were a rigid solid) at an angular velocity ω_0. A constant uniform magnetic field $\mathbf{B}_0$ is present parallel to the axis of rotation.

(a) Assuming that the rigid rotation can be described by $\mathbf{v}_E = \omega_0 \times \mathbf{r}$, where $\mathbf{v}_E = \mathbf{E} \times \mathbf{B}/B^2$, compute the electric field $\mathbf{E}$ in the plasma column. Use cylindrical (ρ, ϕ, z) coordinates.

(b) Is there a polarization charge $\rho_q = \epsilon_0 \nabla \cdot \mathbf{E}$ associated with this electric field? If so, how does ρ_q depend on the distance from the central axis?

(c) Find the electrostatic potential Φ such that $\mathbf{E} = -\nabla\Phi$.

(d) How would you induce a motion of this type in a magnetized plasma column?

3.3. A particle of mass m and charge q is initially orbiting around the origin in the x, y plane in response to the forces produced by a magnetic field $\mathbf{B} = \hat{\mathbf{z}} B_0$. At $t = 0$ a linearly increasing electric field $\mathbf{E} = \hat{\mathbf{y}} kt$ is applied perpendicular to the magnetic field, where k is constant. Using the equations for the electric field drift, $\mathbf{v}_E = \mathbf{E} \times \mathbf{B}/B^2$, and the polarization drift, $\mathbf{v}_P = (m/qB^2)\mathrm{d}\mathbf{E}/\mathrm{d}t$, show that the motion of the guiding center is a parabola, as shown in Figure 3.23.

3.4. A plasma with a constant electric field in the x direction and a constant magnetic field in the y direction is undergoing a uniform $\mathbf{E} \times \mathbf{B}$ drift in the z direction. A neutral particle of mass m at rest in this system is suddenly ionized giving it a charge $+q$.

(a) By using appropriate sketches, describe the resulting motion of the particle in the x, y, z frame and in the plasma rest frame (where $E' = 0$).

(b) What is the resulting kinetic energy of the particle in the plasma rest frame?

(c) If a large number of uniformly distributed neutral particles were to be ionized at random times, what would be the velocity distribution function of the resulting ions?

3.5. A planetary magnetosphere has a dipole magnetic field with a dipole moment $\mathbf{M}$. The dipole axis is aligned along the rotational axis of the planet. All of the plasma in the magnetosphere rotates with the planet at an angular velocity ω_0. Using spherical coordinates (r, θ, ϕ) derive equations for the following:

(a) the magnetic field, $\mathbf{B}$;

(b) the $\mathbf{E} \times \mathbf{B}$ drift velocity, $\mathbf{v}_E = \omega_0 \times \mathbf{r}$;

(c) the electric field $\mathbf{E}$;

(d) the electrostatic potential, Φ, such that $\mathbf{E} = -\nabla\Phi$.

3.6. For a slowly varying electric field that is perpendicular to a constant static magnetic field, we showed that the plasma has an effective dielectric constant ($\epsilon = K\epsilon_0$):

$$K = 1 + \frac{n_0(m_\mathrm{i} + m_\mathrm{e})}{\epsilon_0 B^2}.$$

For a small-amplitude low-frequency electromagnetic wave propagating along the magnetic field, show that the propagation velocity, $V_A = 1/\sqrt{\epsilon\mu_0}$, often called the Alfvén velocity, is given to a good approximation by

$$V_A = \frac{B}{\sqrt{\mu_0 \rho_m}},$$

where $\rho_m = n_0(m_i + m_e)$ is the plasma density and the density is such that $K \gg 1$.

3.7. The equation for a dipole magnetic field in spherical coordinates is given by

$$\mathbf{B} = \frac{\mu_0 M}{4\pi}\frac{1}{r^3}[\hat{\mathbf{r}}\,2\cos\theta + \hat{\boldsymbol{\theta}}\,\sin\theta],$$

where M is the magnetic moment.

(a) Show that the equation for a magnetic field line is $r = R\sin^2\theta$, where R is the radius of the magnetic field line at the equator ($\theta = \pi/2$).
(b) Show that the curvature of the magnetic field line at the equator ($\theta = \pi/2$) is $R_C = R/3$.
(c) Compute the curvature drift of a particle with charge q and parallel energy $w_\parallel$ at a radial distance R at the equator.
(d) Compute the gradient drift of a particle with charge q and perpendicular energy $w_\perp$ at a radial distance R at the equator.
(e) Compare the equations for the curvature drift and gradient drift at the equator.

3.8. A charged particle of mass m and charge q has its guiding center at a radius ρ from the center of a long straight wire carrying a current I_0. A uniform electric field E_0 exists parallel to the axis of the wire.

(a) Using cylindrical coordinates (ρ, ϕ, z) derive equations for the $\mathbf{E}\times\mathbf{B}$ drift velocity, the gradient drift velocity, and the curvature drift velocity in terms of the quantities I_0 and E_0.
(b) Solve for the ρ component of the guiding-center motion as a function of time.

3.9. The ratio of the maximum to minimum magnetic field in a magnetic mirror machine is called the mirror ratio, $r = B_{max}/B_0$. Assume that the particles initially have an isotropic velocity distribution at B_0. What fraction of the particles at B_0 is trapped in the machine?

3.10. A particle is trapped in a magnetic mirror field given by

$$B_z = B_0\left[1 + \left(\frac{z}{L}\right)^2\right]$$

and has a total kinetic energy w and pitch angle α at $z = 0$. Find the oscillation frequency in terms of L, w, and α.

3.11. Demonstrate by direct calculation that the WKB solution

$$X_{\rm WKB} = \frac{1}{\sqrt{\omega(t)}} \exp\left[\pm \mathrm{i} \int^{t} \omega(t')\,\mathrm{d}t'\right]$$

satisfies the differential equation

$$\frac{\mathrm{d}^2 X}{\mathrm{d}t^2} + \left[\omega^2 + \frac{1}{2}\left(\frac{\ddot{\omega}}{\omega}\right) - \frac{3}{4}\left(\frac{\dot{\omega}}{\omega}\right)^2\right] X = 0.$$

3.12. A particle is trapped in an axially symmetric magnetic mirror field. The field has a uniform value B_0 in the region $-L < z < L$, and increases to a very large value at $z = \pm L$ as shown below.

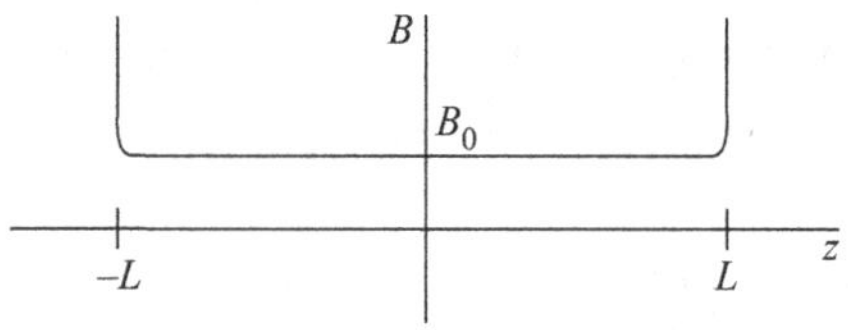

The initial pitch angle of a particle at $z = 0$ is $\alpha_0 = 45°$. The ends of the mirror are now slowly moved into $\pm L/2$ while holding B_0 constant. Assume that the first and second adiabatic invariants are constant. By what factor $(w_{\rm f}/w_{\rm i})$ does the total energy of the particle increase during this process?

3.13. A particle is trapped in a magnetic mirror field given by

$$B_z = B_0\left[1 + \left(\frac{z}{L}\right)^2\right].$$

Initially, the mirror points of the particle are located at $z = \pm L$.

(a) B_0 is now slowly increased to $2B_0$. Using the second adiabatic invariant, find the new mirror point locations and the new mirror field $B_{\rm m}$.

(b) L is then slowly decreased to $L/2$, while holding $2B_0$ constant. Using the second adiabatic invariant, find the new mirror point locations and the new mirror field $B_{\rm m}$.

3.14. A charged particle of mass m and charge q moves in the magnetic field of a magnetic monopole with a magnetic charge $Q_{\rm M}$.

(a) Assuming that the magnetic field is given by $\mathbf{B} = (\mu_0 Q_{\rm M}/4\pi r^2)\hat{\mathbf{r}}$, show that a vector potential that generates this field is $A_r = 0, A_\theta = 0$, and

$$A_\phi = -\frac{\mu_0 Q_{\rm M} \cos\theta}{4\pi r \sin\theta}.$$

(b) Show that the Hamiltonian in spherical coordinates is given by

$$H = \frac{(p_r - qA_r)^2}{2m} + \frac{(p_\theta - qrA_\theta)^2}{2mr^2} + \frac{(p_\phi - qr\sin\theta A_\phi)^2}{2mr^2\sin^2\theta},$$

where $p_r = m\dot{r} + qA_r$, $p_\theta = mr^2\dot{\theta} + q\dot{r}A_\theta$, and $p_\phi = mr^2\sin^2\theta\dot{\phi} + qr\sin\theta A_\phi$.

(c) By using the Hamiltonian formalism and identifying appropriate constants of the motion, show that the magnetic moment, $\mu = w_\perp/B$, is constant, where $w_\perp = \frac{1}{2}m(\dot{\theta}r\sin\theta)^2$.

3.15 (a) Show that for the toroidal magnetic field in a tokamak, the gradient-B and curvature drifts add up to give

$$\mathbf{v}_G + \mathbf{v}_C = \frac{v_\parallel^2 + (v_\perp^2/2)}{\rho\omega_s}\hat{\mathbf{z}}.$$

(b) Show that the distance travelled by a barely trapped particle executing a banana orbit is given by $2\pi q_0\rho$, and hence the bounce period is approximately

$$t_b \approx \frac{2\pi q_0\rho}{\sqrt{2\varepsilon_t}}C_s,$$

where ε_t is the aspect ratio, a/ρ_0.

(c) Assuming that parallel and perpendicular speeds of charged particles are typically equal to the thermal speed, show that the banana orbit width can be estimated to be

$$\Delta_b \approx \frac{\pi q_0\rho_c}{\sqrt{2\varepsilon_t}}.$$

3.16 Consider the three constants of motion W, P_z, and C_x given by Eqs. (3.10.2)–(3.10.4). Demonstrate that not all of them are independent and obey the commutation relation (3.10.7).

References

Biskamp, D. 2000. *Magnetic Reconnection in Plasmas*. Cambridge: Cambridge University Press, p. 254.

Brillouin, L. 1926. La mécanique ondulatoire de Schrödinger, une méthode générale de résolution par approximations successives. *C. R. Acad. Sci.* **183**, 24–26.

Büchner, J., and Zelenyi, L. M. 1989. Regular and chaotic charged particle motion in magnetotaillike field reversals: 1. Basic theory of trapped motion. *J. Geophys. Res.* **94**, 11821–11842.

Chen, J., and Palmadesso, P. J. 1986. Chaos and nonlinear dynamics of single-particle orbits in a magnetotail-like magnetic field. *J. Geophys. Res.* **91**, 1499–1508.

Dragt, A. J., and Finn, J. M. 1976. Insolubility of trapped particle motion in a magnetic dipole field. *J. Geophys. Res.* **81**, 2327–2340.

Dungey, J. W. 1961. Interplanetary magnetic field and the auroral zones. *Phys. Rev. Lett.* **6**, 47–48.

Fermi, E. 1949. On the origin of the cosmic radiation. *Phys. Rev.* **75**, 1169–1174.

Gleick, J. 1987. *Chaos*. New York: Penguin, pp. 142–144.

Goldstein, H. 1959. *Classical Mechanics*. Reading, MA: Addison-Wesley, p. 217.

Griffiths, D. J. 2013. *Introduction to Electrodynamics*, Fourth Edition. Glenview, IL: Pearson, p. 172.

Hundhausen, A. J. 1972. *Coronal Expansion and the Solar Wind*. New York: Springer-Verlag, p. 17.

Kivelson, M. G., and Russell, C. T. 1995. *Introduction to Space Physics*. Cambridge: Cambridge University Press, p. 243.

Kramers, H. A. 1926. Wellenmechanik und halbzahlige Quantisierung. *Z. Phys.* **39**, 828–840.

Kruskal, M. D. 1962. Asymptotic theory of Hamiltonian and other systems with all solutions nearly periodic. *J. Math. Phys.* **3**, 806–828.

Longmire, C. L. 1963. *Elementary Plasma Physics*. New York: Wiley, p. 53.

McIlwain, C. E. 1961. Coordinates for mapping the distribution of magnetically trapped particles. *J. Geophys. Res.* **66**, 3681–3691.

Northrop, T. G. 1963. *The Adiabatic Motion of Charged Particles*. New York: Interscience.

Parker, E. N. 1963. The solar-flare phenomenon and the theory of reconnection and annihilation of magnetic fields. *Astrophys. J. Suppl.* **8**, 177–211.

Priest, E. R., and Forbes, T. G. 2000. *Magnetic Reconnection: MHD Theory and Applications*. Cambridge: Cambridge University Press, p. 304.

Sagdeev, R. Z., Usikov, D. A., and Zaslavsky, G. M. 1988. *Nonlinear Physics: From the Pendulum to Turbulence and Chaos*. Char, Switzerland: Harwood Academic Publishers, Chapter 6.

Störmer, C. 1907. Sur des trajectoires des corpuscles electrisés dans l'éspace sous l'action du magnétisme terrestre, Chapitre 4. *Arch. Sci. Phys. Nat.* **24**, 317–364.

Van Allen, J. A. 1996. Kuiper prize lecture: Electrons, protons, and planets. *Icarus* **122**, 209–232.

Wentzel, G. 1926. Eine Verallgemeinerung der Quantenbedingung für die Zwecke der Wellenmechanik. *Z. Phys.* **38**, 518–529.

Further Reading

Baumjohann, W., and Treumann, R. A. 1997. *Basic Space Plasma Physics*. London: Imperial College Press, Chapter 2.

Bellan, P. M. 2008. *Fundamentals of Plasma Physics*. Cambridge: Cambridge University Press, Chapter 3.

Boyd, T. J. M., and Sanderson, J. J. 2003. *The Physics of Plasmas*. Cambridge: Cambridge University Press, Chapter 2.

Chen, F. F. 1990. *Introduction to Plasma Physics and Controlled Fusion Volume 1: Plasma Physics*. New York: Plenum Press, Chapter 2. Originally published in 1983.

Northrop, T. G. 1963. *The Adiabatic Motion of Charged Particles*. New York: Interscience, Chapters 2 and 3.

Parks, G. K. 2000. *Physics of Space Plasmas: An Introduction*. Redwood City, CA: Addison-Wesley, Chapter 4. Originally published in 1991.

Schmidt, G. 1979. *Physics of High Temperature Plasmas: An Introduction*. New York: Academic Press, Chapter 2. Originally published in 1966.

4

Waves in a Cold Plasma

Due to the presence of long-range forces, various types of waves can exist in a plasma that have no counterpart in ordinary gases or dielectric media. In this chapter we consider the special case of small-amplitude waves propagating in a cold plasma. The assumption that the amplitude is small greatly simplifies the analysis by allowing the equations of motion to be linearized. The assumption that the plasma is cold means that the particles are initially at rest, with no thermal motions, thereby allowing us to ignore the Doppler shifts caused by thermal motions, which provides a considerable simplification.

Since a cold plasma has no pressure, waves that depend on pressure, such as sound waves, cannot occur in a cold plasma. Also, since there is no source of free energy, instabilities are absent. Despite these limitations, the cold plasma model gives a remarkably accurate description of many types of waves that occur in real plasmas.

4.1 Fourier Representation of Waves

For reasons that will be apparent shortly, it is convenient to use Fourier transforms for analyzing linearized wave equations. Under very general conditions it can be shown that a function $f(x)$ can be represented by the integral

$$f(x) = \frac{1}{\sqrt{2\pi}} \int_{-\infty}^{\infty} \tilde{f}(k) e^{ikx} \, dk, \tag{4.1.1}$$

where $\tilde{f}(k)$ is called the Fourier transform of $f(x)$. The Fourier transform can be determined by inverting the above integral, and is given by

$$\tilde{f}(k) = \frac{1}{\sqrt{2\pi}} \int_{-\infty}^{\infty} f(x) e^{-ikx} \, dx. \tag{4.1.2}$$

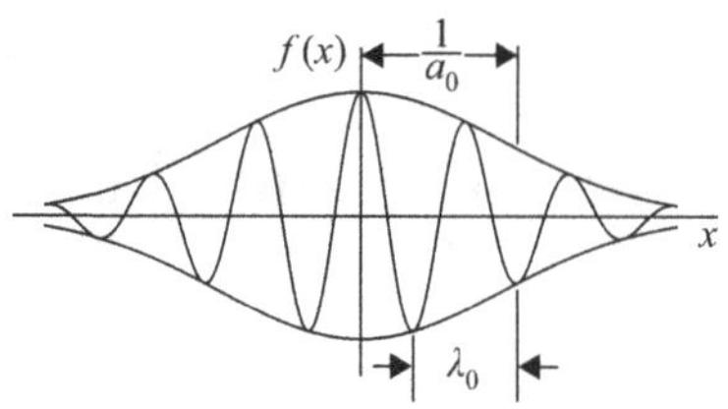

Figure 4.1 The Gaussian harmonic wave packet.

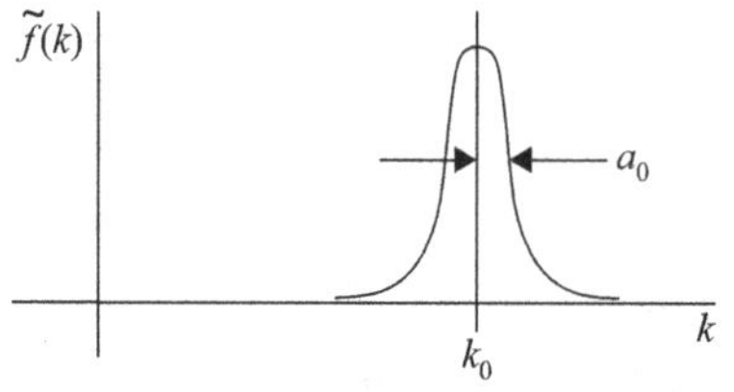

Figure 4.2 The Fourier transform of the Gaussian harmonic wave packet.

The functions $f(x)$ and $\tilde{f}(k)$ are often referred to as a Fourier transform pair, and are tabulated in mathematics handbooks for most commonly used functions. If x represents a spatial coordinate then $k = 2\pi/\lambda$ is called the wave number, where λ is the wavelength.

A specific example that illustrates many of the general properties of Fourier transforms is the Gaussian harmonic packet

$$f(x) = \mathrm{e}^{-a_0^2 x^2/2}\, \mathrm{e}^{\mathrm{i}k_0 x}. \tag{4.1.3}$$

The parameter a_0 controls the width of the Gaussian envelope and $k_0 = 2\pi/\lambda_0$ is the wave number of the harmonic term. The real part of $f(x)$ is shown in Figure 4.1. It is straightforward to show that the Fourier transform of the Gaussian harmonic packet is given by

$$\tilde{f}(k) = \frac{1}{a_0} \mathrm{e}^{-(k-k_0)^2/2a_0^2}, \tag{4.1.4}$$

which is also a Gaussian function (see Problem 4.1). As shown in Figure 4.2, the Fourier transform has a single peak at wave number k_0. The width of this peak, given by the parameter a_0, is inversely proportional to the width of the wave packet. The product of the mean-square width of the function, $\langle \Delta x^2 \rangle$, and the mean-square width of the Fourier transform, $\langle \Delta k^2 \rangle$, obeys the inequality

$$\langle \Delta x^2 \rangle \langle \Delta k^2 \rangle \geq \frac{1}{4}. \tag{4.1.5}$$

The above inequality is often referred to as the uncertainty theorem for Fourier transforms, and is analogous to the well-known uncertainty theorem of quantum mechanics (Townsend, 1992). The Gaussian harmonic packet is unique because $\langle \Delta x^2 \rangle \langle \Delta k^2 \rangle$ is exactly equal to 1/4, which is the smallest allowed value.

The uncertainty theorem simply expresses the fact that a sharply peaked transform corresponds to a wave packet that is spread out over a broad region in space, and vice versa. An extreme example is obtained by taking the limit of Eqs. (4.1.3) and (4.1.4) as the parameter a_0 goes to zero. In this case, the wave packet (4.1.3) becomes a plane wave,

$$f(x) = \mathrm{e}^{\mathrm{i}k_0 x}, \tag{4.1.6}$$

which has a constant amplitude extending from $-\infty$ to $+\infty$, and the Fourier transform (4.1.4) becomes a Dirac delta function,

$$\tilde{f}(k) = \sqrt{2\pi}\,\delta(k - k_0), \tag{4.1.7}$$

which is sharply localized in k-space, being non-zero only when $k = k_0$.

By a simple four-dimensional extension of expression (4.1.1), we can represent any function of space and time, $f(\mathbf{r},t)$, by the four-dimensional integral

$$f(\mathbf{r},t) = \frac{1}{(2\pi)^2} \int_{-\infty}^{\infty} \tilde{f}(\mathbf{k},\omega)\,\mathrm{e}^{\mathrm{i}(\mathbf{k}\cdot\mathbf{r} - \omega t)}\,\mathrm{d}^3k\,\mathrm{d}\omega, \tag{4.1.8}$$

where $\mathrm{d}^3k = \mathrm{d}k_x\,\mathrm{d}k_y\,\mathrm{d}k_z$ and $\mathbf{k}\cdot\mathbf{r} = k_x x + k_y y + k_z z$. The quantity ω is the frequency and is related to the time coordinate t in the same way that the wave number components, k_x, k_y, and k_z, are related to the spatial coordinates, x, y, and z. The function $\exp[\mathrm{i}(\mathbf{k}\cdot\mathbf{r} - \omega t)]$ in the integrand represents an elementary plane wave. As shown in Figure 4.3, the vector $\mathbf{k}$ is perpendicular to the planes of constant phase ($\mathbf{k}\cdot\mathbf{r} - \omega t$ = constant). The minus sign is placed in front of ω so that the planes of constant phase move in the $\mathbf{k}$ direction. The vector $\mathbf{k}$ is often called the wave vector or propagation vector, since it indicates the direction of motion of the planes of constant phase. The velocity with which the planes of constant phase move is called the phase velocity and is given by

$$\mathbf{v}_\mathrm{p} = \frac{\omega}{k}\,\hat{\mathbf{k}}. \tag{4.1.9}$$

4.1.1 The Dispersion Relation

A wave $f(\mathbf{r},t)$ propagating in some medium usually satisfies a differential equation that determines the possible solutions for $f(\mathbf{r},t)$ in that medium. If the differential equation is linear, Fourier analysis provides a simple method of determining the

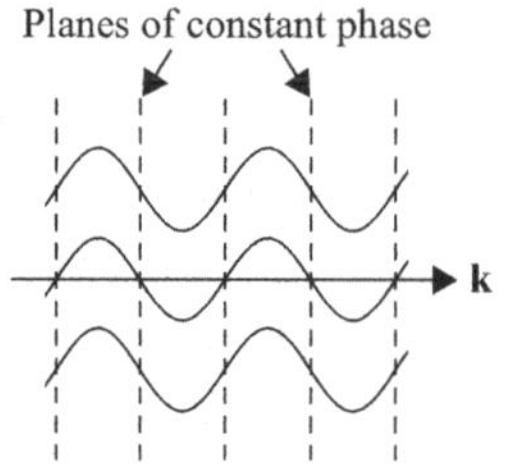

Figure 4.3 An elementary plane wave with wave vector **k**.

allowed solutions. This simplification arises because the Fourier transform of the derivative $\partial f/\partial t$ can be obtained by simply multiplying the transform of f by $\mathrm{i}\omega$. This property can be demonstrated by differentiating both sides of Eq. (4.1.8) with respect to t, which gives

$$\frac{\partial f}{\partial t} = \frac{1}{(2\pi)^2}\int_{-\infty}^{\infty}(-\mathrm{i}\omega\tilde{f})\,\mathrm{e}^{\mathrm{i}(\mathbf{k}\cdot\mathbf{r}-\omega t)}\,\mathrm{d}^3k\,\mathrm{d}\omega. \tag{4.1.10}$$

The Fourier transform of $\partial f/\partial t$ is thus seen to be $(-\mathrm{i}\omega\tilde{f})$. A similar derivation can be performed to show that the Fourier transform of ∇f is $\mathrm{i}\mathbf{k}\tilde{f}$. Therefore, any combination of derivatives can be represented by the following operator substitutions:

$$\begin{aligned}\frac{\partial}{\partial t} &\to -\mathrm{i}\omega\\ \nabla &\to \mathrm{i}\mathbf{k}.\end{aligned} \tag{4.1.11}$$

It is readily shown that these operator substitutions hold for any sequence of operations, including all operations involving vector calculus. Thus, $\partial^n f/\partial t^n \to (-\mathrm{i}\omega)^n\tilde{f}$, $\nabla^2 f \to (\mathrm{i}k)^2\tilde{f}, \nabla\cdot\mathbf{f}\to \mathrm{i}\mathbf{k}\cdot\tilde{\mathbf{f}}, \nabla\times\mathbf{f}\to\mathrm{i}\mathbf{k}\times\tilde{\mathbf{f}}$, etc. It follows then that the Fourier transform of the homogeneous linear differential equation

$$D(\nabla,\partial/\partial t)f = 0 \tag{4.1.12}$$

is simply

$$D(\mathrm{i}\mathbf{k},-\mathrm{i}\omega)\tilde{f} = 0. \tag{4.1.13}$$

The net effect is to convert the original differential equation (4.1.12) into an algebraic equation (4.1.13) involving ω, $\mathbf{k}$, and $\tilde{f}$. The resulting algebraic equation is usually much easier to solve than the original differential equation. For the homogeneous equation (4.1.13) to have a non-trivial solution for $\tilde{f}$ the quantity $D(\mathrm{i}\mathbf{k},-\mathrm{i}\omega)$ must be zero. This condition, written as $D(\mathbf{k},\,\omega) = 0$, is called the dispersion relation. The dispersion relation provides a relationship between ω and

k. In many cases of physical interest, the dispersion relation has multiple, discrete roots for ω, represented by

$$\omega = \omega_\alpha(\mathbf{k}), \qquad \alpha = 1, 2, \ldots, N. \tag{4.1.14}$$

These roots are called the normal modes. Except for degenerate cases, the number of normal modes is equal to the order of the differential equation. For example, the simple wave equation

$$\frac{\partial^2 f}{\partial x^2} - \frac{1}{c^2}\frac{\partial^2 f}{\partial t^2} = 0, \tag{4.1.15}$$

where c is the speed of light, can be Fourier-transformed to yield

$$\left[-k^2 + \frac{\omega^2}{c^2}\right]\tilde{f} = 0, \tag{4.1.16}$$

which has a dispersion relation

$$D(\mathbf{k}, \omega) = -k^2 + \frac{\omega^2}{c^2} = 0 \tag{4.1.17}$$

with roots $\omega = \pm\, ck$. The two normal modes in this case represent waves traveling to the right and left with phase velocities $v_\mathrm{p} = \pm\, c$.

Once the roots of the dispersion relation are known, the principle of superposition allows us to find the general solution to the linear differential equation (4.1.12) by integrating over **k** and summing over all the normal modes. Thus, the general solution can be represented by

$$f(\mathbf{r}, t) = \frac{1}{(2\pi)^{3/2}} \sum_{\alpha=1}^{N} \int_{-\infty}^{\infty} \tilde{f}_\alpha(\mathbf{k})\, \mathrm{e}^{\mathrm{i}[\mathbf{k}\cdot\mathbf{r} - \omega_\alpha(\mathbf{k})t]}\, \mathrm{d}^3k. \tag{4.1.18}$$

The roots $\omega_\alpha(\mathbf{k})$, specified by Eq. (4.1.14), determine the propagation speed of the individual plane waves that make up the wave packet $f(\mathbf{r}, t)$. Because the plane waves usually propagate at different speeds, the wave packet generally tends to spread out in space with increasing time. This spreading is called dispersion.

For a vector wave field **f**, it is usually necessary to specify a system of linear homogeneous differential equations that can be written in the following matrix form:

$$\overleftrightarrow{\mathbf{D}}(\boldsymbol{\nabla}, \partial/\partial t) \cdot \mathbf{f} = 0. \tag{4.1.19}$$

Using the operator substitution given by Eq. (4.1.11), the corresponding Fourier transform is then

$$\overleftrightarrow{\mathbf{D}}(\mathrm{i}\mathbf{k}, -\mathrm{i}\omega) \cdot \tilde{\mathbf{f}} = 0. \tag{4.1.20}$$

It is well known that a homogeneous system of linear equations has a non-trivial solution if and only if the determinant of the matrix is zero, which gives the dispersion relation

$$Ð(\mathbf{k},\omega) = \mathrm{Det}[\overleftrightarrow{\mathbf{Ð}}(\mathrm{i}\mathbf{k},-\mathrm{i}\omega)] = 0. \tag{4.1.21}$$

Each root of the dispersion relation has an associated vector $\tilde{\mathbf{f}}$, that is called the eigenvector for that mode. The eigenvector characterizes the field geometry associated with that mode of propagation.

4.1.2 Group Velocity

Usually the integrals involved in the Fourier representation of a wave are quite complicated, since the individual plane waves can be propagating in different directions at different speeds. However, there is one particular situation for which the waveform has a simple representation. If the Fourier transform is sharply peaked around a particular wave number (such as the Gaussian harmonic wave packet with $a_0 \ll k_0$), the shape of the envelope of the wave packet is preserved to a first approximation. The velocity at which the envelope moves is called the group velocity.

To obtain an expression for the group velocity, we note that if the Fourier transform is sharply peaked at $\mathbf{k} = \mathbf{k}_0$ for one of the modes, only waves near $\mathbf{k}_0$ contribute significantly to the wave packet for that mode. We can then obtain a good approximation to the integral in Eq. (4.1.8) by performing a Taylor series expansion of the dispersion relation around $\mathbf{k}_0$ and keeping only the first two terms

$$\omega(\mathbf{k}) = \omega_0 + \nabla_{\mathbf{k}}\omega \cdot (\mathbf{k} - \mathbf{k}_0) + \cdots, \tag{4.1.22}$$

where ω_0 and $\nabla_{\mathbf{k}}\omega = \hat{\mathbf{x}}\,\partial\omega/\partial k_x + \hat{\mathbf{y}}\,\partial\omega/\partial k_y + \hat{\mathbf{z}}\,\partial\omega/\partial k_z$ are evaluated at $\mathbf{k} = \mathbf{k}_0$. As shown in Figure 4.4, the first two terms represent a straight line approximation to the dispersion relation in the vicinity of the peak in $\tilde{f}(\mathbf{k})$. Substituting Eq. (4.1.22)

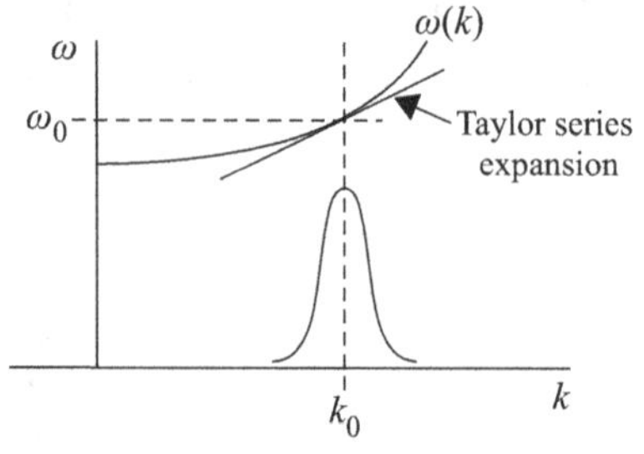

Figure 4.4 Taylor series expansion of $\omega(k)$ around the point k_0, ω_0.

into the integral representation (4.1.18) and neglecting the higher-order terms, we obtain

$$f(\mathbf{r}, t) = \frac{1}{(2\pi)^{3/2}} \int_{-\infty}^{\infty} \tilde{f}(\mathbf{k})\, \mathrm{e}^{\mathrm{i}[\mathbf{k}\cdot\mathbf{r} - \omega_0 t - \nabla_{\mathbf{k}}\omega\cdot(\mathbf{k}-\mathbf{k}_0)t]}\, \mathrm{d}^3k. \tag{4.1.23}$$

After rearranging terms, the integral on the right-hand side of the equation can be rewritten

$$f(\mathbf{r}, t) = \left[\frac{1}{(2\pi)^{3/2}} \int_{-\infty}^{\infty} \tilde{f}(\mathbf{k})\, \mathrm{e}^{\mathrm{i}(\mathbf{k}-\mathbf{k}_0)\cdot(\mathbf{r}-\mathbf{v}_{\mathrm{g}}t)}\, \mathrm{d}^3k\right] \mathrm{e}^{\mathrm{i}(\mathbf{k}_0\cdot\mathbf{r} - \omega_0 t)}, \tag{4.1.24}$$

where $\mathbf{v}_{\mathrm{g}} = \nabla_{\mathbf{k}}\omega$ is called the group velocity. The above equation can be written in the form

$$f(\mathbf{r}, t) = A(\mathbf{r} - \mathbf{v}_{\mathrm{g}}t)\, \mathrm{e}^{\mathrm{i}(\mathbf{k}_0\cdot\mathbf{r} - \omega_0 t)}, \tag{4.1.25}$$

where $A(\mathbf{r} - \mathbf{v}_{\mathrm{g}}t)$ is the term in the bracket. This equation shows that the waveform for a given mode consists of a plane wave of wave vector $\mathbf{k}_0$ and frequency ω_0 modulated by an amplitude factor $A(\mathbf{r} - \mathbf{v}_{\mathrm{g}}t)$. A typical waveform is shown in Figure 4.5. The wave packet consists of a plane wave that moves at the phase velocity $\mathbf{v}_{\mathrm{p}} = (\omega_0/k_0)\hat{\mathbf{k}}$, and an outer envelope that moves at the group velocity $\mathbf{v}_{\mathrm{g}} = \nabla_{\mathbf{k}}\omega$ evaluated at $\mathbf{k} = \mathbf{k}_0$. Since energy is carried by the wave packet, the group velocity is the velocity of energy propagation.

Several interesting features can be noted about the group velocity. Since the group velocity and phase velocity are in general different, the planes of constant phase move through the wave packet. A given wave crest starts with a small amplitude at one end of the packet, grows in amplitude as it reaches the center of the packet, and finally disappears at the other end. An exception to this behavior occurs for waves that obey the simple wave equation

$$\frac{\partial^2 f}{\partial x^2} - \frac{1}{c^2}\frac{\partial^2 f}{\partial t^2} = 0, \tag{4.1.26}$$

for which the phase and group velocities are the same, i.e., $v_{\mathrm{p}} = v_{\mathrm{g}} = c$. In this special case, the shape of the waveform stays constant in time. Such waves are

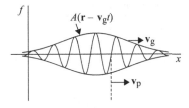

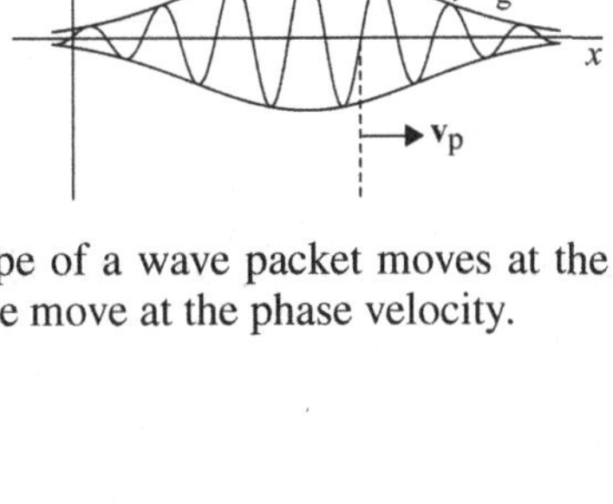

Figure 4.5 The envelope of a wave packet moves at the group velocity and the planes of constant phase move at the phase velocity.

said to be non-dispersive, since the wave packet does not tend to spread out with increasing time. Electromagnetic waves in free space and sound waves in air have this property.

If the medium is anisotropic the group velocity and phase velocity may not even be in the same direction. In such an anisotropic medium the wave packet moves at an angle relative to the wave vector, as shown in Figure 4.6. This unusual behavior occurs in doubly refracting optical crystals and in plasmas with a static magnetic field. In order to follow the direction of wave energy flow in an inhomogeneous medium it is useful to introduce the concept of a ray path. The ray path is a sequence of connected line elements that are everywhere in the direction of the local group velocity. To construct the ray path it is useful to have a simple method of visualizing the direction of the group velocity. The group velocity direction can be visualized by noting that $\mathbf{v}_g$ is simply the gradient of $\omega(\mathbf{k})$ in $\mathbf{k}$-space:

$$\mathbf{v}_g = \nabla_{\mathbf{k}}\omega(\mathbf{k}) = \hat{\mathbf{x}}\frac{\partial\omega}{\partial k_x} + \hat{\mathbf{y}}\frac{\partial\omega}{\partial k_y} + \hat{\mathbf{z}}\frac{\partial\omega}{\partial k_z}. \tag{4.1.27}$$

From the properties of the gradient operator it follows that the group velocity is perpendicular to a contour of constant ω in $\mathbf{k}$-space. This relationship is illustrated by the geometric construction in Figure 4.7, which shows that the group velocity is perpendicular to a contour of constant ω. In practice, it is often more convenient to deal with a dimensionless index of refraction vector, $\mathbf{n} = c\mathbf{k}/\omega$, rather than the wave vector $\mathbf{k}$. Let us assume that the index of refraction is given as a

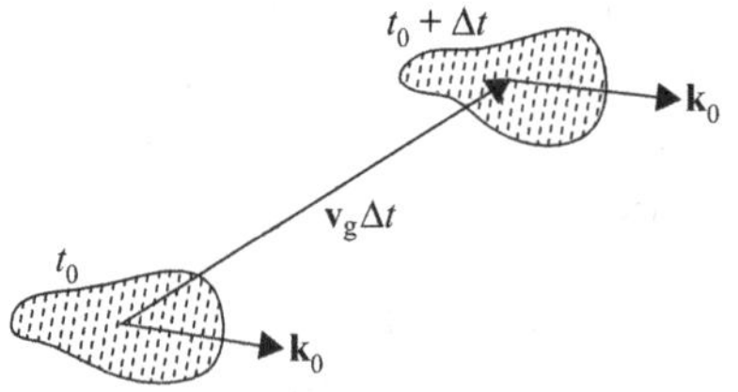

Figure 4.6 In an anisotropic medium the group velocity, $\mathbf{v}_g$, may be in a direction totally different from the wave vector, $\mathbf{k}_0$.

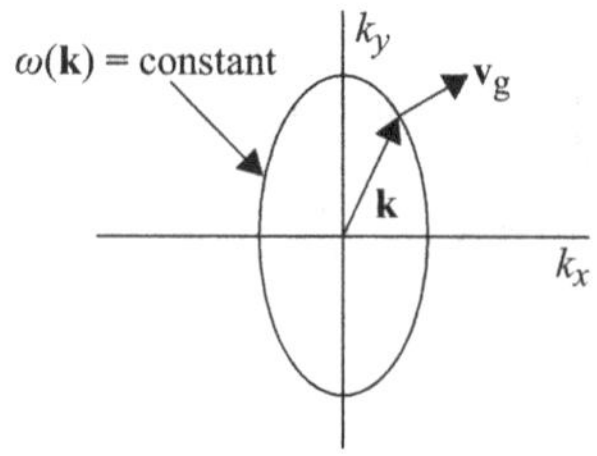

Figure 4.7 In $\mathbf{k}$-space, the group velocity is perpendicular to a surface of constant ω.

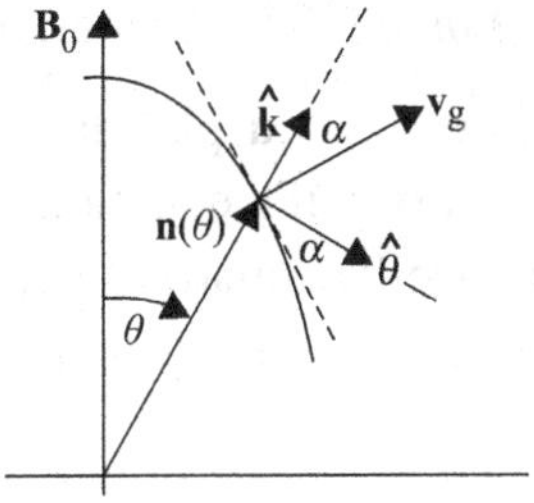

Figure 4.8 Geometric construction used to compute the angle α between the group velocity and the wave vector.

function of the angle θ, called the wave normal angle, between the wave vector and some preferred axis of symmetry, typically the magnetic field, $\mathbf{B}_0$. The index of refraction for constant ω can then be represented as a curve $\mathbf{n}(\theta)$ in polar form, as shown in Figure 4.8. The surface generated by rotating $\mathbf{n}(\theta)$ around the symmetry axis $\mathbf{B}_0$ is called the index of refraction surface. Since ω is constant along this surface, it follows from our previous discussion that the group velocity is perpendicular to the index of refraction surface, i.e., in the direction of the gradient $\nabla_{\mathbf{n}}\omega(\mathbf{n})$. Using $\hat{\mathbf{k}}$ and $\hat{\boldsymbol{\theta}}$ unit vectors, it can be shown that the group velocity can be written in terms of this gradient as

$$\mathbf{v}_g = \frac{c}{\omega}\nabla_{\mathbf{n}}\omega(\mathbf{n}) = \frac{c}{\omega}\left[\hat{\mathbf{k}}\frac{\partial\omega}{\partial n} + \hat{\theta}\frac{1}{n}\frac{\partial\omega}{\partial\theta}\right]. \tag{4.1.28}$$

The angle α between the group velocity and the wave vector is then given by

$$\tan\alpha = \frac{\dfrac{1}{n}\left(\dfrac{\partial\omega}{\partial\theta}\right)_n}{\left(\dfrac{\partial\omega}{\partial n}\right)_\theta}, \tag{4.1.29}$$

where the subscripts n and θ indicate that the derivatives are to be computed with n and θ held constant. Using the following theorem relating the partial derivatives of three implicit functions ω, n, and θ (Kaplan, 1952),

$$\left(\frac{\partial\omega}{\partial\theta}\right)_n\left(\frac{\partial\theta}{\partial n}\right)_\omega\left(\frac{\partial n}{\partial\omega}\right)_\theta = -1, \tag{4.1.30}$$

Eq. (4.1.29) simplifies to

$$\tan\alpha = -\frac{1}{n}\left(\frac{\partial n}{\partial\theta}\right)_\omega. \tag{4.1.31}$$

This relationship provides a convenient method of calculating the group velocity direction if the index of refraction is known as a function of θ.

4.2 General Form of the Dispersion Relation

To analyze the waves that can exist in a plasma, the basic equations that must be solved self-consistently are Maxwell's equations and the particle equations of motion. As is well known, Maxwell's equations can be written in two equivalent forms, the so-called "microscopic" and "macroscopic" forms. These two forms are summarized below:

Microscopic	Macroscopic
$\nabla \times \mathbf{B} = \mu_0 \mathbf{J} + \epsilon_0 \mu_0 \frac{\partial \mathbf{E}}{\partial t}$	$\nabla \times \mathbf{H} = \mathbf{J}_{\mathrm{r}} + \frac{\partial \mathbf{D}}{\partial t}$
$\nabla \times \mathbf{E} = -\frac{\partial \mathbf{B}}{\partial t}$	$\nabla \times \mathbf{E} = -\frac{\partial \mathbf{B}}{\partial t}$
$\nabla \cdot \mathbf{E} = \frac{\rho_q}{\epsilon_0}$	$\nabla \cdot \mathbf{D} = \rho_{\mathrm{r}}$
$\nabla \cdot \mathbf{B} = 0$	$\nabla \cdot \mathbf{B} = 0.$ (4.2.1)

In the macroscopic approach the charge is divided into a real charge and a polarization charge, $\rho_q = \rho_{\mathrm{r}} + \rho_{\mathrm{p}}$, and the current is divided into a real current and a magnetization current, $\mathbf{J} = \mathbf{J}_{\mathrm{r}} + \mathbf{J}_{\mathrm{m}}$. This division is completely arbitrary, depending on how we define real charges and polarization charges, and real currents and magnetization currents. For our purposes it is useful to count all of the charges in the plasma as polarization charges, so that the displacement field is given by $\mathbf{D} = \epsilon_0 \mathbf{E} + \mathbf{P}$, where $\mathbf{P}$ is the dipole moment per unit volume. All of the currents in the plasma are then included in the displacement current $\partial \mathbf{D}/\partial t$. Since the magnetic moment of the individual particles is normally negligible, in a plasma we usually assume that $\mathbf{B} = \mu_0 \mathbf{H}$.

4.2.1 The Conductivity and Dielectric Tensors

To obtain a closed system of equations using the macroscopic approach, it is necessary to specify the relationship between $\mathbf{D}$ and $\mathbf{E}$. This specification must ultimately come from the equations of motion of the particles. The procedure is first to solve the equations of motion for some assumed wave field. Since the equations of motion are assumed to be linear, this solution is most easily obtained by using Fourier transforms, which gives the Fourier transform of the velocity, $\tilde{\mathbf{v}}$, as a linear function of the Fourier transform of the electric field, $\tilde{\mathbf{E}}$. The solution for $\tilde{\mathbf{v}}$ can then be substituted into the equation for the current density,

$$\tilde{\mathbf{J}} = \sum_s n_s e_s \tilde{\mathbf{v}}_s, \tag{4.2.2}$$

and organized into an equation of the form

$$\tilde{\mathbf{J}} = \overleftrightarrow{\sigma} \cdot \tilde{\mathbf{E}}, \tag{4.2.3}$$

where the matrix $\overleftrightarrow{\sigma}$ is a tensor called the conductivity tensor.

Once the conductivity tensor has been determined, the next step is to relate the electric field $\tilde{\mathbf{E}}$ to the displacement vector $\tilde{\mathbf{D}}$. Since the system is assumed to be linear, in the macroscopic approach the displacement vector $\tilde{\mathbf{D}}$ must be a linear function of the electric field $\tilde{\mathbf{E}}$. Since we anticipate that anisotropic effects are likely to occur, the linear relationship between $\tilde{\mathbf{D}}$ and $\tilde{\mathbf{E}}$ is assumed to be of the form

$$\tilde{\mathbf{D}} = \epsilon_0 \overleftrightarrow{\mathbf{K}} \cdot \tilde{\mathbf{E}}, \tag{4.2.4}$$

where the matrix $\overleftrightarrow{\mathbf{K}}$ is called the dielectric tensor. The above equation represents a generalization of the relation $\mathbf{D} = \epsilon_0 K \mathbf{E}$, commonly used in the description of isotropic media.

The relationship between the dielectric tensor and the conductivity tensor can be obtained by comparing the microscopic and macroscopic forms of Ampère's law, which in Fourier transformation form can be written

Microscopic	Macroscopic
$\mathrm{i}\mathbf{k} \times \tilde{\mathbf{B}} = \mu_0 \overleftrightarrow{\sigma} \cdot \tilde{\mathbf{E}} + \epsilon_0 \mu_0 (-\mathrm{i}\omega) \tilde{\mathbf{E}},$	$\mathrm{i}\mathbf{k} \times \tilde{\mathbf{B}} = \epsilon_0 \mu_0 (-\mathrm{i}\omega) \overleftrightarrow{\mathbf{K}} \cdot \tilde{\mathbf{E}}.$ (4.2.5)

Equating the right-hand sides of the above two equations gives the following relationship between the dielectric tensor and the conductivity tensor:

$$\overleftrightarrow{\mathbf{K}} = \overleftrightarrow{\mathbf{1}} - \frac{\overleftrightarrow{\sigma}}{\mathrm{i}\omega\epsilon_0}, \tag{4.2.6}$$

where $\overleftrightarrow{\mathbf{1}}$ is the unit tensor.

The procedure for computing the dielectric tensor is first to compute the conductivity tensor using the particle equation of motion and Eq. (4.2.2) for the current density. The dielectric tensor is then computed using Eq. (4.2.6). This procedure is quite general and works for any plasma, hot or cold, provided only that the plasma can be described by a linear system of equations.

4.2.2 The Homogeneous Equation

To obtain the dispersion relation, we need to look for a solution to Maxwell's equations, ignoring external sources. A homogeneous (source-free) system of equations can be obtained by eliminating either **E** or **B** from Faraday's law and Ampère's law (4.2.1). After Fourier transforming, Faraday's and Ampère's laws become (using $\mu_0\epsilon_0 = 1/c^2$)

$$\mathrm{i}\mathbf{k}\times\tilde{\mathbf{E}} = -(-\mathrm{i}\omega)\tilde{\mathbf{B}} \qquad \text{and} \qquad \mathrm{i}\mathbf{k}\times\tilde{\mathbf{B}} = \frac{-\mathrm{i}\omega}{c^2}\overleftrightarrow{\mathbf{K}}\cdot\tilde{\mathbf{E}}. \tag{4.2.7}$$

Eliminating $\tilde{\mathbf{B}}$ between these two equations gives a homogeneous equation for the electric field:

$$\mathbf{k}\times(\mathbf{k}\times\tilde{\mathbf{E}}) + \frac{\omega^2}{c^2}\overleftrightarrow{\mathbf{K}}\cdot\tilde{\mathbf{E}} = 0. \tag{4.2.8}$$

Using the definition of the index of refraction, $\mathbf{n} = c\mathbf{k}/\omega$, the above equation can be expressed in the following, somewhat simpler, form:

$$\mathbf{n}\times(\mathbf{n}\times\tilde{\mathbf{E}}) + \overleftrightarrow{\mathbf{K}}\cdot\tilde{\mathbf{E}} = 0. \tag{4.2.9}$$

Either of the above two homogeneous equations (4.2.8) or (4.2.9) can be written in matrix form as $\overleftrightarrow{\mathbf{Đ}}\cdot\tilde{\mathbf{E}} = 0$. A non-trivial solution for $\tilde{\mathbf{E}}$ is then possible if and only if the determinant of the matrix is zero, which gives the dispersion relation, $Đ(\mathbf{k},\omega)$. The electric field eigenvector associated with each root of the dispersion relation can be obtained from the homogeneous equation (4.2.8) or (4.2.9), and the corresponding magnetic field eigenvector can be obtained from Faraday's law, $\tilde{\mathbf{B}} = (\mathbf{k}/\omega)\times\tilde{\mathbf{E}}$ or $c\tilde{\mathbf{B}} = \mathbf{n}\times\tilde{\mathbf{E}}$.

4.3 Waves in a Cold Uniform Unmagnetized Plasma

To illustrate the above procedure, we first consider the special case of waves propagating in a cold uniform unmagnetized plasma. To obtain a linear set of equations, all of the dependent variables ($\mathbf{v}_s$, n_s, **E**, and **B**) are assumed to consist of a constant uniform zero-order term plus a small first-order perturbation. For example, the number density is assumed to be a constant uniform density plus a small perturbation, $n_s = n_{s0} + n_{s1}$. Since the plasma is cold (i.e., zero temperature), the zero-order velocities are assumed to be zero. The zero-order electric field must also be zero, otherwise at zero order the particles would not remain at rest. Since the plasma is unmagnetized, the zero-order magnetic field is also assumed to be zero. Based on these assumptions we can then write

$$\begin{aligned} n_s &= n_{s0} + n_{s1} \\ \mathbf{v}_s &= \mathbf{v}_{s1} \\ \mathbf{E} &= \mathbf{E}_1 \\ \mathbf{B} &= \mathbf{B}_1. \end{aligned} \tag{4.3.1}$$

In these equations the subscript 0 identifies the zero-order quantities and the subscript 1 identifies a first-order quantity. To assure charge neutrality for the unperturbed system, we must also require that $\sum_s e_s n_{s0} = 0$.

Next we solve the equation of motion and compute the current for an assumed first-order wave field. The equation of motion for a particle of mass m_s and charge e_s is

$$m_s \frac{\mathrm{d}\mathbf{v}_{s1}}{\mathrm{d}t} = e_s[\mathbf{E}_1 + \mathbf{v}_{s1} \times \mathbf{B}_1] \tag{4.3.2}$$

and the equation for the current density is

$$\mathbf{J} = \sum_s n_{s0} e_s \mathbf{v}_{s1} + \sum_s n_{s1} e_s \mathbf{v}_{s1}. \tag{4.3.3}$$

In these equations the nonlinear terms $\mathbf{v}_{s1} \times \mathbf{B}_1$ and $n_{s1}\mathbf{v}_{s1}$ are of second order because they represent the product of two first-order terms. If the amplitude of the first-order perturbation is assumed to be small, so that $|\mathbf{v}_{s1} \times \mathbf{B}_1| \ll |\mathbf{E}_1|$ and $|\mathbf{n}_{s1}\mathbf{v}_{s1}| \ll |\mathbf{n}_{s0}\mathbf{v}_{s1}|$, then the second-order terms can be neglected compared with the first-order terms. The term $\mathrm{d}\mathbf{v}_{s1}/\mathrm{d}t$ represents the acceleration evaluated in the particle frame of reference. Because the particles are initially at rest, $\mathrm{d}\mathbf{v}_{s1}/\mathrm{d}t$ can be replaced by $\partial\mathbf{v}_{s1}/\partial t$. The linearized equations then become

$$m_s \frac{\partial \mathbf{v}_s}{\partial t} = e_s \mathbf{E} \tag{4.3.4}$$

and

$$\mathbf{J} = \sum_s n_{s0} e_s \mathbf{v}_s, \tag{4.3.5}$$

where for notational simplicity we have dropped the subscript 1 on the first-order terms. The subscript 0 on the zero-order term can always be used to distinguish the zero-order terms from the first-order terms.

Next we compute the conductivity tensor. The Fourier transforms of the equation of motion (4.3.4) and the current density (4.3.5) are

$$m_s(-\mathrm{i}\omega)\tilde{\mathbf{v}}_s = e_s\tilde{\mathbf{E}} \tag{4.3.6}$$

and

$$\tilde{\mathbf{J}} = \sum_s n_{s0} e_s \tilde{\mathbf{v}}_s. \tag{4.3.7}$$

Solving Eq. (4.3.6) for the velocity and substituting into the right-hand side of Eq. (4.3.7) gives

$$\tilde{\mathbf{J}} = \sum_s \frac{n_{s0} e_s^2}{(-\mathrm{i}\omega) m_s} \tilde{\mathbf{E}}. \tag{4.3.8}$$

The conductivity tensor, defined by Eq. (4.2.3), can now be identified, and is given by

$$\overleftrightarrow{\sigma} = \overleftrightarrow{\mathbf{1}} \sum_s \frac{n_{s0} e_s^2}{(-\mathrm{i}\omega) m_s}. \tag{4.3.9}$$

In this case, the conductivity tensor is diagonal, which indicates an isotropic scalar conductivity. The conductivity tensor is also purely imaginary, which indicates that **J** and **E** are shifted in phase by $\pi/2$. This phase shift arises from the $\pi/2$ phase shift between the velocity and the electric field implied by the "i" in Eq. (4.3.6). The conductivity also decreases with increasing frequency. This frequency dependence occurs because at higher frequencies the particles have a shorter time to respond to the electric field, thereby causing smaller velocities and currents. Because of the very small mass of the electron, the dominant contribution to the conductivity tensor comes from the electrons.

Having determined the conductivity tensor, the dielectric tensor can be computed using Eq. (4.2.6), which gives

$$\overleftrightarrow{\mathbf{K}} = \overleftrightarrow{\mathbf{1}} \left(1 - \frac{\omega_{\mathrm{p}}^2}{\omega^2} \right), \tag{4.3.10}$$

where $\omega_{\mathrm{p}}^2 = \sum_s \omega_{\mathrm{p}s}^2$ and $\omega_{\mathrm{p}s}^2 = n_{s0} e^2 / \epsilon_0 m_s$. Note that since the electron mass is much smaller than the ion mass, the sum over species is dominated by the electron term, so to a good approximation $\omega_{\mathrm{p}}^2 \simeq \omega_{\mathrm{pe}}^2$. Thus, electrons play the dominant role in determining the dielectric properties of the plasma.

To find the dispersion relation and the electric field eigenvectors, we must next analyze the homogeneous equation for the electric field. Substituting the dielectric tensor (4.3.10) into the homogeneous equation (4.2.8), we obtain the equation

$$c^2 \mathbf{k} \times (\mathbf{k} \times \tilde{\mathbf{E}}) + \left(\omega^2 - \omega_{\mathrm{p}}^2 \right) \tilde{\mathbf{E}} = 0. \tag{4.3.11}$$

Since the plasma is isotropic, without loss of generality we can choose a coordinate system with the z axis along the **k** vector, so that $\mathbf{k} = (0, 0, k)$ and $\tilde{\mathbf{E}} = (\widetilde{E}_x, \widetilde{E}_y, \widetilde{E}_z)$.

Equation (4.3.11) can then be rewritten in the form

$$\begin{bmatrix} -c^2k^2+\omega^2-\omega_{\rm p}^2 & 0 & 0 \\ 0 & -c^2k^2+\omega^2-\omega_{\rm p}^2 & 0 \\ 0 & 0 & \omega^2-\omega_{\rm p}^2 \end{bmatrix} \begin{bmatrix} \widetilde{E}_x \\ \widetilde{E}_y \\ \widetilde{E}_z \end{bmatrix} = 0. \tag{4.3.12}$$

The dispersion relation is obtained by setting the determinant of the matrix to zero, which gives

$$D(k,\omega) = (-c^2k^2+\omega^2-\omega_{\rm p}^2)^2(\omega^2-\omega_{\rm p}^2) = 0. \tag{4.3.13}$$

The roots of the dispersion relation and the corresponding eigenvectors are easily shown to be

$$\omega^2 = \omega_{\rm p}^2 + c^2k^2, \quad \tilde{\mathbf{E}} = (\widetilde{E}_x, \widetilde{E}_y, 0), \tag{4.3.14}$$

and

$$\omega^2 = \omega_{\rm p}^2, \quad \tilde{\mathbf{E}} = (0, 0, \widetilde{E}_z). \tag{4.3.15}$$

Since $\mathbf{k}$ is in the z direction, the electric fields associated with the above two roots are, respectively, perpendicular and parallel to the direction of propagation. These roots are called the transverse ($\mathbf{k}\cdot\tilde{\mathbf{E}} = 0$) and longitudinal ($\mathbf{k}\times\tilde{\mathbf{E}} = 0$) modes.

4.3.1 The Transverse Mode

The electric field eigenvector for the transverse mode is $\tilde{\mathbf{E}} = (\widetilde{E}_x, \widetilde{E}_y, 0)$. As can be seen, the electric field eigenvector has two independent components, $\widetilde{E}_x$ and $\widetilde{E}_y$. Each of these electric field components has an associated magnetic field component given by $\tilde{\mathbf{B}} = (\mathbf{k}/\omega) \times \tilde{\mathbf{E}}$. For the $\widetilde{E}_x$ component, the $\mathbf{k}$ vector, the electric field $\tilde{\mathbf{E}}$, and the magnetic field $\tilde{\mathbf{B}}$ form an orthogonal triad as shown in Figure 4.9. A similar triad exists for the $\widetilde{E}_y$ component, but rotated by $\pi/2$ with respect to the z axis. Since the field geometry is such that $\mathbf{k}\cdot\tilde{\mathbf{E}} = 0$, Gauss' law, $\tilde{\rho}_{\rm q} = \epsilon_0(\mathrm{i}\mathbf{k})\cdot\tilde{\mathbf{E}}$, shows that there is no charge density fluctuation associated with the transverse mode. Thus, even though this mode generates an electrical current, the current does not cause a charge density fluctuation.

The dispersion relation for the transverse mode, $\omega^2 = \omega_{\rm p}^2 + c^2k^2$, is plotted in Figure 4.10. As can be seen, for large ω and k the dispersion relation is asymptotic to the straight lines $\omega = \pm ck$, which are the free space branches of the electromagnetic dispersion relation. Thus, at high frequencies the transverse mode reduces to the free space electromagnetic mode. No real solution for ω exists at frequencies below the plasma frequency $\omega_{\rm p}$. Therefore, propagation cannot occur

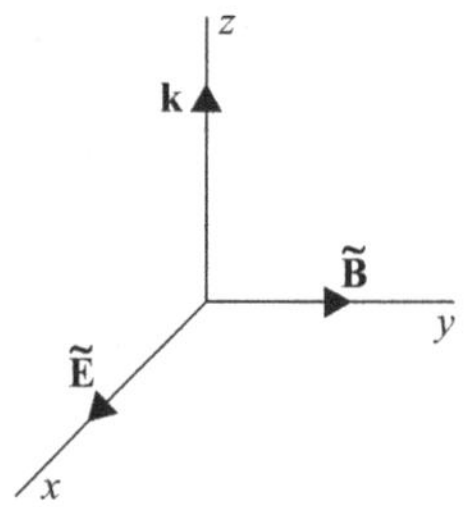

Figure 4.9 The electric and magnetic field eigenvectors for the transverse electromagnetic mode.

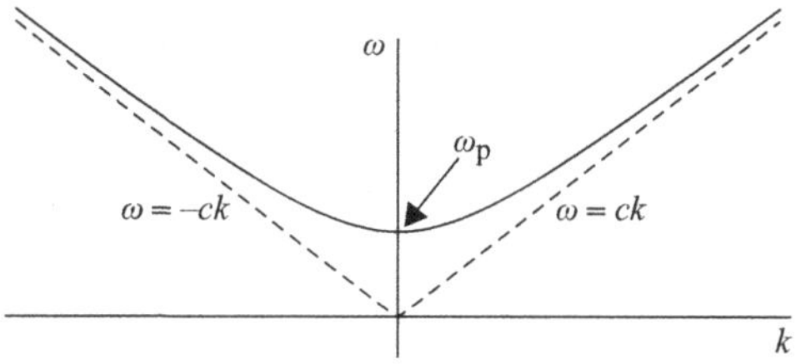

Figure 4.10 The dispersion relation for the transverse electromagnetic mode.

for frequencies below the plasma frequency. The existence of this propagation cutoff can be demonstrated by solving the dispersion relation for the wave number

$$k = \pm\frac{1}{c}\sqrt{\omega^2 - \omega_{\mathrm{p}}^2}. \tag{4.3.16}$$

For $|\omega| > \omega_{\mathrm{p}}$ the wave number is real, which corresponds to a propagating wave. For $|\omega| < \omega_{\mathrm{p}}$ the wave number is purely imaginary, $k = \mathrm{i}k_{\mathrm{i}}$, which corresponds to a pure exponential decay with no spatial oscillations. Such a wave is said to be *evanescent* or non-propagating. The reason for the cutoff at the plasma frequency can be traced to the relationship between the conduction current and the displacement current. It can be shown that the ratio of the conduction current to the displacement current is given by

$$\frac{\text{conduction current}}{\text{displacement current}} = \frac{\sum_s n_{s0}e^2/(-\mathrm{i}\omega)m_s}{\epsilon_0(-\mathrm{i}\omega)} = -\frac{\omega_{\mathrm{p}}^2}{\omega^2}. \tag{4.3.17}$$

The minus sign in the above equation indicates that the conduction current opposes the displacement current. At high frequencies, $\omega \gg \omega_{\mathrm{p}}$, the conduction current has little effect on the free space electromagnetic mode. However, because of the $1/\omega^2$ dependence, as the frequency decreases the conduction current increases, until at the plasma frequency it just cancels the displacement current. At and below this

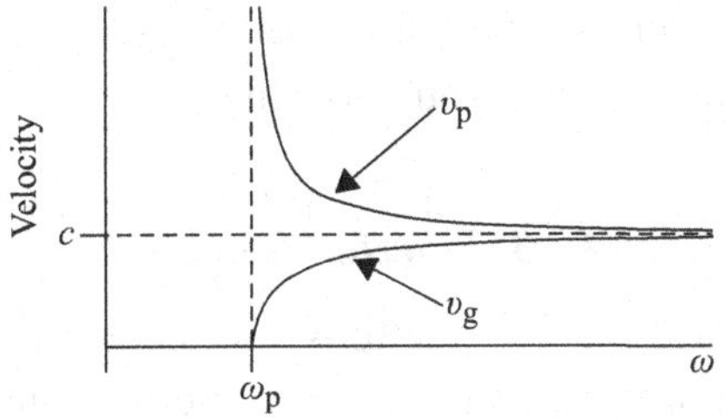

Figure 4.11 The phase velocity (top plot) and group velocity (bottom plot) of the transverse electromagnetic mode.

frequency propagation cannot occur, since the current on the right-hand side of Ampère's law (4.2.5) has the wrong sign.

The phase and group velocities can be computed from the dispersion relation and are given by

$$v_p = \frac{\omega}{k} = \frac{c}{\sqrt{1 - (\omega_p^2/\omega^2)}} \quad \text{and} \quad v_g = \frac{\partial \omega}{\partial k} = c\sqrt{1 - (\omega_p^2/\omega^2)}. \tag{4.3.18}$$

As shown in Figure 4.11, the phase velocity goes to infinity at the plasma frequency, and the group velocity goes to zero. Note that the group velocity, which is the velocity of energy propagation, is always less than the speed of light. It can also be verified that the group velocity and phase velocity satisfy the relationship $v_p v_g = c^2$. The phase and group velocities are also the same as the phase and group velocities for an electromagnetic wave propagating in a waveguide, with the plasma frequency playing the same role as the cutoff frequency of a waveguide.

4.3.2 The Longitudinal Mode

For the longitudinal mode, Eq. (4.3.15) shows that the dispersion relation is $\omega = \pm\omega_p$, independent of k. This mode corresponds to the electron plasma oscillations analyzed in Section 2.3 using a simple slab model. The wave magnetic field of this mode is zero because $\tilde{\mathbf{B}} = (\mathbf{k}/\omega) \times \tilde{\mathbf{E}} = 0$. However, the perturbed charge density is not zero, since $\tilde{\rho}_q = \epsilon_0(i\mathbf{k}) \cdot \tilde{\mathbf{E}} \neq 0$. Thus, the longitudinal mode is associated with a charge density fluctuation. Since $\mathbf{k} \times \tilde{\mathbf{E}} = 0$ (or $\nabla \times \mathbf{E} = 0$), it also follows that the electric field of the longitudinal mode can be written as the gradient of an electrostatic potential, $\mathbf{E} = -\nabla\Phi$. Thus, the oscillations are purely electrostatic. It is also easy to see that the group velocity, $\partial\omega/\partial k$, is zero, since the oscillation frequency is independent of the wave number. Thus, the envelope of the wave packet does not propagate. Since the wave magnetic field is zero, the Poynting flux, $\mathbf{S} = (1/\mu_0)\mathbf{E} \times \mathbf{B}$, which represents the flow of electromagnetic energy, is also zero, consistent with the fact that the group velocity is zero. In the next chapter, when non-zero temperatures are considered, we will show that the longitudinal mode has a non-zero group velocity.

4.3.3 External Sources

The discussions in the previous two sections have been mainly concerned with the wave modes that can exist in a plasma with no external source. An external source can be incorporated by adding an external current density, $\mathbf{J}_{\text{ext}}$, to Ampère's law so that

$$\nabla \times \mathbf{B} = \mu_0 \frac{\partial \mathbf{D}}{\partial t} + \mu_0 \mathbf{J}_{\text{ext}}. \tag{4.3.19}$$

After Fourier transforming, eliminating $\tilde{\mathbf{B}}$ using Faraday's law, and introducing the dielectric tensor via $\tilde{\mathbf{D}} = \epsilon_0 \overleftrightarrow{\mathbf{K}} \cdot \tilde{\mathbf{E}}$, the above equation can be written

$$\mathbf{k} \times (\mathbf{k} \times \tilde{\mathbf{E}}) + \frac{\omega^2}{c^2} \overleftrightarrow{\mathbf{K}} \cdot \tilde{\mathbf{E}} = -\mathrm{i}\omega\mu_0 \tilde{\mathbf{J}}_{\text{ext}}, \tag{4.3.20}$$

where $\tilde{\mathbf{J}}_{\text{ext}}$ is the Fourier transform of the external current distribution. Splitting the electric field and current density into longitudinal (parallel to $\mathbf{k}$) and transverse (perpendicular to $\mathbf{k}$) components, $\tilde{\mathbf{E}} = \tilde{\mathbf{E}}_{\|} + \tilde{\mathbf{E}}_{\perp}$ and $\tilde{\mathbf{J}} = \tilde{\mathbf{J}}_{\|} + \tilde{\mathbf{J}}_{\perp}$, and using the dielectric tensor given by Eq. (4.3.10), it can be shown that the above equation separates nicely into two equations:

$$(-c^2k^2 + \omega^2 - \omega_{\mathrm{p}}^2)^2 \tilde{\mathbf{E}}_{\perp} = -\frac{\mathrm{i}\omega}{\epsilon_0} \tilde{\mathbf{J}}_{\perp\text{ext}} \tag{4.3.21}$$

and

$$(\omega^2 - \omega_{\mathrm{p}}^2) \tilde{\mathbf{E}}_{\|} = -\frac{\mathrm{i}\omega}{\epsilon_0} \tilde{\mathbf{J}}_{\|\text{ext}}. \tag{4.3.22}$$

These two equations show that the transverse component of the external current acts as a source for the transverse electromagnetic waves, and the longitudinal component acts as a source for the longitudinal electrostatic waves.

In practice, it is somewhat difficult to interpret Eqs. (4.3.21) and (4.3.22), since $\tilde{\mathbf{J}}_{\perp}$ and $\tilde{\mathbf{J}}_{\|}$ involve Fourier transforms. The interpretation is simplified somewhat by noting, from the continuity equation $\partial\rho_{\mathrm{q}}/\partial t + \nabla \cdot \mathbf{J} = 0$ (which after Fourier transforming becomes $-\omega\tilde{\rho}_{\mathrm{q}} + k\widetilde{J}_{\|} = 0$), that $\tilde{\mathbf{J}}_{\|}$ is associated with a charge density fluctuation, whereas $\tilde{\mathbf{J}}_{\perp}$ is not. Therefore, longitudinal waves can be excited only if charge density fluctuations are present in the source. For example, a loop antenna, such as that shown in Figure 4.12, can only excite transverse waves, since there are no charge density fluctuations associated with this current system. On the

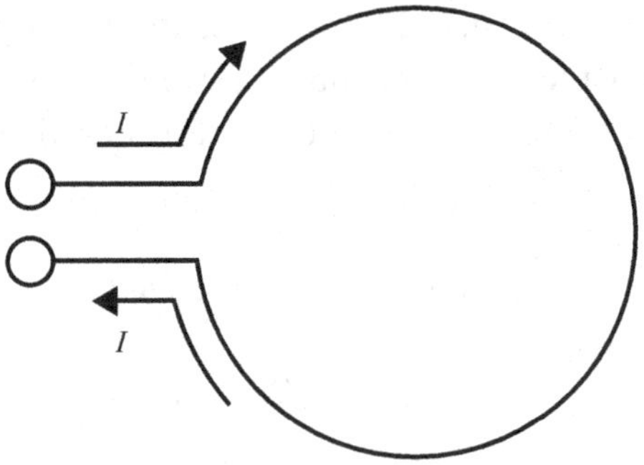

Figure 4.12 A loop antenna has no charge density fluctuations and cannot excite the longitudinal electrostatic mode.

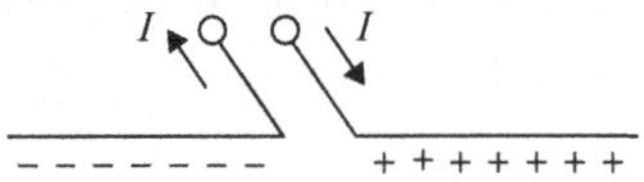

Figure 4.13 An electric dipole antenna has a charge density fluctuation and can excite both longitudinal and transverse modes.

other hand, an electric dipole antenna, such as that shown in Figure 4.13, has both a transverse current and a longitudinal current (i.e., charge fluctuation) and can excite both transverse and longitudinal waves.

Once the Fourier transform of the source current is known, Eqs. (4.3.21) and (4.3.22) can be solved for the electric fields by computing the inverse Fourier transforms of $\tilde{\mathbf{E}}_{\perp}$ and $\tilde{\mathbf{E}}_{\parallel}$. Note that the terms $-c^2k^2 + \omega^2 - \omega_{\rm p}^2$ and $\omega^2 - \omega_{\rm p}^2$ occur in the denominators of the inverse Fourier transforms. These denominators play a crucial role in determining the properties of the inverse Fourier transform by introducing poles in the integrals. The procedure for computing integrals of this type using Laplace transforms will be discussed in Chapter 9.

4.4 Waves in a Cold Uniform Magnetized Plasma

We next consider the case of a cold plasma with an externally imposed static uniform magnetic field. Using the same linearization procedure as in Section 4.3, the magnetic field must now be written $\mathbf{B} = \mathbf{B}_0 + \mathbf{B}_1$, where $\mathbf{B}_0$ is the externally imposed magnetic field and $\mathbf{B}_1$ is the first-order perturbation. The linearized equation of motion is then

$$m_s \frac{\partial \mathbf{v}_{s1}}{\partial t} = e_s[\mathbf{E}_1 + \mathbf{v}_{s1} \times \mathbf{B}_0], \tag{4.4.1}$$

where the term $\mathbf{v}_{s1} \times \mathbf{B}_1$ has been dropped because it is a product of two small first-order terms. Without loss of generality we can let the static magnetic field be in the z direction, $\mathbf{B}_0 = (0, 0, B_0)$. After Fourier transforming, the above equation becomes

$$
\begin{aligned}
-\mathrm{i}\omega m_s \tilde{v}_{sx} &= e_s[\widetilde{E}_x + \tilde{v}_{sy} B_0], \\
-\mathrm{i}\omega m_s \tilde{v}_{sy} &= e_s[\widetilde{E}_y - \tilde{v}_{sx} B_0], \\
-\mathrm{i}\omega m_s \tilde{v}_{sz} &= e_s \widetilde{E}_z,
\end{aligned}
\tag{4.4.2}
$$

where for notational simplicity the subscript 1 on the first-order terms has again been dropped. By introducing the cyclotron frequency, $\omega_{cs} = e_s B_0 / m_s$, the above equations can be written in matrix form as

$$
\begin{bmatrix} -\mathrm{i}\omega & -\omega_{cs} & 0 \\ \omega_{cs} & -\mathrm{i}\omega & 0 \\ 0 & 0 & -\mathrm{i}\omega \end{bmatrix}
\begin{bmatrix} \tilde{v}_{sx} \\ \tilde{v}_{sy} \\ \tilde{v}_{sz} \end{bmatrix}
= \frac{e_s}{m_s}
\begin{bmatrix} \widetilde{E}_x \\ \widetilde{E}_y \\ \widetilde{E}_z \end{bmatrix}.
\tag{4.4.3}
$$

This set of linear equations can be solved for the velocity components $\tilde{v}_{sx}$, $\tilde{v}_{sy}$, and $\tilde{v}_{sz}$ by inverting the matrix on the left-hand side, which gives

$$
\begin{bmatrix} \tilde{v}_{sx} \\ \tilde{v}_{sy} \\ \tilde{v}_{sz} \end{bmatrix}
= \frac{e_s}{m_s}
\begin{bmatrix}
\dfrac{-\mathrm{i}\omega}{\omega_{cs}^2 - \omega^2} & \dfrac{\omega_{cs}}{\omega_{cs}^2 - \omega^2} & 0 \\
\dfrac{-\omega_{cs}}{\omega_{cs}^2 - \omega^2} & \dfrac{-\mathrm{i}\omega}{\omega_{cs}^2 - \omega^2} & 0 \\
0 & 0 & \dfrac{\mathrm{i}}{\omega}
\end{bmatrix}
\begin{bmatrix} \widetilde{E}_x \\ \widetilde{E}_y \\ \widetilde{E}_z \end{bmatrix}.
\tag{4.4.4}
$$

Following the same procedure as in the previous section, the conductivity tensor can be determined by computing the current $\tilde{\mathbf{J}} = \sum_s n_{s0} e_s \tilde{\mathbf{v}}_s$, which in matrix form becomes

$$
\begin{bmatrix} \widetilde{J}_x \\ \widetilde{J}_y \\ \widetilde{J}_z \end{bmatrix}
= \sum_s \frac{n_{s0} e_s^2}{m_s}
\begin{bmatrix}
\dfrac{-\mathrm{i}\omega}{\omega_{cs}^2 - \omega^2} & \dfrac{\omega_{cs}}{\omega_{cs}^2 - \omega^2} & 0 \\
\dfrac{-\omega_{cs}}{\omega_{cs}^2 - \omega^2} & \dfrac{-\mathrm{i}\omega}{\omega_{cs}^2 - \omega^2} & 0 \\
0 & 0 & \dfrac{\mathrm{i}}{\omega}
\end{bmatrix}
\begin{bmatrix} \widetilde{E}_x \\ \widetilde{E}_y \\ \widetilde{E}_z \end{bmatrix}.
\tag{4.4.5}
$$

The conductivity tensor, defined by Eq. (4.2.3), can now be identified and is given by

$$\overleftrightarrow{\sigma} = \sum_s \frac{n_{s0}e_s^2}{m_s}\begin{bmatrix} \dfrac{-\mathrm{i}\omega}{\omega_{\mathrm{c}s}^2-\omega^2} & \dfrac{\omega_{\mathrm{c}s}}{\omega_{\mathrm{c}s}^2-\omega^2} & 0 \\ \dfrac{-\omega_{\mathrm{c}s}}{\omega_{\mathrm{c}s}^2-\omega^2} & \dfrac{-\mathrm{i}\omega}{\omega_{\mathrm{c}s}^2-\omega^2} & 0 \\ 0 & 0 & \dfrac{\mathrm{i}}{\omega} \end{bmatrix}. \tag{4.4.6}$$

Finally, the dielectric tensor, defined by Eq. (4.2.6), can be computed and has the form

$$\overleftrightarrow{\mathbf{K}} = \begin{bmatrix} S & -\mathrm{i}D & 0 \\ -\mathrm{i}D & S & 0 \\ 0 & 0 & P \end{bmatrix}, \tag{4.4.7}$$

where

$$S = 1 - \sum_s \frac{\omega_{\mathrm{p}s}^2}{\omega^2-\omega_{\mathrm{c}s}^2}, \qquad D = \sum_s \frac{\omega_{\mathrm{c}s}\,\omega_{\mathrm{p}s}^2}{\omega(\omega^2-\omega_{\mathrm{c}s}^2)}, \tag{4.4.8}$$

and

$$P = 1 - \sum_s \frac{\omega_{\mathrm{p}s}^2}{\omega^2}. \tag{4.4.9}$$

The term P is seen to be identical to the dielectric tensor elements for an unmagnetized plasma; see Eq. (4.3.10). This similarity occurs because for motions along the magnetic field the $\mathbf{v}\times\mathbf{B}$ force is zero, so the magnetic field has no effect on the $z-z$ element of the dielectric tensor. Following Stix (1992), the terms S and D can be decomposed into a sum and difference using the relations

$$S = \frac{1}{2}(R+L) \quad \text{and} \quad D = \frac{1}{2}(R-L), \tag{4.4.10}$$

where R and L are defined by

$$R = 1 - \sum_s \frac{\omega_{\mathrm{p}s}^2}{\omega(\omega+\omega_{\mathrm{c}s})} \quad \text{and} \quad L = 1 - \sum_s \frac{\omega_{\mathrm{p}s}^2}{\omega(\omega-\omega_{\mathrm{c}s})}. \tag{4.4.11}$$

As will be shown, the equation R is associated with a right-hand polarized mode, and the equation L is associated with a left-hand polarized mode. Note that the signs of the cyclotron frequencies in Eq. (4.4.11) are determined by the signs of

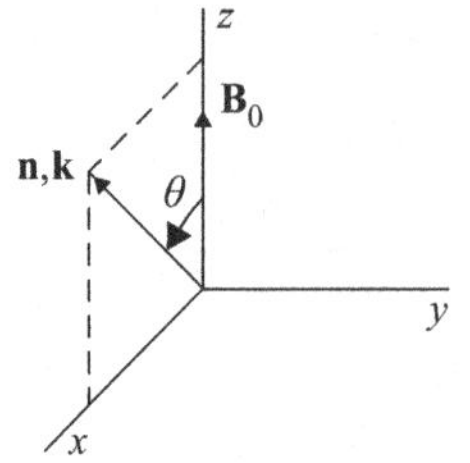

Figure 4.14 The coordinate system used for analyzing wave propagation in a cold magnetized plasma.

the charges (i.e., negative for electrons and positive for positively charged ions); see Eq. (2.4.3).

To obtain the dispersion relation, it is convenient to express the homogeneous equation (4.2.9) in matrix form. Without loss of generality we can rotate the coordinate system so that the index of refraction vector **n** (which is parallel to **k**) is in the x,z plane at an angle θ relative to the magnetic field $\mathbf{B}_0$ as shown in Figure 4.14, so that

$$\mathbf{n} = (n\sin\theta,\ 0,\ n\cos\theta). \tag{4.4.12}$$

Computing the various cross-products in Eq. (4.2.9) and organizing the results in matrix form, the homogeneous equation for the electric field can be written

$$\begin{bmatrix} -n^2\cos^2\theta & 0 & n^2\sin\theta\cos\theta \\ 0 & -n^2 & 0 \\ n^2\sin\theta\cos\theta & 0 & -n^2\sin^2\theta \end{bmatrix}\begin{bmatrix} \widetilde{E}_x \\ \widetilde{E}_y \\ \widetilde{E}_z \end{bmatrix} + \begin{bmatrix} S & -\mathrm{i}D & 0 \\ \mathrm{i}D & S & 0 \\ 0 & 0 & P \end{bmatrix}\begin{bmatrix} \widetilde{E}_x \\ \widetilde{E}_y \\ \widetilde{E}_z \end{bmatrix} = 0 \tag{4.4.13}$$

which reduces to

$$\begin{bmatrix} S - n^2\cos^2\theta & -\mathrm{i}D & n^2\sin\theta\cos\theta \\ \mathrm{i}D & S - n^2 & 0 \\ n^2\sin\theta\cos\theta & 0 & P - n^2\sin^2\theta \end{bmatrix}\begin{bmatrix} \widetilde{E}_x \\ \widetilde{E}_y \\ \widetilde{E}_z \end{bmatrix} = 0. \tag{4.4.14}$$

Following the procedure in the previous section, the homogeneous equation has a non-trivial solution if and only if the determinant of the matrix is zero, which gives the dispersion relation

$$\begin{aligned} Đ(\mathbf{n},\omega) = n^2\sin\theta\cos\theta\,[-(S-n^2)n^2\sin\theta\cos\theta] \\ + [P - n^2\sin^2\theta][(S-n^2)(S-n^2\cos^2\theta) - D^2] = 0. \end{aligned} \tag{4.4.15}$$

Note that the n^6 terms cancel. This is fortunate because otherwise the dispersion relation would be a cubic in n^2, which would be more difficult to solve. Using the identity $S^2 - D^2 = RL$, the dispersion relation can be simplified to the form

$$Ð(\mathbf{n},\omega) = An^4 - Bn^2 + RLP = 0, \tag{4.4.16}$$

where $A = S\sin^2\theta + P\cos^2\theta$ and $B = RL\sin^2\theta + PS(1+\cos^2\theta)$. This equation is a quadratic in n^2 and can be solved using the quadratic formula, which gives

$$n^2 = \frac{B \pm F}{2A}, \tag{4.4.17}$$

where the term F can be written in the positive definite form

$$F^2 = (RL - PS)^2\sin^4\theta + 4P^2D^2\cos^2\theta. \tag{4.4.18}$$

This form is useful because it shows that F must always be real. Since A and B are real, it follows that the index of refraction must be either purely real ($n^2 > 0$), which corresponds to a propagating wave, or purely imaginary ($n^2 < 0$), which corresponds to an evanescent wave. A complex index of refraction, with non-zero real and imaginary parts, cannot occur because such an index of refraction would imply absorption of wave energy by the plasma. In the absence of collisions, energy absorption cannot occur in a cold plasma because there is no mechanism for converting the ordered motions associated with the wave into random thermal motions.

An equivalent form of the dispersion relation can be obtained by multiplying the term RLP in Eq. (4.4.16) by $\sin^2\theta + \cos^2\theta (= 1)$ and sorting out the $\sin^2\theta$ and $\cos^2\theta$ terms. The result is

$$\tan^2\theta = \frac{-P(n^2 - R)(n^2 - L)}{(Sn^2 - RL)(n^2 - P)}. \tag{4.4.19}$$

This relation is sometimes referred to as the "tangent" form of the dispersion relation, and is useful in some circumstances.

4.4.1 Propagation Parallel to the Magnetic Field

For propagation parallel to the magnetic field, $\theta = 0$, Eq. (4.4.19) shows that the dispersion relation has three roots: $P = 0, n^2 = R$, and $n^2 = L$. To understand the nature of the waves associated with each of these roots, it is necessary to examine the homogeneous equation for the electric field.

For $\theta = 0$, the homogeneous equation (4.4.14) becomes

$$\begin{bmatrix} S - n^2 & -\mathrm{i}D & 0 \\ \mathrm{i}D & S - n^2 & 0 \\ 0 & 0 & P \end{bmatrix} \begin{bmatrix} \widetilde{E}_x \\ \widetilde{E}_y \\ \widetilde{E}_z \end{bmatrix} = 0. \tag{4.4.20}$$

The electric field eigenvectors associated with each of the three roots are as follows:

$$P = 0, \qquad \tilde{\mathbf{E}} = (0, 0, E_0), \tag{4.4.21}$$

$$n^2 = R, \qquad \tilde{\mathbf{E}} = (E_0, \mathrm{i}E_0, 0), \tag{4.4.22}$$

$$n^2 = L, \qquad \tilde{\mathbf{E}} = (E_0, -\mathrm{i}E_0, 0). \tag{4.4.23}$$

Since the electric field eigenvector for the first root, $P = 0$, is parallel to the z axis, and therefore to $\mathbf{k}$, this root corresponds to a longitudinal wave. This is the same longitudinal electrostatic mode described by Eq. (4.3.15) for a cold unmagnetized plasma. The magnetic field has no effect on this mode because the particles oscillate along the magnetic field and the $\mathbf{v} \times \mathbf{B}$ force vanishes. As can be seen from Eq. (4.4.9), the condition $P = 0$ implies that the oscillation occurs at the plasma frequency, $\omega = \pm\omega_{\mathrm{p}}$.

Since the electric field eigenvectors for the second and third roots, $n^2 = R$ and $n^2 = L$, are perpendicular to the wave vector, $\mathbf{k}$, which for $\theta = 0$ is along the z axis, these roots correspond to transverse waves. These modes are electromagnetic, since Faraday's law, $\tilde{\mathbf{B}} = \mathbf{k} \times \tilde{\mathbf{E}}/\omega$, shows that the wave magnetic field is non-zero. Since the field geometry is such that $\mathbf{k} \cdot \tilde{\mathbf{E}} = 0$, Gauss' law, $\tilde{\rho}_q = \epsilon_0(\mathrm{i}\mathbf{k}) \cdot \tilde{\mathbf{E}}$, shows that there are no charge density fluctuations associated with these modes. When computing the eigenvectors, note that for $n^2 = R$ one has $S - n^2 = -(1/2)(R - L) = -D$, whereas for $n^2 = L$ one has $S - n^2 = D$. Thus, the only difference between the eigenvectors for these two modes is the sign of $\widetilde{E}_y$. The $+\mathrm{i}$ in the $\widetilde{E}_y$ component for the $n^2 = R$ mode indicates a $\pi/2$ phase shift with respect to the $\widetilde{E}_x$ component. It is easily verified that this phase shift corresponds to a right-hand sense of rotation with respect to the static magnetic field. For example, if E_0 is a real number, then at $z = 0$,

$$E_x = \mathrm{Re}\{E_0\, \mathrm{e}^{-\mathrm{i}\omega t}\} = E_0 \cos\omega t \tag{4.4.24}$$

and

$$E_y = \mathrm{Re}\{\mathrm{i}E_0\, \mathrm{e}^{-\mathrm{i}\omega t}\} = E_0 \sin\omega t, \tag{4.4.25}$$

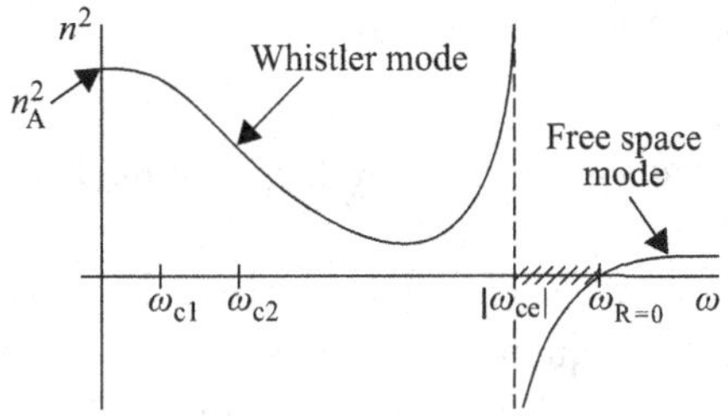

Figure 4.15 A plot of $n^2 = R$ as a function of frequency.

which represents a wave rotating in the right-hand sense with respect to the static magnetic field. Similarly, the $-i$ in the E_y component of the electric field eigenvector for the $n^2 = L$ mode indicates a left-hand polarized wave. Therefore, the R and L notation for the two modes gives the sense of rotation (right-hand and left-hand) with respect to the externally imposed static magnetic field. Note that this convention does *not* correspond to the convention used in optics. In optics the polarization is defined with respect to the wave vector, **k**, rather than the magnetic field vector $\mathbf{B}_0$.

It is useful next to consider the frequency dependence of the index of refraction for the R and L modes. For the R mode the index of refraction is given by

$$n^2 = R = 1 - \sum_s \frac{\omega_{ps}^2}{\omega(\omega + \omega_{cs})}. \qquad (4.4.26)$$

A plot of n^2 as a function of frequency is shown in Figure 4.15 for a plasma consisting of electrons and two types of positively charged ions, with cyclotron frequencies ω_{ce}, ω_{c1}, and ω_{c2}. Note that since ω_{ce} is negative for electrons, the function $n^2 = R$ goes to infinity at $\omega = |\omega_{ce}|$. A condition where the index of refraction goes to infinity is called a *resonance*. In this case the resonance is called the electron cyclotron resonance. Since electrons rotate around the static magnetic field in the same (right-hand) sense as the rotation of the wave electric field, for frequencies near the electron cyclotron frequency the wave interacts very strongly with the electrons. A comparable resonant interaction does not occur at the ion cyclotron frequencies ω_{c1} and ω_{c2}, since the wave field and the ions have opposite senses of rotation.

At a frequency somewhat above the electron cyclotron frequency the index of refraction goes to zero. A condition where the index of refraction goes to zero is called a *cutoff*. The frequency at which this cutoff occurs is called the $R = 0$ cutoff. To a very good approximation, ion motions can be ignored at these high frequencies. Omitting the ion terms in Eq. (4.4.26), it can be shown by setting

$R = 0$ that the cutoff frequency is given by

$$\omega_{R=0} = \frac{|\omega_{ce}|}{2} + \sqrt{\left(\frac{\omega_{ce}}{2}\right)^2 + \omega_{pe}^2}. \tag{4.4.27}$$

Since n^2 is negative between $|\omega_{ce}|$ and $\omega_{R=0}$, right-hand polarized waves cannot propagate in this range of frequencies, since the index of refraction is imaginary. The wave becomes *evanescent*. The evanescent region is indicated by the hatched lines in Figure 4.15. The branch of the dispersion relation below $|\omega_{ce}|$ is called the *whistler mode*, after a type of magnetospheric radio wave called a whistler, which will be discussed shortly. The branch of the dispersion relation above $\omega_{R=0}$ is called the free space mode because at high frequencies, this branch becomes the right-hand polarized free space electromagnetic mode. Note that as the frequency approaches infinity the index of refraction approaches one, its free space value.

As the wave frequency approaches zero, the function R appears to diverge because of the $1/\omega$ terms in Eq. (4.4.26). However, careful analysis (Problem 4.9) shows that the coefficients of the $1/\omega$ terms cancel if the plasma is electrically neutral ($\sum_i n_{0i} = n_{0e}$). In the limit of zero frequency, it can be shown that the index of refraction approaches a constant value called the Alfvén index of refraction, n_A, which is given by

$$n_A^2 = 1 + \sum_s \frac{\omega_{ps}^2}{\omega_{cs}^2}. \tag{4.4.28}$$

Waves obeying this dispersion relation are called Alfvén waves, after Alfvén (1942) who first predicted their existence. This same index of refraction can be obtained from the dielectric constant given by Eq. (3.6.9) based on polarization drift. A more detailed discussion of Alfvén waves is given in Chapter 6.

The index of refraction for the left-hand polarized mode is given by

$$n^2 = L = 1 - \sum_s \frac{\omega_{ps}^2}{\omega(\omega - \omega_{cs})}. \tag{4.4.29}$$

The frequency dependence of the L mode is similar to the R mode, except that the resonances are now at ion cyclotron frequencies rather than at the electron cyclotron frequency (again assuming positively charged ions). These resonances are called ion cyclotron resonances. They occur because positively charged ions rotate around the static magnetic field in the same left-hand sense as the rotation of the wave electric field, thereby causing strong interactions near the ion cyclotron frequencies. A plot of n^2 as a function of frequency is shown in Figure 4.16.

The frequency variation consists of an alternating series of resonances and cutoffs that separate regions of propagation and evanescence. The evanescent

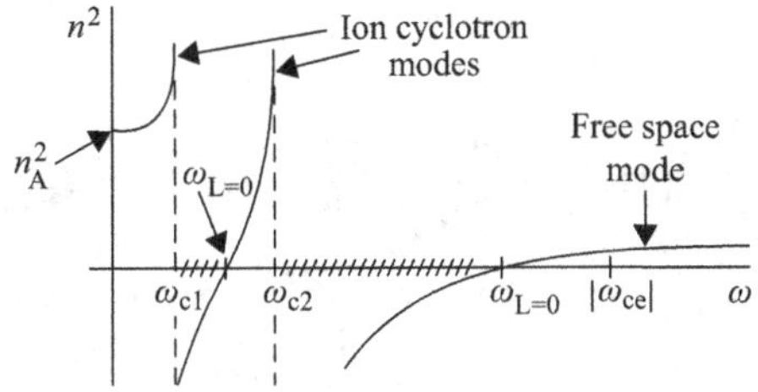

Figure 4.16 A plot of $n^2 = L$ as a function of frequency.

regions are indicated by hatched lines. The branches below each of the ion cyclotron frequencies are called the ion cyclotron modes. As the frequency approaches zero, the index of refraction approaches the Alfvén index of refraction, just as in the case of the R mode. At high frequencies the index of refraction approaches the free space value ($n \to 1$ as $\omega \to \infty$). This branch is called the left-hand polarized free space mode. The low-frequency cutoff of the free space branch can be obtained by setting $L = 0$ in Eq. (4.4.11), and is called the $L = 0$ cutoff. Usually ion motions can be ignored when computing this cutoff, in which case the cutoff frequency is given by

$$\omega_{L=0} = -\frac{|\omega_{ce}|}{2} + \sqrt{\left(\frac{\omega_{ce}}{2}\right)^2 + \omega_{pe}^2}. \tag{4.4.30}$$

This approximation fails at very low densities, since in this limit $\omega_{L=0}$ becomes comparable to the ion cyclotron frequencies. The cutoff frequency must then be calculated numerically. Cutoff frequencies also occur between the ion cyclotron frequencies. These cutoffs are called the ion–ion cutoffs, and must usually be computed numerically.

Numerous examples of the above modes of propagation exist in both laboratory and space plasmas. Some of these examples are discussed below.

Whistlers

In 1918, Barkhausen (1919), while listening to signals from an antenna connected to a rudimentary vacuum tube amplifier, reported hearing unusual whistling tones that decrease, in frequency with increasing time. These signals came to be known as whistlers. Over thirty years later, in 1953, Storey (1953) explained the origin of whistlers. He showed that whistlers are produced by lightning and that the whistling tone is caused by the dispersive propagation of the lightning signal as it travels along a magnetic field line from one hemisphere to the other in the right-hand polarized mode now known as the whistler mode.

A frequency–time spectrogram of several whistlers detected on the ground is shown in Figure 4.17. Dark regions in the spectrogram indicate higher intensities. According to currently accepted ideas, whistlers are guided almost exactly

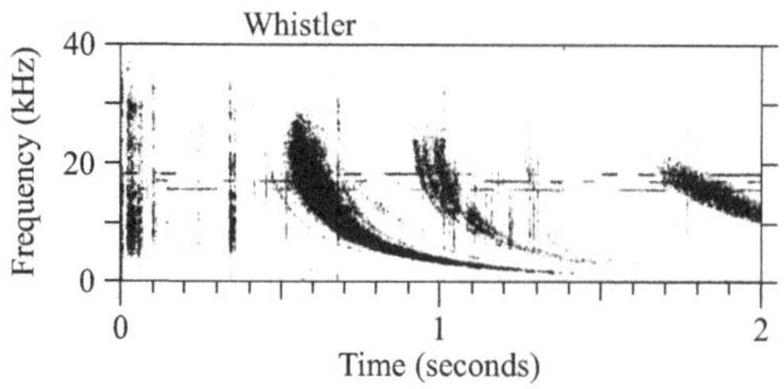

Figure 4.17 A frequency–time spectrogram of several whistlers (adapted from Helliwell (1965)).

along the magnetic field line ($\theta = 0$) by small magnetic field-aligned density irregularities that exist in the magnetized plasma surrounding the Earth, known as the magnetosphere. The guiding of the wave is thought to be similar to the guiding of light in an optical fiber. Later, we will show that even in the absence of such field-aligned irregularities, the whistler energy still follows the magnetic field to a good approximation because of the anisotropy introduced by the magnetic field. The whistling character of the signal arises because of the dispersive nature of the propagation. At the low (audio) frequencies where whistlers are typically observed, the higher frequencies propagate faster than lower frequencies. Since the lightning stroke simultaneously emits a broad range of frequencies, the higher frequencies arrive first, thereby dispersing the signal from the lightning stroke into a tone that decreases in frequency with increasing time. Because the wave can reflect at the base of the ionosphere, whistlers sometimes bounce back and forth along the magnetic field line from one hemisphere to the other. A periodic sequence of whistlers is then observed, with each successive whistler having proportionally more dispersion.

For the conditions that exist in Earth's magnetosphere, the index of refraction of the whistler mode is usually much greater than one, so the first term in Eq. (4.4.26) can be discarded. Since the frequencies of interest are usually well above the ion cyclotron frequencies, the ion terms can also be discarded. The index of refraction is then given by

$$n^2 = \frac{\omega_{\rm p}^2}{\omega(\omega_{\rm c} - \omega)}, \tag{4.4.31}$$

where, for notational convenience, we write $\omega_{\rm p}$ for $\omega_{\rm pe}$ and $\omega_{\rm c}$ for $|\omega_{\rm ce}|$. Using the above equation, it can be shown that the group velocity is given by

$$v_{\rm g} = 2c\frac{\omega^{1/2}(\omega_{\rm c} - \omega)^{3/2}}{\omega_{\rm c}\omega_{\rm p}}. \tag{4.4.32}$$

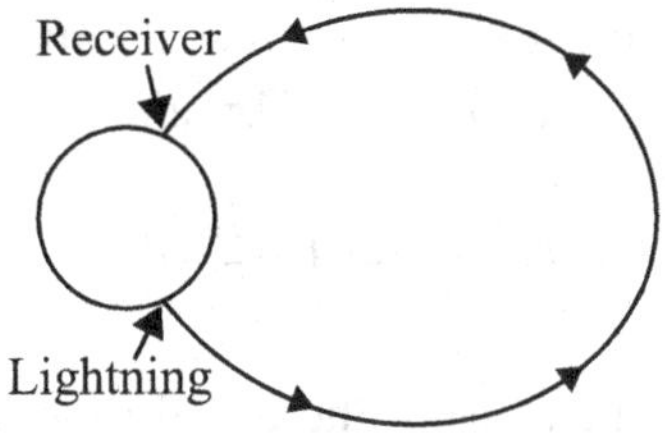

Figure 4.18 Whistlers observed on the ground are guided along the magnetic field line from one hemisphere to the other by small field-aligned density irregularities.

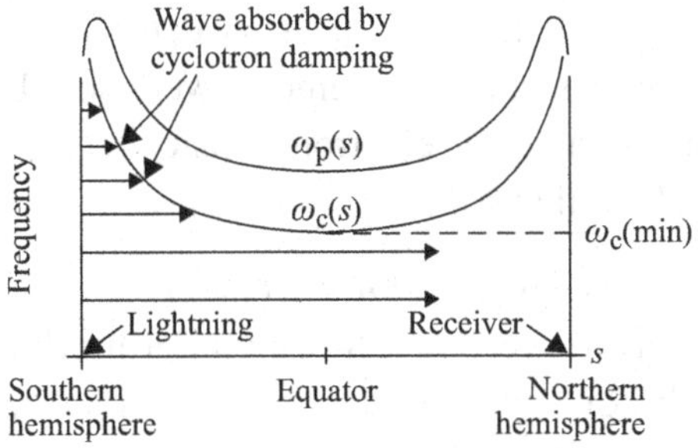

Figure 4.19 A representative plot of the electron plasma frequency, $\omega_p(s)$, and the cyclotron frequency, $\omega_c(s)$, as a function of the path length, s, along the magnetic field line.

The travel time $t(\omega)$ for a given frequency ω is then given by

$$t(\omega) = \frac{1}{2c\omega^{1/2}} \int \frac{\omega_p \omega_c}{(\omega_c - \omega)^{3/2}} \mathrm{d}s, \tag{4.4.33}$$

where the integration is carried out along the magnetic field line from the lightning source to the receiver, as shown in Figure 4.18. In the above equation, both ω_c and ω_p are functions of the path length s along the magnetic field line. Representative plots of $\omega_c(s)$ and $\omega_p(s)$ as a function of the path length, s, from one hemisphere to the other are shown in Figure 4.19. Note from Eq. (4.4.33) that the travel time goes to infinity at zero frequency and at the minimum electron cyclotron frequency along the path, $\omega_c(\text{min})$. Frequencies above $\omega_c(\text{min})$ are absorbed at the electron cyclotron frequency during the upgoing portion of the ray path by a process called cyclotron damping (see Chapter 10) and cannot reach the opposite hemisphere.

Since the travel time depends on the electron density (via the plasma frequency, ω_p) it is obvious that the whistler dispersion gives information on the electron density. Since ω_c is known, the integral equation for $t(\omega)$ can in principle be inverted to give $\omega_p(s)$. However, in practice small errors in $t(\omega)$ usually lead to large errors in the computed electron density, so certain simplifying assumptions are usually made. At frequencies well below the electron cyclotron frequency,

$\omega \ll \omega_c$, where whistlers are usually observed, it can be shown that Eq. (4.4.33) simplifies to $t(f) = D_W/f^{1/2}$, where $f = \omega/2\pi$, and

$$D_W = \frac{1}{2c}\int \frac{f_p}{f_c^{1/2}} ds \tag{4.4.34}$$

is called the dispersion. By measuring the travel time as a function of frequency it is easy to determine the dispersion. From the above equation one can see that the dispersion is controlled by the line integral of $(n_e/B)^{1/2}$ along the magnetic field line. Since the magnetic field B reaches a minimum near the equator, the dispersion is mainly controlled by the electron density near the equator. In the pre-space era, before direct in situ measurements were possible, whistlers provided the only method of measuring the electron density at high altitudes, in Earth's magnetosphere; see Helliwell (1965).

Ion Cyclotron Whistlers

When plasma wave receivers were first flown on Earth-orbiting spacecraft, a new type of whistler was discovered that cannot be observed from the ground. This type of whistler propagates in the left-hand polarized ion cyclotron mode, and is called an ion cyclotron whistler (Gurnett et al., 1965). Ion cyclotron whistlers occur immediately after an upward propagating right-hand polarized whistler and consist of one or more slowly rising tones lasting several seconds. The frequencies of the tones asymptotically approach the ion cyclotron frequencies of the local plasma. A frequency–time spectrogram showing a pair of ion cyclotron whistlers observed at an altitude of about 1300 km over a lightning storm is shown in Figure 4.20. The long tones occur because the group velocity of the ion cyclotron mode goes to zero at the ion cyclotron frequencies, H^+ and He^+ in this case. Ion cyclotron whistlers exhibit an unusual feature called polarization reversal. If plots of $n^2 = R$ and $n^2 = L$ versus frequency are superposed on the same diagram, as in Figure 4.21, the plots cross at a series of frequencies called the crossover frequencies, ω_x. The crossover frequencies are solutions of the equation $D = 0$, which can be written in the form

$$D = \sum_s \frac{\omega_{ps}^2 \omega_{cs}}{\omega_x(\omega_x^2 - \omega_{cs}^2)} = 0. \tag{4.4.35}$$

Inspection of the above equation shows that one crossover frequency occurs between each pair of adjacent ion cyclotron frequencies.

Because the index of refraction curves for $\theta = 0$ intersect at the crossover frequencies, it is not immediately obvious how the two branches of the dispersion relation are connected at these intersections. This question can be answered by numerically computing the index of refraction slightly away from $\theta = 0$ using the general dispersion relation (4.4.17). The R and L branches are found to connect as

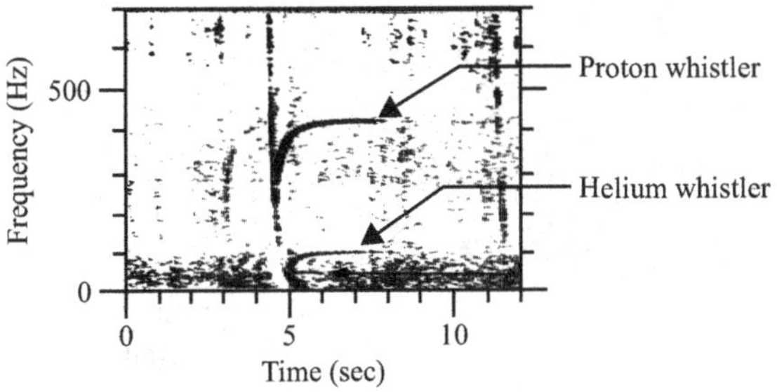

Figure 4.20 A frequency–time spectrogram of two ion cyclotron whistlers detected in the ionosphere over a lightning storm.

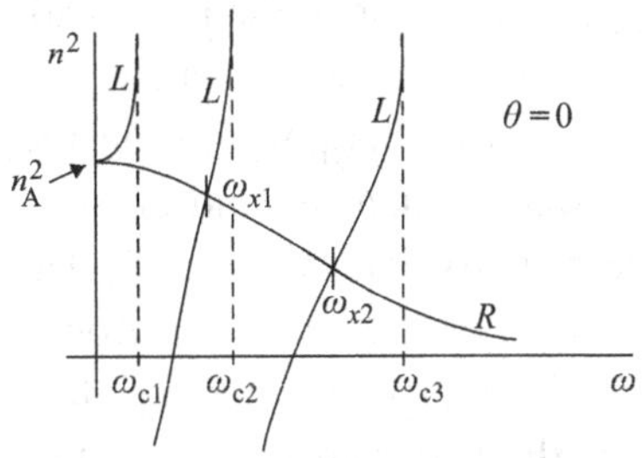

Figure 4.21 Superimposed plots of $n^2 = R$ and $n^2 = L$. The plots intersect at the crossover frequencies, ω_{x1} and ω_{x2}.

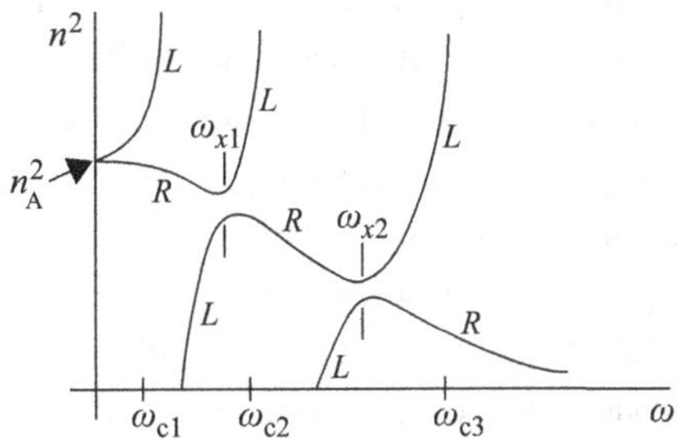

Figure 4.22 The index of refraction of the R and L modes at a small but non-zero wave normal angle.

shown in Figure 4.22. The polarization of a given branch is determined by the ratio of $\widetilde{E}_y$ to $\widetilde{E}_x$, which from Eq. (4.4.20) is given by

$$\frac{\widetilde{E}_y}{\widetilde{E}_x} = \frac{-\mathrm{i}D}{S - n^2}. \tag{4.4.36}$$

From the above equation one can see that the polarization, $\widetilde{E}_y/\widetilde{E}_x$, changes sign as D goes through zero (i.e., at the crossover frequencies). Thus, for a wave

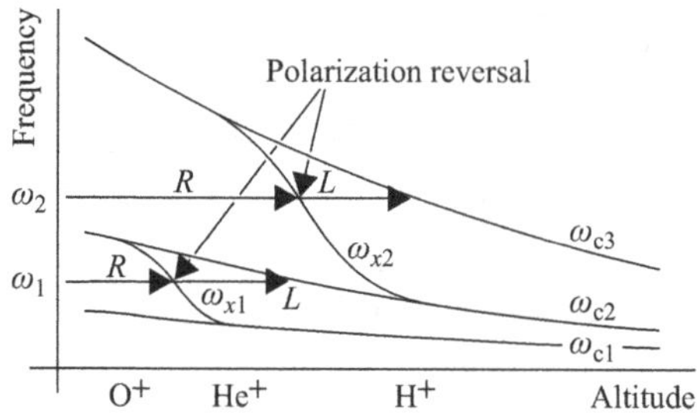

Figure 4.23 A representative plot of the crossover frequencies ω_{x1} and ω_{x2} as a function of altitude.

propagating in an inhomogeneous plasma, the polarization reverses whenever the crossover frequency moves across the wave frequency.

To illustrate in more detail how this polarization reversal occurs, it is useful to consider how the cyclotron frequencies and crossover frequencies vary as a function of altitude. Three primary ion species, O^+, He^+, and H^+ are present in Earth's ionosphere. The corresponding ion cyclotron frequencies are ω_{c1}, ω_{c2}, and ω_{c3}. Because of their different molecular weights these ions are separated into distinct height bands, with O^+ dominant at low altitudes, He^+ at intermediate altitudes, and H^+ at high altitudes. These concentration variations cause the crossover frequency to vary with altitude, as shown in Figure 4.23. One can see from this plot that an upward propagating right-hand polarized wave, R, from a lightning stroke at frequency ω_2 is converted to a left-hand polarized ion cyclotron wave, L, at crossover frequency ω_{x2}. This wave becomes a proton (H^+) whistler. At frequency ω_1 a similar polarization reversal also occurs at crossover frequency ω_{x1}. This wave becomes a helium (He^+) whistler.

After the upgoing right-hand polarized wave from the lightning stroke has been converted to the left-hand polarized ion cyclotron mode, it continues to propagate upward, approaching closer and closer to the ion cyclotron frequency. The wave energy is eventually absorbed at the ion cyclotron frequency by a process called cyclotron damping (see Chapter 10). The above geometry, where a wave approaches resonance in a region of decreasing magnetic field, is called a magnetic beach, in analogy with the dissipation of ocean waves propagating into a beach. The magnetic beach geometry has been used for heating ions in fusion machines and is called ion cyclotron heating.

Faraday Rotation

As can be seen by comparing Figures 4.15 and 4.16, both the left-hand and right-hand polarized free space modes can propagate at frequencies above the right-hand cutoff, $\omega_{R=0}$. In this frequency range, above the right-hand cutoff, a wave of any arbitrary polarization (linear, elliptical, or circular) can be constructed from a linear superposition of the right-hand and left-hand polarized waves. However, unlike in free space, in a plasma the phase velocities of the right-hand and left-hand polarized modes are not the same. The difference in the phase velocities causes the principal axis of the electric field polarization to rotate as the wave propagates through the plasma. This effect is called Faraday rotation, and occurs in a variety of astrophysical and laboratory situations.

To illustrate the basic effect, consider the special case of a wave that is constrained to be linearly polarized along the x axis at $z = 0$, with $\mathbf{k}$ and $\mathbf{B}_0$ aligned along the z axis. This boundary condition can be satisfied by superposing left- and right-hand circularly polarized waves of the form

$$E_y = \frac{1}{2} E_0 [\sin(k_{\mathrm{L}} z - \omega t) - \sin(k_{\mathrm{R}} z - \omega t)] \tag{4.4.37}$$

and

$$E_x = \frac{1}{2} E_0 [\cos(k_{\mathrm{L}} z - \omega t) + \cos(k_{\mathrm{R}} z - \omega t)], \tag{4.4.38}$$

where k_{R} and k_{L} are the wave numbers of the right-hand and left-hand polarized modes. It is easily verified that the electric field is aligned along the x axis at $z = 0$. Using well-known trigonometric identities, the above two equations can be rewritten in the form

$$E_y = E_0 \sin\left[\frac{1}{2}(k_{\mathrm{L}} - k_{\mathrm{R}})z\right] \cos\left[\frac{1}{2}(k_{\mathrm{L}} + k_{\mathrm{R}})z - \omega t\right] \tag{4.4.39}$$

and

$$E_x = E_0 \cos\left[\frac{1}{2}(k_{\mathrm{L}} - k_{\mathrm{R}})z\right] \cos\left[\frac{1}{2}(k_{\mathrm{L}} + k_{\mathrm{R}})z - \omega t\right]. \tag{4.4.40}$$

Taking the ratio of E_y to E_x, it can be shown that

$$\frac{E_y}{E_x} = \tan\psi, \text{ where } \psi = \frac{1}{2}(k_{\mathrm{L}} - k_{\mathrm{R}})z. \tag{4.4.41}$$

The above equation shows that the wave remains linearly polarized at all points along the z axis, and that the plane of polarization rotates through an angle ψ relative to the plane of polarization at $z = 0$. Since ψ increases linearly with z, the electric field vector lies on a helix, as shown in Figure 4.24. A similar rotation also occurs for an elliptically polarized wave.

If the wave frequency is much greater than both the electron cyclotron frequency and the electron plasma frequency, a simple formula can be derived for the Faraday rotation angle. Dropping the ion terms in Eqs. (4.4.26) and (4.4.29) and assuming

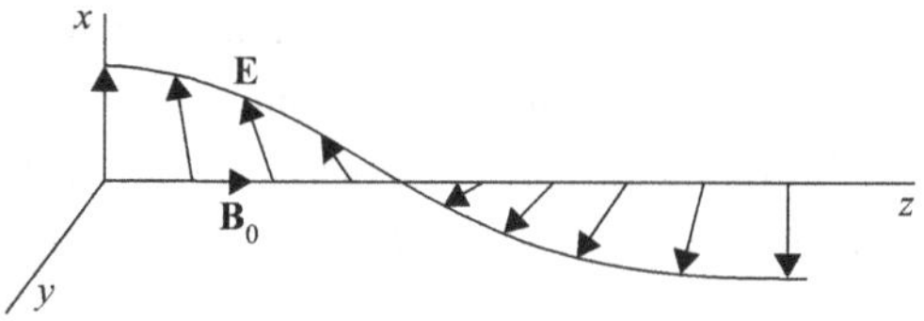

Figure 4.24 The electric field of a linearly polarized free space electromagnetic wave rotates as the wave propagates through a magnetized plasma.

that $\omega \gg \omega_c$ and $\omega \gg \omega_p$, the indices of refraction of the R and L modes can be approximated by

$$n_R \simeq 1 - \frac{1}{2}\frac{\omega_p^2}{\omega^2}\left(1 + \frac{\omega_c}{\omega}\right) \tag{4.4.42}$$

and

$$n_L \simeq 1 - \frac{1}{2}\frac{\omega_p^2}{\omega^2}\left(1 - \frac{\omega_c}{\omega}\right). \tag{4.4.43}$$

Converting n to k using the relation $k = \omega n/c$, and using Eq. (4.4.41), it can be shown that the Faraday rotation angle is given by

$$\psi \simeq \frac{1}{2c}\left(\frac{\omega_p^2\omega_c}{\omega^2}\right)z. \tag{4.4.44}$$

This equation shows that ψ increases in direct proportion to the product of the electron density and the magnetic field strength. If the electron density and magnetic field strength vary along the ray path, the rotation angle is given by the integral

$$\psi = \frac{1}{2c}\left(\frac{e^3}{\epsilon_0 m_e^2}\right)\frac{1}{\omega^2}\int n_e B_z \, d\ell, \tag{4.4.45}$$

where the integration is performed along the ray path. If the magnetic field is at an angle to the direction of propagation, it can be shown that Eq. (4.4.45) is still valid, provided B_z is taken to be the component of the magnetic field along the direction of propagation.

Measurements of the Faraday rotation angle can be used to provide information on the electron density or magnetic field strength along the ray path. Since the rotation angle involves the product of the electron density and the magnetic field, one of these parameters must be known to get useful results. For example, in a laboratory plasma, the magnetic field is usually known, so the line integral of the electron density can be obtained by measuring the Faraday rotation angle. In practice, this is done by generating a linearly polarized wave with a microwave transmitter on one side of a plasma chamber and measuring the polarization angle with a suitable antenna on the other side of the chamber. Similarly, ground-based measurements of the Faraday rotation angle of signals from a satellite transmitter can be used to give information on the electron density in Earth's ionosphere.

Because of the $1/\omega^2$ dependence in expression (4.4.45), it is not necessary to know the polarization of the source if measurements are made at two or more frequencies. Such multi-frequency Faraday rotation measurements have been used to provide estimates of the magnetic field strength in the interstellar medium. By measuring the Faraday rotation of signals from a distant radio source, such as a pulsar, the integral $\int n_{\mathrm{e}} B_z \,\mathrm{d}\ell$ can be obtained. The problem is then to unscramble the n_{e} and B_z dependences. For a pulsar, the line integral $\int n_{\mathrm{e}} \,\mathrm{d}\ell$ along the ray path from the pulsar to the Earth can be determined from the dispersion of the pulsar signal. The ratio of the two integrals can then be used to give a measure of the average magnetic field component along the propagation path. The sign of B_z can also be determined from the sense of the Faraday rotation. This technique has been used by Manchester and Taylor (1977) to determine the strength and orientation of the magnetic field in our Galaxy.

4.4.2 Propagation Perpendicular to the Magnetic Field

For propagation perpendicular to the magnetic field, $\theta = \pi/2$, Eq. (4.4.19) shows that the dispersion relation has two roots: $n^2 = P$ and $n^2 = RL/S$. The electric field eigenvectors associated with these two roots can be obtained from the homogeneous equation (4.4.14), which for $\theta = \pi/2$ becomes

$$\begin{bmatrix} S & -\mathrm{i}D & 0 \\ \mathrm{i}D & S-n^2 & 0 \\ 0 & 0 & P-n^2 \end{bmatrix} \begin{bmatrix} \widetilde{E}_x \\ \widetilde{E}_y \\ \widetilde{E}_z \end{bmatrix} = 0. \tag{4.4.46}$$

The electric field eigenvectors associated with the two roots are as follows:

$$n^2 = P, \qquad \tilde{\mathbf{E}} = (0, 0, E_0), \tag{4.4.47}$$

and

$$n^2 = \frac{RL}{S}, \qquad \tilde{\mathbf{E}} = \left(\frac{\mathrm{i}D}{S}E_0,\, E_0, 0\right). \tag{4.4.48}$$

The root $n^2 = P$ is identical to the dispersion relation for transverse waves in a cold unmagnetized plasma, as discussed in Section 4.3.1. The static magnetic field has no effect on this mode because the particle motion is parallel to the magnetic

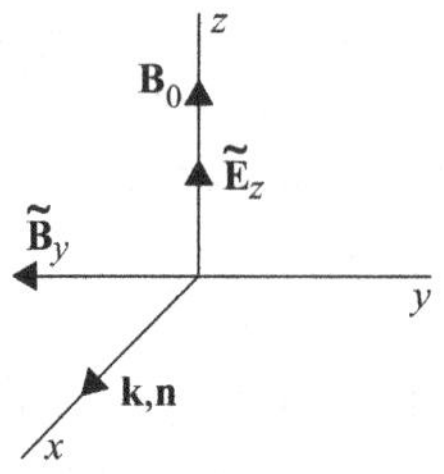

Figure 4.25 The electric and magnetic field eigenvectors for the ordinary (O) mode.

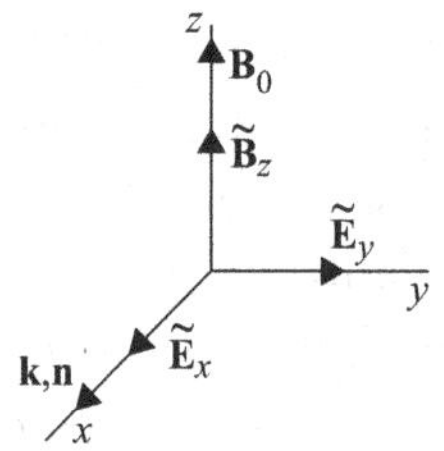

Figure 4.26 The electric and magnetic field eigenvectors for the extraordinary (X) mode.

field, and hence there is no $\mathbf{v} \times \mathbf{B}_0$ force. The relative orientations of the $\mathbf{k}$ vector, the wave electric and magnetic fields, and the static magnetic field are shown in Figure 4.25. Since there is no dependence on the magnetic field strength, this mode of propagation is called the ordinary (O) mode. Note that the electric field in this case is linearly polarized.

The root $n^2 = RL/S$ is more complicated. For this mode the electric field consists of both longitudinal (parallel to $\mathbf{k}$) and transverse (perpendicular to $\mathbf{k}$) components. Therefore, this mode has both electrostatic and electromagnetic characteristics. The relative orientations of the $\mathbf{k}$ vector, the wave electric and magnetic fields, and the static magnetic field are shown in Figure 4.26. Since this mode of propagation depends on the magnetic field strength, this mode is called the extraordinary (X) mode.

The frequency dependence of the index of refraction for the extraordinary mode is determined by the functions R, L, and S. The functions R and L have already been described in Section 4.4.1. The function S, defined by Eq. (4.4.8), consists of alternating poles and zeros, with a pole at each cyclotron frequency. The index of refraction, $n^2 = RL/S$, has zeros at the zeros of R and L and resonances at the zeros of S. Plots of S and $n^2 = RL/S$ for a plasma consisting of electrons and two ion species are shown in Figure 4.27. The resonances at $S = 0$ are called hybrid resonances because they involve a combination (or hybrid) of electric and magnetic forces. From the electric field eigenvector, $\tilde{\mathbf{E}} = [(\mathrm{i}D/S)E_0, E_0, 0]$, one can

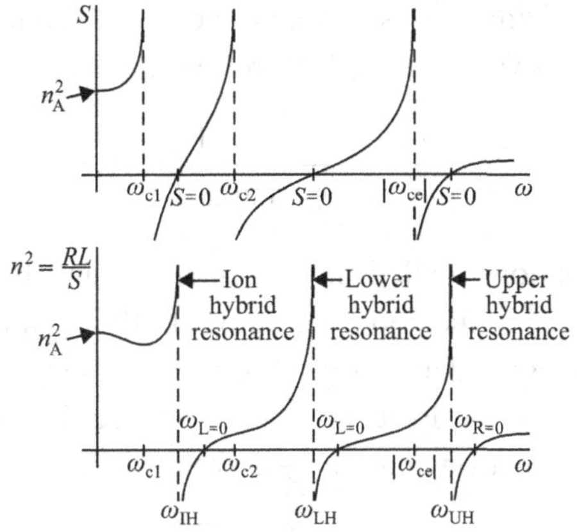

Figure 4.27 Plots of S and $n^2 = RL/S$ as a function of frequency.

see that as S approaches zero the longitudinal component $\widetilde{E}_x = (\mathrm{i}D/S)E_0$ becomes arbitrarily large compared to the transverse component $\widetilde{E}_y = E_0$. Therefore, near the hybrid resonances the wave becomes almost purely electrostatic, similar in many respects to electron plasma oscillations, except magnetic forces are now important.

As can be seen in Figure 4.27, a hybrid resonance is associated with each species present in the plasma. For a plasma consisting of electrons and two ion species, there are three hybrid resonances: one above the electron cyclotron frequency, one between the electron cyclotron frequency and the largest ion cyclotron frequency, and one between the two ion cyclotron frequencies. These resonances are called the upper hybrid resonance, ω_{UH}, the lower hybrid resonance, ω_{LH}, and the ion hybrid resonance, ω_{IH}, respectively. Additional ion hybrid resonances occur if more ion species are present.

Since the upper hybrid resonance frequency is much greater than any of the ion cyclotron frequencies, the upper hybrid resonance mainly involves electron motions. The corresponding solution to $S = 0$ can be computed to a good approximation by omitting the ion terms in Eq. (4.4.8), which gives

$$\omega_{\mathrm{UH}} = \sqrt{\omega_{\mathrm{p}}^2 + \omega_{\mathrm{c}}^2}. \tag{4.4.49}$$

If the magnetic field is reduced to zero ($\omega_{\mathrm{c}} \ll \omega_{\mathrm{p}}$) the oscillation at the upper hybrid resonance reduces to the electron plasma oscillation found in an unmagnetized plasma.

Since the lower hybrid resonance frequency is between the electron cyclotron frequency and the largest ion cyclotron frequency, both the electron and ion terms must be included in Eq. (4.4.8) when solving for the corresponding root of $S = 0$. For simplicity, we consider a plasma consisting of electrons and only one ion species. In order to obtain a simple result, one must make the further assumption that $\omega_{ci} \ll \omega_{LH} \ll |\omega_{ce}|$. With this assumption it can be shown that the lower hybrid resonance frequency is given to a good approximation by

$$\frac{1}{\omega_{LH}^2} = \frac{1}{\omega_{pi}^2} + \frac{1}{|\omega_{ce}|\omega_{ci}}. \tag{4.4.50}$$

Two further approximations, called the high-density and low-density limits, are often employed when using the above equation. In the high-density limit, defined by $|\omega_{ce}|\omega_{ci} \ll \omega_{pi}^2$ (or equivalently $\omega_{ce}^2 \ll \omega_{pe}^2$), the first term on the right can be ignored, in which case the lower hybrid resonance frequency is given by the geometric mean of the electron and ion cyclotron frequencies:

$$\omega_{LH} = \sqrt{|\omega_{ce}|\omega_{ci}}. \tag{4.4.51}$$

The assumption that $\omega_{LH} \ll |\omega_{ce}|$ is clearly satisfied by this solution. Both electron and ion motions are involved in the oscillation. In the low-density limit, which is defined by $|\omega_{ce}|\omega_{ci} \gg \omega_{pi}^2$, the lower hybrid resonance frequency is given by

$$\omega_{LH} = \omega_{pi}. \tag{4.4.52}$$

In this case the electrons are held essentially motionless by the strong magnetic field. The ions then play the dominant role in the oscillation. These ion oscillations are exactly analogous to the electron plasma oscillations described in Section 2.3, except the ions move instead of the electrons.

The ion hybrid resonances involve motions of two or more ion species. An ion hybrid resonance occurs between each adjacent pair of ion cyclotron frequencies. It is not possible to obtain a simple analytic expression for the ion hybrid resonance frequencies because the corresponding roots of $S = 0$ involve equations of third (or higher) degree in ω^2. These roots must be computed numerically.

Numerous examples of hybrid resonances exist in both space and laboratory plasmas. Some of these examples are discussed below.

Ionospheric Sounders

Early satellite studies of Earth's ionosphere utilized a technique called ionospheric sounding. Ionospheric sounding consists of transmitting a short pulse of radio waves at a frequency ω, and then recording the time delay of the signals reflected by the plasma (Franklin and Maclean, 1969). After a suitable listening interval, the transmitter frequency is increased by a small increment and the transmit/receive cycle is repeated. This cycle is repeated again and again until a complete scan is obtained of all the characteristic frequencies of the plasma. Typically, two types of reflected signals are detected: (1) echoes associated with remote reflections of electromagnetic waves by the ionosphere; and (2) electrostatic oscillations excited

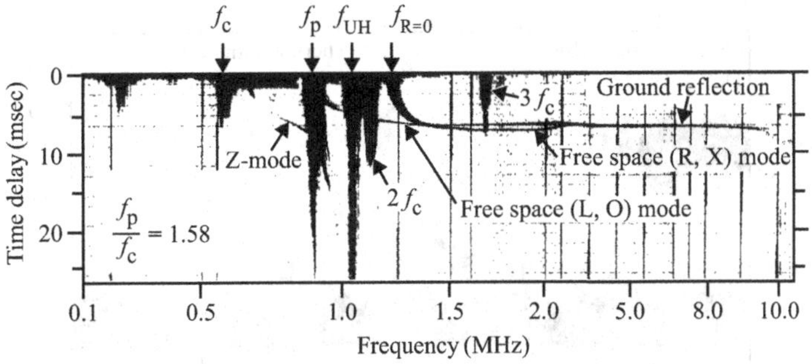

Figure 4.28 An ionogram obtained from an ionospheric sounder on the *Alouette 2* spacecraft at an altitude of 1542 km (adapted from Benson (1982)).

in the plasma near the spacecraft. The reflected signal intensity is typically plotted as a function of the transmitter frequency and the time delay after transmission of the pulse. This type of display is called an ionogram. An example of an ionogram obtained from an ionospheric sounder is shown in Figure 4.28. The darker regions indicate stronger signals. Usually the time delay is plotted vertically downward, since reflections are expected from the ionosphere, which is below the spacecraft. The lines labeled "free space (R, X)" and "free space (L, O)" are ionospheric echoes of the right-hand and left-hand polarized free space modes. The intense vertical "spikes" in the ionogram are local electrostatic resonances. The vertical character of the spikes indicates that a long-duration oscillation is being excited. Very strong spikes occur at the electron plasma frequency, f_p, and the upper hybrid resonance frequency, f_{UH}. These correspond to the electrostatic oscillation at the electron plasma frequency and the upper hybrid resonance frequency. Strong responses are observed at these resonances because the group velocity is zero at these frequencies, which causes the wave to remain in the vicinity of the spacecraft for a long time after the pulse is transmitted. Several other spike-like features are also evident in Figure 4.28, for example, at harmonics of the electron cyclotron frequency. These resonances involve electrostatic oscillations that require a finite temperature, and will be discussed in Chapter 10.

Thermal Excitation of Hybrid Resonances

Early rocket and satellite measurements of waves in Earth's ionosphere revealed narrowband emissions at both the upper and lower hybrid resonance frequencies (Walsh et al., 1964; Brice and Smith, 1965). These emissions are observed in the ionosphere and magnetosphere at geocentric radial distances of up to several Earth radii (R_E). An example of an emission at the upper hybrid resonance frequency is shown in Figure 4.29. The narrowband emission labeled "thermal emission at

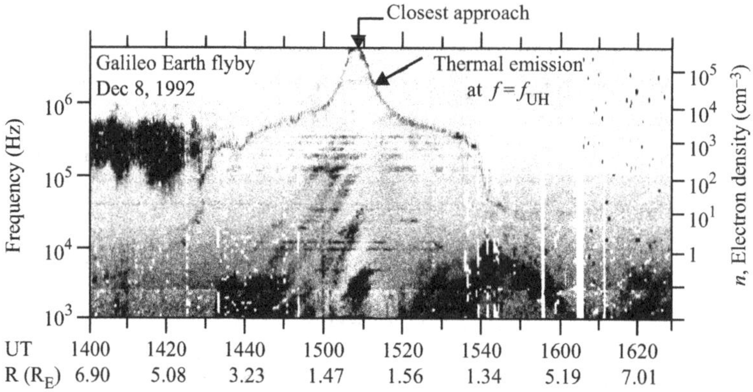

Figure 4.29 A frequency–time spectrogram showing an emission line at the upper hybrid frequency during a close flyby of Earth by the *Galileo* spacecraft. UT is Universal Time.

$f = f_{UH}$" is at the upper hybrid frequency, $f_{UH} = \sqrt{f_c^2 + f_p^2}$. In this case the electron cyclotron frequency is much less than the electron plasma frequency, $f_c \ll f_p$, so the upper hybrid frequency is essentially at the electron plasma frequency, which is proportional to the square root of the electron density; see Eq. (2.3.5). Therefore, the frequency of the upper hybrid emission line provides a direct measurement of the electron density. An electron density scale is given on the right-hand side of the spectrogram. The sharp peak in the electron density around the time of closest approach, indicated by the arrow marked "closest approach," is due to the passage of the spacecraft through the ionosphere. These weak, nearly continuous emissions are believed to be caused by thermal excitation of electrostatic oscillations at the upper hybrid frequency. Occasionally, much more intense emissions are observed at the upper hybrid resonance frequency. These much stronger emissions are believed to be due to plasma instabilities, and will be discussed in Chapter 10.

Plasma Heating via Hybrid Resonances

One of the basic challenges of controlled fusion is to heat the ions to very high temperatures, on the order of 10^8 K. Because of the greater mobility of electrons, simple Ohmic heating, such as by driving a current through the plasma, causes most of the energy to be absorbed by the electrons. What is needed is a mechanism that preferentially heats the ions. Because the lower hybrid and ion hybrid resonances involve primarily ion motions, it has been suggested that these resonances be used for ion heating. The basic idea is to transmit waves into the plasma perpendicular to the magnetic field at a frequency slightly below either the lower hybrid resonance frequency or the ion hybrid resonance frequency. If the density and magnetic field profiles are favorable, the wave can be brought into resonance at the appropriate hybrid resonance frequency inside the plasma, very similar to the magnetic beach geometry discussed for ion cyclotron whistlers. The wave energy is then absorbed at the resonance, heating the ions. The lower hybrid resonance is particularly favored for this purpose because in the high-density limit, the resonance frequency $\omega_{LH} = \sqrt{\omega_{ci}|\omega_{ce}|}$ can be controlled by the magnetic field. This method has been very successful in heating ions in laboratory fusion devices. Attempts have also been made to heat plasmas at the ion hybrid resonance frequency, with more limited success. In both cases the achieved heating rates are limited by the efficiency of the transmitter and the coupling of the radiated power into the plasma.

4.4.3 Oblique Wave Propagation

We next consider the propagation of waves at intermediate wave normal angles, neither parallel nor perpendicular to the magnetic field. The index of refraction at an arbitrary angle $n(\theta)$ is given by Eq. (4.4.17). Simple inspection of this equation shows that the index of refraction surface $n(\theta)$ has several obvious symmetries. Since the plasma has rotational symmetry around the magnetic field, the index of refraction surface has rotational symmetry around the $\mathbf{B}_0$ axis. The index of refraction surface also has mirror symmetry with respect to $\theta = \pi/2$. This symmetry arises because the index of refraction only involves terms of the form $\cos^2\theta$ and $\sin^2\theta$.

The topological shape of the index of refraction surface can be deduced from the general properties of Eq. (4.4.17). First, if the index of refraction goes to zero (i.e., a cutoff), then it must do so simultaneously for all angles. This result can be seen by examining Eq. (4.4.16). If $n = 0$, then either $R = 0, L = 0$, or $P = 0$. Since the condition for a cutoff does not involve the wave normal angle, then a cutoff must occur simultaneously for all wave normal angles. Second, if the index of refraction becomes infinite, then in the range $0 < \theta < \pi/2$ it can do so at only one angle, θ_{Res}, called the resonance cone angle. This result can be seen by examining Eqs. (4.4.16) and (4.4.17). The index of refraction can go to infinity if and only if $A = 0$. However, the condition $A = 0$ can be satisfied if and only if S and P have opposite signs. If S and P have opposite signs then the index of refraction goes to infinity at wave normal angles given by

$$\tan^2\theta_{Res} = -\frac{P}{S}. \tag{4.4.53}$$

There are only two solutions of the above equation: one at θ_{Res} and the other at $\pi - \theta_{Res}$. Therefore, if the index of refraction becomes infinite, then it must do so along two oppositely directed cones, the axes of which are aligned parallel to the

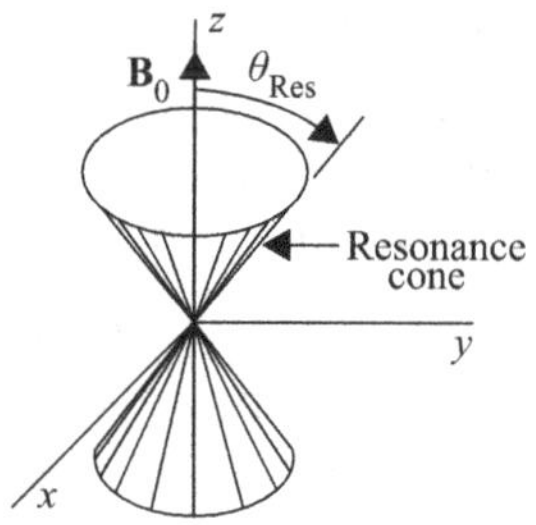

Figure 4.30 The resonance cone.

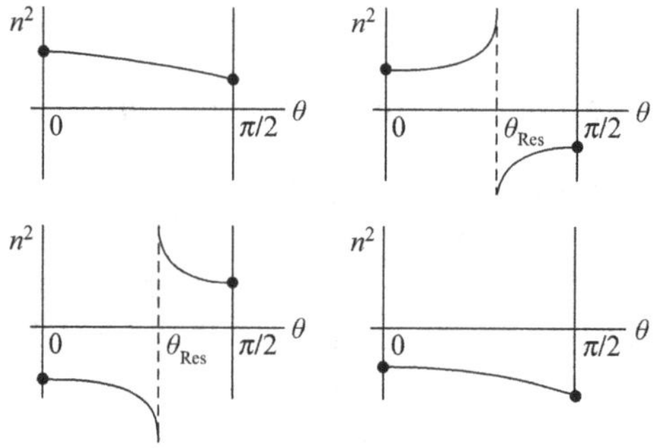

Figure 4.31 The four possible plots of n^2 versus the wave normal angle, θ.

magnetic field, as illustrated in Figure 4.30. Based on the above results, consider how the index of refraction must vary between $\theta = 0$ and $\theta = \pi/2$. Since the index of refraction cannot go to zero (unless it does so for all angles), and since n^2 can be either positive or negative at $\theta = 0$ and at $\theta = \pi/2$, it follows that there are only four possible ways for the index of refraction to vary between $\theta = 0$ and $\theta = \pi/2$. These four possibilities are illustrated in Figure 4.31. Of the four, only three give rise to real index of refraction surfaces. The fourth corresponds to an imaginary index of refraction at all angles, which is physically uninteresting. As shown in Figure 4.32, the index of refraction surfaces have the topological shapes of an ellipsoid, a hyperboloid of two sheets, and a hyperboloid of one sheet. The asymptotes of the hyperboloid-like surfaces are at the resonance cone angle.

The possible combinations of index of refraction surfaces that can occur when both modes of propagation are considered involve two basic conditions. First, except for degenerate cases at $\theta = 0$ and $\theta = \pi/2$, the two modes can never intersect. This follows from the fact that F in Eq. (4.4.18) is positive definite. The only exception occurs when the quantities $RL - PS$ and PD are simultaneously zero, which is physically unlikely. Second, only one of the two modes can have a resonance cone. This follows because if $n = \infty$ for one of the modes, then $A = 0$. Equation (4.4.16) then shows that the index of refraction of the other mode,

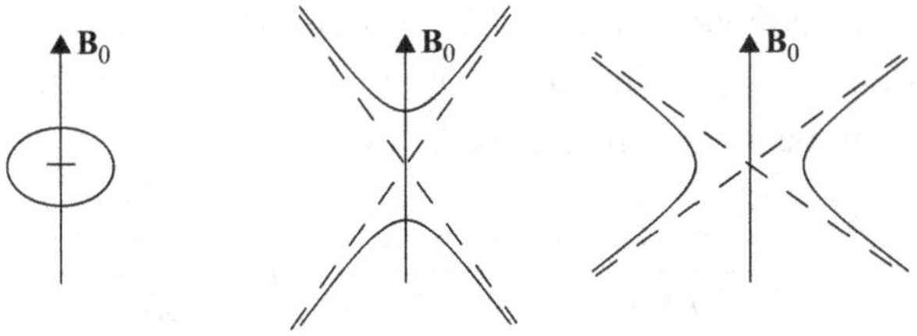

Figure 4.32 The three possible real index of refraction surfaces in a cold magnetized plasma.

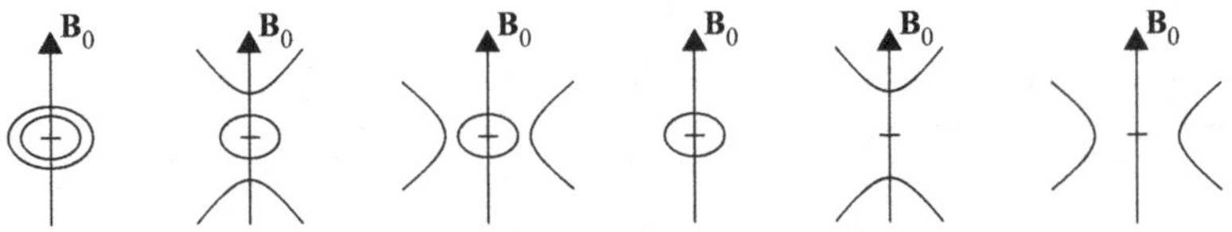

Figure 4.33 The six possible combinations of index of refraction surfaces that can occur for the two electromagnetic modes of propagation in a cold magnetized plasma.

$n^2 = RLP/B$, is finite. Considering the two possible signs of n^2 for each of the two modes at $\theta = 0$ and at $\theta = \pi/2$, a total of sixteen combinations occur. Of these, one is not of interest because both surfaces are imaginary, three represent redundant combinations, two violate the condition that the index of refraction surfaces cannot intersect, and four violate the condition that only one mode can have a resonance cone. After eliminating these possibilities, only six combinations survive. These six combinations are shown in Figure 4.33. The specific combination that occurs for any given set of plasma parameters is completely determined by the indices of refraction at $\theta = 0$ and $\theta = \pi/2$. Thus, using the n^2 plots in Figure 4.31, the topological shape of the index of refraction surfaces can be determined at any frequency, and for any combination of plasma parameters.

The above analysis shows that the index of refraction in a plasma is qualitatively different than in most common dielectric media. Most striking is the highly anisotropic behavior of the index of refraction, especially the occurrence of resonance cones where the index of refraction goes to infinity. As an example that illustrates the highly unusual effects that can occur, we next describe the index of refraction for the whistler mode.

In order to obtain a mathematically tractable equation for the index of refraction of the whistler mode, it is necessary to make some simplifying assumptions. First, we assume that the ion terms can be ignored. Except for waves propagating very nearly perpendicular to the magnetic field, this assumption is normally good down to frequencies on the order of the ion cyclotron frequency. Second, we assume that the wave frequency and the electron cyclotron frequency are both much less than the electron plasma frequency, i.e., $\omega^2 \ll \omega_{\rm p}^2$ and $\omega_{\rm c}^2 \ll \omega_{\rm p}^2$. This assumption is often called the high-density approximation. With these limitations, the expressions for R, L, D, S, and P can be approximated as follows:

$$R \simeq \frac{-\omega_{\rm p}^2}{\omega(\omega-\omega_{\rm c})}, \quad L \simeq \frac{-\omega_{\rm p}^2}{\omega(\omega+\omega_{\rm c})}, \quad S \simeq \frac{-\omega_{\rm p}^2}{\omega^2-\omega_{\rm c}^2}, \quad D \simeq \frac{-\omega_{\rm p}^2\omega_{\rm c}}{\omega(\omega^2-\omega_{\rm c}^2)}, \quad P \simeq -\frac{\omega_{\rm p}^2}{\omega^2}. \tag{4.4.54}$$

Noting that $RL = PS$, the quantities A, B, and F in Eq. (4.4.17) simplify to the following:

$$A \simeq \frac{-\omega_{\rm p}^2(\omega^2-\omega_{\rm c}^2\cos^2\theta)}{\omega^2(\omega^2-\omega_{\rm c}^2)}, \quad B \simeq \frac{2\omega_{\rm p}^4}{\omega^2(\omega^2-\omega_{\rm c}^2)}, \quad F \simeq 2\frac{\omega_{\rm p}^4\omega_{\rm c}}{\omega^3(\omega^2-\omega_{\rm c}^2)}\cos\theta. \tag{4.4.55}$$

Substituting expressions (4.4.54) and (4.4.55) into Eq. (4.4.17) and using the plus sign, which yields the whistler mode, the index of refraction is given by

$$n^2 = \frac{\omega_{\rm p}^2}{\omega(\omega_{\rm c}\cos\theta-\omega)}. \tag{4.4.56}$$

This equation reduces to Eq. (4.4.31) when $\theta = 0$. A polar plot of $n(\theta)$ is shown in Figure 4.34. The index of refraction goes to infinity at the resonance cone angle, which in this case is given by $\cos\theta_{\rm Res} = \omega/\omega_{\rm c}$. Note the frequency dependence of the resonance cone angle. At high frequencies the resonance cone angle goes to zero at the electron cyclotron frequency. Therefore, the cone of wave normal angles within which propagation is allowed shrinks to zero as the wave frequency approaches the electron cyclotron frequency. Above the electron cyclotron frequency, the index of refraction is imaginary. At low frequencies, which is where whistlers are usually observed, the resonance cone angle approaches $\pi/2$.

The index of refraction surface then has the shape shown in Figure 4.35. Since the group velocity is perpendicular to the index of refraction surface, it is evident from Figure 4.35 that the group velocity, $\mathbf{v}_{\rm g}$, is constrained to be within

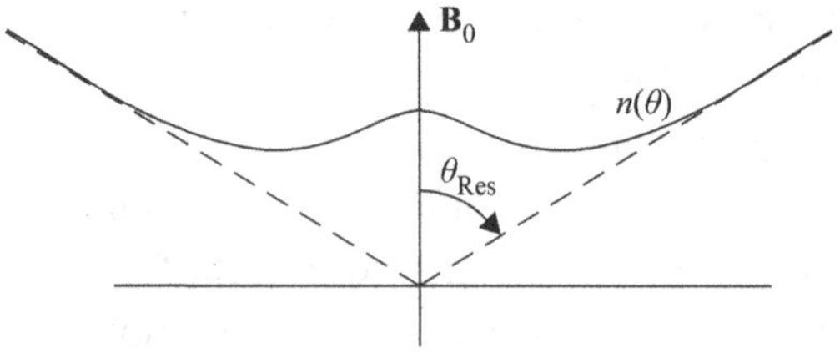

Figure 4.34 The index of refraction for the whistler mode in the high-density limit.

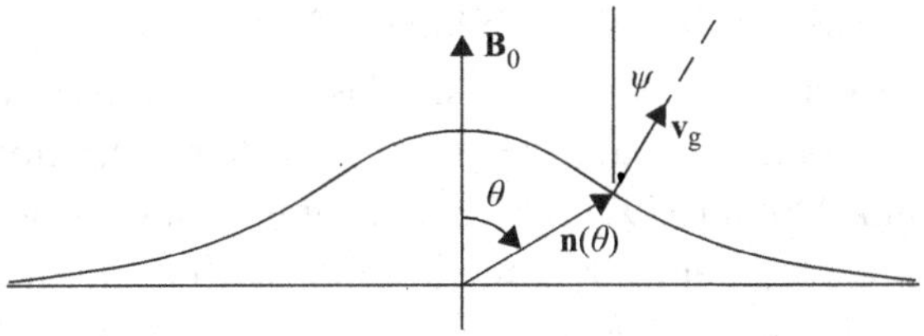

Figure 4.35 The index of refraction for the whistler mode in the low-frequency, high-density limit.

a small cone of angles relative to the magnetic field for all wave normal angles. In this high-density low-frequency limit, it can be shown (see Problem 4.14) that the maximum angle ψ between the group velocity and the magnetic field is $\psi_{\text{max}} = \tan^{-1}(1/\sqrt{8}) = 19°28'$. This result was first shown by Storey (1953). Therefore, even if there are insufficient density irregularities to maintain the wave normal angle at $\theta = 0°$, whistlers are still guided along the magnetic field to a good approximation. Furthermore, since the index of refraction goes to infinity at $\theta = \pi/2$, the wave normal angle cannot rotate through $\theta = \pi/2$.

Thus, once a whistler starts up the magnetic field in one hemisphere, it cannot reverse its direction of motion until it reflects at the base of the ionosphere in the opposite hemisphere. When ion effects are included it can be shown that the wave normal angle can rotate through $\theta = \pi/2$ at frequencies below the lower hybrid resonance frequency. This occurs because the index of refraction at $\theta = \pi/2$ is real for $\omega < \omega_{\text{LH}}$.

4.4.4 The Clemmow–Mullaly–Allis (CMA) Diagram

The dispersion relation derived in this chapter is based on the assumption that the plasma is homogeneous. The results obtained can be applied to inhomogeneous plasmas if the index of refraction varies sufficiently slowly. As a wave propagates through an inhomogeneous medium the plasma parameters can be expected to change. Because of these changes, a wave originating in a particular mode at one point in the plasma may or may not be able to reach some other region of the plasma. This issue is referred to as *accessibility*. Two points A and B are said to be accessible via a particular mode if a ray path can connect the two points. A necessary condition for accessibility is that the index of refraction be real and continuous at all points along some ray path between the two points. Continuity of the index of refraction is a necessary, but not sufficient, condition for accessibility. It is not a sufficient condition because refraction may prevent a ray path from connecting the two points, even if the index of refraction is real and continuous between the two points. The effects of refraction are discussed in the next section.

To answer the question of whether the index of refraction is real and continuous along a path between two points in a plasma, it is useful to consider a multi-dimensional parameter space called a CMA diagram, after Clemmow and Mullaly (1955) and Allis (1959). A CMA diagram consists of a diagram with one coordinate for each parameter of the plasma, such as n_s and B_0. Within this parameter space a set of *bounding surfaces* is constructed defined by the equations

$$R = 0, \quad L = 0, \quad P = 0, \quad S = 0, \quad R = \infty, \quad \text{and} \quad L = \infty.$$

These surfaces define *bounded volumes* in parameter space. Since R, L, S, and P can only change sign at a bounding surface, the topological form of the index of refraction surface remains the same throughout a bounded volume. Therefore, if the topological form is determined at one point in a bounded volume, it is known at all points in that bounded volume.

At a bounding surface the topological form of the index of refraction surface may or may not change. If the index of refraction surface changes from real to imaginary (i.e., non-propagating) at the bounding surface, then the transition across the bounding surface is called a *destructive* transition, since the wave cannot propagate into the region with the imaginary index of refraction. On the other hand, if the refractive index remains real and changes continuously upon crossing a bounding surface, then the transition is called a *non-destructive* transition, since a wave can in principle cross the bounding surface. All the bounded volumes connected by non-destructive transitions are possible accessible regions, subject only to the restrictions imposed by refraction.

The CMA Diagram for a Plasma of Electrons and Immobile Ions

To illustrate these concepts, we construct the CMA diagram for a plasma consisting of electrons and a fixed background of immobile ions. The assumption of immobile ions basically implies that the frequencies must be sufficiently high that ion motions are unimportant, i.e., well above the ion cyclotron frequencies. Dropping the ion terms, the expressions for R, L, S, and P become

$$R = 1 - \frac{\omega_{\rm p}^2}{\omega(\omega - \omega_{\rm c})}, \; L = 1 - \frac{\omega_{\rm p}^2}{\omega(\omega + \omega_{\rm c})}, \; S = 1 - \frac{\omega_{\rm p}^2}{\omega^2 - \omega_{\rm c}^2}, \; P = 1 - \frac{\omega_{\rm p}^2}{\omega^2}. \tag{4.4.57}$$

To proceed further, it is convenient to define the normalized, dimensionless parameters $Y = \omega_{\rm c}/\omega$ and $X = \omega_{\rm p}^2/\omega^2$. Note that X is proportional to the electron density and Y is proportional to the magnetic field strength. In terms of X and Y, the above equations can be rewritten as

$$R = 1 - \frac{X}{1 - Y}, \; L = 1 - \frac{X}{1 + Y}, \; S = 1 - \frac{X}{1 - Y^2}, \; \text{and } P = 1 - X. \tag{4.4.58}$$

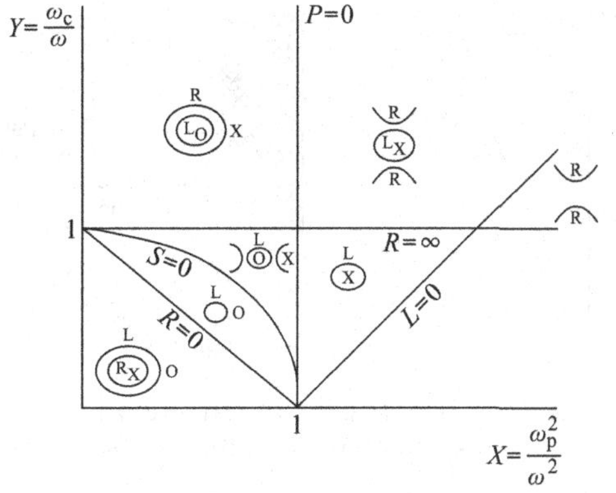

Figure 4.36 The CMA diagram for a plasma of cold electrons and immobile ions.

Using these equations, it is easy to show that the bounding surfaces are given by

$R = 0,$	$X = 1 - Y$	cutoff,
$L = 0,$	$X = 1 + Y$	cutoff,
$P = 0,$	$X = 1$	cutoff,
$S = 0,$	$X = 1 - Y^2$	hybrid resonance,
$R = \infty,$	$Y = 1$	cyclotron resonance,
$L = \infty,$		no solution.

The CMA diagram can then be constructed by plotting the bounding surfaces as a function of X and Y, as shown in Figure 4.36. The topological shapes of the index of refraction surfaces are summarized by the sketches in each bounded volume. Each sketch is labeled by the polarization (R, L) at $\theta = 0$ and the mode of propagation (O, X) at $\theta = \pi/2$.

In order to determine the accessible regions for each mode, the continuity of the index of refraction surfaces must be studied at each bounding surface. In some cases this is relatively easy, whereas in other cases it is quite difficult. For example, it is easy to see that the (L, O) mode is continuous throughout the region $X < 1$. On the other hand, the analysis of what happens to the extraordinary mode in the region around $X = 1$ and $Y = 1$ is quite complicated. In such complicated regions, the best approach is to numerically investigate the continuity of the index of refraction across the boundary in question. Using such an approach, it can be shown that there are only four continuously connected branches of the index of refraction. These four branches are indicated by the four shaded regions in Figures 4.37 and 4.38. Figure 4.37 is for the case $\omega_p > \omega_c$, and Figure 4.38 is for the case $\omega_p < \omega_c$.

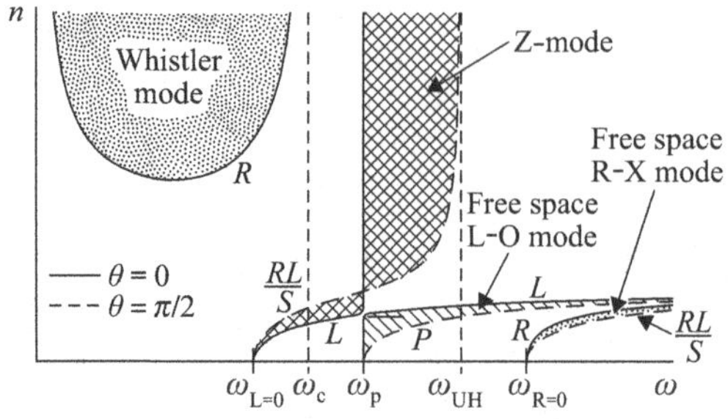

Figure 4.37 The boundaries of the index of refraction for the four continuously connected branches of the dispersion relation for the case $\omega_p > \omega_c$.

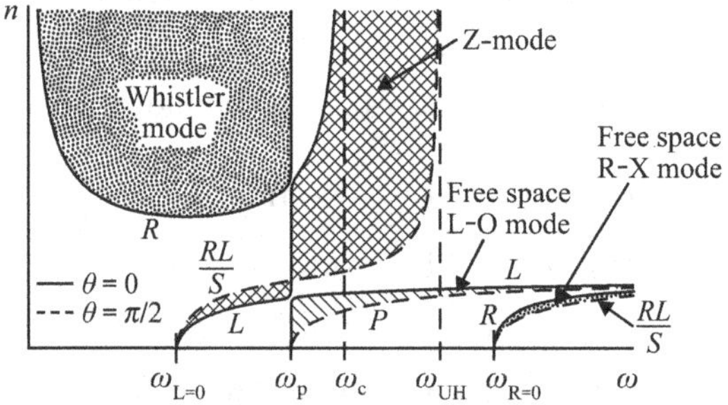

Figure 4.38 The boundaries of the index of refraction for the four continuously connected branches of the dispersion relation for the case $\omega_p < \omega_c$.

The boundaries of the shaded regions are determined by the indices of refraction at $\theta = 0$ (solid line) and $\theta = \pi/2$ (dashed line). At intermediate wave normal angles, the index of refraction remains in the shaded regions. In both cases note that for $\theta = 0$ the (L, O) mode has a gap at $\omega = \omega_p$. Although the index of refraction at $\theta = 0$ is continuous across $\omega = \omega_p$, at any angle $\theta \neq 0$ the (L, O) mode splits into two branches at the plasma frequency. The high-frequency branch is the free space (L, O) mode, and the low-frequency branch is the Z-mode, named after a characteristic Z-shaped feature first identified in ground-based ionograms (Ratcliffe, 1962). The Z-mode is bounded at low frequencies by the $\omega_{L=0}$ cutoff, and at high frequencies by the upper hybrid resonance ω_{UH}. The whistler mode is bounded at high frequencies by either the electron cyclotron frequency, ω_c, or the electron plasma frequency, ω_p, whichever is smaller. The remaining mode, the free space R-X mode is bounded by $R = 0$. The regions of the CMA diagram that are accessible via the four continuously connected branches of the dispersion relation are summarized in Figure 4.39. The four branches can be grouped into modes that are accessible from free space ($X = 0, Y = 0$), and modes that are internally trapped (not accessible from free space). The modes accessible

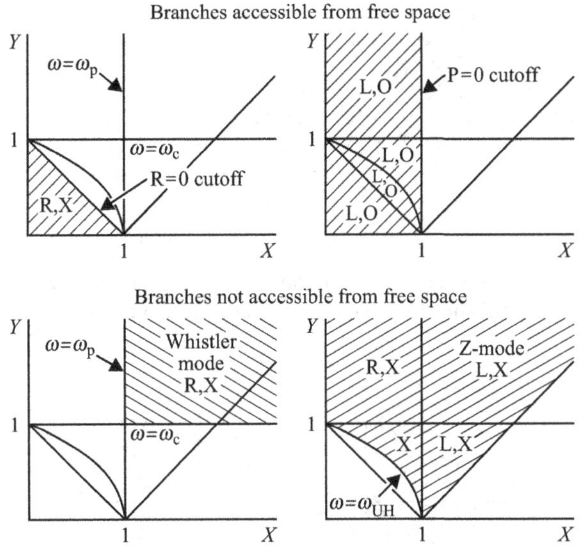

Figure 4.39 The regions of the CMA diagram for the four continuously connected branches of the dispersion relation for a plasma of cold electrons and immobile ions.

from free space are the right-hand polarized extraordinary mode (R, X) and the left-hand polarized ordinary mode (L, O). These two modes are of particular interest to radio astronomers because they are the only modes that can escape from an astrophysical radio source. In the ionosphere the cutoff frequencies of these two modes also give the low-frequency limit for ground-based radio astronomy. Typical plots of the cutoff frequencies, f_{p} and $f_{\mathrm{R}=0}$, as a function of altitude in Earth's ionosphere, are shown in Figure 4.40. Left-hand and right-hand polarized radio waves arriving from a distant source are reflected from the top side of the ionosphere at f_{p} and $f_{\mathrm{R}=0}$. Such waves can only reach the ground at frequencies above the maximum values for f_{p} and $f_{\mathrm{R}=0}$, which are typically about 5 to 8 MHz. The same reflection process also occurs for waves incident on the ionosphere from below. Reflections from the bottom side of the ionosphere are important for long-range, over-the-horizon radio communication.

The remaining two modes are the whistler and the Z-modes. These two modes are internally trapped and cannot reach free space ($X = 0$ and $Y = 0$). The CMA diagram shows that the whistler mode is bounded by $X > 1$ and $Y > 1$. The index of refraction of the whistler mode has the same topological form throughout the region where it exists. The Z-mode is more complicated and changes topological form at $X = 1$ and $Y = 1$. The Z-mode has a resonance cone within the region bounded by the lines $X = 1, Y = 1$, and $X = 1 - Y^2$. Elsewhere, the Z-mode has a quasi-ellipsoidal index of refraction surface. The boundary at $X = 1 - Y^2$

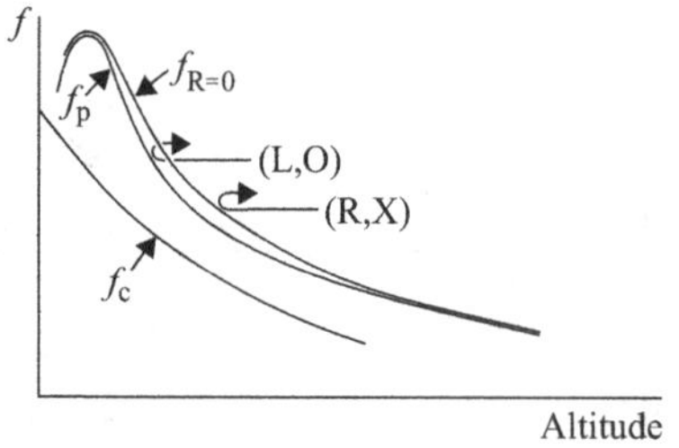

Figure 4.40 A representative plot of the cutoff frequencies, f_p and $f_{R=0}$, as a function of altitude for the free space (R, X) and (L, O) modes.

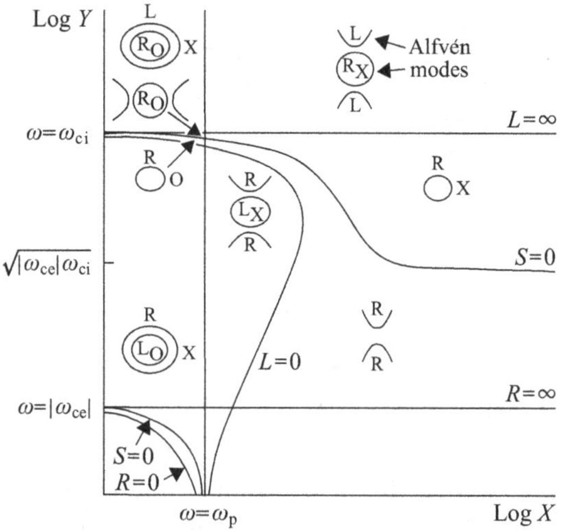

Figure 4.41 The CMA diagram for a cold plasma of electrons and ions.

corresponds to the upper hybrid resonance. Along this boundary the Z-mode becomes almost purely electrostatic and the allowed range of wave normal angles shrinks to zero at $\theta = \pi/2$.

The CMA Diagram for Electrons and One Ion Species

The inclusion of a single type of positive ion with finite mass does not appreciably alter the results presented in the previous section provided the wave frequency is above the lower hybrid resonance frequency. However, at frequencies below the lower hybrid resonance frequency, the effects of the finite ion mass become quite important.

Since charge neutrality demands that $n_{i0} = n_{e0}$, the CMA diagram for a two-component plasma still has only two parameters, just as when the ions were considered immobile. A convenient set of parameters for describing a two-component plasma are the normalized electron cyclotron frequency $Y =$

$|\omega_{ce}|/\omega$ and the normalized plasma frequency $X = (\omega_{pe}^2 + \omega_{pi}^2)/\omega^2$. The corresponding CMA diagram is shown in Figure 4.41. The two modes at large X and Y (in the upper right-hand corner) are the Alfvén modes described earlier. One of these modes is isotropic, and the other is highly anisotropic (i.e., it has a resonance cone). The isotropic Alfvén mode transforms smoothly and continuously into the whistler mode, and the anisotropic Alfvén mode transforms smoothly and continuously into the ion cyclotron mode. The ion cyclotron mode has a resonance at the ion cyclotron frequency and does not propagate above this frequency. At the ion cyclotron frequency, the ion cyclotron mode undergoes a destructive transition very similar to the destructive transition that occurs for the whistler mode at the electron cyclotron frequency.

4.5 Ray Paths in Inhomogeneous Plasmas

Next we return to the important question of refraction. Because the index of refraction changes with spatial position in an inhomogeneous medium, the ray path is usually not a straight line. The bending of the ray path by inhomogeneous effects is called refraction. The problem of determining the ray path of a wave propagating in an inhomogeneous anisotropic medium is difficult. Meaningful solutions are possible only if the index of refraction varies slowly on a spatial scale comparable to the wavelength.

The problem of an electromagnetic wave propagating in a slowly varying medium can be solved by assuming a solution of the form $Ae^{i\phi}$, where the amplitude A is a slowly varying function of position. The function ϕ, which describes the wave front, is called the eikonal. Starting with Maxwell's equations and assuming that the dielectric properties are slowly varying functions of position, it is possible to derive an approximate set of equations ordered according to the magnitude of the inhomogeneity. These equations were first derived for isotropic optical media in the late nineteenth century and were extended to anisotropic media by Weinberg (1962). Because of the complexity of the subject, it is our intention to give only a brief outline of the theory and focus on some simple applications.

For an inhomogeneous time-dependent medium, Maxwell's equations can be written as a set of linear homogeneous partial differential equations of the form

$$\overleftrightarrow{\mathbf{Đ}}(\boldsymbol{\nabla}, \partial/\partial t, \mathbf{r}, t) \cdot \mathbf{f} = 0, \tag{4.5.1}$$

where $\mathbf{f}$ is a generalized vector that includes all field components. In the eikonal approach it is assumed that $\overleftrightarrow{\mathbf{Đ}}$ depends so weakly on $\mathbf{r}$ that all the rapid spatial dependence in $\mathbf{f}$ can be represented by a common multiplicative factor of the form

$e^{i\phi}$. With this assumption one can replace ∇ in $\overleftrightarrow{\mathbf{D}}$ by $i\mathbf{k}$. If the temporal variations of the medium are slow, one can further assume that $\mathbf{f}$ depends on time through a factor $e^{-i\omega t}$, in which case $\partial/\partial t$ can be replaced by $-i\omega$. The system of linear differential equations given by Eqs. (4.5.1) then becomes

$$\overleftrightarrow{\mathbf{D}}(i\mathbf{k}, -i\omega, \mathbf{r}, t) \cdot \mathbf{f} = 0, \tag{4.5.2}$$

where $i\mathbf{k} = \nabla\phi$.

To solve the above equation for $\phi(\mathbf{r})$, one further assumes that $\mathbf{k}$ must satisfy the dispersion relation for a homogeneous medium evaluated at $\mathbf{r}$, i.e., $D(\mathbf{k}, \omega, \mathbf{r}, t) = 0$. The dispersion relation by itself does not give $\mathbf{k}$. To find $\mathbf{k}$ at all points along the ray path, one introduces a one-parameter family of trajectories in $\mathbf{r}, \mathbf{k}$ space defined by

$$\mathbf{r} = \mathbf{r}(\tau) \quad \text{and} \quad \mathbf{k} = \mathbf{k}(\tau). \tag{4.5.3}$$

A requirement is then imposed that the variation in the dispersion relation be zero along the actual ray path, i.e., $\delta D = 0$. Therefore, if the dispersion relation is satisfied at the initial point $\tau = 0$, it is satisfied at all points along the ray path. The variation in D is given by

$$\delta D = \nabla_{\mathbf{k}} D \cdot \frac{d\mathbf{k}}{d\tau} d\tau + \frac{\partial D}{\partial \omega}\frac{d\omega}{d\tau} d\tau + \nabla D \cdot \frac{d\mathbf{r}}{d\tau} d\tau + \frac{\partial D}{\partial t}\frac{dt}{d\tau} d\tau = 0. \tag{4.5.4}$$

The above equation is satisfied if we require that $\mathbf{r}$, $\mathbf{k}$, and ω satisfy the following system of equations:

$$\frac{d\mathbf{r}}{d\tau} = \nabla_{\mathbf{k}} D, \tag{4.5.5}$$

$$\frac{d\mathbf{k}}{d\tau} = -\nabla D, \tag{4.5.6}$$

$$\frac{d\omega}{d\tau} = \frac{\partial D}{\partial t}, \tag{4.5.7}$$

and

$$\frac{dt}{d\tau} = -\frac{\partial D}{\partial \omega}. \tag{4.5.8}$$

The last equation can be regarded as the defining equation for the parameter τ.

To interpret the above system of equations, note that Eqs. (4.5.5) and (4.5.8) can be combined to give

$$\frac{d\mathbf{r}}{dt} = -\frac{\nabla_{\mathbf{k}} D}{\partial D/\partial \omega} = \nabla_{\mathbf{k}}\omega = \mathbf{v}_g. \tag{4.5.9}$$

Thus, the space–time trajectory is parallel to the group velocity, which by definition is the ray path. If we further identify $\mathcal{D}$ as the Hamiltonian, $\mathbf{k}$ as the momentum, ω as the total energy, and τ as the time, the first three equations are seen to be identical to Hamilton's equations of classical mechanics. Thus, the ray path problem is formally identical to the problem of solving for the motion of a particle in a system with a Hamiltonian $\mathcal{D}(\mathbf{k}, \omega, \mathbf{r}, t)$. The close correspondence between the motion of a wave packet and the motion of a particle should not be surprising, since the duality between waves and particles is a well-known feature of physics. The correspondence between the ray path equations and Hamilton's equations also means that one can now make use of the many powerful techniques available in classical mechanics to solve particle trajectories.

Even though the ray path problem can in principle be solved like a problem in classical mechanics, in all but the simplest cases one must resort to numerical computations. One such case of special interest is a planar geometry in which the parameters of the medium depend on only one coordinate, z for example. The coordinates x and y are then cyclic coordinates, since $\mathcal{D}(\mathbf{k}, \omega)$ does not depend on x and y. As in classical mechanics, it follows that the "conjugate momenta," k_x and k_y, are constants. By a suitable coordinate rotation we can then arrange for $\mathbf{k}$ to be in the x,z plane, so that we need only concern ourselves with the condition $k_x =$ constant. This condition is equivalent to Snell's law, which can be written

$$n_x = n \sin\theta = \text{constant}. \tag{4.5.10}$$

For a linearly polarized wave with the electric field in the y direction, it is straightforward to show that Maxwell's equations can be combined to give the following equation for E_y:

$$\frac{\mathrm{d}^2 E_y}{\mathrm{d}z^2} + k_z^2(z) E_y = 0. \tag{4.5.11}$$

This equation is seen to be of the same form as the time varying harmonic oscillator discussed in Section 3.8.1. If k_z varies sufficiently slowly, an approximate solution is given by the Wentzel–Kramers–Brillouin (WKB) equation

$$E_y = \frac{A}{\sqrt{k_z}} \exp\left[\pm \mathrm{i} \int^z k_z(z')\,\mathrm{d}z'\right], \tag{4.5.12}$$

which is valid if the condition

$$k_z^2 \gg \left| \frac{3}{4}\left(\frac{1}{k_z}\frac{\mathrm{d}k_z}{\mathrm{d}z}\right)^2 - \frac{1}{2}\frac{1}{k_z}\frac{\mathrm{d}^2 k_z}{\mathrm{d}z^2} \right| \tag{4.5.13}$$

is satisfied. A similar solution can be obtained for E_x. Since k_x and k_y are determined by the initial conditions and are constant, k_z can be computed as a

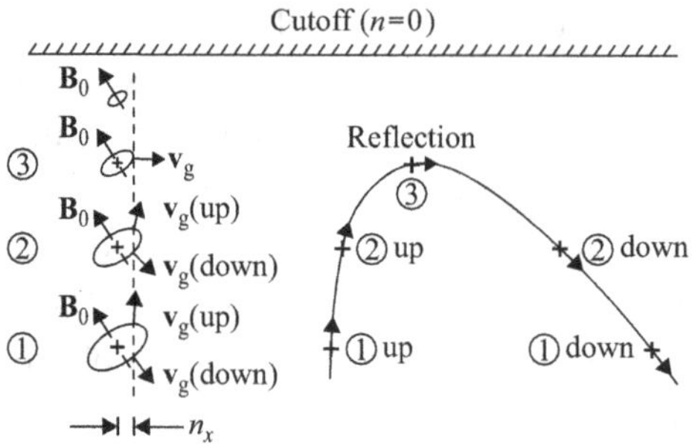

Figure 4.42 A Poeverlein construction illustrating the ray path of a wave approaching a cutoff in a horizontally stratified plasma.

function of z from the dispersion relation. The integral in the WKB solution then gives the phase ϕ of the electric field at any arbitrary height, i.e., the eikonal.

The ray path for a planar geometry can be determined using a graphical construction introduced by Poeverlein (1950). This construction makes use of the facts that the x component of the index of refraction vector is constant and that the group velocity is perpendicular to the index of refraction surface. To make a Poeverlein construction, the index of refraction surface is drawn at a sequence of heights, as shown in Figure 4.42. The wave is assumed to be approaching a cutoff, indicated by the horizontal hashed line labeled $n = 0$. The vertical coordinate is z and the horizontal coordinate is x. A vertical (dashed) line is drawn through the index of refraction surfaces at $n_x = n \sin\theta_0$, where θ_0 is the initial angle of incidence with respect to the vertical. The intersection of the dashed line and the index of refraction surfaces then gives the index of refraction vector at each height. Note that two solutions occur, one corresponding to an upgoing wave and the other to a downgoing wave. The group velocities can be determined for each of these solutions from the shape of the index of refraction surface (see Figure 4.7). Once the group velocity vectors are known, the ray path can be determined by drawing a smooth curve parallel to the group velocity at each altitude. Starting with the upgoing solution the ray path proceeds upward and becomes horizontal at point 3. At this point, the group velocity is horizontal and the upgoing and downgoing solutions merge. This is a reflection point. The wave energy is then transferred to the downgoing solution, and the ray path continues downward. Note that the wave does not reach the cutoff surface where $n = 0$. This is because the wave is refracted away from the region where the index of refraction goes to zero. Thus, even though the index of refraction remains real and varies smoothly up to the cutoff, the region above the reflection point is not accessible. Accessibility in this case is limited by refraction effects. The reflection point can be moved closer to the cutoff surface by decreasing θ_0. For $\theta_0 = 0$, the wave can propagate all the way up to the cutoff. However, in the vicinity of the cutoff the WKB solution (4.5.11) fails because k_z goes to zero. In this region, a so-called full-wave solution is required to solve Maxwell's equations. Typically these solutions involve Airy functions; see Budden (1961) and Stix (1992).

If a cyclic coordinate does not exist, as often happens in two- or three-dimensional geometries, then the ray paths must be computed numerically using Eqs. (4.5.5) through (4.5.8). Various numerical ray tracing codes are available for solving these equations, usually based on techniques developed by Haselgrove (1955). For a more rigorous and modern treatment of geometric optics, the reader is referred to the recent monograph by Tracy et al. (2014).

Problems

4.1. Compute the Fourier transform of a Gaussian harmonic wave packet

$$f(x) = \mathrm{e}^{-a_0^2 x^2/2} \mathrm{e}^{i k_0 x},$$

where a_0 and k_0 are constants.

4.2. Show for the Gaussian harmonic wave packet that $\langle \Delta x^2 \rangle \langle \Delta k^2 \rangle = 1/4$.

4.3. A waveguide consists of two parallel conducting plates separated by a distance L. The dispersion relation of electromagnetic waves propagating in this waveguide is given by

$$\omega^2 = \omega_n^2 + c^2 k^2,$$

where ω_n is the cutoff frequency of the nth mode ($n = 1, 2, 3, \ldots$, etc.) and $\omega_n = \pi c n / L$.

(a) Show that the phase velocity is

$$v_{\mathrm{p}} = \frac{\omega}{k} = \frac{c}{\sqrt{1 - (\omega_n^2/\omega^2)}}.$$

(b) Show that the group velocity is

$$v_{\mathrm{g}} = \frac{\partial \omega}{\partial k} = c \sqrt{1 - (\omega_n^2/\omega^2)}.$$

(c) Plot v_{p} and v_{g} as a function of ω.

(d) Compare the above results to the propagation of electromagnetic waves in an unmagnetized plasma.

4.4. Waves are incident on a slab of plasma at an angle θ as shown below. The plasma has a uniform plasma frequency ω_{p} and the index of refraction is given by $n = \sqrt{1-\omega_\mathrm{p}^2/\omega^2}$. Using Snell's law, find the critical angle θ_c for total internal reflection (within the free space region) as a function of frequency.

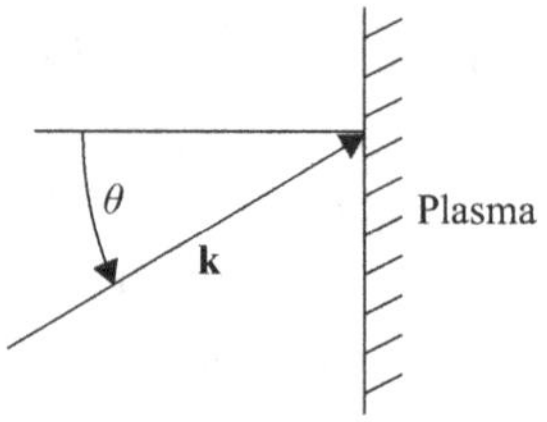

4.5. The effect of collisions can be included in the dispersion relation for waves in a cold plasma by adding a drag force $\nu_s m_s \mathbf{v}_s$ to the equation of motion (ν_s is called the collision frequency):

$$m_s \frac{\mathrm{d}\mathbf{v}_s}{\mathrm{d}t} + \nu_s m_s \mathbf{v}_s = e_s[\mathbf{E} + \mathbf{v}_s \times \mathbf{B}].$$

(a) Show that the effect of collisions can be obtained by making the substitution

$$m_s \to m_s\left(1 + \frac{\mathrm{i}\nu_s}{\omega}\right)$$

in the collisionless dispersion relation.

(b) For transverse waves in a cold plasma, show that if $\nu_s \ll \omega$ and $\omega_\mathrm{p} \ll \omega$ (the high-frequency approximation), the real and imaginary parts of the wave number are approximately

$$k_\mathrm{r} = \frac{\omega}{c}\left[1 - \frac{\omega_\mathrm{p}^2}{2\omega^2}\right] \quad \text{and} \quad k_\mathrm{i} = \frac{1}{2c}\sum_s \frac{\nu_s \omega_{\mathrm{p}s}^2}{\omega^2}.$$

(c) Find the dispersion relation for longitudinal electron plasma oscillations including collisions. Briefly discuss the damping of these waves (i.e., what is the damping decrement and how does it depend on ν_e).

4.6. In the derivation of the index of refraction for waves propagating in a cold plasma with a magnetic field, we defined the quantities $D = (1/2)(R - L)$ and $S = (1/2)(R + L)$. Prove the following identity:

$$S^2 - D^2 = RL.$$

4.7. Show that the determinant of the matrix

$$\mathrm{Det}\begin{bmatrix} S-n^2\cos^2\theta & -\mathrm{i}D & n^2\sin\theta\cos\theta \\ \mathrm{i}D & S-n^2 & 0 \\ n^2\sin\theta\cos\theta & 0 & P-n^2\sin^2\theta \end{bmatrix} = 0$$

can be written in the form

$$Ð(\mathbf{n},\omega) = An^4 - Bn^2 + C = 0,$$

where

$$A = S\sin^2\theta + P\cos^2\theta,$$
$$B = RL\sin^2\theta + PS(1+\cos^2\theta),$$

and

$$C = RLP.$$

4.8. Show that $F^2 = B^2 - 4AC$ can be rewritten in the positive definite form

$$F^2 = (RL - PS)^2\sin^4\theta + 4P^2D^2\cos^2\theta.$$

4.9. Show in the limit $\omega \to 0$ that

$$R = L = S = 1 + \sum_s \frac{\omega_{\mathrm{p}s}^2}{\omega_{\mathrm{c}s}^2},$$

$$D = 0 \quad \text{and} \quad P = -\sum_s \frac{\omega_{\mathrm{p}s}^2}{\omega^2}.$$

Hint: You will have to make use of the condition $\sum_s e_s n_{s0} = 0$.

4.10. Assuming that the ions are infinitely massive, derive Eqs. (4.4.27), (4.4.30), and (4.4.49) for the following frequencies:
(a) the right-hand cutoff $\omega_{\mathrm{R}=0}$;
(b) the left-hand cutoff $\omega_{\mathrm{L}=0}$;
(c) the upper hybrid resonance ω_{UH}.

4.11. For a plasma consisting of electrons and an arbitrary number of ion species, show that the lower hybrid resonance frequency is given by the following equation if $\omega_{\mathrm{ci}} \ll \omega_{\mathrm{LH}} \ll \omega_{\mathrm{ce}}$:

$$\omega_{\mathrm{LH}}^2 = \frac{1}{\frac{1}{\omega_{\mathrm{ce}}^2} + \frac{1}{\omega_{\mathrm{pe}}^2}} \sum_i \left(\frac{m_\mathrm{e}}{m_i}\right)\alpha_i, \quad \text{where} \quad \alpha_i = \frac{n_i}{n_\mathrm{e}}.$$

4.12. Assuming the frequency is sufficiently high that the ions do not move and that $\omega \ll \omega_p, \omega_c \ll \omega_p$, show that the index of refraction of the whistler mode is approximately

$$n^2 = \frac{\omega_p^2}{\omega(\omega_c \cos\theta - \omega)}.$$

Plot $n(\theta)$ for $\omega \ll \omega_c$, $\omega = (0.25)\omega_c$, and $\omega = (0.5)\omega_c$.
Hint: Use the approximations

$$R \simeq -\omega_p^2/\omega(\omega - \omega_c), \;\; L \simeq -\omega_p^2/\omega(\omega + \omega_c),$$

$$S \simeq -\omega_p^2/(\omega^2 - \omega_c^2), \;\; D \simeq -\omega_p^2\omega_c/\omega(\omega^2 - \omega_c^2), \text{ and}$$

$$P = -\omega_p^2/\omega^2.$$

4.13. Using the above formula for the index of refraction, show that for parallel propagation ($\theta = 0$), the group velocity of the whistler mode is given by

$$v_g = 2c\frac{\omega^{1/2}(\omega_c - \omega)^{3/2}}{\omega_c \omega_p}.$$

4.14. For frequencies $\omega \ll \omega_c$, one can see from Problem 4.12 that the index of refraction of the whistler mode is given by

$$n^2 = \frac{\omega_p^2}{\omega\omega_c \cos\theta}.$$

Show that the group velocity makes an angle ψ relative to the magnetic field that is given by

$$\tan\psi = \frac{\sin\theta\cos\theta}{1+\cos^2\theta}.$$

From the above equation, show that the group velocity makes a maximum angle of $\psi_{max} = \tan^{-1}(1/\sqrt{8}) = 19°28'$ with respect to the magnetic field.
Hint: In the text, it is shown that the angle ψ is given by

$$\psi = \theta + \alpha, \quad \text{where} \quad \tan\alpha = -\frac{1}{n}\frac{\partial n}{\partial\theta}.$$

Evaluate $\partial\psi/\partial\theta = 0$ to find the maximum value of ψ.

4.15. A plasma consists of electrons and one ion species. Assuming that $\omega \ll \omega_{pe}$ and $\omega_{pi} \gg \omega_{ci}$, show that near resonance ($\omega \sim \omega_{ci}$) the index of refraction of the ion cyclotron mode is given by

$$n^2 \simeq n_A^2 \left(\frac{\omega_{ci}}{\omega_{ci} - \omega}\right) \frac{1+\cos^2\theta}{2\cos^2\theta},$$

where n_A is the Alfvén index of refraction.
Hint: Assume that P is sufficiently large that one can make the approximations $A \simeq P\cos^2\theta$ and $B \simeq PS(1+\cos^2\theta)$. Near resonance one can assume that $L \gg R$ so that $S \simeq L/2$. Since $n_e = n_i$, one can also use the relationship $\omega_{pe}^2/|\omega_{ce}| = \omega_{pi}^2/\omega_{ci}$. Finally, note that $n_A^2 = \omega_{pi}^2/\omega_{ci}^2$ when $\omega_{pi} \gg \omega_{ci}$.

4.16. Using the results of Problem 4.15 above and the same procedure as in Problem 4.14, show that the maximum group velocity angle of the ion cyclotron mode is $\psi_{max} = 12.3°$.
Hint: In this case

$$\tan\psi = \frac{\sin\theta\cos^3\theta}{1+\cos^4\theta}.$$

Because of the complicated form of this equation, the best way to find ψ_{max} is to use a computer to calculate ψ as a function of θ.

4.17 For a sufficiently large plasma density, show that the Alfvén velocity $V_A = c/n_A$ is given to a good approximation by the following equation:

$$V_A = \sqrt{\frac{B^2}{\mu_0\rho_m}},$$

where $\rho_m = \sum_s m_s n_s$ is the mass density.

4.18. A transverse wave on a wire of linear mass density λ and tension T propagates at a velocity $V = \sqrt{T/\lambda}$. Show that the Alfvén velocity can be obtained from the above formula if the magnetic field has a tension (per unit cross-sectional area) given by B^2/μ_0.

4.19. Waves are sometimes classified as compressible or non-compressible, according to whether there is or is not a density fluctuation associated with the wave. By examining the electric field eigenvector for the Alfvén modes and the particle equations of motion, show that the isotropic Alfvén mode, $n^2 = n_A^2$, is compressible, whereas the anisotropic Alfvén mode, $n^2 = n_A^2/\cos^2\theta$, is incompressible.

4.20. A plasma consists of equal densities of electrons and positrons (equal masses, $m_e^- = m_e^+$) in a uniform, static magnetic field.
(a) Write down and simplify the equations for R, L, S, D, and P using only the parameters ω_p and ω_c, where $\omega_p = \omega_{pe}^- = \omega_{pe}^+$ and $\omega_c = |\omega_{ce}^-| = |\omega_{ce}^+|$.
(b) Using the parameters $X = \omega_p/\omega$ and $Y = \omega_c/\omega$, construct a CMA diagram showing all bounding surfaces.
(c) Show that the indices of refraction of the two modes are given by

$$n^2 = S \quad \text{and} \quad n^2 = \frac{PS}{S\sin^2\theta + P\cos^2\theta}.$$

(d) Sketch the shape of the index of refraction surfaces in each region of the CMA diagram.
(e) Using appropriate diagrams, summarize the accessible region of the CMA diagram for each continuous branch of the index of refraction.

References

Alfvén, H. 1942. Existence of electromagnetic-hydrodynamic waves. *Nature* **150**, 405.

Allis, W. P. 1959. Waves in *Plasmas, Sherwood Conference on Controlled Fusion, April 27–28, 1959, Gatlinburg, TN,* p. 32.

Barkhausen, H. 1919. Zwei mit Hilfe der neuen Verstarker entdeckte Ersheinungen. *Phys. Z.* **20**, 401–403.

Benson, R. F. 1982. Stimulated plasma instability and nonlinear phenomena in the ionosphere. *Radio Sci.* **17** (6), 1637–1659.

Brice, N. M., and Smith, R. L. 1965. Lower hybrid resonance emissions. *J. Geophys. Res.* **78** (1), 71–80.

Budden, K. G. 1961. *Radio Waves in the Ionosphere: The Mathematical Theory of the Reflection of Radio Waves from Stratified Ionised Layers.* London: Cambridge University Press.

Clemmow, P. C., and Mullaly, R. F. 1955. Dependence of the refractive index in magneto-ionic theory on the direction of the wave normal. *Physics of the Ionosphere: Rep. Phys. Soc. Conf.* London: London Physical Society, p. 340.

Franklin, C. A., and Maclean, M. A. 1969. The design of swept-frequency topside sounders. *Proc. IEEE* **57**, 897–929.

Gurnett, D. A., Shawhan, S. D., Brice, N. M., and Smith, R. L. 1965. Ion cyclotron whistlers. *J. Geophys. Res.* **70**, 1665–1688.

Haselgrove, J. 1955. Ray theory and a new method for ray tracing. *Physics of the Ionosphere: Rep. Phys. Soc. Conf.* London: London Physical Society, pp. 355–364.

Helliwell, R. A. 1965. *Whistlers and Related Ionospheric Phenomena.* Stanford, CA: Stanford University Press, p. 103.

Kaplan, W. 1952. *Advanced Calculus.* Reading, MA: Addision-Wesley, p. 99.

Manchester, R. N., and Taylor, J. H. 1977. *Pulsars.* San Francisco, CA: Freeman and Company, p. 134.

Poeverlein, H. 1950. Strahlwege von radiowellen in der ionosphere. Z. *Angew. Phys.* **2** (4), 152–160.

Ratcliffe, J. A. 1962. *The Magneto-Ionic Theory and its Application to the Ionosphere.* London: Cambridge University Press, p. 126.

Stix, T. H. 1992. *Waves in Plasmas.* New York: American Institute of Physics, p. 343.

Storey, L. R. O. 1953. An investigation of whistling atmospherics. *Philos. Trans. R. Soc. London, Ser. A* **246**, 113–141.

Townsend, J. S. 1992. *A Modern Approach to Quantum Mechanics.* New York: McGraw-Hill, p. 79.

Tracy, E. R., Brizard, A. J., Richardson, A. S., and Kaufman, A. N. 2014. *Ray Tracing and Beyond: Phase Space Methods in Plasma Wave Theory.* New York: Cambridge University Press.

Walsh, D., Haddock, T. F., and Schultz, H. F. 1964. Cosmic radio intensities at 1.225 and 2.0 MC measured up to an altitude of 1700 km. *Space Sci.* **4**, 935–959.

Weinberg, S., 1962. Eikonal methods in magnetohydrodynamics. *Phys. Rev.* **126**, 1899–1909.

Further Reading

Goertz, C. K., and Strangeway, R. J. 1997. *Introduction to Space Physics,* eds. Kivelson, M. G., and Russell, C. T. Cambridge: Cambridge University Press. Originally published in 1995.

Parks, G. K. 1997. *Physics of Space Plasmas: An Introduction.* Redwood City, CA: Addison-Wesley, Chapter 9. Originally published in 1991.

Stix, T. H. 1992. *Waves in Plasmas.* New York: American Institute of Physics.

Swanson, D. G. 1989. *Plasma Waves.* San Diego, CA: Academic Press, Harcourt Brace, Chapter 2.

5

Kinetic Theory and the Moment Equations

In the previous chapter we considered the idealized case of a cold plasma in which there were no random thermal motions. In this chapter we introduce a more general framework for analyzing plasmas in which thermal motions must be considered. For a system with a large number of particles it is neither possible nor desirable to determine the motion of each and every particle. Instead, we will use a statistical approach to compute the average motion of a large number of particles. This approach is called kinetic theory. It is not our intention in this chapter to present an exhaustive treatment of plasma kinetic theory. Instead we will introduce the basic concepts of kinetic theory and derive a system of equations known as the moment equations. These equations will then be used to analyze various simple applications that are of interest.

5.1 The Distribution Function

To carry out a statistical description of a plasma, it is convenient to introduce a six-dimensional space, called *phase space,* that consists of the position coordinates x, y, and z, and the velocity coordinates v_x, v_y, and v_z. At any given time the dynamical state of a particle can be represented by a point in phase space. For a system of many particles, the dynamical state of the entire system can then be represented by a collection of points in phase space, with one point for each particle. If the number of particles is very large, it is meaningful to define the average number density of particles in a small volume element of phase space. This density is called the distribution function $f(\mathbf{r}, \mathbf{v}, t)$ and is defined such that

$$\mathrm{d}N = f(\mathbf{r}, \mathbf{v}, t)\,\mathrm{d}^3x\,\mathrm{d}^3v \qquad (5.1.1)$$

is the probable number of particles in the phase-space volume element $\mathrm{d}^3x\,\mathrm{d}^3v = \mathrm{d}x\,\mathrm{d}y\,\mathrm{d}z\,\mathrm{d}v_x\,\mathrm{d}v_y\,\mathrm{d}v_z$ at time t. We shall assume that the number of particles is sufficiently large that $f(\mathbf{r},\mathbf{v},t)$ can be regarded as a continuous function of $\mathbf{r}$ and $\mathbf{v}$. The total number of particles, N, in the system is obtained by integrating $f(\mathbf{r},\mathbf{v},t)$ over all of phase space (assuming a spatially infinite plasma)

$$N = \int_{-\infty}^{\infty} f(\mathbf{r},\mathbf{v},t)\, \mathrm{d}^3x\, \mathrm{d}^3v. \tag{5.1.2}$$

Having defined the distribution function, we can then use the standard rules of statistics for computing various macroscopic averages. The average (expectation) value of any dynamical quantity $g(\mathbf{r},\mathbf{v})$ in a region R of phase space is given by

$$\langle g(\mathbf{r},\mathbf{v})\rangle = \frac{1}{N}\int_R g(\mathbf{r},\mathbf{v})\, f(\mathbf{r},\mathbf{v},t)\, \mathrm{d}^3x\, \mathrm{d}^3v, \tag{5.1.3}$$

where the angle brackets, $\langle\,\rangle$, mean "average."

Since several species may be present in a plasma, it is necessary to define a distribution function $f_s(\mathbf{r},\mathbf{v},t)$ for each particle species s in the plasma. By making suitable choices for $g(\mathbf{r},\mathbf{v})$ we can then compute various types of averages. Some of the most commonly used averages are the following:

$$\text{number density, } n_s = \int_V f_s(\mathbf{r},\mathbf{v},t)\, \mathrm{d}^3v, \tag{5.1.4}$$

$$\text{average velocity, } \mathbf{U}_s = \frac{1}{n_s}\int_V \mathbf{v} f_s(\mathbf{r},\mathbf{v},t)\, \mathrm{d}^3v, \tag{5.1.5}$$

$$\text{kinetic energy density, } W_s = \int_V \frac{1}{2} m_s v^2 f_s(\mathbf{r},\mathbf{v},t)\, \mathrm{d}^3v, \tag{5.1.6}$$

$$\text{pressure tensor, } \overleftrightarrow{\mathbf{P}}_s = \int_V m_s (\mathbf{v}-\mathbf{U}_s)(\mathbf{v}-\mathbf{U}_s)\, f_s(\mathbf{r},\mathbf{v},t)\, \mathrm{d}^3v, \tag{5.1.7}$$

where the integrations are over velocity space, V. The number density, n_s, is the average number of particles of type s per unit volume. The velocity, $\mathbf{U}_s$, is the average velocity of particles of type s. The kinetic energy density, W_s, is the average kinetic energy, $(1/2)m_s v^2$, of particles of type s per unit volume. In index notation, the pressure tensor, $\overleftrightarrow{\mathbf{P}}_s = [P_{ij}]$, is the average rate at which momentum is transported in the i direction across surface j in a frame of reference moving at the average velocity, $\mathbf{U}_s$. The quantity $(\mathbf{v}-\mathbf{U}_s)(\mathbf{v}-\mathbf{U}_s)$ in Eq. (5.1.7) is defined by the following matrix:

$$(\mathbf{v} - \mathbf{U}_s)(\mathbf{v} - \mathbf{U}_s)$$

$$= \begin{bmatrix} (v_x - U_{sx})(v_x - U_{sx}), (v_x - U_{sx})(v_y - U_{sy}), (v_x - U_{sx})(v_z - U_{sz}) \\ (v_y - U_{sy})(v_x - U_{sx}), (v_y - U_{sy})(v_y - U_{sy}), (v_y - U_{sy})(v_z - U_{sz}) \\ (v_z - U_{sz})(v_x - U_{sx}), (v_z - U_{sz})(v_y - U_{sy}), (v_z - U_{sz})(v_z - U_{sz}) \end{bmatrix}. \tag{5.1.8}$$

The average number density, n_s, and the average velocity, $\mathbf{U}_s$, can be used to compute the charge density, ρ_q, and the mass density, ρ_m, using the equations

$$\rho_\mathrm{q} = \sum_s e_s n_s \tag{5.1.9}$$

and

$$\rho_\mathrm{m} = \sum_s m_s n_s. \tag{5.1.10}$$

Similarly, the current density is given by

$$\mathbf{J} = \sum_s e_s n_s \mathbf{U}_s. \tag{5.1.11}$$

To illustrate the methods used to compute the various averages defined above, it is instructive to work out the integrals involved for certain specific choices of the distribution function. For example, consider a cold beam of electrons all moving at a velocity $\mathbf{v} = \mathbf{U}_0$ and of ions all at rest, $\mathbf{v} = 0$. The velocity distribution functions in this case can be written

$$f_\mathrm{e} = n_0 \delta(\mathbf{v} - \mathbf{U}_0) \quad \text{and} \quad f_\mathrm{i} = n_0 \delta(\mathbf{v}), \tag{5.1.12}$$

where δ is the Dirac delta function. For this simple case, it is easy to show using Eq. (5.1.4) that $n_\mathrm{e} = n_\mathrm{i} = n_0$. The charge density is then $\rho_q = 0$. The average velocity of the electrons, also sometimes called the drift velocity, is $\mathbf{U}_\mathrm{e} = \mathbf{U}_0$, and the average velocity of the ions is $\mathbf{U}_\mathrm{i} = 0$. Since the electrons are drifting with respect to the ions, a current is present. The corresponding current density is given by $\mathbf{J} = -en_0\mathbf{U}_0$. Since there is no spread in velocity with respect to the average velocity, the pressure tensors $\overleftrightarrow{\mathbf{P}}_\mathrm{e}$ and $\overleftrightarrow{\mathbf{P}}_\mathrm{i}$ are zero.

Next, consider the more complicated case of a Maxwellian velocity distribution function given by

$$f = n_0 \left(\frac{m}{2\pi\kappa T} \right)^{3/2} \exp\left[-\frac{m(\mathbf{v} - \mathbf{U}_0)^2}{2\kappa T} \right], \tag{5.1.13}$$

where for simplicity the subscript s has been omitted. This distribution function has a peak at velocity $\mathbf{v} = \mathbf{U}_0$ and is spherically symmetric around this velocity. The average velocity, defined by Eq. (5.1.5), is simply $\mathbf{U}_0$. The integrals in the pressure tensor can also be computed and are

$$P_{ij} = n_0 \kappa T \quad \text{for} \quad i = j \tag{5.1.14}$$

and

$$P_{ij} = 0 \quad \text{for} \quad i \neq j. \tag{5.1.15}$$

The pressure tensor is diagonal, with equal elements, which indicates an isotropic pressure. The pressure is also directly proportional to the number density, n_0, and the temperature, T, which is simply the ideal gas law, $P = n_0 \kappa T$.

If the distribution function is anisotropic, the diagonal elements of the pressure tensor are in general not the same. An example of an anisotropic distribution function is given by the bi-Maxwellian

$$f = n_0 \left(\frac{m}{2\pi \kappa T_\perp} \right) \left(\frac{m}{2\pi \kappa T_\parallel} \right)^{1/2} \exp\left[-\frac{m v_\perp^2}{2\kappa T_\perp} \right] \exp\left[-\frac{m v_\parallel^2}{2\kappa T_\parallel} \right], \tag{5.1.16}$$

which is often used in a magnetized plasma. The symbols $\perp$ and $\parallel$ refer to directions perpendicular and parallel to the magnetic field (i.e., cylindrical coordinates). If we choose the coordinate system such that the magnetic field is aligned with the z axis, then the pressure tensor, defined by Eq. (5.1.7), is diagonal and has the simple form

$$\overleftrightarrow{\mathbf{P}} = \begin{bmatrix} P_\perp & 0 & 0 \\ 0 & P_\perp & 0 \\ 0 & 0 & P_\parallel \end{bmatrix}, \tag{5.1.17}$$

where $P_\perp = n_0 \kappa T_\perp$ and $P_\parallel = n_0 \kappa T_\parallel$. In the absence of collisions, the parallel and perpendicular pressures are decoupled. Later we will discuss an equation of state, called the Chew–Goldberger–Low equation of state, that describes how the perpendicular and parallel pressures respond to dynamical changes.

5.2 The Boltzmann and Vlasov Equations

The kinetic equation that governs the evolution of the distribution function in phase space is a differential equation called the *Boltzmann equation*. To derive this equation, one starts by assuming that the forces acting on the particles can be classified into two types: long range and short range. Forces that are the same for all particles in a given phase-space volume element are long-range forces, and forces that are not the same are short-range forces. Long-range forces originate from the collective effects of a large number of particles acting over relatively long distances, and short-range forces originate from a small number of particles (usually only two) acting over relatively short distances. Short-range forces correspond to what are commonly called collisions. Collisions produce large, nearly instantaneous changes in the velocities of the interacting particles.

To derive the Boltzmann equation, we first assume that there are no short-range forces (i.e., no collisions). All particles in an infinitesimal volume of phase space, $\mathrm{d}^3x\,\mathrm{d}^3v$, then experience the same long-range force $\mathbf{F}$. Since the force on a representative particle in phase space is given by Newton's law

$$\mathbf{F} = m\frac{\mathrm{d}\mathbf{v}}{\mathrm{d}t}, \tag{5.2.1}$$

it follows that the particle trajectories can never intersect. If the trajectories did intersect, two different values of $\mathrm{d}\mathbf{v}/\mathrm{d}t$ would exist for the same $\mathbf{F}$ and $\mathbf{v}$, which cannot occur. Therefore, if we follow the evolution of a volume element that moves with the particles, then all of the particles within the volume element $\mathrm{d}^3x\,\mathrm{d}^3v$ at time t map into the corresponding volume element $\mathrm{d}^3x'\mathrm{d}^3v'$ at time $t' = t + \mathrm{d}t$. The evolution of a typical phase-space volume element is illustrated in Figure 5.1. Since the number of particles in the phase-space volume element is conserved, we can write

$$f(\mathbf{r},\mathbf{v},t)\,\mathrm{d}^3x\,\mathrm{d}^3v = f'(\mathbf{r}',\mathbf{v}',t')\,\mathrm{d}^3x'\,\mathrm{d}^3v', \tag{5.2.2}$$

where f is the distribution function at time t, and f' is the distribution function at time t'. The equations of motion for the particles in phase space are given by

$$x_r' = x_r + v_r\,\mathrm{d}t \text{ and } v_r' = v_r + \frac{F_r}{m}\mathrm{d}t, \tag{5.2.3}$$

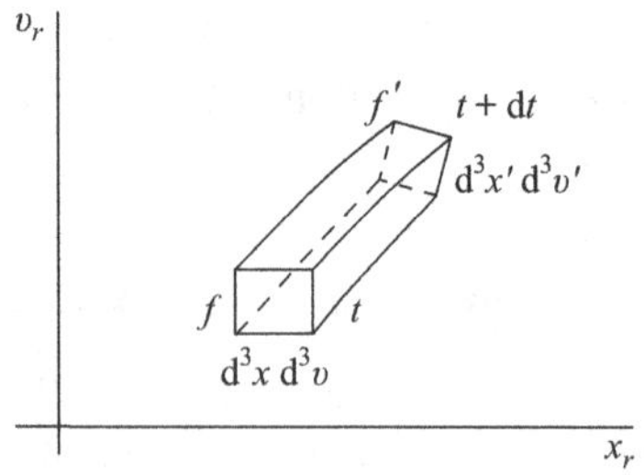

Figure 5.1 The evolution of a volume element $\mathrm{d}^3x\mathrm{d}^3v$ in phase space from time t to $t + \mathrm{d}t$.

where we use index notation $r = (1, 2, 3)$ to represent $x_r = (x, y, z)$ and $v_r = (v_x, v_y, v_z)$. These equations can be viewed as simply a transformation from phase-space coordinates $(\mathbf{r}, \mathbf{v})$ to coordinates $(\mathbf{r}', \mathbf{v}')$. The infinitesimal volume elements involved in this transformation are determined by

$$\mathrm{d}^3x'\,\mathrm{d}^3v' = J\,\mathrm{d}^3x\,\mathrm{d}^3v, \tag{5.2.4}$$

where J is the Jacobian of the transformation (Kaplan, 1952). Since there are six transformation equations, the Jacobian is given by the 6×6 determinant

$$J = \mathrm{Det}\begin{bmatrix} \frac{\partial x'}{\partial x} & \frac{\partial x'}{\partial y} & \cdots \\ \frac{\partial y'}{\partial x} & & \\ \cdot & & \\ \cdot & & \\ \cdot & & \frac{\partial v_z'}{\partial v_z} \end{bmatrix}. \tag{5.2.5}$$

After computing the appropriate derivatives and sorting out the lowest-order terms in $\mathrm{d}t$, Eq. (5.2.4) can be written in the form

$$\mathrm{d}^3x'\,\mathrm{d}^3v' = \left[1 + \sum_r \frac{\partial v_r}{\partial x_r}\mathrm{d}t + \sum_r \frac{\partial}{\partial v_r}\left(\frac{F_r}{m}\right)\mathrm{d}t\right]\mathrm{d}^3x\,\mathrm{d}^3v, \tag{5.2.6}$$

where the summation extends over $r = 1$, 2, and 3 (i.e., x, y, and z). To complete the derivation, the function f' in Eq. (5.2.2) is expanded in a Taylor series using the chain rule, with $v_r = \mathrm{d}x_r/\mathrm{d}t$ and $\mathrm{d}v_r/\mathrm{d}t = F_r/m$, which gives

$$f' = f + \frac{\partial f}{\partial t}\mathrm{d}t + \sum_r v_r \frac{\partial f}{\partial x_r}\mathrm{d}t + \sum_r \left(\frac{F_r}{m}\right)\frac{\partial f}{\partial v_r}\mathrm{d}t. \tag{5.2.7}$$

Substituting $\mathrm{d}^3x'\,\mathrm{d}^3v'$ from Eq. (5.2.6) and f' from Eq. (5.2.7) into Eq. (5.2.2) then gives

$$f = f + \frac{\partial f}{\partial t}\mathrm{d}t + \sum_r v_r \frac{\partial f}{\partial x_r}\mathrm{d}t + \sum_r \left(\frac{F_r}{m}\right)\frac{\partial f}{\partial v_r}\mathrm{d}t + \sum_r f\frac{\partial v_r}{\partial x_r}\mathrm{d}t + \sum_r f\frac{\partial}{\partial v_r}\left(\frac{F_r}{m}\right)\mathrm{d}t, \tag{5.2.8}$$

which simplifies to the first-order partial differential equation

$$\frac{\partial f}{\partial t} + \sum_r \frac{\partial}{\partial x_r}(v_r f) + \sum_r \frac{\partial}{\partial v_r}\left(\frac{F_r}{m}f\right) = 0. \tag{5.2.9}$$

Comparing the above equation with the familiar three-dimensional mass continuity equation of fluid mechanics,

$$\frac{\partial \rho_{\mathrm{m}}}{\partial t} + \sum_r \frac{\partial}{\partial x_r}(\upsilon_r \rho_{\mathrm{m}}) = 0, \tag{5.2.10}$$

we see that it is simply a continuity equation in six-dimensional phase space. Equation (5.2.9) is valid for both relativistic and non-relativistic velocities.

Two simplifications can be made to Eq. (5.2.9). First, note that since υ_r and x_r are independent variables, it follows that $\partial \upsilon_r/\partial x_r$ is zero. The independence of υ_r and x_r may not be immediately obvious, since in particle mechanics $\mathbf{v}(t)$ is obtained as a time derivative of $\mathbf{r}(t)$ and hence depends on $\mathbf{r}(t)$. However, the kinetic description is not based on following an individual particle in phase space. Instead, it is based on the average number of particles in a volume element of phase space at any given time. Since the spatial coordinates of the volume element can be chosen independent of the velocity, x_r and υ_r are independent variables. Second, note that if we restrict our consideration to non-relativistic velocities, for which the mass m is independent of velocity, the term $(\partial/\partial \upsilon_r)(F_r/m)$ is zero. This result can be proven for electromagnetic forces by simply working out the appropriate derivatives. For $r = x$, it follows that

$$\frac{\partial}{\partial \upsilon_x}\left(\frac{F_x}{m}\right) = \frac{q}{m}\frac{\partial}{\partial \upsilon_x}[E_x + \upsilon_y B_z - \upsilon_z B_y] = 0. \tag{5.2.11}$$

A similar result holds for $r = y$ and $r = z$. These two simplifications imply that the Jacobian is $J = 1$. The phase-space flow is then volume preserving; that is, $\mathrm{d}^3x'\,\mathrm{d}^3\upsilon' = \mathrm{d}^3x\mathrm{d}^3\upsilon$. With these simplifications Eq. (5.2.9) reduces to

$$\frac{\partial f}{\partial t} + \sum_r \upsilon_r \frac{\partial f}{\partial x_r} + \sum_r \frac{F_r}{m}\frac{\partial f}{\partial \upsilon_r} = 0, \tag{5.2.12}$$

or in vector notation

$$\frac{\partial f}{\partial t} + \mathbf{v}\cdot\nabla f + \frac{\mathbf{F}}{m}\cdot\nabla_{\mathbf{v}} f = 0, \tag{5.2.13}$$

where $\nabla_{\mathbf{v}}$ is the gradient operator in velocity space, $\nabla_{\mathbf{v}} = \hat{\mathbf{x}}(\partial/\partial \upsilon_x) + \hat{\mathbf{y}}(\partial/\partial \upsilon_y) + \hat{\mathbf{z}}(\partial/\partial \upsilon_z)$. This result does not hold for relativistic velocities, since the mass in Eq. (5.2.11) depends on the velocity, i.e., $m = m_0/(1 - \upsilon^2/c^2)^{1/2}$, where m_0 is the rest mass.

Next, consider the effect of short-range forces (i.e., collisions). Because short-range forces are not the same for all particles in a volume element, collisions act to scatter particles into and out of the phase-space volume element. As mentioned earlier, the main effect of short-range interactions is to introduce a

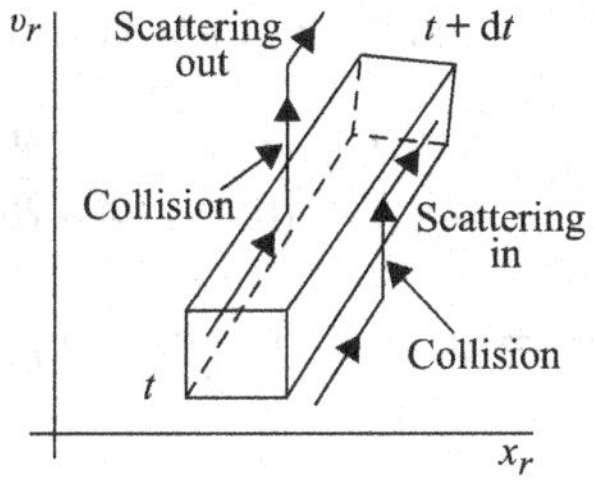

Figure 5.2 Short-range forces (i.e., collisions) scatter particles out of and into the phase-space volume element as it evolves.

nearly discontinuous change in the particle velocity. Some typical trajectories in phase space are shown in Figure 5.2. Changes to the distribution function due to collisions as the phase-space volume element evolves from time t to time $t + \mathrm{d}t$ are taken into account by adding a collision operator, $\delta_c f/\delta t$, not yet specified, to the right-hand side of Eq. (5.2.13), i.e.,

$$\frac{\partial f}{\partial t} + \mathbf{v} \cdot \nabla f + \frac{\mathbf{F}}{m} \cdot \nabla_{\mathbf{v}} f = \frac{\delta_c f}{\delta t} . \tag{5.2.14}$$

This equation is called the Boltzmann equation, after Boltzmann (1995), who first used this equation to analyze particle transport phenomena in ordinary gases. The left-hand side of the Boltzmann equation describes the response of the particles to long-range collective forces, and the right-hand side describes the response to short-range collisional forces. As discussed earlier, if the number of electrons per Debye cube is large, $n\lambda_{\mathrm{D}}^3 \gg 1$, then collective forces are much more important than collisional forces. Under these conditions the collision term can be ignored, i.e., $\delta_c f/\delta t = 0$. The resulting equation, with the force given by the Lorentz force, $\mathbf{F} = q[\mathbf{E} + \mathbf{v} \times \mathbf{B}]$, is

$$\frac{\partial f}{\partial t} + \mathbf{v} \cdot \nabla f + \frac{q}{m}[\mathbf{E} + \mathbf{v} \times \mathbf{B}] \cdot \nabla_{\mathbf{v}} f = 0. \tag{5.2.15}$$

This equation is called the Vlasov equation, after Vlasov (1945), who first applied this equation to the study of plasmas.

5.3 Solutions Based on Constants of the Motion

If a system has one or more known constants of the motion, then it is possible to immediately write down a general form of the solution to the Vlasov equation. These solutions are based on the fact that the distribution function is constant along a representative trajectory in phase space. To prove this result, compute the total derivative of f along a representative trajectory, $\mathbf{r}(t)$ and $\mathbf{v}(t)$, using the chain rule

$$\mathrm{d}f_s = \frac{\partial f_s}{\partial t}\mathrm{d}t + \sum_r \frac{\partial f_s}{\partial x_r}\frac{\mathrm{d}x_r}{\mathrm{d}t}\mathrm{d}t + \sum_r \frac{\partial f_s}{\partial v_r}\frac{\mathrm{d}v_r}{\mathrm{d}t}\mathrm{d}t. \tag{5.3.1}$$

After recognizing that $\mathrm{d}x_r/\mathrm{d}t = v_r$ and $\mathrm{d}v_r/\mathrm{d}t = F_r/m$, and dividing by $\mathrm{d}t$ one obtains

$$\frac{\mathrm{d}f_s}{\mathrm{d}t} = \frac{\partial f_s}{\partial t} + \mathbf{v}\cdot\boldsymbol{\nabla} f_s + \frac{\mathbf{F}}{m}\cdot\boldsymbol{\nabla}_{\mathbf{v}} f_s = 0, \tag{5.3.2}$$

which is zero, since the right-hand side is just the Vlasov equation. Since $\mathrm{d}f_s/\mathrm{d}t = 0$, it follows that f_s is constant along the trajectory. The possible trajectories are constrained by the constants of the motion, $C_{\mathrm{m}}(\mathbf{r},\mathbf{v})$. The constants of the motion can be obtained from the adiabatic invariants, as in Section 3.8, or from Hamilton's equations, as in Section 3.9. Any function of the constants of the motion

$$f_s[C_1(\mathbf{r},\mathbf{v}),\, C_2(\mathbf{r},\mathbf{v}),\ldots] \tag{5.3.3}$$

is then a solution of the Vlasov equation.

As an example of this procedure, consider the motion of a particle of mass m and charge q in a time-stationary electrostatic potential $\Phi(\mathbf{r})$. In this system the total energy

$$W = \frac{1}{2}mv^2 + q\,\Phi(\mathbf{r}) \tag{5.3.4}$$

is a constant of the motion. Any function $f_s[W(\mathbf{r},\mathbf{v})]$ is then a solution of the Vlasov equation. The exact form of $f_s[W]$ is determined by the boundary conditions. For example, if $f(\mathbf{r},\mathbf{v})$ is known to be a Maxwellian of number density n_0 and temperature T_0 as $r \to \infty$, where $\Phi(\infty) = 0$, then the general solution at any arbitrary point is

$$f(\mathbf{r},\mathbf{v}) = n_0\left(\frac{m}{2\pi\kappa T_0}\right)^{3/2} e^{-\left[\frac{1}{2}mv^2 + q\Phi(\mathbf{r})\right]\Big/\kappa T_0}. \tag{5.3.5}$$

This distribution function can be recognized as the function used in Eq. (2.2.2) to determine the potential in the plasma sheath around a point charge.

When constructing solutions of the Vlasov equation using constants of the motion, one must be careful about the boundary conditions. If the electrostatic potential is not monotonic, such as the example illustrated in Figure 5.3, then certain regions of phase space (x, v) may not be accessible from the boundary point x_0. An example of such an inaccessible region is shown by the cross-hatching.

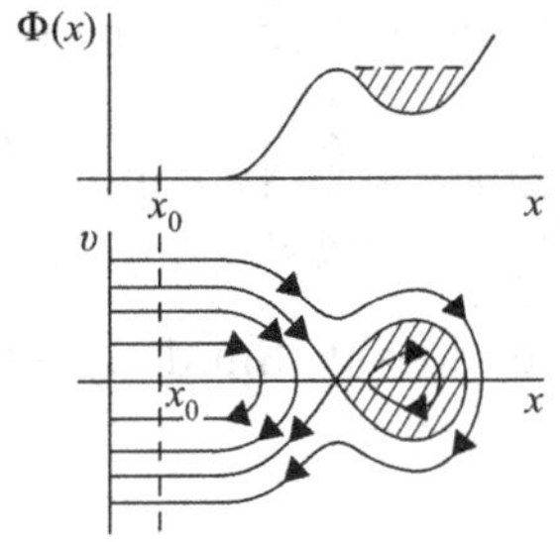

Figure 5.3 The bottom (x, v) diagram shows the phase-space trajectories of the particles moving in the one-dimensional potential $\Phi(x)$ shown in the top diagram.

Since the trajectories in the cross-hatched region do not go through the point x_0, the boundary conditions on f at x_0 do not constrain the distribution function in the inaccessible region. Some other boundary condition is then necessary to specify the distribution function. If no source is present in the inaccessible region, then the distribution function is zero in this region. In practice, collisions and various wave–particle interactions often scatter particles into the inaccessible region. An exact determination of the distribution function in such cases is usually quite difficult and must take into account these scattering effects.

5.4 The Moment Equations

Often in analyzing problems in plasma physics, we are more interested in certain macroscopic averages, such as the density and pressure, than the detailed form of the distribution function. It is then convenient to use a system of equations called the moment equations. The moment equations are obtained by multiplying the Boltzmann equation by powers of the velocity and integrating over all of velocity space. This procedure generates one moment equation for each power of the velocity. We illustrate the procedure involved by calculating the first three moments of the Boltzmann equation.

5.4.1 Zeroth Moment

To obtain the zeroth moment, we multiply the Boltzmann equation (5.2.14) by $v^0 = 1$ and integrate over all of velocity space, V, which gives

$$\int_V \frac{\partial f_s}{\partial t} \mathrm{d}^3 v + \int_V \mathbf{v} \cdot \nabla f_s \, \mathrm{d}^3 v + \frac{1}{m_s} \int_V \mathbf{F} \cdot \nabla_{\mathrm{v}} f_s \, \mathrm{d}^3 v = \int_V \frac{\delta_c f_s}{\delta t} \mathrm{d}^3 v. \tag{5.4.1}$$

The first integral on the left of this equation can be written

$$\int_V \frac{\partial f_s}{\partial t} \mathrm{d}^3 v = \frac{\partial}{\partial t} \int_V f_s \, \mathrm{d}^3 v = \frac{\partial n_s}{\partial t}, \tag{5.4.2}$$

where n_s is the average number density. Similarly, the second integral on the left can be written

$$\int_V \mathbf{v} \cdot \boldsymbol{\nabla} f_s \, \mathrm{d}^3 v = \boldsymbol{\nabla} \cdot \int_V \mathbf{v} f_s \, \mathrm{d}^3 v = \boldsymbol{\nabla} \cdot (n_s \mathbf{U}_s), \tag{5.4.3}$$

where $\mathbf{U}_s$ is the average velocity. Noting that $\boldsymbol{\nabla}_\mathbf{v} \cdot (\mathbf{v} \times \mathbf{B}) f_s = (\mathbf{v} \times \mathbf{B}) \cdot \boldsymbol{\nabla}_\mathbf{v} f_s$, the third integral on the left can be rewritten

$$\int_V \mathbf{F} \cdot \boldsymbol{\nabla}_\mathbf{v} f_s \, \mathrm{d}^3 v = \int_V \boldsymbol{\nabla}_\mathbf{v} \cdot (\mathbf{F} f_s) \mathrm{d}^3 v = \int_S (\mathbf{F} f_s) \cdot \mathrm{d}\mathbf{s}_v = 0, \tag{5.4.4}$$

where we have used Gauss' theorem to convert the volume integral to a surface integral at infinity. The surface integral is assumed to vanish if $f(\mathbf{v})$ decays sufficiently rapidly at infinite velocity. The integral on the right side of Eq. (5.4.1) involving the collision term is zero,

$$\int_V \frac{\delta_\mathrm{c} f_s}{\delta t} \mathrm{d}^3 v = \frac{\delta_\mathrm{c} n_s}{\delta t} = 0, \tag{5.4.5}$$

since collisions are assumed to not change the density. Combining Eqs. (5.4.2)–(5.4.5), the zeroth moment of the Boltzmann equation (5.4.1) yields

$$\frac{\partial n_s}{\partial t} + \boldsymbol{\nabla} \cdot (n_s \mathbf{U}_s) = 0. \tag{5.4.6}$$

This equation is simply the continuity equation for the sth species. Multiplying the above equation by the charge e_s and summing over all species, one obtains the charge continuity equation

$$\frac{\partial \rho_\mathrm{q}}{\partial t} + \boldsymbol{\nabla} \cdot \mathbf{J} = 0, \tag{5.4.7}$$

which is a fundamental equation of electricity and magnetism.

5.4.2 First Moment

The first moment is obtained by multiplying the Boltzmann equation (5.2.14) by $\mathbf{v}$ and integrating over all of velocity space. For reasons that will soon become apparent, it is more convenient to multiply by the momentum $m_s\mathbf{v}$, which gives

$$m_s \int_V \mathbf{v} \frac{\partial f_s}{\partial t} \mathrm{d}^3 v + m_s \int_V \mathbf{v}(\mathbf{v} \cdot \nabla f_s)\, \mathrm{d}^3 v + \int_V \mathbf{v}(\mathbf{F} \cdot \nabla_{\mathbf{v}} f_s)\, \mathrm{d}^3 v$$
$$= m_s \int_V \mathbf{v} \frac{\delta_{\mathrm{c}} f_s}{\delta t} \mathrm{d}^3 v. \tag{5.4.8}$$

The first integral on the left of the above equation can be written

$$m_s \int_V \mathbf{v} \frac{\partial f_s}{\partial t}\, \mathrm{d}^3 v = \frac{\partial}{\partial t}\left(m_s \int_V \mathbf{v} f_s\, \mathrm{d}^3 v\right) = \frac{\partial}{\partial t}(m_s n_s \mathbf{U}_s). \tag{5.4.9}$$

The second integral is more complicated and can be written

$$m_s \int_V \mathbf{v}(\mathbf{v} \cdot \nabla f_s)\, \mathrm{d}^3 v = \nabla \cdot \left[m_s \int_V \mathbf{v}\,\mathbf{v} f_s\, \mathrm{d}^3 v\right]. \tag{5.4.10}$$

It is convenient to rewrite the **vv** tensor by making use of the following identity:

$$\mathbf{vv} = (\mathbf{v} - \mathbf{U}_s)(\mathbf{v} - \mathbf{U}_s) + \mathbf{U}_s\mathbf{v} + \mathbf{v}\,\mathbf{U}_s - \mathbf{U}_s\mathbf{U}_s. \tag{5.4.11}$$

The integral in Eq. (5.4.10) then simplifies to

$$m_s \int_V \mathbf{vv} f_s\, \mathrm{d}^3 v = m_s \int_V (\mathbf{v} - \mathbf{U}_s)(\mathbf{v} - \mathbf{U}_s) f_s\, \mathrm{d}^3 v + m_s n_s \mathbf{U}_s\mathbf{U}_s. \tag{5.4.12}$$

The first term on the right in the above equation is just the pressure tensor, defined by Eq. (5.1.7). Equation (5.4.10) can then be written

$$m_s \int_V \mathbf{v}(\mathbf{v} \cdot \nabla f_s) \mathrm{d}^3 v = \nabla \cdot \overset{\leftrightarrow}{\mathbf{P}}_s + \nabla \cdot (m_s n_s \mathbf{U}_s\mathbf{U}_s). \tag{5.4.13}$$

The third integral on the left of Eq. (5.4.8) is of the form

$$\int_V \mathbf{v}(\mathbf{F} \cdot \nabla_{\mathbf{v}} f_s) \mathrm{d}^3 v = \int_V (v_x\hat{\mathbf{x}} + v_y\hat{\mathbf{y}} + v_z\hat{\mathbf{z}}) \left[\frac{\partial}{\partial v_x}(F_x f_s)\right.$$
$$\left. + \frac{\partial}{\partial v_y}(F_y f_s) + \frac{\partial}{\partial v_z}(F_z f_s)\right] \mathrm{d}^3 v, \tag{5.4.14}$$

where we have used the fact that for the Lorentz force

$$\mathbf{F} \cdot \nabla_{\mathbf{v}} f_s = \nabla_{\mathbf{v}} \cdot (\mathbf{F} f_s). \tag{5.4.15}$$

Of the nine integrals that must be computed, six are of the type

$$\int_V v_y \frac{\partial}{\partial v_z}(F_z f_s)\, \mathrm{d}^3 v = \int_{-\infty}^{\infty}\int_{-\infty}^{\infty} \mathrm{d}v_x\, \mathrm{d}v_y\, v_y \int_{-\infty}^{\infty} \frac{\partial}{\partial v_z}(F_z f_s)\, \mathrm{d}v_z, \tag{5.4.16}$$

which can be integrated directly to give

$$\int_{-\infty}^{\infty} \frac{\partial}{\partial v_z}(F_z f_s)\, \mathrm{d}v_z = [F_z f_s]_{-\infty}^{\infty}. \tag{5.4.17}$$

If we assume that f_s goes to zero sufficiently rapidly at infinite velocity, these integrals are all zero. The remaining three integrals are of the type

$$\int_{-\infty}^{\infty}\int_{-\infty}^{\infty} \mathrm{d}v_x\, \mathrm{d}v_y \int_{-\infty}^{\infty} \mathrm{d}v_z\, v_z \frac{\partial}{\partial v_z}(F_z f_s), \tag{5.4.18}$$

which, after integrating by parts once, gives

$$\int_{-\infty}^{\infty} \mathrm{d}v_z\, v_z \frac{\partial}{\partial v_z}(F_z f_s) = [v_z F_z f_s]_{-\infty}^{\infty} - \int_{-\infty}^{\infty} F_z f_s\, \mathrm{d}v_z. \tag{5.4.19}$$

If we again assume that f_s goes to zero sufficiently rapidly at infinite velocity, the first term on the right is zero. Combining all of these integrals, Eq. (5.4.14) can be written

$$\int_V \mathbf{v}(\mathbf{F}\cdot\nabla_{\mathrm{v}} f_s)\mathrm{d}^3v = -\int_V \mathbf{F} f_s\, \mathrm{d}^3v. \tag{5.4.20}$$

After substituting the Lorentz force for $\mathbf{F}$ and integrating, the above equation yields the simple result

$$\int_V \mathbf{v}(\mathbf{F}\cdot\nabla_{\mathrm{v}} f_s)\mathrm{d}^3v = -e_s n_s(\mathbf{E} + \mathbf{U}_s \times \mathbf{B}). \tag{5.4.21}$$

Finally, the first moment of the collision term on the right of Eq. (5.4.8) can be written

$$m_s \int_V \mathbf{v}\frac{\delta_{\mathrm{c}} f_s}{\delta t}\mathrm{d}^3v = \frac{\delta_{\mathrm{c}}\mathbf{p}_s}{\delta t}, \tag{5.4.22}$$

where $\delta_{\mathrm{c}}\mathbf{p}_s/\delta t$ is the average rate of change of the momentum per unit volume due to collisions. This term, which represents the drag force due to collisions, will be discussed in greater detail in Section 5.6 and again in Chapter 12.

Using Eqs. (5.4.9), (5.4.13), (5.4.21), and (5.4.22), Eq. (5.4.8) becomes

$$\frac{\partial}{\partial t}(m_s n_s \mathbf{U}_s) + \nabla\cdot(m_s n_s \mathbf{U}_s \mathbf{U}_s) = n_s e_s(\mathbf{E} + \mathbf{U}_s \times \mathbf{B}) - \nabla\cdot\overleftrightarrow{\mathbf{P}}_s + \frac{\delta_{\mathrm{c}}\mathbf{p}_s}{\delta t}. \tag{5.4.23}$$

The two terms on the left of this equation can be written in a more useful form by expanding the time derivative and using the identity $\nabla\cdot(\mathbf{AB}) = (\nabla\cdot\mathbf{A})\mathbf{B} + (\mathbf{A}\cdot\nabla)\mathbf{B}$, which gives

$$\begin{aligned}&\frac{\partial}{\partial t}(m_s n_s \mathbf{U}_s) + \nabla \cdot (m_s n_s \mathbf{U}_s \mathbf{U}_s)\\&= m_s n_s \frac{\partial \mathbf{U}_s}{\partial t} + m_s \mathbf{U}_s \left[\frac{\partial n_s}{\partial t} + \nabla \cdot (n_s \mathbf{U}_s)\right] + m_s n_s (\mathbf{U}_s \cdot \nabla)\mathbf{U}_s.\end{aligned} \tag{5.4.24}$$

From the continuity equation (5.4.6), the term in the brackets on the right side of the above equation is seen to be zero. The first moment of the Boltzmann equation then becomes

$$m_s n_s \left[\frac{\partial \mathbf{U}_s}{\partial t} + (\mathbf{U}_s \cdot \nabla)\mathbf{U}_s\right] = n_s e_s [\mathbf{E} + \mathbf{U}_s \times \mathbf{B}] - \nabla \cdot \overleftrightarrow{\mathbf{P}}_s + \frac{\delta_c \mathbf{p}_s}{\delta t}. \tag{5.4.25}$$

The quantity in the brackets on the left contains an operator that occurs frequently in fluid mechanics and is called the *convective derivative*. The term $\partial \mathbf{U}_s/\partial t$ gives the rate of change of the velocity due to explicit time variations, and the term $(\mathbf{U}_s \cdot \nabla)\mathbf{U}_s$ gives the rate of change of the velocity due to spatial variations. If we follow a given volume element moving at velocity $\mathbf{U}_s$, the convective derivative is simply the total derivative, $\mathrm{d}\mathbf{U}_s/\mathrm{d}t$, evaluated along the trajectory of the fluid element; that is,

$$\frac{\mathrm{d}\mathbf{U}_s}{\mathrm{d}t} = \frac{\partial \mathbf{U}_s}{\partial t} + (\mathbf{U}_s \cdot \nabla)\mathbf{U}_s. \tag{5.4.26}$$

Equation (5.4.25) can then be written in the form

$$m_s n_s \frac{\mathrm{d}\mathbf{U}_s}{\mathrm{d}t} = n_s e_s [\mathbf{E} + \mathbf{U}_s \times \mathbf{B}] - \nabla \cdot \overleftrightarrow{\mathbf{P}}_s + \frac{\delta_c \mathbf{p}_s}{\delta t}. \tag{5.4.27}$$

This equation is called the *momentum equation* and can be recognized as Newton's law for the sth species of the plasma. The term on the left is the rate of change of the momentum per unit volume evaluated at the volume element, as though it were a particle. The first term on the right is the electromagnetic force per unit volume, the second term is the force per unit volume due to pressure gradients, and the third term, $\delta_c \mathbf{p}_s/\delta t$, is the collisional drag force per unit volume.

5.4.3 Second Moment

The second moment is obtained by multiplying the Boltzmann equation by the tensor $\mathbf{vv}$ and integrating over all of velocity space. This procedure leads to a symmetric matrix with nine elements, only six of which are independent, since the off-diagonal elements of a symmetric matrix are equal to each other. Here we restrict our discussion to the trace of this matrix, which is obtained by multiplying the Boltzmann equation by the kinetic energy (1/2) $m_s v^2$ and integrating over velocity space, which gives

$$\frac{m_s}{2}\int_V v^2 \frac{\partial f_s}{\partial t}\,\mathrm{d}^3v + \frac{m_s}{2}\int_V v^2(\mathbf{v}\cdot\boldsymbol{\nabla} f_s)\mathrm{d}^3v$$
$$+\frac{1}{2}\int_V v^2(\mathbf{F}\cdot\boldsymbol{\nabla}_{\mathbf{v}} f_s)\mathrm{d}^3v = \frac{m_s}{2}\int_V v^2\frac{\delta_{\mathrm{c}} f_s}{\delta t}\mathrm{d}^3v. \quad (5.4.28)$$

The first integral on the left of the above equation can be written

$$\frac{m_s}{2}\int_V v^2 \frac{\partial f_s}{\partial t}\,\mathrm{d}^3v = \frac{\partial}{\partial t}\left(\int_V \frac{1}{2} m_s v^2 f_s\,\mathrm{d}^3v\right) = \frac{\partial W_s}{\partial t}, \quad (5.4.29)$$

where W_s is the kinetic energy density. The second integral can be written

$$\frac{m_s}{2}\int_V v^2(\mathbf{v}\cdot\boldsymbol{\nabla} f_s)\mathrm{d}^3v = \boldsymbol{\nabla}\cdot\left(\int_V \frac{1}{2} m_s \mathbf{v} v^2 f_s\,\mathrm{d}^3v\right) = \boldsymbol{\nabla}\cdot\mathbf{Q}_s, \quad (5.4.30)$$

where

$$\mathbf{Q}_s = \int_V \frac{1}{2} m_s v^2 \mathbf{v}\, f_s\,\mathrm{d}^3v \quad (5.4.31)$$

is the kinetic energy flux. The third integral can be simplified by integrating by parts which, after discarding the contribution at infinite velocity, gives

$$\frac{1}{2}\int_V v^2(\mathbf{F}\cdot\boldsymbol{\nabla}_{\mathbf{v}} f_s)\mathrm{d}^3v = -\frac{1}{2}\int_V f_s(\mathbf{F}\cdot\boldsymbol{\nabla}_{\mathbf{v}} v^2)\mathrm{d}^3v. \quad (5.4.32)$$

Introducing the Lorentz force and noting that $\boldsymbol{\nabla}_{\mathbf{v}} v^2 = 2\mathbf{v}$, and that $\mathbf{v}\cdot(\mathbf{v}\times\mathbf{B}) = 0$, the integral on the right side of the above equation can be written

$$\frac{1}{2}\int_V v^2(\mathbf{F}\cdot\boldsymbol{\nabla}_{\mathbf{v}} f_s)\,\mathrm{d}^3v = -e_s\mathbf{E}\cdot\int_V \mathbf{v} f_s\,\mathrm{d}^3v = -\mathbf{E}\cdot(n_s e_s \mathbf{U}_s) = -\mathbf{E}\cdot\mathbf{J}_s, \quad (5.4.33)$$

where $\mathbf{J}_s$ is the current contributed by the species s. Using Eqs. (5.4.29), (5.4.30), and (5.4.33), the second moment of the Boltzmann equation becomes

$$\frac{\partial W_s}{\partial t} + \boldsymbol{\nabla}\cdot\mathbf{Q}_s - \mathbf{E}\cdot\mathbf{J}_s = \int_V \frac{1}{2} m_s v^2 \frac{\delta_{\mathrm{c}} f_s}{\delta t}\mathrm{d}^3v. \quad (5.4.34)$$

This equation is called the *energy equation* because it is simply a statement of conservation of energy for the sth species. The term $\mathbf{E}\cdot\mathbf{J}_s$ is the contribution to the Joule heating by the sth species, and the collision term on the right gives the rate of transfer of energy to the sth species due to collisions with other species. Summing the above equation over all species then gives an energy conservation equation for the entire plasma:

$$\frac{\partial W}{\partial t} + \nabla \cdot \mathbf{Q} = \mathbf{E} \cdot \mathbf{J}, \tag{5.4.35}$$

where $W = \sum_s W_s$ is the total kinetic energy density, $\mathbf{Q} = \sum_s \mathbf{Q}_s$ is the total kinetic energy flux, and $\mathbf{E} \cdot \mathbf{J}$ is the Joule heating. In a fully ionized plasma, the collision terms sum to zero because for Coulomb collisions the effect of collisions is simply to transfer energy from one species to another without changing the overall energy.

5.4.4 The Closure Problem

It is evident by inspection of the first three moment equations that the nth moment equation always involves the $(n + 1)$st moment of the distribution function. For example, the zeroth moment of the Boltzmann equation involves the first moment $\mathbf{U}_s$, etc. This hierarchy occurs because the $\mathbf{v} \cdot \nabla f_s$ term in the Boltzmann equation always increases the order of the moments by one. Therefore, for any finite number of moment equations, the number of unknown moments always exceeds the number of equations. There are never enough equations to solve for all of the unknown moments. This is called the closure problem. To obtain a closed set of equations it is then necessary to specify the $(n + 1)$st moment in terms of the first n moments. For example, if only the continuity and momentum equations are used, a closed system of equations can be obtained by specifying an equation of state that relates the pressure tensor, $\overleftrightarrow{\mathbf{P}}_s$, to the number density, n_s. Unfortunately, it is impossible to do this in a completely rigorous manner for plasmas of arbitrary collisionality. Typically, an equation of state is assumed that is based on one of the well-known equations of state used in thermodynamics. This approach works well when there are sufficient collisions to maintain the plasma in a state of local thermodynamic equilibrium, but often fails in a collisionless plasma. Next, we discuss some of the commonly used equations of state.

The Cold Plasma Equation of State

In this simple approximation, the temperature is assumed to be so low that the pressure is negligible. The equation of state is simply $\overleftrightarrow{\mathbf{P}}_s = 0$ for any n_s. The momentum equation then becomes

$$m_s n_s \frac{\mathrm{d}\mathbf{U}_s}{\mathrm{d}t} = n_s e_s [\mathbf{E} + \mathbf{U}_s \times \mathbf{B}] + \frac{\delta_c \mathbf{p}_s}{\delta t}. \tag{5.4.36}$$

If the collisional drag term is assumed to be zero, the number density, n_s, can be factored out of this equation. The momentum equation then reduces to the equation of motion for a single particle,

$$m_s \frac{\mathrm{d}\mathbf{U}_s}{\mathrm{d}t} = e_s[\mathbf{E} + \mathbf{U}_s \times \mathbf{B}], \tag{5.4.37}$$

which is identical to the equation of motion (4.3.2) used in the analysis of waves in a cold plasma (Chapter 4). However, there is one subtle difference. In the above equation, $\mathbf{U}_s$ is a fluid velocity field which is a function of the independent coordinates $\mathbf{r}$ and t, whereas $\mathbf{v}_s$ in Eq. (4.3.2) is the velocity of a single particle which is tagged and followed as a function of t. In some cases, this distinction can be important because the convective derivative, $\mathrm{d}\mathbf{U}_s/\mathrm{d}t$, consists of two terms, i.e.,

$$\frac{\mathrm{d}\mathbf{U}_s}{\mathrm{d}t} = \frac{\partial \mathbf{U}_s}{\partial t} + (\mathbf{U}_s \cdot \nabla)\mathbf{U}_s. \tag{5.4.38}$$

In the case of small-amplitude waves in a fluid at rest (i.e., $\mathbf{U}_{s0} = 0$), the second term on the right can be ignored, since it involves the product of two first-order quantities, $(\mathbf{U}_{s1} \cdot \nabla)\mathbf{U}_{s1}$, which is small compared with the first-order quantity $\partial \mathbf{U}_{s1}/\partial t$. We then have $\mathrm{d}\mathbf{U}_{s1}/\mathrm{d}t \simeq \partial \mathbf{U}_{s1}/\partial t$. The operator substitution $\mathrm{d}/\mathrm{d}t \rightarrow (-\mathrm{i}\omega)$ is then a good first-order approximation. On the other hand, if the fluid is moving, i.e., $\mathbf{U}_{s0} \neq 0$, then the correct linearized form for the operator $\mathrm{d}\mathbf{U}_{s1}/\mathrm{d}t$ is

$$\frac{\mathrm{d}\mathbf{U}_{s1}}{\mathrm{d}t} = \frac{\partial \mathbf{U}_{s1}}{\partial t} + (\mathbf{U}_{s0} \cdot \nabla)\mathbf{U}_{s1}. \tag{5.4.39}$$

After Fourier transforming, this equation becomes

$$(-\mathrm{i}\omega')\tilde{\mathbf{U}}_{s1} = (-\mathrm{i}\omega + \mathrm{i}\mathbf{k} \cdot \mathbf{U}_{s0})\tilde{\mathbf{U}}_{s1}, \tag{5.4.40}$$

where ω' is the frequency in a frame of reference moving with the particle. From the above equation it follows that the frequency in the particle frame of reference is related to the rest frame frequency by the equation $\omega' = \omega - \mathbf{k} \cdot \mathbf{U}_{s0}$. The term $\mathbf{k} \cdot \mathbf{U}_{s0}$ is the Doppler shift introduced by the change in the frame of reference. In general, the correct linearized operator substitution for $\mathrm{d}/\mathrm{d}t$ is $(-\mathrm{i}\omega + \mathrm{i}\mathbf{k} \cdot \mathbf{U}_{s0})$, and not $(-\mathrm{i}\omega)$.

The Adiabatic Equation of State

The concept of an adiabatic equation of state is borrowed from thermodynamics. Using the ideal gas law, $PV = N\kappa T$, and the first law of thermodynamics, $\mathrm{d}Q = \mathrm{d}U + P\,\mathrm{d}V$, it can be shown that if no heat flows, then the pressure, P, and volume, V, are given by $PV^{\gamma} = \text{constant}$, where $\gamma = C_\mathrm{P}/C_\mathrm{V}$ is the ratio of the heat capacity at constant pressure, C_P, to the heat capacity at constant volume, C_V; see Halliday and Resnick (1978). The adiabatic equation of state applies to situations where the gas is compressed so rapidly that there is not enough time for heat to flow.

The compression of a gas by a sound wave or by a rapidly moving piston are typical examples where the adiabatic equation of state can be used. Although the compression must occur sufficiently rapidly that the heat flow can be neglected, it must not occur so rapidly that collisions cannot maintain the gas in a state of local thermodynamic equilibrium. For example, the adiabatic equation of state cannot be used to describe the free expansion of a gas into an evacuated container.

Since local thermodynamic equilibrium is maintained by collisions, the question arises as to whether an adiabatic equation of state can be applied to a collisionless plasma. The obvious answer is no. However, microscopic wave–particle interactions sometimes play a role similar to collisions in an ordinary gas. The adiabatic equation of state may then provide an adequate approximation, even in a collisionless plasma.

When the adiabatic equation of state is used, it is usually assumed that collisions (or wave–particle interactions) maintain an isotropic velocity distribution. Under these conditions, the pressure tensor becomes diagonal, with equal diagonal elements, i.e.,

$$\overleftrightarrow{\mathbf{P}}_s = \overleftrightarrow{\mathbf{1}} P_s, \tag{5.4.41}$$

where $\overleftrightarrow{\mathbf{1}}$ is the unit tensor. Since we normally use the number density, n_s, rather than the volume, V, as the independent variable, it can be seen that the adiabatic equation of state can be written in the following alternative form:

$$\frac{\mathrm{d}}{\mathrm{d}t}(P_s/n_s^{\gamma}) = 0, \tag{5.4.42}$$

where the pressure obeys the ideal gas law, $P_s = n_s \kappa T_s$. From statistical mechanics it is well known that γ is determined by the number of degrees of freedom, f, and is given by

$$\gamma = \frac{f+2}{f}. \tag{5.4.43}$$

Thus, for a gas with three degrees of freedom ($f = 3$), it follows that $\gamma = 5/3$.

If the collision rate is too low to maintain an isotropic distribution then the number of degrees of freedom is controlled by the geometry of the compression process. Later, we will show that in a collisionless plasma certain types of longitudinal waves behave as if the compression is one-dimensional, in which case $\gamma = 3$.

The power law index γ is also sometimes considered to be an adjustable parameter that can be selected to represent non-adiabatic equations of state. For example, $\gamma = 1$ represents an isothermal equation of state, and $\gamma = \infty$ represents an incompressible equation of state. In the fluid dynamics and astrophysics literature, the parameter γ is sometimes called the polytrope index (Fetter and Walecka, 1980).

The Chew–Goldberger–Low Equation of State

If the collision rate in a magnetized plasma is too small to transfer momentum effectively between the parallel and perpendicular directions, an isotropic velocity distribution cannot be maintained. To account for such anisotropic effects, Chew, Goldberger, and Low (1956), developed a modified adiabatic equation of state in which the pressures parallel and perpendicular to the magnetic field are not the same, known as the Chew–Goldberger–Low (CGL) equation of state. They showed that in a coordinate system in which the magnetic field is parallel to the z axis, the pressure tensor is diagonal and can be written in the form

$$\overleftrightarrow{\mathbf{P}}_s = \begin{bmatrix} P_{s\perp} & 0 & 0 \\ 0 & P_{s\perp} & 0 \\ 0 & 0 & P_{s\|} \end{bmatrix}, \tag{5.4.44}$$

where the perpendicular and parallel pressures, $P_{s\perp}$ and $P_{s\|}$, are related to the magnetic field and number density by the equations

$$\frac{\mathrm{d}}{\mathrm{d}t}\left(\frac{P_{s\perp}}{n_s B}\right) = 0 \quad \text{and} \quad \frac{\mathrm{d}}{\mathrm{d}t}\left(\frac{P_{s\|}B^2}{n_s^3}\right) = 0. \tag{5.4.45}$$

The bi-Maxwellian velocity distribution given by Eq. (5.1.16) has a pressure tensor of the type described by Eq. (5.4.44).

The CGL equations of state (5.4.45) can be derived heuristically by making use of the first and second adiabatic invariants discussed in Chapter 3. In a coordinate system at rest with respect to the plasma, the perpendicular pressure is given by the product of the number density and the average perpendicular kinetic energy, i.e., $P_{s\perp} = n_s\langle w_{s\perp}\rangle$. It then follows, from the first adiabatic invariant and the relation $w_{s\perp} = \mu_s B$, that $P_{s\perp}/n_s B = \langle \mu_s \rangle$ is constant, which corresponds to the first equation in (5.4.45). The second equation can be derived similarly by making use of the second adiabatic invariant. The parallel pressure is given by the product of the number density and the average parallel kinetic energy, i.e., $P_{s\|} = n_s\langle w_{s\|}\rangle$. Writing the second adiabatic invariant in the form $v_\| L =$ constant, where L is the distance between mirror points, it follows that $P_{s\|}$ is proportional to n_s/L^2. The distance L between mirror points can next be eliminated by considering a volume with cross-sectional area A, perpendicular to the magnetic field, and length L, along the magnetic field. Conservation of mass implies $n_s AL =$ constant, and conservation of magnetic flux implies $BA =$ constant. Making use of these simple algebraic relations gives $P_{s\|}B^2/n_s^3 =$ constant, which is the second equation in (5.4.45). Since the adiabatic invariants are constant only if the magnetic field varies slowly on a time-scale compared to the parallel bounce period, the CGL equations of state can only be expected to be valid for time-scales that are long compared to the bounce period.

5.5 Electron and Ion Pressure Waves

To illustrate the use of the adiabatic equation of state, we next consider the propagation of small-amplitude waves in a plasma with a finite pressure. Since the temporal fluctuations associated with a wave are relatively rapid, it is reasonable to assume that an adiabatic equation of state is valid. If we restrict our analysis to a plasma without a magnetic field, then the adiabatic equation of state can be written as a diagonal pressure tensor of the form

$$\overleftrightarrow{\mathbf{P}}_s = P_{s0}\left(\frac{n_s}{n_{s0}}\right)^{\gamma}\overleftrightarrow{\mathbf{1}}, \tag{5.5.1}$$

where P_{s0} and n_{s0} are the undisturbed pressure and density. For a small-amplitude wave, we assume, as in Chapter 4, that these quantities consist of a constant spatially uniform zero-order term plus a small first-order perturbation. The relevant equations are then

$$P_s = P_{s0} + P_{s1}, \tag{5.5.2}$$

$$n_s = n_{s0} + n_{s1}, \tag{5.5.3}$$

$$\mathbf{U}_s = \mathbf{U}_{s1} \qquad (\text{assumes } \mathbf{U}_{s0} = 0), \tag{5.5.4}$$

and

$$\mathbf{E} = \mathbf{E}_1 \qquad (\text{assumes } \mathbf{E}_0 = 0), \tag{5.5.5}$$

where the zero-order velocity and electric field are assumed to be zero.

The equations that must be solved to determine the fluid motions are the continuity equation, the momentum equation, and the adiabatic equation of state:

$$\frac{\partial n_s}{\partial t} + \boldsymbol{\nabla}\cdot(n_s\mathbf{U}_s) = 0, \tag{5.5.6}$$

$$m_s n_s\left[\frac{\partial \mathbf{U}_s}{\partial t} + (\mathbf{U}_s\cdot\boldsymbol{\nabla})\mathbf{U}_s\right] = n_s e_s\mathbf{E} - \boldsymbol{\nabla}P_s, \tag{5.5.7}$$

and

$$P_s = P_{s0}\left(\frac{n_s}{n_{s0}}\right)^{\gamma}. \tag{5.5.8}$$

Linearizing and Fourier transforming Eqs. (5.5.6)–(5.5.8), one obtains

$$(-\mathrm{i}\omega)\tilde{n}_s + n_{s0}\mathrm{i}\mathbf{k}\cdot\tilde{\mathbf{U}}_s = 0, \tag{5.5.9}$$

$$m_s n_{s0}(-\mathrm{i}\omega)\tilde{\mathbf{U}}_s = n_{s0}e_s\tilde{\mathbf{E}} - \mathrm{i}\mathbf{k}\tilde{P}_s, \tag{5.5.10}$$

and

$$\tilde{P}_s = m_s\gamma_s C_s^2\tilde{n}_s, \tag{5.5.11}$$

where for notational convenience we have dropped the subscript 1 on the first-order terms and introduced the thermal speed, C_s, of the sth species via the definition $C_s^2 = P_{s0}/m_s n_{s0}$. Note that since $P_{s0} = n_{s0}\kappa T_{s0}$, it follows that $C_s^2 = \kappa T_{s0}/m_s$, which is consistent with the thermal speed defined by Eq. (2.1.3). Using Eqs. (5.5.9) and (5.5.11), we can eliminate the first-order pressure from Eq. (5.5.10) to obtain

$$m_s(-\mathrm{i}\omega)\tilde{\mathbf{U}}_s + m_s\gamma_s C_s^2\frac{\mathrm{i}\mathbf{k}}{\omega}(\mathbf{k}\cdot\tilde{\mathbf{U}}_s) = e_s\tilde{\mathbf{E}}. \tag{5.5.12}$$

It is useful to rewrite the above equation in matrix form:

$$\left(\overset{\leftrightarrow}{\mathbf{1}} - \gamma_s C_s^2\frac{\mathbf{kk}}{\omega^2}\right)\cdot\tilde{\mathbf{U}}_s = \frac{e_s}{(-\mathrm{i}\omega)m_s}\tilde{\mathbf{E}}. \tag{5.5.13}$$

Since there is no preferred direction in this isotropic plasma, without loss of generality we can choose the **k** vector to be aligned with the z axis. The above equation can then be written

$$\begin{bmatrix} 1 & 0 & 0 \\ 0 & 1 & 0 \\ 0 & 0 & 1-\gamma_s C_s^2\dfrac{k^2}{\omega^2} \end{bmatrix}\begin{bmatrix} \tilde{U}_{sx} \\ \tilde{U}_{sy} \\ \tilde{U}_{sz} \end{bmatrix} = \frac{e_s}{(-\mathrm{i}\omega)m_s}\begin{bmatrix} \tilde{E}_x \\ \tilde{E}_y \\ \tilde{E}_z \end{bmatrix}, \tag{5.5.14}$$

which can be solved for $\tilde{\mathbf{U}}_s$ to give

$$\begin{bmatrix} \tilde{U}_{sx} \\ \tilde{U}_{sy} \\ \tilde{U}_{sz} \end{bmatrix} = \frac{e_s}{(-\mathrm{i}\omega)m_s}\begin{bmatrix} 1 & 0 & 0 \\ 0 & 1 & 0 \\ 0 & 0 & \dfrac{1}{1-\gamma_s C_s^2(k^2/\omega^2)} \end{bmatrix}\begin{bmatrix} \tilde{E}_x \\ \tilde{E}_y \\ \tilde{E}_z \end{bmatrix}. \tag{5.5.15}$$

The conductivity tensor, defined by Eq. (4.2.3), can then be obtained by computing the current density $\tilde{\mathbf{J}} = \sum_s n_{s0}e_s\tilde{\mathbf{U}}_s$. Once the conductivity tensor has been determined, it is relatively simple to compute the dielectric tensor (4.2.6) and write down the homogeneous equation (4.2.8) for the electric field, which in matrix form is given by

$$\begin{bmatrix} -c^2k^2+\omega^2-\omega_{\rm p}^2 & 0 & 0 \\ 0 & -c^2k^2+\omega^2-\omega_{\rm p}^2 & 0 \\ 0 & 0 & \omega^2-\sum_s \dfrac{\omega_{{\rm p}s}^2}{1-\gamma_s C_s^2(k^2/\omega^2)} \end{bmatrix} \begin{bmatrix} \widetilde{E}_x \\ \widetilde{E}_y \\ \widetilde{E}_z \end{bmatrix} = 0. \tag{5.5.16}$$

The dispersion relation is then obtained by computing the determinant of the matrix, which is

$$Ð(k,\omega) = \left(-c^2k^2+\omega^2-\omega_{\rm p}^2\right)^2 \left(\omega^2 - \sum_s \frac{\omega_{{\rm p}s}^2}{1-\gamma_s C_s^2(k^2/\omega^2)}\right) = 0. \tag{5.5.17}$$

It is evident from the above equation that there are two types of roots. The roots associated with the first term in parentheses correspond to transverse electromagnetic waves. The roots associated with the term in the large brackets correspond to longitudinal electrostatic waves. It is interesting to note that the transverse electromagnetic mode is not affected by the pressure. This occurs because the electric field, $\tilde{\mathbf{E}} = (\widetilde{E}_x, \widetilde{E}_y, 0)$, and particle velocity, $\tilde{\mathbf{U}}_s = (\widetilde{U}_{sx}, \widetilde{U}_{sy}, 0)$, for the transverse electromagnetic mode are perpendicular to $\mathbf{k}$. From the continuity equation (5.5.9), it follows that the first-order number density, $\tilde{n}_s$, is zero. Since there is no density perturbation, there is no pressure perturbation associated with this mode. On the other hand, the longitudinal electrostatic mode is strongly affected by the pressure. For the longitudinal mode $\tilde{\mathbf{U}}_s$ and $\mathbf{k}$ are parallel, so there is a non-zero fluctuation in the number density as well as the pressure. By factoring out ω^2, the dispersion relation for the longitudinal mode can be written in the form

$$Ð_\ell(k,\omega) = 1 - \sum_s \frac{\omega_{{\rm p}s}^2}{\omega^2 - \gamma_s C_s^2 k^2} = 0. \tag{5.5.18}$$

A plot of the function $Ð_\ell(k,\omega)$ is shown in Figure 5.4.

From this plot, it can be seen that the dispersion equation has one root associated with each species in the plasma. The root associated with the electrons is a modification of the electron plasma oscillation found in cold plasma theory and is often called the Langmuir mode, after Tonks and Langmuir (1929) who first studied this mode of propagation. As shown below, the electron pressure produces a shift in the frequency of the electron plasma oscillations. The roots associated with the ions have no analog in cold plasma theory and are called the ion acoustic modes. The properties of each of these modes are discussed below.

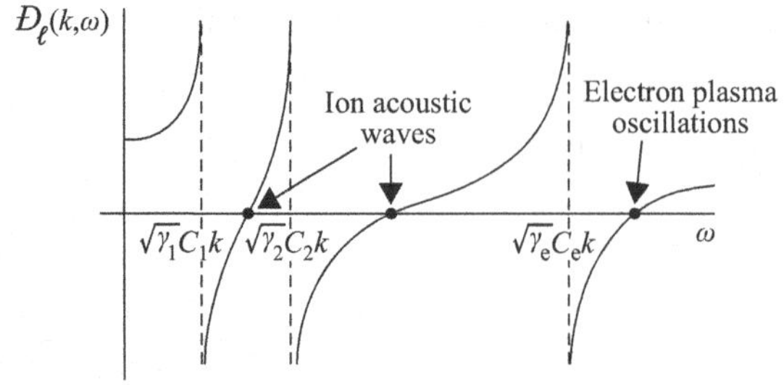

Figure 5.4 A plot of $Đ_\ell(k, \omega)$ as a function of frequency for the longitudinal mode.

5.5.1 The Langmuir Mode

Since the electron plasma frequency is much greater than the characteristic frequencies of the ions, for the Langmuir mode we can essentially ignore the motion of the ions. This is achieved by omitting the ion terms in the general dispersion relation, Eq. (5.5.18). The Langmuir wave dispersion relation then becomes

$$1 - \frac{\omega_{pe}^2}{\omega^2 - \gamma_e C_e^2 k^2} = 0, \tag{5.5.19}$$

which simplifies to

$$\omega^2 = \omega_{pe}^2 + \gamma_e C_e^2 k^2. \tag{5.5.20}$$

This equation is called the Bohm–Gross dispersion relation, after Bohm and Gross (1949), who first derived it. If the electron temperature is zero, so that $C_e = 0$, the oscillation frequency reduces to the cold plasma result, $\omega^2 = \omega_{pe}^2$. It can be shown that the term $\gamma_e C_e^2 k^2$ comes from the electron pressure gradient term in the momentum equation (5.5.7). For long wavelengths ($k \simeq 0$), the pressure gradient force is small compared with the electric field force. The oscillation frequency is then identical to the cold plasma result, i.e., $\omega^2 = \omega_{pe}^2$. As the wavelength decreases (and k increases), the pressure gradient force increases and eventually becomes larger than the electric field force. The pressure gradient force then causes the oscillation frequency to shift upwards relative to the cold plasma result.

Using the relation $\omega_{pe}\lambda_{De} = C_e$, we can rewrite the Bohm–Gross dispersion relation (5.5.20) in the form

$$\omega^2 = \omega_{pe}^2 \left[1 + \gamma_e \lambda_{De}^2 k^2\right]. \tag{5.5.21}$$

The Debye length is now seen to be a basic length-scale in the dispersion relation. The variation of the oscillation frequency with wave number is shown in Figure 5.5. As the wavelength approaches the Debye length, i.e., $k \approx 1/\lambda_{De}$,

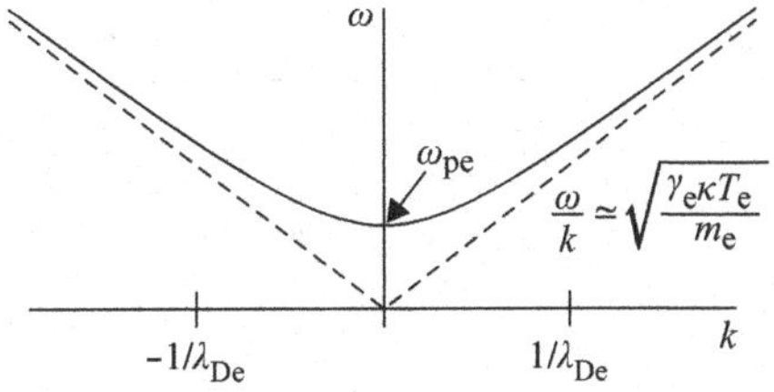

Figure 5.5 The dispersion relation for electron plasma oscillations, including the effects of electron pressure.

the oscillation frequency asymptotically approaches the dashed lines, which correspond to a phase velocity given by

$$\frac{\omega}{k} = \pm\sqrt{\frac{\gamma_e \kappa T_e}{m_e}}. \tag{5.5.22}$$

This phase velocity is formally identical to the phase velocity of an acoustic wave in an ordinary gas, except the temperature is the electron temperature and the mass is the electron mass. Because of this close similarity, Langmuir waves are sometimes referred to as electron acoustic waves. However, this comparison is somewhat misleading. Later (in Chapter 9) we show that Langmuir waves are very strongly damped when $k\lambda_{De} \gtrsim 1$, due to a collisionless damping process called Landau damping. The failure of the moment equations to reveal this damping at large wave numbers is one of the shortcomings of the moment equation approach. The moment equations give a good description for $k\lambda_{De} \ll 1$, but fail when $k\lambda_{De} \gtrsim 1$.

5.5.2 The Ion Acoustic Mode

To investigate the roots of the dispersion relation associated with the ions, we start by making the simplifying assumption that the ion temperature is small (i.e., $C_i = 0$), which we shall later justify. The dispersion relation (5.5.18) can then be written

$$D_\ell(k,\omega) = 1 - \frac{\omega_{pi}^2}{\omega^2} - \frac{\omega_{pe}^2}{\omega^2 - \gamma_e C_e^2 k^2} = 0. \tag{5.5.23}$$

For the ion roots, we expect that the phase velocity will be much less than the electron thermal velocity, so we assume that $\omega^2 \ll \gamma_e C_e^2 k^2$. The dispersion relation (5.5.23) can then be solved for ω^2 to yield

$$\omega^2 = \frac{1}{1+\gamma_e \lambda_{De}^2 k^2}\left(\frac{\gamma_e \kappa T_e}{m_i}\right)k^2, \tag{5.5.24}$$

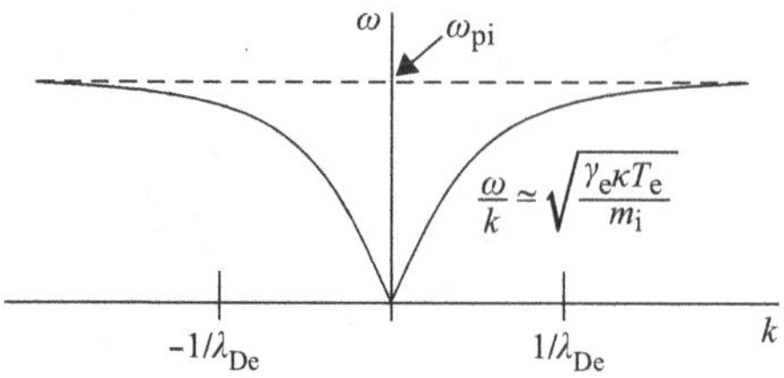

Figure 5.6 The dispersion relation for the ion acoustic mode.

where we have again introduced $\lambda_{\rm De}$ by means of the relation $\omega_{\rm pe}\lambda_{\rm De} = C_{\rm e}$, and substituted $m_{\rm e}/m_{\rm i}$ for $\omega_{\rm pi}^2/\omega_{\rm pe}^2$. This mode of propagation is called the ion acoustic mode. Figure 5.6 shows a plot of the frequency, ω, of the ion acoustic mode as a function of the wave number, k. For long wavelengths $k \ll 1/\lambda_{\rm De}$, the frequency ω varies linearly with k. In this region, $k\lambda_{\rm De} \ll 1$, the phase velocity of the wave is essentially independent of wavelength and is given by

$$\frac{\omega}{k} = \pm\sqrt{\frac{\gamma_{\rm e}\kappa T_{\rm e}}{m_{\rm i}}}. \tag{5.5.25}$$

Note that the phase velocity involves the *electron* temperature and the *ion* mass. This combination of parameters is interpreted as indicating that the electrons provide the pressure while the ions provide the inertia. As the wave number increases, the frequency begins to decrease relative to the linear relation given by Eq. (5.5.25) and asymptotically approaches the ion plasma frequency, $\omega_{\rm pi}$, for $k\lambda_{\rm De} \gg 1$. It can be verified that the phase velocity of the ion acoustic mode is always much less than the electron thermal velocity if the ion temperature is zero. Therefore, the initial assumption that $\omega^2 \ll \gamma_{\rm e}C_{\rm e}^2k^2$ is always satisfied.

It is instructive to explore the electron and ion motions for the Langmuir mode and for the ion acoustic mode. From the momentum equation (5.5.10), it can be shown that the velocities of the electrons and ions are given, respectively, by

$$\widetilde{U}_{\rm e} = \left(\frac{-e}{-{\rm i}\omega m_{\rm e}}\right)\frac{\widetilde{E}}{1-\gamma_{\rm e}C_{\rm e}^2(k^2/\omega^2)} \tag{5.5.26}$$

and

$$\widetilde{U}_{\rm i} = \left(\frac{+e}{-{\rm i}\omega m_{\rm i}}\right)\widetilde{E}, \tag{5.5.27}$$

where we have again assumed that $C_{\rm i} = 0$ in the ion equation. Taking the ratio of the electron-to-ion velocities, we obtain

$$\frac{\widetilde{U}_{\rm e}}{\widetilde{U}_{\rm i}} = -\left(\frac{m_{\rm i}}{m_{\rm e}}\right)\frac{1}{1-\gamma_{\rm e}C_{\rm e}^2(k^2/\omega^2)}. \tag{5.5.28}$$

For the Langmuir mode it can be verified using Eq. (5.5.21) that in the long-wavelength limit the term $\gamma_e C_e^2 k^2/\omega^2$ is much less than one. It follows then that $\widetilde{U}_e/\widetilde{U}_i \simeq -m_i/m_e$. The minus sign indicates that the electrons and ions move in opposite directions. Also, since the ion mass is much greater than the electron mass, the amplitude of the electron motion is much greater than the amplitude of the ion motion. The ions are essentially motionless. In contrast, for the ion acoustic mode it can be shown that in the long-wavelength limit the term $\gamma_e C_e^2 k^2/\omega^2$ is approximately m_i/m_e. It follows that $\widetilde{U}_e/\widetilde{U}_i \simeq 1 + m_e/m_i$. The electrons and ions then move in the same direction at nearly the same velocity, causing a fluid-like motion that is very similar to an acoustic wave in an ordinary gas. The small difference in the electron and ion velocities caused by the m_e/m_i term produces the charge separation that is responsible for the electric field of the wave. It is this electric field that couples the electron and ion motions.

If the electron and ion temperatures are equal ($T_e = T_i$) it can be shown that the phase velocity of the ion acoustic mode is comparable to the ion thermal speed. The ion acoustic mode then suffers heavy Landau damping due to absorption of wave energy by the ions, very similar to the situation for electron plasma oscillations when $k\lambda_{De} \gtrsim 1$. Again, this damping is not predicted by the moment equations. If $T_e \gg T_i$, Eq. (5.5.24) shows that the phase velocity of the ion acoustic wave is much larger than the ion thermal speed. Later, in Chapter 9, we show that it is only when $T_e \gg T_i$ that the ion acoustic wave can propagate with relatively little damping. This justifies our earlier assumption that the ion temperature must be small (i.e., $C_i = 0$).

5.6 Collisional Drag Force

So far we have ignored the effect of collisions. The effect of collisions enters into the momentum equations (5.4.27) via the collisional drag force, $\delta_c \mathbf{p}_s/\delta t$. For a collisionless plasma this term is often assumed to be zero. If collisions cannot be neglected, then an equation is needed for the collisional drag force. Collisional effects are quite complicated and a full discussion is not presented until Chapter 12. However, for some applications a simple model called the Lorentz gas model can be used. This model is discussed below.

5.6.1 The Lorentz Gas Model

In the Lorentz gas model it is assumed that the plasma particles are scattered by fixed immobile scattering centers. If, in addition, we assume that the scattering is isotropic, the average momentum change per collision is simply $-m\mathbf{v}$, since the average momentum after a collision is zero for isotropic scattering. The average drag force per unit volume for a particle of type s is then given by

$$\frac{\delta_c \mathbf{p}_s}{\delta t} = \nu_s \int_V (-m_s \mathbf{v}) f_s \mathrm{d}^3 v \tag{5.6.1}$$

or

$$\frac{\delta_c \mathbf{p}_s}{\delta t} = -\nu_s m_s n_s \mathbf{U}_s, \tag{5.6.2}$$

where ν_s is an effective collision frequency. This equation shows that the collisional drag force, $\delta_c \mathbf{p}_s/\delta t$, is directly proportional to the average momentum density, $m_s n_s \mathbf{U}_s$, with the collision frequency, ν_s, acting as the constant of proportionality.

The Lorentz gas model is particularly well suited for electron scattering by neutral particles. Neutral particles essentially act as hard-sphere scattering centers, thereby causing nearly isotropic scattering. Since the electron mass is much less than the mass of the neutral particles, very little momentum is transferred to the neutral particle, so the assumption of fixed immobile scattering centers is valid. For ion-neutral collisions the isotropic scattering approximation is not as good, since an appreciable amount of momentum is transferred to the neutral atom during the collision, thereby introducing a non-isotropic component to the scattering. Nevertheless, the Lorentz gas model is still sometimes used for ion-neutral collisions. Even though the scattering is not completely isotropic, it can be shown that the drag force is, to a first approximation, proportional to the relative velocity between the ions and the neutral atoms. For scattering by charged particles, the isotropic assumption is not valid since charged-particle scattering is highly anisotropic; see Section 2.5.2. For a more detailed discussion of collisions between charged particles, see Chapter 12.

5.6.2 Ohm's Law

When a steady electric field is applied to a plasma, the electrons and ions are accelerated in opposite directions by the electric field force, thereby producing a current. In the absence of collisions the acceleration would continue indefinitely. It would then not be possible to define a conductivity, since a steady-state current is never achieved. However, if a collisional drag force is present, an equilibrium velocity is eventually reached at which the drag force balances the electric field force. If the equilibrium velocity is proportional to the electric field, then a conductivity, σ, can be defined by means of Ohm's law, $\mathbf{J} = \sigma \mathbf{E}$. To evaluate this conductivity, consider a homogeneous plasma with no magnetic field. The time required to reach equilibrium can be determined by assuming that a uniform electric field, $\mathbf{E}_0$, is turned on at $t = 0$. Using Eq. (5.6.2) for the collisional drag force, the momentum equation then becomes

$$m_s n_s \frac{\partial \mathbf{U}_s}{\partial t} = e_s n_s \mathbf{E}_0 - \nu_s m_s n_s \mathbf{U}_s, \tag{5.6.3}$$

which simplifies to

$$\frac{\partial \mathbf{U}_s}{\partial t} + \nu_s \mathbf{U}_s = \frac{e_s}{m_s} \mathbf{E}_0. \tag{5.6.4}$$

This linear differential equation is easily solved for the drift velocity $\mathbf{U}_s$. The solution is

$$\mathbf{U}_s = \frac{e_s \mathbf{E}_0}{m_s \nu_s} \left(1 - e^{-\nu_s t}\right). \tag{5.6.5}$$

The drift velocity starts at zero and exponentially approaches a steady-state equilibrium on a time-scale of $\tau = 1/\nu_s$. The initial phase of the transient response is controlled by inertia and the asymptotic final phase is controlled by the collisional drag force. For times $t \gg \tau$ one can see from Eq. (5.6.5) that the steady-state drift velocity is given by $\mathbf{U}_{s0} = e_s \mathbf{E}_0 / m_s \nu_s$. The resulting current is then given by

$$\mathbf{J}_s = n_s e_s \mathbf{U}_{s0} = \left(\frac{n_s e_s^2}{m_s \nu_s}\right) \mathbf{E}_0, \tag{5.6.6}$$

where the quantity in parentheses is the conductivity due to the sth species. The total conductivity is then the sum of the conductivities of each species:

$$\sigma = \sum_s \sigma_s, \tag{5.6.7}$$

where

$$\sigma_s = \frac{n_s e_s^2}{m_s \nu_s}. \tag{5.6.8}$$

Because the electron mass is much smaller than the ion mass, electrons usually provide the main contribution to the conductivity, so $\sigma \simeq \sigma_{\mathrm{e}}$.

An Ohm's law conductivity, σ, such as that given by Eq. (5.6.8), arises whenever the drift velocity is linearly proportional to the electric field. Note that in contrast to the conductivity encountered in the analysis of waves in a cold plasma, which was always imaginary, the conductivity caused by collisions is real, which implies that energy is being dissipated. The energy dissipation occurs because the ordered motion produced by the applied electric field is converted into random thermal motion (i.e., heat) by collisions. Because of their small mass, electrons are heated more rapidly than ions during such collisional interactions. For fusion plasmas this is unfortunate, since it is the ions that must be heated to a high temperature in order to initiate a fusion reaction. Collisional or "Ohmic" heating is a relatively ineffective method of heating a fusion plasma.

5.6.3 Pedersen and Hall Conductivities

If a magnetic field is present, the conductivity is modified because of the $e_s\mathbf{U}_s \times \mathbf{B}$ force. For a homogeneous plasma with a magnetic field $\mathbf{B}_0$ and an electric field $\mathbf{E}_0$, the momentum equation in steady state is given by

$$0 = n_s e_s[\mathbf{E}_0 + \mathbf{U}_s \times \mathbf{B}_0] - \nu_s m_s n_s \mathbf{U}_s. \tag{5.6.9}$$

Assuming that the magnetic field is parallel to the z axis, the above equation can be written in matrix form as follows:

$$\begin{bmatrix} 1 & -\frac{\omega_{cs}}{\nu_s} & 0 \\ \frac{\omega_{cs}}{\nu_s} & 1 & 0 \\ 0 & 0 & 1 \end{bmatrix} \begin{bmatrix} U_{sx} \\ U_{sy} \\ U_{sz} \end{bmatrix} = \frac{e_s}{m_s \nu_s} \begin{bmatrix} E_{0x} \\ E_{0y} \\ E_{0z} \end{bmatrix}, \tag{5.6.10}$$

where $\omega_{cs} = e_s B_0 / m_s$. This equation can be inverted to give

$$\begin{bmatrix} U_{sx} \\ U_{sy} \\ U_{sz} \end{bmatrix} = \frac{e_s}{m_s \nu_s \left(1 + \omega_c s^2/\nu_s^2\right)} \begin{bmatrix} 1 & \frac{\omega_{cs}}{\nu_s} & 0 \\ -\frac{\omega_{cs}}{\nu_s} & 1 & 0 \\ 0 & 0 & 1 + (\omega_c s^2/\nu_s^2) \end{bmatrix} \begin{bmatrix} E_{0x} \\ E_{0y} \\ E_{0z} \end{bmatrix}. \tag{5.6.11}$$

By computing the current using $\mathbf{J} = \sum_s n_s e_s \mathbf{U}_s$ and expressing the current in the form $\mathbf{J} = \overleftrightarrow{\sigma} \cdot \mathbf{E}$, the conductivity tensor can be identified and is given by

$$\overleftrightarrow{\sigma} = \begin{bmatrix} \sigma_\perp & \sigma_{\mathrm{H}} & 0 \\ -\sigma_{\mathrm{H}} & \sigma_\perp & 0 \\ 0 & 0 & \sigma_\parallel \end{bmatrix}, \tag{5.6.12}$$

where

$$\sigma_\perp = \sum_s \frac{\sigma_s}{\left(1 + \omega_{cs}^2/\nu_s^2\right)}, \quad \sigma_{\mathrm{H}} = \sum_s \frac{\sigma_s(\omega_{cs}/\nu_s)}{\left(1 + \omega_{cs}^2/\nu_s^2\right)}, \tag{5.6.13}$$

and $\sigma_\parallel = \sum_s \sigma_s$.

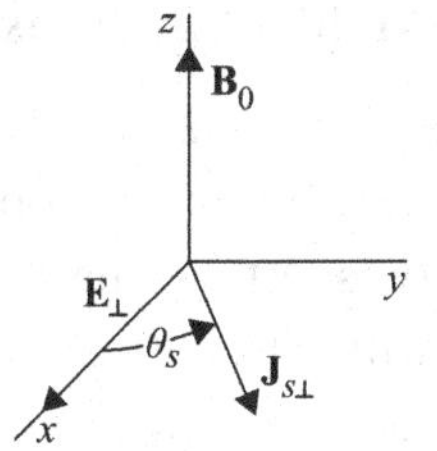

Figure 5.7 In a magnetized plasma, collisions cause the current to flow at an angle to the electric field.

The conductivity tensor elements $\sigma_\parallel$ and $\sigma_\perp$ are called the parallel and perpendicular (Pedersen) conductivities, respectively, and the element $\sigma_{\rm H}$ is called the Hall conductivity. For currents flowing parallel to the magnetic field, the magnetic field has no effect on the conductivity since the $e_s\mathbf{U}_s \times \mathbf{B}$ force is zero. Therefore, the parallel conductivity is the same as in a plasma with no magnetic field. As can be seen, the perpendicular conductivity gives the conductivity perpendicular to the magnetic field. The magnetic field has a strong effect on the perpendicular conductivity, particularly when $\omega_{cs} \gg \nu_s$. The Hall conductivity is the off-diagonal element in the conductivity tensor. For an electric field applied perpendicular to the magnetic field, the Hall conductivity causes a current to flow perpendicular to both $\mathbf{E}_\perp$ and $\mathbf{B}_0$. This component of the current is called the Hall current. The Hall current is caused by the $e_s\mathbf{U}_s \times \mathbf{B}$ force in the momentum equation, which is in the opposite direction for positive and negative charges, thereby causing a current to flow perpendicular to the applied electric field. Because of the Hall current, the total current flows at an angle to the electric field, as shown in Figure 5.7.

It is relatively straightforward to show that the angle between the perpendicular electric field $\mathbf{E}_\perp$ and the perpendicular component of the current for the sth species $\mathbf{J}_{s\perp}$ is given by

$$\tan\theta_s = \frac{J_{sy}}{J_{sx}} = -\frac{\omega_{cs}}{\nu_s}. \tag{5.6.14}$$

From the above equation one can see that the ratio of the cyclotron frequency to the collision frequency plays an important role in determining the direction of current flow. If $\omega_{cs} \ll \nu_s$ for all species, then the conductivity tensor becomes diagonal with $\overleftrightarrow{\boldsymbol{\sigma}} = \overleftrightarrow{\mathbf{1}}\sigma$, where σ is given by Eq. (5.6.7). On the other hand, if $\omega_{cs} \gg \nu_s$, then $\sigma_\perp$ and $\sigma_{\rm H}$ go to zero and $\sigma_\parallel$ becomes infinite. The only current that can flow under these conditions is along the magnetic field.

5.7 Ambipolar Diffusion

Frequently in laboratory and space applications one finds situations where plasma is produced by an ionization process in one region and transported to some other region where it is lost. If the transport is caused by random collisional forces, the transport process is called diffusion. Diffusion causes particles to move from a region of high density to a region of low density. In this section we consider the diffusion of electrons and one species of positive ions using the Lorentz gas model. This model is often applied to the diffusion of electrons and ions through a neutral gas.

If the diffusion is a steady-state process ($\partial/\partial t \simeq 0$), the momentum equation for the electrons and ions can be written

$$m_s n_s (\mathbf{U}_s \cdot \nabla)\mathbf{U}_s = n_s e_s [\mathbf{E} + \mathbf{U}_s \times \mathbf{B}] - \nabla P_s - \nu_s m_s n_s \mathbf{U}_s. \quad (5.7.1)$$

Since diffusion is associated with a density gradient, the inhomogeneous terms cannot be neglected. Usually the drift velocity is much smaller than the root–mean–square thermal velocity (i.e., $U_s^2 \ll \kappa T_s/m_s$). Under these conditions the inertial term on the left-hand side of the above equation is small compared to the pressure gradient term. Dropping the inertial term and assuming that the magnetic field is zero, the momentum equation can be solved for the particle flux, $n_s \mathbf{U}_s$, which is given by

$$n_s \mathbf{U}_s = -\frac{1}{\nu_s m_s} \nabla P_s + \frac{e_s n_s}{\nu_s m_s} \mathbf{E}. \quad (5.7.2)$$

Substituting the ideal gas law, $P_s = n_s \kappa T_s$, for the pressure term, the above equation can be written

$$n_s \mathbf{U}_s = -\nabla(\bar{\kappa}_s n_s) + \bar{\mu}_s n_s \mathbf{E}, \quad (5.7.3)$$

where the quantities $\bar{\kappa}_s = \kappa T_s/\nu_s m_s$ and $\bar{\mu}_s = e_s/\nu_s m_s$ are called the diffusion and mobility coefficients. The diffusion coefficient provides a measure of the transport rate caused by the gradient in the number density, and the mobility coefficient provides a measure of the transport rate caused by the electric field.

To compute the particle flux, we must also take into account the constraint imposed by the continuity equation. If we avoid source or loss regions, there are two continuity equations, one for the electrons,

$$\frac{\partial n_{\mathrm{e}}}{\partial t} + \nabla \cdot (n_{\mathrm{e}} \mathbf{U}_{\mathrm{e}}) = 0, \quad (5.7.4)$$

and one for the ions,

$$\frac{\partial n_{\mathrm{i}}}{\partial t} + \nabla \cdot (n_{\mathrm{i}} \mathbf{U}_{\mathrm{i}}) = 0. \quad (5.7.5)$$

Subtracting Eq. (5.7.5) from Eq. (5.7.4) gives

$$\nabla \cdot [n(\mathbf{U}_e - \mathbf{U}_i)] = -\frac{\partial}{\partial t}(n_e - n_i), \tag{5.7.6}$$

where $n = n_e \simeq n_i$. If the plasma is quasi-neutral, so that $n_e \simeq n_i$, it follows that the right-hand side of the above equation is very small, so that, to a good approximation,

$$\nabla \cdot [n(\mathbf{U}_e - \mathbf{U}_i)] \simeq 0. \tag{5.7.7}$$

In many cases it can be shown that $\mathbf{U}_e = \mathbf{U}_i$ provides the only solution of the above equation. For example, for a cylindrically symmetric plasma where all the parameters are independent of the ϕ and z coordinates, Eq. (5.7.7) becomes

$$\frac{1}{\rho}\frac{\mathrm{d}}{\mathrm{d}\rho}[\rho n(U_{\rho e} - U_{\rho i})] = 0, \tag{5.7.8}$$

where ρ is the radial distance from the central axis. The above equation implies that the term $\rho n(U_{\rho e} - U_{\rho i})$ must be constant. If there is no current source at $\rho = 0$ then it follows that $U_{\rho e} = U_{\rho i}$. These drift velocities depend on the axial current. If the boundary conditions at each end of the cylinder are such that no current can flow along the z axis, then it follows that $\mathbf{U}_e = \mathbf{U}_i$. A similar argument can be made for a spherical geometry. A process for which $\mathbf{U}_e = \mathbf{U}_i$ is called ambipolar diffusion.

If ambipolar diffusion is assumed, so that $\mathbf{U}_e = \mathbf{U}_i = \mathbf{U}$, then the following two equations must be satisfied:

$$n\mathbf{U} = -\nabla(\bar{\kappa}_e n) - |\bar{\mu}_e| n\mathbf{E} \tag{5.7.9}$$

and

$$n\mathbf{U} = -\nabla(\bar{\kappa}_i n) + \bar{\mu}_i n\mathbf{E}. \tag{5.7.10}$$

The negative sign in the electron mobility term has been used to emphasize the fact that the electric field force on the electrons is directed opposite to the electric field. The electric field in these equations is intimately related to the ambipolar diffusion process. In the absence of an electric field, the diffusion rate of the electrons would be much higher than the diffusion rate of the ions because of the lighter mass and higher thermal velocity of the electrons. However, as soon as the electrons start to diffuse, a polarization charge develops. This polarization charge produces an electric field that reduces the electron flux and enhances the ion flux. Equilibrium is achieved when the electron and ion drift velocities are equal. Equations (5.7.9)

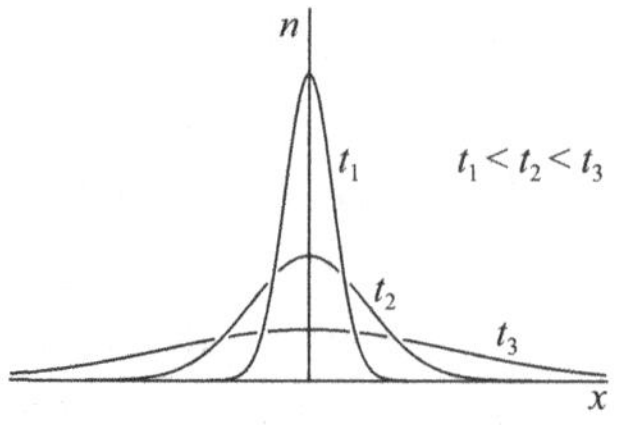

Figure 5.8 Solutions to the diffusion equations at a sequence of times: t_1, t_2, and t_3.

and (5.7.10) are self-consistent relations that must be satisfied for this equilibrium to be achieved. The electric field can be eliminated from these equations to give

$$n\mathbf{U} = -\nabla(\bar{\kappa}_a n), \tag{5.7.11}$$

where

$$\bar{\kappa}_a = \frac{\bar{\kappa}_e \bar{\mu}_i + \bar{\kappa}_i |\bar{\mu}_e|}{\bar{\mu}_i + |\bar{\mu}_e|} \tag{5.7.12}$$

is called the ambipolar diffusion coefficient. Equation (5.7.11) can then be substituted into the continuity equation to obtain

$$\frac{\partial n}{\partial t} = \nabla^2(\bar{\kappa}_a n). \tag{5.7.13}$$

This equation is of a general form called the diffusion equation.

The magnitude of the ambipolar diffusion coefficient can be estimated by evaluating the various terms in $\bar{\kappa}_a$. Because the electron mobility is always much greater than the ion mobility, $|\bar{\mu}_e| \gg \bar{\mu}_i$, it can be shown that to a good approximation

$$\bar{\kappa}_a \simeq \bar{\kappa}_i \left(1 + \frac{T_e}{T_i}\right). \tag{5.7.14}$$

For equal electron and ion temperatures, the ambipolar diffusion coefficient is approximately twice the ion diffusion coefficient, which is much smaller than the electron diffusion coefficient. Thus, even though the electrons tend to diffuse much more rapidly than the ions, the ambipolar electric field restricts the electrons to diffuse at the same rate as ions. Therefore, the ion diffusion effectively controls the electron transport.

The diffusion equation (5.7.13) has the property of eliminating peaks in the density distribution. This is because the flux of particles is always away from regions of higher density. A time sequence of solutions with a constant diffusion coefficient is shown in Figure 5.8. From simple scaling considerations, it can be shown that the time-scale for the decay is $\tau_D = L^2/\bar{\kappa}_a$, where L is a typical length-scale. In steady state ($\partial/\partial t = 0$), the diffusion equation reduces to Laplace's equation,

$$\nabla^2(\bar{\kappa}_a n) = 0, \tag{5.7.15}$$

the solutions of which are discussed in most textbooks on electricity and magnetism.

Problems

5.1. The general form of a Maxwellian velocity distribution drifting at velocity $\mathbf{U}_0$ is

$$f(\mathbf{v}) = n_0\left(\frac{m}{2\pi\kappa T}\right)^{3/2}\exp\left[-\frac{m(\mathbf{v}-\mathbf{U}_0)^2}{2\kappa T}\right].$$

(a) Show that

$$\int_{-\infty}^{\infty} f(\mathbf{v})d^3v = n_0.$$

(b) Show that

$$\langle\mathbf{v}\rangle = \frac{1}{n}\int_{-\infty}^{\infty}\mathbf{v}f(\mathbf{v})d^3v = \mathbf{U}_0.$$

(c) Show that the pressure tensor is

$$\overleftrightarrow{\mathbf{P}} = \begin{vmatrix} P & 0 & 0 \\ 0 & P & 0 \\ 0 & 0 & P \end{vmatrix}, \quad \text{where } P = n_0\kappa T.$$

5.2. Later we will use a one-dimensional distribution function of the form

$$F_0 = \frac{C}{\pi}\frac{1}{C^2+v_z^2},$$

called the Cauchy distribution. Show that $\int_{-\infty}^{\infty} F_0 dv_z = 1$.

5.3. The distribution function used in Problem 5.2 above has the disadvantage that the second moment is infinite, which implies an infinite temperature. To avoid this problem, one can use a function of the form $1/(C^2+v_z^2)^2$, so that

$$F_0(v_z) = \frac{2C^3}{\pi}\frac{1}{(C^2+v_z^2)^2}.$$

(a) Show that

$$\int_{-\infty}^{\infty} F_0(v_z)\mathrm{d}v_z = 1.$$

(b) Show that the root–mean–square value of v_z is C; in other words, that

$$\langle v_z^2\rangle^{1/2} = C.$$

5.4. To represent a loss-cone distribution, we sometimes use a distribution function of the form

$$f_0(v_\perp, v_\parallel) = \frac{n_0}{(2\pi)^{3/2} C_\perp^2 C_\parallel \ell!}\left(\frac{v_\perp^2}{2C_\perp^2}\right)^\ell \exp\left(-\frac{v_\perp^2}{2C_\perp^2}\right)\exp\left(-\frac{v_\parallel^2}{2C_\parallel^2}\right),$$

where $C_\perp^2 = \kappa T_\perp/m$ and $C_\parallel^2 = \kappa T_\parallel/m$, and ℓ is an integer.
(a) Show that

$$\int_{-\infty}^{\infty} f_0(\mathbf{v})d^3v = n_0.$$

(b) Show that

$$\overleftrightarrow{\mathbf{P}} = \begin{vmatrix} P_\perp & 0 & 0 \\ 0 & P_\perp & 0 \\ 0 & 0 & P_\parallel \end{vmatrix},$$

where $P_\perp = (\ell+1)n_0\kappa T_\perp$ and $P_\parallel = n_0\kappa T_\parallel$.

5.5. As an example of a solution of the Vlasov equation based on a constant of the motion, we discussed a Maxwellian

$$f(\mathbf{v}) = n_0\left(\frac{m}{2\pi\kappa T_0}\right)^{3/2} \exp\left[\frac{-(\frac{1}{2}mv^2 + q\Phi)}{\kappa T_0}\right],$$

where Φ is the electrostatic potential, $\mathbf{E} = -\nabla\Phi$. Explicitly show that this distribution function satisfies the Vlasov equation. In other words, show that

$$\frac{\partial f}{\partial t} + \mathbf{v}\cdot\nabla f + \frac{q\mathbf{E}}{m}\cdot\nabla_{\mathbf{v}} f = 0.$$

5.6. As an example of the use of the above solution, consider a plasma of electrons and ions that have the distribution functions

$$f_e = n_0\left(\frac{m_e}{2\pi\kappa T_e}\right)^{3/2} \exp\left[\frac{-(\frac{1}{2}m_e v^2 - e\Phi)}{\kappa T_e}\right]$$

and

$$f_{\rm i} = n_0\left(\frac{m_{\rm i}}{2\pi\kappa T_{\rm i}}\right)^{3/2} \exp\left[\frac{-(\frac{1}{2}m_{\rm i}v^2 + e\Phi)}{\kappa T_{\rm i}}\right].$$

(a) By integrating over velocity space to obtain the number densities $n_{\rm e}$ and $n_{\rm i}$, show that the potential must satisfy

$$\nabla^2\Phi = -\frac{n_0 e}{\epsilon_0}\left[\exp\left(\frac{-e\Phi}{\kappa T_{\rm i}}\right) - \exp\left(\frac{e\Phi}{\kappa T_{\rm e}}\right)\right].$$

(b) By assuming that $e\Phi/\kappa T_{\rm i} \ll 1$ and $e\Phi/\kappa T_{\rm e} \ll 1$, show that

$$\nabla^2\Phi = \frac{n_0 e^2}{\epsilon_0 \kappa}\left[\frac{1}{T_{\rm i}} + \frac{1}{T_{\rm e}}\right]\Phi.$$

(c) A charge Q is introduced at $r = 0$. Show that the potential is given by

$$\Phi = \frac{1}{4\pi\epsilon_0}\frac{Q}{r}{\rm e}^{-r/\lambda_{\rm D}},$$

where $(1/\lambda_{\rm D})^2 = (1/\lambda_{\rm De})^2 + (1/\lambda_{\rm Di})^2$.

(d) In the process of integrating $f_{\rm e}$ and $f_{\rm i}$ over all of velocity space, we have made an error. Identify and discuss the nature of this error. Two cases must be considered, $Q > 0$ and $Q < 0$.

5.7. A plasma with a magnetic field $\mathbf{B}_0(s)$ has a parallel electric field that can be described by a monotonic electrostatic potential $\Phi(s)$, where s is the distance along the magnetic field line. Suppose that an absorbing boundary (the wall of the plasma chamber or the atmosphere) exists at $s = s_a$ [i.e., s_a].

(a) By assuming conservation of energy and conservation of the first adiabatic invariant, show that, at an arbitrary point s, the loss cone in velocity space $(v_\parallel, v_\perp)$ is given by

$$v_\parallel^2 - \left(\frac{Bs_a}{B} - 1\right)v_\perp^2 = \frac{2q}{m}(\Phi_{\rm m} - \Phi).$$

(b) Sketch the shape of the loss cone for electrons and ions assuming that $\Phi_{\rm m} > \Phi$ and $B_{\rm m} > B$.

(c) Suppose ions are now emitted at $s = s_a$. What region of $(v_\parallel, v_\perp)$ is accessible to these ions, again assuming that $\Phi s_a > \Phi$ and $Bs_a > B$.

(d) Find the ion distribution function at the arbitrary point if the distribution function at $s = s_a$ is a Maxwellian,

$$f_{\rm i} = n_0\left(\frac{m}{2\pi\kappa T}\right)^{3/2} \exp\left[-\frac{mv^2}{2\kappa T}\right].$$

5.8. Show that the Vlasov equation in cylindrical phase-space coordinates $(\rho, \phi, z, \dot{\rho}, \dot{\phi}, \dot{z})$ is given by

$$\frac{\partial f}{\partial t} + \dot{\rho}\frac{\partial f}{\partial \rho} + \dot{\phi}\frac{\partial f}{\partial \phi} + \dot{z}\frac{\partial f}{\partial z} + \rho\dot{\phi}^2\frac{\partial f}{\partial \dot{\rho}} - \frac{2\dot{\rho}\dot{\phi}}{\rho}\frac{\partial f}{\partial \dot{\phi}} + \frac{1}{m}\left(F_\rho\frac{\partial f}{\partial \dot{\rho}} + \frac{F_\phi}{\rho}\frac{\partial f}{\partial \dot{\phi}} + F_z\frac{\partial f}{\partial \dot{z}}\right) = 0.$$

5.9. Show that the following distribution function

$$f\left(\frac{1}{2}mv^2, m\rho^2\dot{\phi} + q\rho A_\phi\right)$$

is a time-stationary solution of the Vlasov equation for an azimuthally symmetric magnetic field of the form

$$\mathbf{B} = \left(-\frac{\partial A_\phi}{\partial z}\right)\hat{\boldsymbol{\rho}} + \frac{1}{\rho}\frac{\partial}{\partial \rho}(\rho A_\phi)\,\hat{\mathbf{z}}.$$

5.10. By eliminating $\mathbf{E}$ from Eqs. (5.7.9) and (5.7.10), show that the ambipolar diffusion coefficient is given by

$$\bar{\kappa}_{\rm a} = \frac{\bar{\kappa}_{\rm e}\bar{\mu}_{\rm i} + \bar{\kappa}_{\rm i}|\bar{\mu}_{\rm e}|}{\bar{\mu}_{\rm i} + |\bar{\mu}_{\rm e}|},$$

which, when $|\bar{\mu}_{\rm e}| >> \bar{\mu}_{\rm i}$, is given to a good approximation by

$$\bar{\kappa}_{\rm a} \simeq \bar{\kappa}_{\rm i}\left(1 + \frac{T_{\rm e}}{T_{\rm i}}\right).$$

References

Bohm, D., and Gross, E. P. 1949. Theory of plasma oscillations. A. Origin of medium-like behavior; B. Excitation and damping of oscillations. *Phys. Rev.* **75**, 1851–1876.

Boltzmann, L. 1995. *Vorlesungen Über Gastheorie, 1896–1898.* Translated by S. Brush, *Lectures on Gas Theory, 1896-1898.* New York: Dover Publications. Originally published in 1896–1898 by University of California Press.

Chew, G. F., Goldberger, M. L., and Low, F. E. 1956. The Boltzmann equation and the one-fluid hydrodynamic equations in the absence of collisions. *Proc. R. Soc. London, Ser. A* **236**, 112–118.

Fetter, A. L., and Walecka, J. D. 1980. *Theoretical Mechanics of Particles and Continua.* New York: McGraw-Hill, p. 345.

Halliday, D., and Resnick, R. 1978. *Physics, Parts 1 and 2 Combined*, Third Edition. New York: Wiley, pp. 509–510.

Kaplan, W. 1952. *Advanced Calculus.* Reading, MA: Addison-Wesley, p. 99.

Tonks, L., and Langmuir, I. 1929. Oscillations in ionized gases. *Phys. Rev.* **33**, 195–210.

Vlasov, A. A. 1945. On the kinetic theory of an assembly of particles with collective interaction. *J. Phys. (USSR)* **9**, 25–40.

Further Reading

Chen, F. F. 1990. *Introduction to Plasma Physics and Controlled Fusion, Volume 1: Plasma Physics*. New York: Plenum Press, Chapter 7. Originally published in 1983.

Montgomery, D. C., and Tidman, D. A. 1964. *Plasma Kinetic Theory*. New York: McGraw-Hill, Chapter 1.

Nicholson, D. R. 1992. *Introduction to Plasma Theory*. Malabar, FL: Krieger Publishing, Chapters 3, 4, 5, and 6. Originally published in 1983 by Wiley.

Parks, G. K. 2000. *Physics of Space Plasmas: An Introduction*. Redwood City, CA: Addison-Wesley, Chapter 2. Originally published in 1991.

6

Magnetohydrodynamics

Historically, magnetohydrodynamics (MHD) preceded the development of modern plasma physics. The original intent of MHD was to treat a plasma as a conducting fluid (see Cowling, 1957). The governing equations were adapted from fluid mechanics with appropriate modifications to account for electrical forces. To obtain a complete set of equations, it was necessary to specify the current as a function of the applied electric field. This was accomplished by using a linear Ohm's law, such as is often used to describe conducting media. Since, to a first approximation, plasmas are electrically neutral, the net charge density was assumed to be negligible. Also, since fluid motions tend to be slow compared to the characteristic time-scales of a plasma, the displacement current was assumed to be small compared to the conduction current. These assumptions, together with an appropriate equation of state, were sufficient to obtain a closed system of equations.

Although it would be adequate simply to write down the equations of MHD as they were originally postulated, it is useful to try to derive the equations from first principles, using the moment equations developed in Chapter 5. Although this process is not entirely mathematically rigorous, it has the advantage of revealing more clearly the underlying assumptions and the range of applicability of MHD. It also has the advantage of providing a theoretical basis for precisely defining certain quantities, such as the fluid velocity and the plasma pressure.

6.1 The Basic Equations of MHD

To derive the MHD equations from the moment equations, it is convenient to define the mass density as the sum of the mass densities of the individual species

$$\rho_{\mathrm{m}} = \sum_s \rho_{\mathrm{m}s}, \tag{6.1.1}$$

where $\rho_{ms} = m_s n_s$, and the fluid velocity as the mass weighted average of the velocities of the individual species is

$$\mathbf{U} = \frac{1}{\rho_m} \sum_s \rho_{ms} \mathbf{U}_s, \tag{6.1.2}$$

where $\mathbf{U}_s$ is defined by Eq. (5.1.5). Note that since the ions have a much greater mass than the electrons, the fluid velocity $\mathbf{U}$ is heavily weighted by the velocity of the ions.

6.1.1 The Mass Continuity Equation

To derive the mass continuity equation for an MHD fluid, we start with the mass continuity equation for the sth species,

$$\frac{\partial \rho_{ms}}{\partial t} + \nabla \cdot (\rho_{ms} \mathbf{U}_s) = 0, \tag{6.1.3}$$

which follows by multiplying Eq. (5.4.6) by m_s and using the definition $\rho_{ms} = m_s n_s$. Summing the above equation over all species s and making use of the definitions for the mass density (6.1.1) and the fluid velocity (6.1.2), we obtain the equation

$$\frac{\partial \rho_m}{\partial t} + \nabla \cdot (\rho_m \mathbf{U}) = 0. \tag{6.1.4}$$

This equation is identical to the well-known mass continuity equation of fluid dynamics; see Batchelor (1967). Note that the above derivation depends critically on the fact that the fluid velocity is the mass weighted average of the flow velocities of the individual species. A simple average of the flow velocities of the individual species would not give this result.

6.1.2 The Momentum Equation

To derive an equation for the time rate of change of the fluid momentum, we start with the momentum equation (5.4.23) for the sth species. Summing this equation over all species, we obtain

$$\sum_s \frac{\partial}{\partial t}(\rho_{ms} \mathbf{U}_s) + \sum_s \nabla \cdot (\rho_{ms} \mathbf{U}_s \mathbf{U}_s)$$
$$= \rho_q \mathbf{E} + \mathbf{J} \times \mathbf{B} - \nabla \cdot \left(\sum_s \overleftrightarrow{\mathbf{P}}_s \right) + \sum_s \frac{\delta_c \mathbf{p}_s}{\delta t}, \tag{6.1.5}$$

where we have used $\rho_q = \sum_s n_s e_s$ and $\mathbf{J} = \sum_s n_s e_s \mathbf{U}_s$.

Since the total momentum is conserved for any collision process, it follows that the collision term $\sum_s \delta_c \mathbf{p}_s/\delta t$ must sum to zero. In order to simplify the remaining terms, it is convenient to introduce a new pressure tensor,

$$\overleftrightarrow{\mathbf{P}}_{0s} = m_s \int_{-\infty}^{\infty} (\mathbf{v}-\mathbf{U})(\mathbf{v}-\mathbf{U}) f_s \, \mathrm{d}^3 v, \tag{6.1.6}$$

that gives the pressure in a frame of reference moving at the fluid velocity. It is easily verified that $\overleftrightarrow{\mathbf{P}}_{0s}$ is related to $\overleftrightarrow{\mathbf{P}}_s$, defined earlier by Eq. (5.1.7), via the equation

$$\overleftrightarrow{\mathbf{P}}_{0s} = \overleftrightarrow{\mathbf{P}}_s + \rho_{\mathrm{m}s} \mathbf{W}_s \mathbf{W}_s, \tag{6.1.7}$$

where $\mathbf{W}_s = \mathbf{U}_s - \mathbf{U}$ is the average velocity of the sth species relative to the fluid velocity, otherwise known as the *diffusion velocity* for species s. Writing the total pressure as $\overleftrightarrow{\mathbf{P}}_0 = \sum_s \overleftrightarrow{\mathbf{P}}_{0s}$, the momentum equation (6.1.5) becomes

$$\sum_s \frac{\partial}{\partial t}(\rho_{\mathrm{m}s}\mathbf{U}_s) + \sum_s \nabla \cdot (\rho_{\mathrm{m}s}\mathbf{U}_s\mathbf{U}_s)$$
$$= \rho_{\mathrm{q}} \mathbf{E} + \mathbf{J} \times \mathbf{B} - \nabla \cdot \overleftrightarrow{\mathbf{P}}_0 + \sum_s \nabla \cdot (\rho_{\mathrm{m}s}\mathbf{W}_s\mathbf{W}_s). \tag{6.1.8}$$

With the substitution $\mathbf{U}_s = \mathbf{W}_s + \mathbf{U}$, the left-hand side of this equation can be written

$$\frac{\partial}{\partial t}(\rho_{\mathrm{m}s}\mathbf{U}) + \frac{\partial}{\partial t}(\rho_{\mathrm{m}s}\mathbf{W}_s) + \nabla \cdot (\rho_{\mathrm{m}s}\mathbf{W}_s\mathbf{W}_s) + \nabla \cdot (\rho_{\mathrm{m}s}\mathbf{W}_s\mathbf{U})$$
$$+ \nabla \cdot (\rho_{\mathrm{m}s}\mathbf{U}\mathbf{W}_s) + \nabla \cdot (\rho_{\mathrm{m}s}\mathbf{U}\mathbf{U}). \tag{6.1.9}$$

Summing over all species and using the identity $\sum_s \rho_{\mathrm{m}s} \mathbf{W}_s = 0$ (implied by the definition of $\mathbf{W}_s$), this expression simplifies to

$$\sum_s \frac{\partial}{\partial t}(\rho_{\mathrm{m}s}\mathbf{U}) + \sum_s \nabla \cdot (\rho_{\mathrm{m}s}\mathbf{U}\mathbf{U}) + \sum_s \nabla \cdot (\rho_{\mathrm{m}s}\mathbf{W}_s\mathbf{W}_s). \tag{6.1.10}$$

Substituting this result into the left-hand side of Eq. (6.1.8), the momentum equation simplifies further to

$$\frac{\partial}{\partial t}(\rho_{\mathrm{m}}\mathbf{U}) + \nabla \cdot (\rho_{\mathrm{m}}\mathbf{U}\mathbf{U}) = \rho_q \mathbf{E} + \mathbf{J} \times \mathbf{B} - \nabla \cdot \overleftrightarrow{\mathbf{P}}_0. \tag{6.1.11}$$

Using the continuity equation (6.1.4) and following the same procedure used to rewrite the left-hand side of Eq. (5.4.23) in the form given by Eq. (5.4.25), the left-hand side of the above equation can be rewritten as

$$\rho_{\mathrm{m}} \left[\frac{\partial \mathbf{U}}{\partial t} + (\mathbf{U} \cdot \nabla)\mathbf{U} \right] = \rho_{\mathrm{q}} \mathbf{E} + \mathbf{J} \times \mathbf{B} - \nabla \cdot \overleftrightarrow{\mathbf{P}}_0, \tag{6.1.12}$$

where the quantity in the brackets on the left is the convective derivative, $d\mathbf{U}/dt$; see Eq. (5.4.26). This equation is called the *momentum equation* and has the simple interpretation that the rate of change of the momentum of a fluid element is equal to the sum of the electric field, magnetic field, and pressure forces acting on that fluid element. In what follows, we will drop the subscript 0 on the plasma pressure tensor, $\overleftrightarrow{\mathbf{P}}_0$, with the understanding that the pressure is always defined relative to the fluid frame of reference.

It is important to note that no assumptions, other than the validity of the Boltzmann equation, have been used in the derivation of the mass continuity and momentum equations, both of which are exact equations.

6.1.3 Generalized Ohm's Law

Since a plasma is a conducting medium, it is necessary to determine how the current density, $\mathbf{J}$, depends on the electric field, $\mathbf{E}$. In the historical approach it was assumed that the current density can be represented by a simple Ohm's law of the form $\mathbf{J}' = \sigma\mathbf{E}'$, where σ is the conductivity, and $\mathbf{J}'$ and $\mathbf{E}'$ are the current density and electric field in the rest frame of the fluid. For non-relativistic velocities, the current density and the electric field transform from the rest frame of the fluid to the laboratory frame according to the relations $\mathbf{J}' = \mathbf{J}$ and $\mathbf{E}' = \mathbf{E} + \mathbf{U} \times \mathbf{B}$, respectively. In the laboratory frame of reference Ohm's law then becomes

$$\mathbf{J} = \sigma(\mathbf{E} + \mathbf{U} \times \mathbf{B}). \tag{6.1.13}$$

Unfortunately, this equation is not completely rigorous. As will be shown below, numerous assumptions must be made to derive it, some of which are questionable. Nevertheless, Ohm's law is widely used in MHD calculations, and is a reasonable approximation if the collisional mean-free path is short compared with the characteristic length-scale of the system.

In an attempt to derive Ohm's law, let us consider a two-component plasma consisting of electrons (e) and one species of positively charged ions (i). Let us assume further that the drag force on the sth species is proportional to the velocity difference between the sth and rth species and is given by the equation

$$\frac{\delta_c \mathbf{p}_s}{\delta t} = -n_s m_s \nu_{sr}(\mathbf{U}_s - \mathbf{U}_r), \tag{6.1.14}$$

where ν_{sr} is the effective collision frequency between the two species. The above equation is simply a generalized version of the collisional drag force given by Eq. (5.6.2), and is valid only if the velocity difference $\mathbf{U}_s - \mathbf{U}_r$ is small compared to the thermal velocities of the two species.

Next consider the moment equations for the electrons and the ions. The momentum equation for the ions is given by

$$\frac{\partial}{\partial t}(m_i n_i \mathbf{U}_i) + \nabla \cdot (m_i n_i \mathbf{U}_i \mathbf{U}_i) = n_i e[\mathbf{E} + \mathbf{U}_i \times \mathbf{B}] - \nabla \cdot \overleftrightarrow{\mathbf{P}}_i - n_i m_i \nu_{ie}(\mathbf{U}_i - \mathbf{U}_e), \tag{6.1.15}$$

and the momentum equation for the electrons is given by

$$\frac{\partial}{\partial t}(m_e n_e \mathbf{U}_e) + \nabla \cdot (m_e n_e \mathbf{U}_e \mathbf{U}_e) = -n_e e[\mathbf{E} + \mathbf{U}_e \times \mathbf{B}] - \nabla \cdot \overleftrightarrow{\mathbf{P}}_e - n_e m_e \nu_{ei}(\mathbf{U}_e - \mathbf{U}_i). \tag{6.1.16}$$

Since the current density is proportional to $(\mathbf{U}_i - \mathbf{U}_e)$, we start by multiplying the above equations by m_e and m_i, respectively, and then subtracting the two equations in order to generate a $(\mathbf{U}_i - \mathbf{U}_e)$ term, which gives

$$\begin{aligned} m_e m_i \left\{ \frac{\partial}{\partial t}[n(\mathbf{U}_i - \mathbf{U}_e)] + \nabla \cdot [n(\mathbf{U}_i \mathbf{U}_i - \mathbf{U}_e \mathbf{U}_e)] \right\} \\ = e\rho_m \mathbf{E} + en(m_e \mathbf{U}_i + m_i \mathbf{U}_e) \times \mathbf{B} - m_e \nabla \cdot \overleftrightarrow{\mathbf{P}}_i + m_i \nabla \cdot \overleftrightarrow{\mathbf{P}}_e \\ - m_e m_i n \nu_{ie}(\mathbf{U}_i - \mathbf{U}_e) + m_i m_e n \nu_{ei}(\mathbf{U}_e - \mathbf{U}_i), \end{aligned} \tag{6.1.17}$$

where $\rho_m = n(m_e + m_i)$ is the mass density and we have assumed that $n_e = n_i = n$. Since the drag force on the ions must be equal and opposite to the drag force on the electrons, it follows from Eq. (6.1.14) that $m_i \nu_{ie} = m_e \nu_{ei}$. The last two terms in Eq. (6.1.17) then simplify to

$$-m_e m_i n \nu_{ie}(\mathbf{U}_i - \mathbf{U}_e) + m_i m_e n \nu_{ei}(\mathbf{U}_e - \mathbf{U}_i) = -e\rho_m \left(\frac{m_e \nu_{ei}}{ne^2} \right) \mathbf{J}, \tag{6.1.18}$$

where $\mathbf{J} = en(\mathbf{U}_i - \mathbf{U}_e)$ is the current density. Comparing the term in the parentheses on the right-hand side of the above equation with Eq. (5.6.8), one can see that this term is simply the inverse of the conductivity

$$\frac{m_e \nu_{ei}}{ne^2} = \frac{1}{\sigma} = \eta, \tag{6.1.19}$$

which is called the resistivity. Next, consider the $m_e \mathbf{U}_i + m_i \mathbf{U}_e$ term on the right-hand side of Eq. (6.1.17). By simple algebraic manipulation of the equations for the fluid velocity, $\mathbf{U} = (m_e \mathbf{U}_e + m_i \mathbf{U}_i)/(m_e + m_i)$, and the current density, $\mathbf{J} = ne(\mathbf{U}_i - \mathbf{U}_e)$, this term can be written in the form

$$m_e \mathbf{U}_i + m_i \mathbf{U}_e = (m_e + m_i)\mathbf{U} - (m_i - m_e)\frac{\mathbf{J}}{en}. \tag{6.1.20}$$

Finally, consider the left-hand side of Eq. (6.1.17). Using the equation for the current density, $\mathbf{J} = en(\mathbf{U}_\mathrm{i} - \mathbf{U}_\mathrm{e})$, the $\partial/\partial t$ term can be written

$$\frac{\partial}{\partial t}[n(\mathbf{U}_\mathrm{i} - \mathbf{U}_\mathrm{e})] = \frac{1}{e}\frac{\partial \mathbf{J}}{\partial t}. \tag{6.1.21}$$

The ∇ term is more complicated, but can be simplified by noting that Eq. (6.1.14) is valid only if the velocity difference $\delta\mathbf{U} = \mathbf{U}_\mathrm{e} - \mathbf{U}_\mathrm{i}$ is small. The $(\mathbf{U}_\mathrm{i}\mathbf{U}_\mathrm{i} - \mathbf{U}_\mathrm{e}\mathbf{U}_\mathrm{e})$ term can then be expanded in powers of the small quantity $\delta\mathbf{U}$ to give, to a first approximation,

$$\nabla \cdot [n(\mathbf{U}_\mathrm{i}\mathbf{U}_\mathrm{i} - \mathbf{U}_\mathrm{e}\mathbf{U}_\mathrm{e})] = \frac{1}{e}\nabla \cdot (\mathbf{J}\mathbf{U} + \mathbf{U}\mathbf{J}), \tag{6.1.22}$$

where we have made use of the fact that $\mathbf{U} \simeq \mathbf{U}_\mathrm{i}$ for $m_\mathrm{e} \ll m_\mathrm{i}$, and $\mathbf{J} = -en\delta\mathbf{U}$. Substituting Eqs. (6.1.18)–(6.1.22) into Eq. (6.1.17), making use of the fact that $m_\mathrm{e} \ll m_\mathrm{i}$, and omitting the ion pressure term since it is reduced by a factor of $m_\mathrm{e}/m_\mathrm{i}$ relative to the electron pressure term, we obtain the equation

$$\mathbf{E} + \mathbf{U} \times \mathbf{B} - \frac{\mathbf{J}}{\sigma} = \frac{1}{en}\mathbf{J} \times \mathbf{B} - \frac{1}{en}\nabla \cdot \overleftrightarrow{\mathbf{P}}_\mathrm{e} + \frac{m_\mathrm{e}}{ne^2}\left[\frac{\partial \mathbf{J}}{\partial t} + \nabla \cdot (\mathbf{J}\mathbf{U} + \mathbf{U}\mathbf{J})\right]. \tag{6.1.23}$$

This equation is called the *generalized Ohm's law*. Comparing with Eq. (6.1.13), it can be seen that if all of the terms on the right-hand side are sufficiently small, the equation reduces to the simple form of Ohm's law. The last term on the left-hand side of the above equation as well as the terms on the right are often important in boundary layers where the current density, **J**, is large. However, away from such regions, these terms can usually be neglected. A more detailed discussion of the validity of Ohm's law is given in Section 6.6.

As the collision frequency goes to zero, the conductivity σ tends to infinity. Ohm's law then reduces to $\mathbf{E} + \mathbf{U} \times \mathbf{B} = 0$. A plasma that obeys this equation is called an *ideal MHD plasma.* The fluid velocity component perpendicular to **B** is then given by $\mathbf{U}_\perp = \mathbf{E} \times \mathbf{B}/B^2$, which is identical to the $\mathbf{E} \times \mathbf{B}$ drift velocity encountered in single-particle orbit theory; see Eq. (3.2.8). Also, note that for an ideal MHD plasma, the electric field component parallel to the magnetic field, $\mathbf{E}_\parallel$, is identically zero. The magnetic field lines are then equipotentials (i.e., all points along a given magnetic field line are at the same potential) if $\partial\mathbf{A}/\partial t = 0$.

In collisional plasmas, it is often assumed that all of the terms on the right-hand side of Eq. (6.1.23) are negligible. This leads to the resistive form of Ohm's law,

$$\mathbf{E} + \mathbf{U} \times \mathbf{B} = \frac{1}{\sigma}\mathbf{J} = \eta\mathbf{J}, \tag{6.1.24}$$

which is widely used, not because it is a rigorous model, but because it represents the simplest and qualitatively most important deviation from the ideal MHD model.

6.1.4 The Equation of State

Since the moment equations do not define a closed system of equations, we must choose an equation of state in order to close the system of equations. The equation of state specifies the plasma pressure as a function of the temperature and density, and its form depends on various assumptions that must be made concerning the effect of collisions.

If there are sufficient collisions to establish an isotropic Maxwellian velocity distribution, then the plasma pressure is isotropic. The pressure gradient force can then be represented by the gradient of a scalar pressure, $\nabla \cdot \overleftrightarrow{\mathbf{P}} = \nabla P$, where we have dropped the subscript 0 on the pressure, which is now understood to be defined by Eq. (6.1.6). Under these conditions, the equation of state is commonly assumed to be a power law of the form

$$\frac{\mathrm{d}}{\mathrm{d}t}(P\rho_{\mathrm{m}}^{-\gamma}) = 0, \tag{6.1.25}$$

where the exponent γ is called the polytrope index (Fetter and Walecka, 1980). By choosing various values for γ, a variety of situations can be represented. For example, if the compression is so rapid that no heat can flow, i.e., $\mathrm{d}Q = 0$, then following the usual thermodynamic derivation, it follows that $\gamma = C_{\mathrm{P}}/C_{\mathrm{V}}$, where C_{P} is the heat capacity at constant pressure, and C_{V} is the heat capacity at constant volume. Equation (6.1.25) is then called the adiabatic equation of state. Note that the word "adiabatic" in this context means no heat flow, which is different from the meaning of adiabatic used in Chapter 3, which means slowly varying. Note also that although the compression must be rapid, as in a sound wave, it must not be so rapid that the process is irreversible (i.e., not a succession of equilibrium states). This difficulty will be encountered later when we discuss shock waves. In statistical mechanics it can be shown that for the adiabatic equation of state the polytrope index is given by $\gamma = (f+2)/f$, where f is the number of degrees of freedom. An MHD flow that is adiabatic is also sometimes called an isentropic flow, since the entropy, $S = \int d\,Q/T$, is constant along the streamlines. Other equations of state that can be represented by Eq. (6.1.25) include an isothermal equation of state ($\gamma = 1$), and an incompressible equation of state ($\gamma \to \infty$).

If there are insufficient collisions to maintain an isotropic velocity distribution, then the pressure is anisotropic and must be represented by a tensor. If the particle motions in the rest frame of the fluid are azimuthally symmetric with respect to the magnetic field, the pressure tensor can be represented as a diagonal matrix of the form

$$\overleftrightarrow{\mathbf{P}} = \begin{pmatrix} P_\perp & 0 & 0 \\ 0 & P_\perp & 0 \\ 0 & 0 & P_\parallel \end{pmatrix}, \tag{6.1.26}$$

where $P_\perp$ and $P_\parallel$ are pressures perpendicular and parallel to the local magnetic field. For a slowly varying adiabatic process in which there is no heat flow, it can be shown that the parallel and perpendicular pressures obey the relations

$$\frac{\mathrm{d}}{\mathrm{d}t}\left(\frac{P_\perp}{\rho_{\mathrm{m}} B}\right) = 0 \quad \text{and} \quad \frac{\mathrm{d}}{\mathrm{d}t}\left(\frac{P_\perp^2 P_\parallel}{\rho_{\mathrm{m}}^5}\right) = 0. \tag{6.1.27}$$

The above system of equations is known as the Chew–Goldberger–Low (CGL) equation of state (see Section 5.4.4). As previously discussed, the first equation is equivalent to conservation of the first adiabatic invariant and the second equation is equivalent to conservation of the second adiabatic invariant (see Section 3.8.2).

6.1.5 The Complete Set of Resistive MHD Equations

In order to obtain a complete and self-consistent description of a resistive MHD plasma, the fluid equations and Ohm's law must be coupled with Maxwell's equations. The resistive MHD model is typically used to describe plasma phenomena that are characterized by low frequencies (much smaller than the electron plasma frequency) and long wavelengths (much larger than the Debye length). Two approximations are made concerning the detailed form of Maxwell's equations. First, since temporal variations are assumed to be slow, the displacement current, $\epsilon_0 \partial \mathbf{E}/\partial t$, is ignored in comparison with the conduction current. This also means that the phase velocity of electromagnetic waves and the characteristic fluid velocities in the resistive MHD model are much smaller than the speed of light. In other words, the resistive MHD model is essentially non-relativistic. Second, the charge density ρ_q is set equal to zero in Eq. (6.1.12), neglecting the effect of charge separation. In addition, we assume that the pressure is a scalar and obeys Eq. (6.1.25). With these assumptions, the complete set of resistive MHD equations becomes

$$\nabla \times \mathbf{B} = \mu_0 \mathbf{J} \quad \text{(Ampère's law)}, \tag{6.1.28}$$

$$\nabla \cdot \mathbf{B} = 0, \tag{6.1.29}$$

$$\nabla \times \mathbf{E} = -\partial \mathbf{B}/\partial t \quad \text{(Faraday's law)}, \tag{6.1.30}$$

$$\frac{\partial \rho_{\rm m}}{\partial t} + \nabla \cdot (\rho_{\rm m}\mathbf{U}) = 0 \quad \text{(mass continuity equation)}, \tag{6.1.31}$$

$$\rho_{\rm m}\frac{\mathrm{d}\mathbf{U}}{\mathrm{d}t} = \mathbf{J}\times\mathbf{B} - \nabla P \quad \text{(momentum equation)}, \tag{6.1.32}$$

$$\mathbf{E} + \mathbf{U}\times\mathbf{B} = \eta\mathbf{J} \quad \text{(Ohm's law)}, \tag{6.1.33}$$

and

$$\frac{\mathrm{d}}{\mathrm{d}t}(P\rho_{\rm m}^{-\gamma}) = 0 \quad \text{(equation of state)}. \tag{6.1.34}$$

If we take the divergence of Ampère's law (6.1.28), we obtain the condition for charge continuity, $\nabla \cdot \mathbf{J} = 0$, which is always valid when it is assumed that $\rho_{\rm q} = 0$. Note that the charge continuity equation is not an independent equation, since it is implied by Ampère's law. Furthermore, the approximation $\rho_q = 0$, which is often referred to as the quasi-neutrality condition, does not imply that $\nabla \cdot \mathbf{E} = 0$. Applying Gauss' law, $\nabla \cdot \mathbf{E} = \rho_{\rm q}/\epsilon_0$, the quasi-neutrality condition $n_{\rm e} = n_{\rm i} \equiv n$ requires that $|\epsilon_0 \nabla \cdot \mathbf{E}|/(ne) \ll 1$. Thus, Gauss' law drops out of the resistive MHD equations, and the electric field $\mathbf{E}$ is now obtained from Ohm's law (6.1.33).

6.2 Magnetic Pressure

Further insight into the physical nature of the interaction of the magnetic field with an MHD plasma can be obtained by using Ampère's law (6.1.28) to eliminate $\mathbf{J}$ from the $\mathbf{J}\times\mathbf{B}$ force term in the momentum equation, which then becomes

$$\mathbf{J}\times\mathbf{B} = \frac{1}{\mu_0}(\nabla\times\mathbf{B})\times\mathbf{B}. \tag{6.2.1}$$

Using the identity $\nabla(\mathbf{F}\cdot\mathbf{G}) = (\mathbf{F}\cdot\nabla)\mathbf{G} + \mathbf{F}\times(\nabla\times\mathbf{G}) + (\mathbf{G}\cdot\nabla)\mathbf{F} + \mathbf{G}\times(\nabla\times\mathbf{F})$ with $\mathbf{F} = \mathbf{G} = \mathbf{B}$, the $\mathbf{J}\times\mathbf{B}$ force can be rewritten in the form

$$\mathbf{J}\times\mathbf{B} = \frac{1}{\mu_0}(\mathbf{B}\cdot\nabla)\mathbf{B} - \nabla\left(\frac{B^2}{2\mu_0}\right). \tag{6.2.2}$$

Since the pressure term in the momentum equation involves the negative of the divergence of the pressure tensor, it is interesting to ask whether the terms on the right-hand side of the above equation can be expressed as the negative of the divergence of a tensor. The answer is that they can. The appropriate tensor $\overleftrightarrow{\mathbf{T}}$ is given in index notation by

$$T_{rs} = -\frac{B_r B_s}{\mu_0} + \delta_{rs}\frac{B^2}{2\mu_0}, \tag{6.2.3}$$

where δ_{rs} is the Kronecker delta ($\delta_{rs} = 1$ for $r = s, \delta_{rs} = 0$ for $r \neq s$), or in vector notation by

$$\overleftrightarrow{\mathbf{T}} = -\frac{\mathbf{BB}}{\mu_0} + \overleftrightarrow{\mathbf{1}}\frac{B^2}{2\mu_0}. \tag{6.2.4}$$

To show this relationship, consider the s component of the negative of the divergence of this tensor:

$$\begin{aligned} -(\nabla \cdot \overleftrightarrow{\mathbf{T}})_s &= -\sum_r \frac{\partial}{\partial x_r}\left[-\frac{B_r B_s}{\mu_0} + \delta_{rs}\frac{B^2}{2\mu_0}\right] \\ &= \frac{1}{\mu_0}\sum_r \left[B_r \frac{\partial B_s}{\partial x_r} + B_s \frac{\partial B_r}{\partial x_r}\right] - \frac{\partial}{\partial x_s}\frac{B^2}{2\mu_0}. \end{aligned} \tag{6.2.5}$$

Since $\nabla \cdot \mathbf{B} = 0$, which in index notation is $\sum_r (\partial B_r/\partial x_r) = 0$, the above equation simplifies to

$$-(\nabla \cdot \overleftrightarrow{\mathbf{T}})_s = \frac{1}{\mu_0}\sum_r B_r \frac{\partial B_s}{\partial x_r} - \frac{\partial}{\partial x_s}\left(\frac{B^2}{2\mu_0}\right), \tag{6.2.6}$$

which in vector notation can be written

$$-\nabla \cdot \overleftrightarrow{\mathbf{T}} = \frac{1}{\mu_0}(\mathbf{B} \cdot \nabla)\mathbf{B} - \nabla\left(\frac{B^2}{2\mu_0}\right). \tag{6.2.7}$$

Comparing the right-hand side of this equation with the right-hand side of Eq. (6.2.2), we see that, as advertised, $\mathbf{J} \times \mathbf{B} = -\nabla \cdot \overleftrightarrow{\mathbf{T}}$. The tensor $\overleftrightarrow{\mathbf{T}}$ is called the magnetic pressure tensor. Using the magnetic pressure tensor, the momentum equation (6.1.32) can then be written in the compact form

$$\rho_{\mathrm{m}}\frac{\mathrm{d}\mathbf{U}}{\mathrm{d}t} = -\nabla \cdot (\overleftrightarrow{\mathbf{T}} + \overleftrightarrow{\mathbf{P}}). \tag{6.2.8}$$

From the structure of the terms on the right-hand side of the above equation, it is apparent that the magnetic field produces an anisotropic "magnetic pressure" that simply adds to the plasma pressure. The anisotropic nature of the magnetic pressure can be illustrated by choosing a coordinate system in which the magnetic field is parallel to the z axis, $\mathbf{B} = (0, 0, B)$. In this coordinate system the magnetic pressure tensor can be written as the sum of two terms

$$\overleftrightarrow{\mathbf{T}} = \begin{bmatrix} 0 & 0 & 0 \\ 0 & 0 & 0 \\ 0 & 0 & -B^2/\mu_0 \end{bmatrix} + \begin{bmatrix} B^2/2\mu_0 & 0 & 0 \\ 0 & B^2/2\mu_0 & 0 \\ 0 & 0 & B^2/2\mu_0 \end{bmatrix}. \tag{6.2.9}$$

The above equation shows that the magnetic pressure is the sum of a negative pressure (i.e., a *tension*) B^2/μ_0 that acts along the magnetic field, plus a positive pressure $B^2/2\mu_0$ that acts equally in all directions.

It often occurs in the above equations that one or other of the two terms, $B^2/2\mu_0$ or P, dominates, in which case the smaller of the two terms can be neglected. To characterize the relative importance of the two terms, it is useful to define a quantity called the plasma β, which is given by

$$\beta = \frac{P}{(B^2/2\mu_0)}. \tag{6.2.10}$$

If $\beta \gg 1$, the plasma pressure force dominates, and the magnetic field pressure can be neglected. On the other hand, if $\beta \ll 1$, the magnetic field force dominates and the plasma pressure can be neglected.

6.3 Magnetic Field Convection and Diffusion

By a relatively simple series of manipulations, the resistive MHD equations can be rearranged in a form that gives information on the time evolution of the magnetic field. Eliminating $\mathbf{J}$ between Ampère's law (6.1.28) and Ohm's law (6.1.33) gives

$$\nabla \times \mathbf{B} = \mu_0 \sigma (\mathbf{E} + \mathbf{U} \times \mathbf{B}). \tag{6.3.1}$$

Taking the curl of the above equation,

$$\nabla \times (\nabla \times \mathbf{B}) = \mu_0 \sigma (\nabla \times \mathbf{E} + \nabla \times (\mathbf{U} \times \mathbf{B})), \tag{6.3.2}$$

using the identity

$$\nabla \times (\nabla \times \mathbf{B}) = \nabla(\nabla \cdot \mathbf{B}) - \nabla^2 \mathbf{B} \tag{6.3.3}$$

with $\nabla \cdot \mathbf{B} = 0$, and using Faraday's law (6.1.30), then gives the equation

$$\frac{\partial \mathbf{B}}{\partial t} = \nabla \times (\mathbf{U} \times \mathbf{B}) + \frac{1}{\mu_0 \sigma} \nabla^2 \mathbf{B}. \tag{6.3.4}$$

The above equation is often called the *induction equation* and shows that the time rate of change of the magnetic field is controlled by two terms. The first term, which involves the fluid velocity, is called the *convection term*, and the second term, which involves the conductivity, is called the *diffusion term*. To determine which term dominates, it is useful to introduce a dimensionless parameter R_{m}, called the magnetic Reynolds number, which is defined as the ratio of the typical magnitude of the convection term to the typical magnitude of the diffusion term,

$$R_{\mathrm{m}} = \frac{|\nabla \times (\mathbf{U} \times \mathbf{B})|}{|\frac{1}{\mu_0 \sigma} \nabla^2 \mathbf{B}|}. \tag{6.3.5}$$

The magnetic Reynolds number is analogous to the Reynolds number commonly used in fluid mechanics; see Batchelor (1967). To make a rough estimate of the magnetic Reynolds number, replace ∇ by $1/L$, where L is a length-scale that characterizes the spatial gradients in the fluid, and ignore the vector character of the equation. The ratio of the magnitude of the convection term to the magnitude of the diffusion term is then given by

$$R_{\mathrm{m}} = \mu_0 \sigma U L. \tag{6.3.6}$$

If $R_{\mathrm{m}} \gg 1$, the convection term dominates, and if $R_{\mathrm{m}} \ll 1$, the diffusion term dominates.

6.3.1 $R_{\mathrm{m}} \gg 1$, the Frozen Field Theorem

Large magnetic Reynolds numbers occur whenever the conductivity, the fluid velocity, and the length-scale are sufficiently large to make the product $\mu_0 \sigma_0 U L \gg 1$. The diffusion term is then much smaller than the convection term and, to a first approximation, can be ignored. When $R_{\mathrm{m}} \gg 1$, Eq. (6.3.4) simplifies to

$$\frac{\partial \mathbf{B}}{\partial t} = \nabla \times (\mathbf{U} \times \mathbf{B}). \tag{6.3.7}$$

The above equation can be used to prove an important result known as the "frozen field theorem" (also known as Alfvén's theorem). Two different, but equivalent, statements of the frozen field theorem can be made, each of which involves a somewhat different proof.

Statement 1. The magnetic flux threading any closed curve moving with the fluid is constant.

To prove Statement 1, consider the magnetic flux through a closed curve C at time t, as shown in Figure 6.1. Suppose that curve C moves with the fluid, with each point on the curve moving to a new point $\mathbf{U}\,\mathrm{d}t$ after a time $\mathrm{d}t$. It can then be seen that the rate of change of the magnetic flux through curve C is given by the sum of two terms:

$$\frac{\mathrm{d}\Phi_{\mathrm{B}}}{\mathrm{d}t} = \int_S \frac{\partial \mathbf{B}}{\partial t} \cdot \mathrm{d}\mathbf{A} + \int_C \mathbf{B} \cdot (\mathbf{U} \times \mathrm{d}\boldsymbol{\ell}). \tag{6.3.8}$$

The first term gives the rate of change of the flux due to the explicit time dependence of $\mathbf{B}$, and the second term gives the rate of change of the flux due

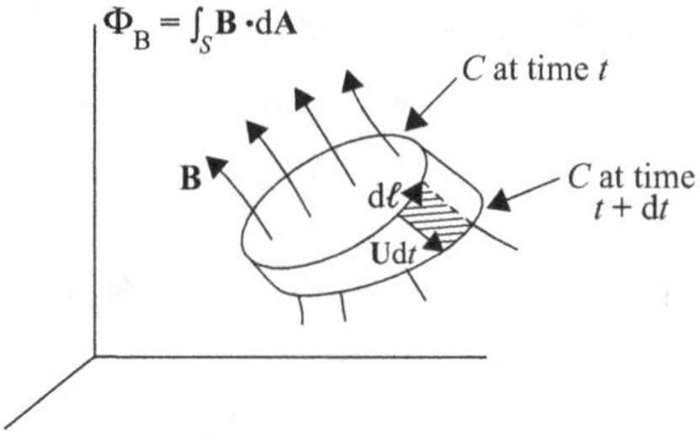

Figure 6.1 The temporal evolution of the magnetic flux through a curve C moving with the fluid from time t to time $t + \mathrm{d}t$.

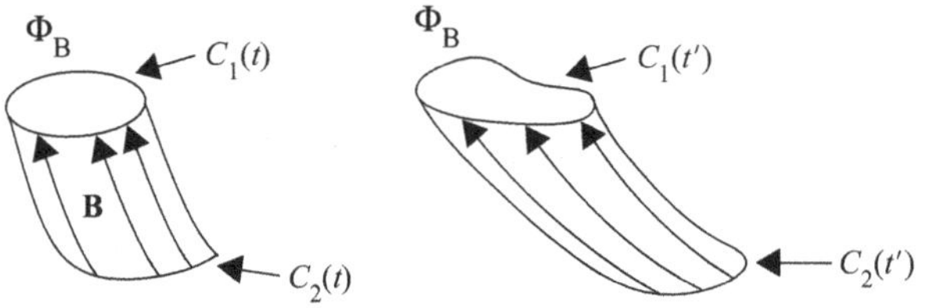

Figure 6.2 For large magnetic Reynolds numbers, the magnetic flux, Φ_B, is "frozen" into the fluid.

to the motion of the curve C. The triple vector product in the second term can be permuted once to give

$$\frac{\mathrm{d}\Phi_B}{\mathrm{d}t} = \int_S \frac{\partial \mathbf{B}}{\partial t} \cdot \mathrm{d}\mathbf{A} + \int_C \mathrm{d}\boldsymbol{\ell} \cdot (\mathbf{B} \times \mathbf{U}). \tag{6.3.9}$$

By Stokes' theorem and Eq. (6.3.7), it follows that

$$\frac{\mathrm{d}\Phi_B}{\mathrm{d}t} = \int_S \left[\frac{\partial \mathbf{B}}{\partial t} - \boldsymbol{\nabla} \times (\mathbf{U} \times \mathbf{B}) \right] \cdot \mathrm{d}\mathbf{A} = 0, \tag{6.3.10}$$

which proves Statement 1. This theorem has the following consequence. Consider two curves, C_1 and C_2, that are connected by magnetic field lines at time t, as shown in Figure 6.2. These field lines form a flux tube with a total magnetic flux Φ_B. As the fluid moves, since $\mathrm{d}\Phi_B/\mathrm{d}t = 0$ on the surfaces that make up the tube, it is easy to see that the tube continues to contain the same total magnetic flux at any later time t'. Since the ends of the flux tube can be shrunk to infinitesimal size, the flux tube becomes a magnetic field line, which then justifies the statement that "the magnetic field lines are frozen into the fluid."

Statement 2. If a line moving with the fluid is a magnetic field line initially, it will be so for all times.

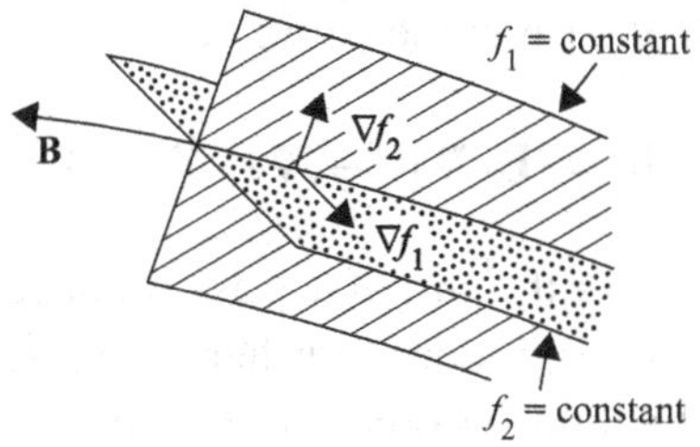

Figure 6.3 Surfaces of constant f_1 and f_2 define a magnetic field line.

To prove Statement 2, we start by introducing the concept of *flux coordinates*. Consider the partial differential equation

$$\mathbf{B}\cdot\nabla f = 0. \tag{6.3.11}$$

Since ∇f is perpendicular to a contour of constant f, it follows that f is a constant along a magnetic field line. Consider two functionally independent families of surfaces, f_1 and f_2, that obey

$$\mathbf{B}\cdot\nabla f_1 = \mathbf{B}\cdot\nabla f_2 = 0. \tag{6.3.12}$$

A magnetic field line can then be specified by the intersection of two such surfaces, as shown in Figure 6.3. Since $\mathbf{B}$ is perpendicular to ∇f_1 and to ∇f_2, we can think of $\mathbf{B}$, ∇f_1, and ∇f_2 as a triad of vectors in three-dimensional space, such that

$$\mathbf{B} = \nabla f_1 \times \nabla f_2, \tag{6.3.13}$$

provided f_1 and f_2 are suitably normalized. The variables f_1 and f_2 are known as flux coordinates. By computing the divergence of the above equation, it can be shown that this representation automatically satisfies the divergence-free condition $\nabla\cdot\mathbf{B} = 0$.

With this preparation, we are now ready to proceed with the proof of Statement 2. The basic approach is to compute the time rate of change of $\mathbf{B}\cdot\nabla f$ along a representative fluid trajectory. If we can show that the rate of change is identically zero, then Statement 2 is proven. To compute the time rate of change, we must use the convective derivative for d/dt, so that

$$\frac{\mathrm{d}}{\mathrm{d}t}(\mathbf{B}\cdot\nabla f) = \frac{\partial}{\partial t}(\mathbf{B}\cdot\nabla f) + (\mathbf{U}\cdot\nabla)(\mathbf{B}\cdot\nabla f). \tag{6.3.14}$$

The first term of the above equation can be expanded as follows:

$$\frac{\partial}{\partial t}(\mathbf{B}\cdot\nabla f) = \frac{\partial \mathbf{B}}{\partial t}\cdot\nabla f + \mathbf{B}\cdot\nabla\frac{\partial f}{\partial t}. \tag{6.3.15}$$

Using Eq. (6.3.7), the $\partial \mathbf{B}/\partial t$ term in the above equation can be written in the form

$$\frac{\partial \mathbf{B}}{\partial t} = \nabla \times (\mathbf{U} \times \mathbf{B}) = (\mathbf{B} \cdot \nabla)\mathbf{U} - (\nabla \cdot \mathbf{U})\mathbf{B} - (\mathbf{U} \cdot \nabla)\mathbf{B}, \tag{6.3.16}$$

where we have used the rule for the curl of a cross-product, $\nabla \times (\mathbf{F} \times \mathbf{G}) = \mathbf{F}(\nabla \cdot \mathbf{G}) - \mathbf{G}(\nabla \cdot \mathbf{F}) + (\mathbf{G} \cdot \nabla)\mathbf{F} - (\mathbf{F} \cdot \nabla)\mathbf{G}$. To proceed further, we must require that the surfaces of constant f move with the fluid. This condition can be imposed by requiring that

$$\frac{\mathrm{d}f}{\mathrm{d}t} = \frac{\partial f}{\partial t} + \mathbf{U} \cdot \nabla f = 0, \tag{6.3.17}$$

where $\mathrm{d}/\mathrm{d}t = \partial/\partial t + \mathbf{U} \cdot \nabla$ is the convective derivative.

Solving the above equation for $\partial f/\partial t$ and substituting the result into Eq. (6.3.15), and then substituting Eqs. (6.3.15) and (6.3.16) into Eq. (6.3.14) gives

$$\begin{aligned}\frac{\mathrm{d}}{\mathrm{d}t}(\mathbf{B} \cdot \nabla f) = {} & [(\mathbf{B} \cdot \nabla)\mathbf{U}] \cdot \nabla f - (\nabla \cdot \mathbf{U})(\mathbf{B} \cdot \nabla f) - [(\mathbf{U} \cdot \nabla)\mathbf{B}] \cdot \nabla f \\ & - \mathbf{B} \cdot \nabla[(\mathbf{U} \cdot \nabla) f] + (\mathbf{U} \cdot \nabla)(\mathbf{B} \cdot \nabla f).\end{aligned} \tag{6.3.18}$$

The first, third, and fourth terms on the right-hand side of the above equation can be simplified by using index notation, as shown below:

$$\begin{aligned}&[(\mathbf{B} \cdot \nabla)\mathbf{U}] \cdot \nabla f - [(\mathbf{U} \cdot \nabla)\mathbf{B}] \cdot \nabla f - \mathbf{B} \cdot \nabla[(\mathbf{U} \cdot \nabla) f] \\ &= \sum_{i,j} \left[B_i \frac{\partial U_j}{\partial x_i} \frac{\partial f}{\partial x_j} - U_i \frac{\partial B_j}{\partial x_i} \frac{\partial f}{\partial x_j} - B_i \frac{\partial}{\partial x_i} \left(U_j \frac{\partial f}{\partial x_j} \right) \right] \\ &= \sum_{i,j} \left[B_i \frac{\partial U_j}{\partial x_i} \frac{\partial f}{\partial x_j} - U_i \frac{\partial B_j}{\partial x_i} \frac{\partial f}{\partial x_j} - B_i \frac{\partial U_j}{\partial x_i} \frac{\partial f}{\partial x_j} - B_i U_j \frac{\partial}{\partial x_i} \frac{\partial f}{\partial x_j} \right] \\ &= -(\mathbf{U} \cdot \nabla)(\mathbf{B} \cdot \nabla f).\end{aligned} \tag{6.3.19}$$

Note that the final expression in the above equation is obtained by combining the second and last terms inside the parentheses, while the first and third terms cancel each other. Substituting the above expression into Eq. (6.3.18) gives the result that we are seeking, which is

$$\frac{\mathrm{d}}{\mathrm{d}t}(\mathbf{B} \cdot \nabla f) = -(\nabla \cdot \mathbf{U})(\mathbf{B} \cdot \nabla f). \tag{6.3.20}$$

The above equation shows that if $\mathbf{B} \cdot \nabla f = 0$ at $t = 0$, then $\mathbf{B} \cdot \nabla f = 0$ for all times. Hence, if the contours of constant f_1 and f_2 label a magnetic field line (i.e., $\mathbf{B} \cdot \nabla f = 0$) at $t = 0$, then they label a magnetic field line for all times, which proves Statement 2.

From the frozen field theorem, we can assert that if at $t = 0$ we label a field line by tagging fluid particles on it, then the same fluid particles will remain frozen to the field line at all subsequent times. It follows that in such a plasma, field lines cannot intersect or break. For if they did, the velocity of the fluid particles at the break-point or the point of intersection would have to change discontinuously. In other words, the "topology" of the magnetic field lines cannot change.

An interesting consequence of the frozen field theorem is that the magnetic field strength can be amplified by changes in the fluid geometry or the mass density. Consider a fluid element that is stretched between two times t and t', as shown in Figure 6.4. The frozen field theorem shows that $B\Delta A = B'\Delta A'$. From mass conservation, one also has the relation $\rho_{\rm m}\Delta A\ell = \rho'_{\rm m}\Delta A'\ell'$. Eliminating ΔA and $\Delta A'$ between these two equations gives

$$B' = B\left(\frac{\rho'_{\rm m}}{\rho_{\rm m}}\right)\left(\frac{\ell'}{\ell}\right). \tag{6.3.21}$$

This equation shows that the magnetic field strength increases whenever either the mass density $\rho'_{\rm m}$ or the length ℓ' increases. An example of magnetic amplification occurs when a star collapses to form a neutron star. As the star collapses, the density ($\rho'_{\rm m} \sim 1/\ell'^3$) increases much faster than the length ℓ' decreases, so the magnetic field is increased by a factor of $(\ell/\ell')^2$, which can be very large. Field strengths as large as 10^{11} to 10^{12} gauss can be produced at the surface of a neutron star from the ~1 gauss surface field of the parent star.

An example of magnetic amplification due to stretching occurs in the solar atmosphere. Because the Sun rotates faster near the equator than near the poles, magnetic field lines near the equator are stretched azimuthally around the Sun, as shown in Figure 6.5. Since the stretching motion is nearly horizontal, the density

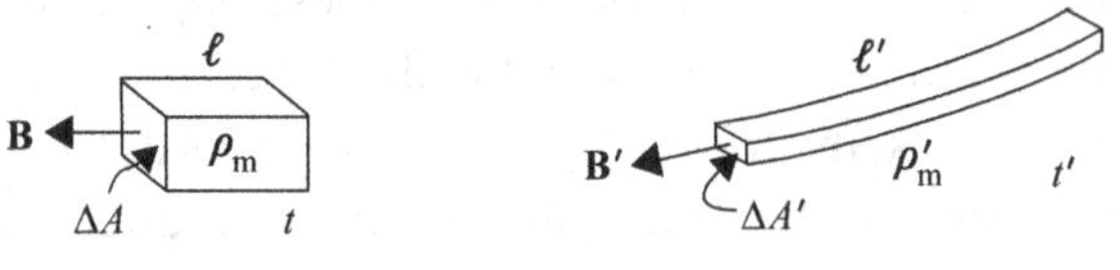

Figure 6.4 Stretching a magnetic flux tube that is frozen into the fluid increases the magnetic field strength.

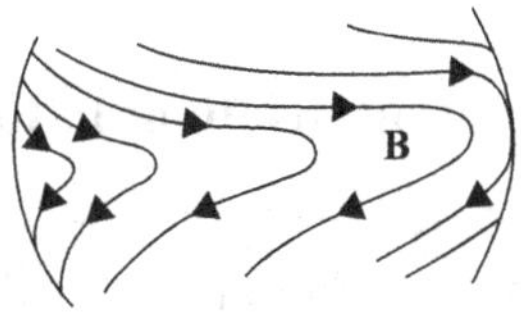

Figure 6.5 Differential rotation at the Sun stretches the magnetic field lines, thereby increasing the magnetic field strength.

remains essentially constant ($\rho' \sim \rho$). As the length of the field line increases, the magnetic field strength increases ($B \sim \ell'$). Eventually the magnetic field becomes so strong that the Rayleigh–Taylor instability (see Section 7.3.7) causes the magnetic field lines to bulge up through the surface of the Sun (Elsasser, 1956; Babcock, 1961). The strong magnetic field in the bulge inhibits the transport of heat from deeper layers in the Sun, causing sunspots, which are cooler and darker than the surrounding gas.

6.3.2 $R_{\mathrm{m}} \ll 1$, Magnetic Diffusion

If the magnetic Reynolds number is much less than one, then the convection term is much smaller than the diffusion term, and to a first approximation can be ignored. Equation (6.3.4) then simplifies to

$$\frac{\partial \mathbf{B}}{\partial t} = \frac{1}{\mu_0 \sigma} \nabla^2 \mathbf{B}. \tag{6.3.22}$$

The above equation has the mathematical form of a diffusion equation. Therefore, the magnetic field diffuses through the fluid in a manner analogous to the diffusion of heat through a solid. The characteristic time-scale for the diffusion, T, can be estimated by substituting $1/T$ for $\partial/\partial t$ and $1/L^2$ for ∇^2, which gives

$$T = \mu_0 \sigma L^2. \tag{6.3.23}$$

Some representative diffusion times are given in Table 6.1. These examples show that for most laboratory applications the diffusion times are quite short, typically on the order of seconds. On the other hand, for most geophysical and astrophysical applications, the diffusion times are extremely long, sometimes even longer than the age of the Universe. Therefore, for most of these applications, the magnetic field can be considered to be "frozen" into the fluid (i.e., $R_{\mathrm{m}} \gg 1$). However, even in such large-scale systems, the frozen field concept breaks down in small localized regions where the scale size becomes very small. Such effects are discussed in more detail in Sections 7.5 and 7.6 under the topics of resistive instabilities and magnetic reconnection.

6.4 Conservation Relations in Ideal MHD

In the limit of ideal MHD (i.e., $\sigma \to \infty$ or $\eta \to 0$), the MHD equations obey rigorous conservation relations, which can be made manifest by rewriting the ideal MHD equations in conservation form. The conservation form of a conserved quantity consists of an equation in which the time rate of change of the density

Table 6.1 *Diffusion times for various conducting systems*

Conductor	T
Copper sphere (10 cm diameter)	10^{-1} s
Fusion machine	10 s
Earth's molten iron core	10^4 years
Interior of the Sun	10^{10} years

of that quantity plus the divergence of the density flux is zero. Note that the mass continuity equation (6.1.31) is already written in conservation form, i.e.,

$$\frac{\partial \rho_{\rm m}}{\partial t} + \boldsymbol{\nabla} \cdot (\rho_{\rm m} \mathbf{U}) = 0. \tag{6.4.1}$$

To write the conservation of momentum in conservation form, we can use the continuity equation together with Eqs. (6.1.32) and (6.2.2) to put the momentum equation in the form

$$\frac{\partial}{\partial t}(\rho_{\rm m} \mathbf{U}) + \boldsymbol{\nabla} \cdot (\rho_{\rm m} \mathbf{U}\mathbf{U}) = \frac{1}{\mu_0}(\mathbf{B} \cdot \boldsymbol{\nabla})\mathbf{B} - \boldsymbol{\nabla}\left(\frac{B^2}{2\mu_0} + P\right), \tag{6.4.2}$$

which can be easily rearranged as follows:

$$\frac{\partial(\rho_{\rm m} \mathbf{U})}{\partial t} + \boldsymbol{\nabla} \cdot (\rho_{\rm m} \mathbf{U}\mathbf{U} + \overset{\leftrightarrow}{\mathbf{T}} + \overset{\leftrightarrow}{\mathbf{P}}) = 0. \tag{6.4.3}$$

To write the conservation of energy in conservation form, we start with the momentum equation (6.1.32). Multiplying the momentum equation by **U** and eliminating **J** via Ampère's law (6.1.28) gives

$$\rho_{\rm m} \mathbf{U} \cdot \left[\frac{\partial \mathbf{U}}{\partial t} + (\mathbf{U} \cdot \boldsymbol{\nabla})\mathbf{U}\right] = -(\mathbf{U} \cdot \boldsymbol{\nabla})P + \frac{\mathbf{U}}{\mu_0} \cdot [(\boldsymbol{\nabla} \times \mathbf{B}) \times \mathbf{B}]. \tag{6.4.4}$$

Using the mass continuity equation (6.1.31), the left-hand side of the above equation can be rewritten in the following form:

$$\begin{aligned} &\rho_{\rm m} \mathbf{U} \cdot \left[\frac{\partial \mathbf{U}}{\partial t} + (\mathbf{U} \cdot \boldsymbol{\nabla})\mathbf{U}\right] \\ &= \frac{1}{2}\rho_{\rm m}\left(\frac{\partial U^2}{\partial t} + \mathbf{U} \cdot \boldsymbol{\nabla} U^2\right) \end{aligned}$$

$$= \frac{\partial}{\partial t}\left(\frac{1}{2}\rho_{\rm m} U^2\right) - \frac{U^2}{2}\frac{\partial \rho_{\rm m}}{\partial t} + \frac{1}{2}\rho_{\rm m}\mathbf{U}\cdot\boldsymbol{\nabla}U^2$$

$$= \frac{\partial}{\partial t}\left(\frac{1}{2}\rho_{\rm m} U^2\right) + \frac{U^2}{2}\boldsymbol{\nabla}\cdot(\rho_{\rm m}\mathbf{U}) + \frac{1}{2}\rho_{\rm m}\mathbf{U}\cdot\boldsymbol{\nabla}U^2$$

$$= \frac{\partial}{\partial t}\left(\frac{1}{2}\rho_{\rm m} U^2\right) + \boldsymbol{\nabla}\cdot\left(\frac{U^2}{2}\rho_{\rm m}\mathbf{U}\right). \tag{6.4.5}$$

Next, consider the $(\mathbf{U}\cdot\boldsymbol{\nabla})P$ term on the right-hand side of Eq. (6.4.4). Since this term involves the pressure, for an equation of state we use the adiabatic equation of state given by Eq. (6.1.34) which, when expanded, becomes

$$\frac{\rm d}{{\rm d}t}\left(\frac{P}{\rho_{\rm m}^{\gamma}}\right) = \left(\frac{{\rm d}P}{{\rm d}t}\right)\rho_{\rm m}^{-\gamma} - \gamma P \rho_{\rm m}^{-(\gamma+1)}\frac{{\rm d}\rho_{\rm m}}{{\rm d}t} = 0. \tag{6.4.6}$$

The above equation simplifies to

$$\frac{{\rm d}P}{{\rm d}t} = \frac{\gamma P}{\rho_{\rm m}}\frac{{\rm d}\rho_{\rm m}}{{\rm d}t}. \tag{6.4.7}$$

To proceed further, the total derivatives in this equation must be replaced, using the convective derivative (${\rm d}/{\rm d}t = \partial/\partial t + \mathbf{U}\cdot\boldsymbol{\nabla}$), which gives

$$\frac{\partial P}{\partial t} + (\mathbf{U}\cdot\boldsymbol{\nabla})P = \frac{\gamma P}{\rho_{\rm m}}\left[\frac{\partial \rho_{\rm m}}{\partial t} + (\mathbf{U}\cdot\boldsymbol{\nabla})\rho_{\rm m}\right]. \tag{6.4.8}$$

Note that the desired term, $(\mathbf{U}\cdot\boldsymbol{\nabla})P$, has now appeared on the left-hand side of the equation. Using the mass continuity equation (6.1.31) to replace the $\partial\rho_{\rm m}/\partial t$ term, the above equation simplifies to

$$\frac{\partial P}{\partial t} + (\mathbf{U}\cdot\boldsymbol{\nabla})P = -\gamma P\,\boldsymbol{\nabla}\cdot\mathbf{U}. \tag{6.4.9}$$

Using the identity $P\boldsymbol{\nabla}\cdot\mathbf{U} = \boldsymbol{\nabla}\cdot(P\mathbf{U}) - \mathbf{U}\cdot\boldsymbol{\nabla}P$, the above equation can be written

$$\frac{\partial P}{\partial t} + (\mathbf{U}\cdot\boldsymbol{\nabla})P = -\gamma\boldsymbol{\nabla}\cdot(P\mathbf{U}) + \gamma\mathbf{U}\cdot\boldsymbol{\nabla}P, \tag{6.4.10}$$

which can be solved for $(\mathbf{U}\cdot\boldsymbol{\nabla})P$ to give

$$(\mathbf{U}\cdot\boldsymbol{\nabla})P = \frac{1}{\gamma-1}\frac{\partial P}{\partial t} + \frac{\gamma}{\gamma-1}\boldsymbol{\nabla}\cdot(P\mathbf{U}). \tag{6.4.11}$$

Returning again to Eq. (6.4.4), the second term on the right can be rewritten as

$$\mathbf{U}\cdot(\boldsymbol{\nabla}\times\mathbf{B})\times\mathbf{B} = -(\mathbf{U}\times\mathbf{B})\cdot(\boldsymbol{\nabla}\times\mathbf{B}). \tag{6.4.12}$$

Noting that $\mathbf{E} = -\mathbf{U} \times \mathbf{B}$ for an ideal MHD fluid, the term on the right can be expanded using the identity $\nabla \cdot (\mathbf{F} \times \mathbf{G}) = \mathbf{G} \cdot (\nabla \times \mathbf{F}) - \mathbf{F} \cdot (\nabla \times \mathbf{G})$ to give

$$\begin{aligned}\mathbf{U} \cdot (\nabla \times \mathbf{B}) \times \mathbf{B} = \mathbf{E} \cdot (\nabla \times \mathbf{B}) &= \mathbf{B} \cdot (\nabla \times \mathbf{E}) - \nabla \cdot (\mathbf{E} \times \mathbf{B}) \\ &= -\mathbf{B} \cdot \frac{\partial \mathbf{B}}{\partial t} - \nabla \cdot (\mathbf{E} \times \mathbf{B}) \\ &= -\frac{\partial}{\partial t}\left(\frac{B^2}{2}\right) - \nabla \cdot (\mathbf{E} \times \mathbf{B}).\end{aligned} \tag{6.4.13}$$

Finally, substituting Eqs. (6.4.5), (6.4.11), and (6.4.13) into Eq. (6.4.4), we obtain the conservation of energy for an ideal MHD fluid in conservation form:

$$\begin{aligned}&\frac{\partial}{\partial t}\left(\frac{1}{2}\rho_{\mathrm{m}} U^2 + \frac{P}{\gamma - 1} + \frac{B^2}{2\mu_0}\right) \\ &\quad + \nabla \cdot \left(\frac{1}{2}\rho_{\mathrm{m}} U^2 \mathbf{U} + \frac{\gamma}{\gamma - 1} P \mathbf{U} + \frac{1}{\mu_0} \mathbf{E} \times \mathbf{B}\right) = 0.\end{aligned} \tag{6.4.14}$$

In thermodynamics, the term

$$\xi = P/(\gamma - 1) \tag{6.4.15}$$

is called the *internal energy*, and the term

$$h = \gamma P/(\gamma - 1) \tag{6.4.16}$$

is called the *enthalpy*. Although Eq. (6.4.14) was derived using the adiabatic equation of state, which is a reversible process, Eqs. (6.4.15) and (6.4.16) for the internal energy and enthalpy are also applicable to irreversible processes, provided we use $\gamma = C_{\mathrm{P}}/C_{\mathrm{V}}$, or equivalently $\gamma = (f + 2)/f$. The reason these equations are valid for irreversible as well as reversible processes is that they are functions of the state of the system and do not depend on the detailed process by which this state is achieved. Other terms recognizable in Eq. (6.4.14) are the magnetic energy density, $B^2/2\mu_0$, the kinetic energy density, $(1/2)\rho_{\mathrm{m}} U^2$, and the Poynting flux, $\mathbf{S} = (1/\mu_0)\mathbf{E} \times \mathbf{B}$.

If we integrate the energy conservation law over all space, the divergence term in Eq. (6.4.14) can be converted to a surface integral at infinity, which gives

$$\frac{\partial}{\partial t} \int_V \left(\frac{1}{2}\rho_{\mathrm{m}} U^2 + \xi + \frac{B^2}{2\mu_0}\right) \mathrm{d}^3 x + \int_S \left(\frac{1}{2}\rho_{\mathrm{m}} U^2 \mathbf{U} + h\mathbf{U} + \frac{1}{\mu_0}\mathbf{E} \times \mathbf{B}\right) \cdot \mathrm{d}\mathbf{A} = 0, \tag{6.4.17}$$

where we have introduced the symbols ξ and h for the internal energy and the enthalpy, respectively. Assuming that the surface integral vanishes at infinity, the above equation simplifies to

$$\frac{\partial}{\partial t}\int_V\left(\frac{1}{2}\rho_{\mathrm{m}}U^2+\xi+\frac{B^2}{2\mu_0}\right)\mathrm{d}^3x=0, \tag{6.4.18}$$

which can be written in the form of an energy conservation equation

$$K+W=\text{constant}, \tag{6.4.19}$$

where

$$K=\int_V\frac{1}{2}\rho_{\mathrm{m}}U^2\,\mathrm{d}^3x \tag{6.4.20}$$

is the kinetic energy, and

$$W=\int_V\left(\frac{P}{\gamma-1}+\frac{B^2}{2\mu_0}\right)\mathrm{d}^3x \tag{6.4.21}$$

is the potential energy. These energy conservation equations will be useful later when we consider the stability of various MHD systems.

6.5 Magnetohydrodynamic Waves

Next we consider the propagation of *small-amplitude* waves in a homogeneous ideal (infinitely conducting) MHD fluid. The relevant equations are Faraday's law (6.1.30), the mass continuity equation (6.1.31), the momentum equation (6.1.32), and an equation of state, which we take to be the adiabatic equation of state (6.1.34). Since waves usually occur on time-scales sufficiently short that no heat flows, the adiabatic equation of state is expected to be a good approximation. The equations are linearized in the usual way by assuming that $\mathbf{U}$,ρ_{m},$\mathbf{B}$, and P are the sum of a spatially uniform time-independent zero-order (0) quantity plus a small first-order (1) perturbation, i.e., $\mathbf{U}=\mathbf{U}_1$, $\rho_{\mathrm{m}}=\rho_{\mathrm{m}0}+\rho_{\mathrm{m}1}$, $\mathbf{B}=\mathbf{B}_0+\mathbf{B}_1$, and $P=P_0+P_1$. The linearized *first-order* equations of motion then become

$$\frac{\partial\rho_{\mathrm{m}1}}{\partial t}+\rho_{\mathrm{m}0}\nabla\cdot\mathbf{U}_1=0, \tag{6.5.1}$$

$$\rho_{\mathrm{m}0}\frac{\partial\mathbf{U}_1}{\partial t}=\frac{1}{\mu_0}(\nabla\times\mathbf{B}_1)\times\mathbf{B}_0-\nabla P_1, \tag{6.5.2}$$

$$\frac{\partial\mathbf{B}_1}{\partial t}=\nabla\times(\mathbf{U}_1\times\mathbf{B}_0), \tag{6.5.3}$$

and

$$P_1 = \gamma\left(\frac{P_0}{\rho_{m0}}\right)\rho_{m1}. \tag{6.5.4}$$

In the above equation, it is convenient to introduce the speed of sound V_S, which is defined by the equation

$$V_S^2 = \gamma\frac{P_0}{\rho_{m0}} = \frac{\gamma\kappa T_0}{m}, \tag{6.5.5}$$

where T_0 is the zero-order temperature, and m is the average molecular mass. Next, we Fourier-analyze the above equations by making the usual operator substitutions, $\nabla \to \mathrm{i}\mathbf{k}$ and $\partial/\partial t \to -\mathrm{i}\omega$. At this point we also drop the subscript 1 on the first-order terms, which can be distinguished from the zero-order terms by the subscript 0 on the zero-order terms. The resulting equations are

$$-\mathrm{i}\omega\tilde{\rho}_m + \mathrm{i}\rho_{m0}\mathbf{k}\cdot\tilde{\mathbf{U}} = 0, \tag{6.5.6}$$

$$-\mathrm{i}\omega\rho_{m0}\tilde{\mathbf{U}} = \frac{\mathrm{i}}{\mu_0}(\mathbf{k}\times\tilde{\mathbf{B}})\times\mathbf{B}_0 - \mathrm{i}\mathbf{k}\widetilde{P}, \tag{6.5.7}$$

$$-\mathrm{i}\omega\tilde{\mathbf{B}} = \mathrm{i}\mathbf{k}\times(\tilde{\mathbf{U}}\times\mathbf{B}_0), \tag{6.5.8}$$

$$\widetilde{P} = V_S^2\tilde{\rho}_m. \tag{6.5.9}$$

Eliminating $\tilde{\rho}_m$ from the above equation by using Eq. (6.5.6), we obtain the following equation for the first-order pressure perturbation:

$$\widetilde{P} = V_S^2\frac{\rho_{m0}}{\omega}\mathbf{k}\cdot\tilde{\mathbf{U}}. \tag{6.5.10}$$

Using the above equation, the pressure $\widetilde{P}$ can be eliminated from Eq. (6.5.7), which, after multiplying by $\mathrm{i}\omega/\rho_{m0}$, becomes

$$\omega^2\tilde{\mathbf{U}} = \frac{-\omega}{\mu_0\rho_{m0}}(\mathbf{k}\times\tilde{\mathbf{B}})\times\mathbf{B}_0 + V_S^2\mathbf{k}(\mathbf{k}\cdot\tilde{\mathbf{U}}). \tag{6.5.11}$$

Finally, $\tilde{\mathbf{B}}$ can be eliminated from the above equation by using Eq. (6.5.8), which gives the homogeneous equation for the fluid velocity:

$$\omega^2\tilde{\mathbf{U}} = \frac{1}{\mu_0\rho_{m0}}\{\mathbf{k}\times(\mathbf{k}\times[\tilde{\mathbf{U}}\times\mathbf{B}_0])\}\times\mathbf{B}_0 + V_S^2\mathbf{k}(\mathbf{k}\cdot\tilde{\mathbf{U}}). \tag{6.5.12}$$

With no loss in generality, we can assume that $\mathbf{B}_0 = (0,0,B_0)$ and $\mathbf{k} = (k\sin\theta, 0, k\cos\theta)$, as shown in Figure 6.6. Working out the cross-products in

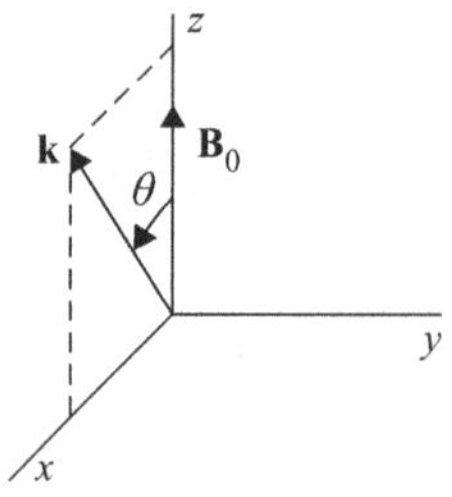

Figure 6.6 The coordinate system used to analyze MHD wave propagation.

Eq. (6.5.12), dividing by k^2, and factoring out B_0^2, we obtain the equation

$$\left(\frac{\omega}{k}\right)^2 \begin{bmatrix} \widetilde{U}_x \\ \widetilde{U}_y \\ \widetilde{U}_z \end{bmatrix} = V_A^2 \begin{bmatrix} \widetilde{U}_x \\ \widetilde{U}_y \cos^2\theta \\ 0 \end{bmatrix} + V_S^2 \begin{bmatrix} \widetilde{U}_x \sin^2\theta + \widetilde{U}_z \sin\theta\cos\theta \\ 0 \\ \widetilde{U}_x \sin\theta\cos\theta + \widetilde{U}_z \cos^2\theta \end{bmatrix}, \tag{6.5.13}$$

where the quantity

$$V_A = B_0 / \sqrt{\mu_0 \rho_{m0}} \tag{6.5.14}$$

has units of velocity and is called the Alfvén velocity, after Alfvén (1942). Substituting $\upsilon_p = \omega/k$ for the phase velocity, the homogeneous equation (6.5.13) can be written in matrix form as

$$\begin{bmatrix} \upsilon_p^2 - V_S^2 \sin^2\theta - V_A^2 & 0 & -V_S^2 \sin\theta\cos\theta \\ 0 & \upsilon_p^2 - V_A^2 \cos^2\theta & 0 \\ -V_S^2 \sin\theta\cos\theta & 0 & \upsilon_p^2 - V_S^2\cos^2\theta \end{bmatrix} \begin{bmatrix} \widetilde{U}_x \\ \widetilde{U}_y \\ \widetilde{U}_z \end{bmatrix} = 0. \tag{6.5.15}$$

This equation has non-trivial solutions for $\widetilde{\mathbf{U}}$ if and only if the determinant of the matrix is zero, which gives the dispersion relation

$$D(k,\omega) = \left(\upsilon_p^2 - V_A^2\cos^2\theta\right)\left[\upsilon_p^4 - \upsilon_p^2\left(V_A^2 + V_S^2\right) + V_A^2 V_S^2 \cos^2\theta\right] = 0. \tag{6.5.16}$$

It can be shown that the dispersion relation has three roots:

$$\upsilon_p^2 = \frac{1}{2}\left(V_A^2 + V_S^2\right) - \frac{1}{2}\left[\left(V_A^2 - V_S^2\right)^2 + 4V_A^2 V_S^2 \sin^2\theta\right]^{1/2}, \tag{6.5.17}$$

$$\upsilon_p^2 = V_A^2 \cos^2\theta, \tag{6.5.18}$$

$$\upsilon_p^2 = \frac{1}{2}\left(V_A^2 + V_S^2\right) + \frac{1}{2}\left[\left(V_A^2 - V_S^2\right)^2 + 4V_A^2 V_S^2 \sin^2\theta\right]^{1/2}. \tag{6.5.19}$$

The roots, Eqs. (6.5.17), (6.5.18), and (6.5.19), are called the slow magnetosonic mode, the transverse Alfvén mode (also called the shear Alfvén mode), and the fast

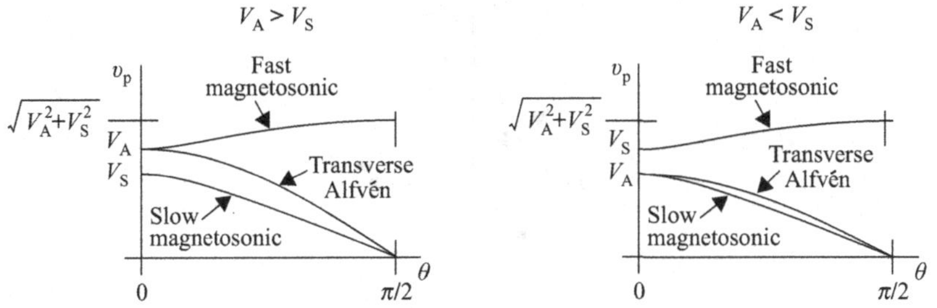

Figure 6.7 Plots of the phase velocities for the three MHD modes as a function of the wave normal angle for two cases, $V_A > V_S$ and $V_A < V_S$.

magnetosonic mode, respectively. In the low-temperature, low-frequency limit, it can be shown that for $\theta = 0$ the transverse Alfvén mode corresponds to the $n^2 = L$ mode, and the fast magnetosonic mode corresponds to the $n^2 = R$ mode (see Section 4.4.1). The slow magnetosonic mode disappears completely in the limit of zero temperature (i.e., $V_S = 0$). The phase velocities of the three modes are shown as a function of the wave normal angle in Figure 6.7. Two cases, $V_A > V_S$ and $V_A < V_S$, must be considered. For $V_A > V_S$, the transverse Alfvén mode connects with the fast magnetosonic mode at $\theta = 0$, whereas for $V_A < V_S$ the transverse Alfvén mode connects with the slow magnetosonic mode at $\theta = 0$.

Next, we discuss the eigenvectors associated with each of these modes.

6.5.1 The Transverse (or Shear) Alfvén Mode

It can be verified that the root for the transverse Alfvén mode, given by Eq. (6.5.18), has eigenvectors

$$\tilde{\mathbf{U}} = (0, \widetilde{U}_y, 0), \tag{6.5.20}$$

$$\tilde{\mathbf{B}} = (0, \widetilde{B}_y, 0), \quad \widetilde{B}_y = -B_0(\widetilde{U}_y/V_A)\ \mathrm{Sign}(\cos\theta), \tag{6.5.21}$$

$$\tilde{\mathbf{E}} = (\widetilde{E}_x, 0, 0), \quad \widetilde{E}_x = -B_0\widetilde{U}_y, \tag{6.5.22}$$

$$\tilde{\rho}_m = 0. \tag{6.5.23}$$

These eigenvectors are shown in Figure 6.8.

As can be seen, the fluid motions for this mode are entirely transverse, with no compressional component (i.e., $\mathbf{k} \cdot \tilde{\mathbf{U}} = 0$). This is the reason the mode is called the transverse Alfvén mode. The propagation velocity is controlled entirely by the Alfvén speed. Since there is no compression, the fluid pressure and temperature play no role in the propagation of this mode. From the direction of the electric and magnetic field, it is easy to see that the Poynting flux, $\mathbf{S} = (1/\mu_0)\,\mathbf{E}\times\mathbf{B}$, is parallel to the static magnetic field $\mathbf{B}_0$. The electromagnetic energy flow is then exactly

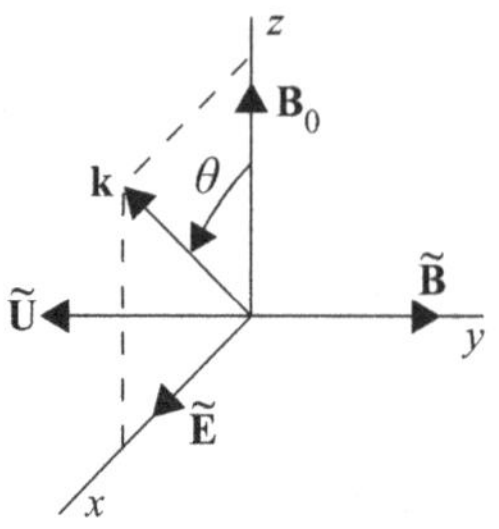

Figure 6.8 The eigenvectors for the transverse Alfvén mode.

along the static magnetic field, *independent* of the wave normal angle. From the dispersion relation for this mode (6.5.18), it can be shown that the group velocity is given by

$$\mathbf{v}_g = \boldsymbol{\nabla}_\mathbf{k}\omega = V_A\hat{\mathbf{z}}, \tag{6.5.24}$$

which is parallel to the static magnetic field, consistent with the fact that the Poynting flux is parallel to the magnetic field.

The propagation of the transverse Alfvén wave has an interesting analogy with the propagation of waves on a taut string. From the magnetic pressure tensor we have seen that the magnetic pressure can be thought of as an isotropic pressure plus a tension force per unit area, B^2/μ_0, along the magnetic field. It is well known that the velocity of propagation of a wave along a taut string is given by

$$v_p = \sqrt{T/\lambda_m}, \tag{6.5.25}$$

where T is the tension and λ_m is the mass per unit length. If we substitute $T = (B^2/\mu_0)\Delta A$ for the tension force and $\lambda_m = \rho_m\Delta A$ for the mass per unit length (where ΔA is the cross-sectional area), the velocity of propagation is given by $v_p = B/\sqrt{\mu_0\rho_m}$, which is just the Alfvén velocity. Thus, the propagation of the transverse (or shear) Alfvén wave can be thought of as being analogous to the propagation of a wave on a taut string with the magnetic field providing the tension and the MHD fluid providing the mass per unit length. This analogy also suggests that waves on different magnetic field lines propagate independently, as though they were on separate strings. Thus, even though the wave vector makes a substantial angle to the magnetic field, as illustrated in Figure 6.9, the wave energy is transported along the magnetic field lines, just as though the wave were propagating on a system of taut strings.

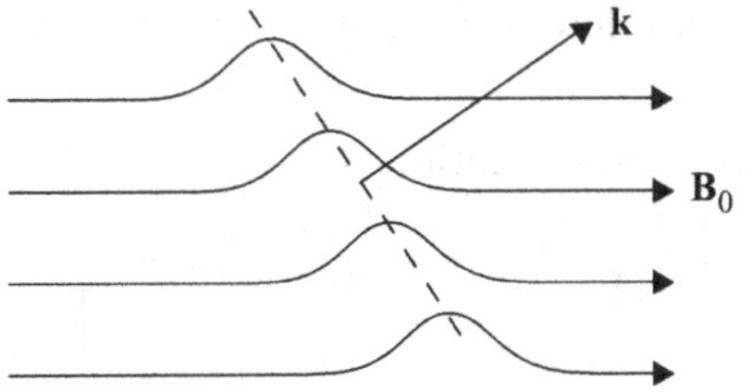

Figure 6.9 The taut string analogy for the propagation of the transverse Alfvén mode.

6.5.2 The Fast and Slow Magnetosonic Modes

As can be seen from the dispersion relations for the slow and fast magnetosonic modes, Eqs. (6.5.17) and (6.5.19) respectively, these modes involve both the magnetic field (via the Alfvén speed) and the plasma pressure (via the sound speed). For this reason these two modes are called the magnetosonic modes. For oblique angles of propagation, it can be shown that the flow velocities have components along the **k** vector. Thus, in contrast to the transverse Alfvén mode, which is purely transverse, the magnetosonic modes have both transverse and longitudinal (i.e., compressional) components. Since the dispersion relation is rather complicated for arbitrary angles of propagation, we first consider the limiting cases of $\theta = 0$ and $\theta = \pi/2$.

For $\theta = 0$ the homogeneous equation is

$$\begin{bmatrix} v_{\rm p}^2 - V_{\rm A}^2 & 0 \\ 0 & v_{\rm p}^2 - V_{\rm S}^2 \end{bmatrix} \begin{bmatrix} \widetilde{U}_x \\ \widetilde{U}_z \end{bmatrix} = 0. \qquad (6.5.26)$$

The dispersion relation can be seen to have two roots, $v_{\rm p}^2 = V_{\rm A}^2$ and $v_{\rm p}^2 = V_{\rm S}^2$. Just which of these two roots corresponds to the fast mode and which corresponds to the slow mode depends on the relative magnitudes of $V_{\rm A}$ and $V_{\rm S}$. For $V_{\rm A} > V_{\rm S}$, $v_{\rm p}^2 = V_{\rm A}^2$ is the fast mode and $v_{\rm p}^2 = V_{\rm S}^2$ is the slow mode. For $V_{\rm S} > V_{\rm A}$, the fast and slow designations are reversed. The eigenvectors corresponding to the $v_{\rm p}^2 = V_{\rm A}^2$ root are $\tilde{\mathbf{U}} = (\widetilde{U}_x, 0, 0), \tilde{\mathbf{B}} = (\widetilde{B}_x, 0, 0), \tilde{\mathbf{E}} = (0, \widetilde{E}_y, 0)$, and $\tilde{\rho}_{\rm m} = 0$. Note that **k** and $\tilde{\mathbf{U}}$ are orthogonal. Since $\mathbf{k} \cdot \tilde{\mathbf{U}} = 0$, there is no density compression. Thus, the eigenvectors for this mode have the properties of a *transverse electromagnetic wave*. (That this mode turns out to be transverse is a special case that only occurs at $\theta = 0$.) In contrast, the eigenvectors corresponding to the $v_{\rm p}^2 = V_{\rm S}^2$ root are $\tilde{\mathbf{U}} = (0, 0, \widetilde{U}_z)$, $\tilde{\mathbf{B}} = (0, 0, 0)$, $\tilde{\mathbf{E}} = (0, 0, 0)$, and $\tilde{\rho}_{\rm m} = \rho_{\rm m0}(\widetilde{U}_z / v_{\rm p})$. Note that the electromagnetic fields are zero and that there is a density compression. These eigenvectors have the properties of a *sound wave*. (That the electromagnetic component disappears completely is also a special case that only occurs at $\theta = 0$.)

For $\theta = \pi/2$ the homogeneous equation is

$$\begin{bmatrix} \upsilon_{\mathrm{p}}^2 - (V_{\mathrm{A}}^2 + V_{\mathrm{S}}^2) & 0 \\ 0 & \upsilon_{\mathrm{p}}^2 \end{bmatrix} \begin{bmatrix} \widetilde{U}_x \\ \widetilde{U}_z \end{bmatrix} = 0. \tag{6.5.27}$$

In this case, the dispersion relation has only one non-trivial root, $\upsilon_{\mathrm{p}}^2 = V_{\mathrm{A}}^2 + V_{\mathrm{S}}^2$. The eigenvectors corresponding to this root are $\tilde{\mathbf{U}} = (\widetilde{U}_{\mathrm{x}}, 0, 0)$, $\tilde{\mathbf{B}} = (0, 0, \widetilde{B}_{\mathrm{z}})$, $\tilde{\mathbf{E}} = (0, \widetilde{E}_y, 0)$, and $\tilde{\rho}_{\mathrm{m}} = \rho_{\mathrm{m0}}(\widetilde{U}_{\mathrm{x}}/\upsilon_{\mathrm{p}})$. These eigenvectors share features of both an electromagnetic wave and a sound wave.

For intermediate wave normal angles, the two magnetosonic modes have both longitudinal (i.e., sound wave) and transverse (i.e., electromagnetic) components. Further insight into the nature of the magnetosonic waves at intermediate wave normal angles can be gained by considering the limit $V_{\mathrm{S}}^2 \ll V_{\mathrm{A}}^2$. Under this condition it can be shown that the magnetosonic part of the dispersion relation is approximately

$$D(k, \omega) = (\upsilon_{\mathrm{p}}^2 - V_{\mathrm{A}}^2)(\upsilon_{\mathrm{p}}^2 - V_{\mathrm{S}}^2 \cos^2\theta) = 0. \tag{6.5.28}$$

As shown in Figure 6.10, the isotropic root $\upsilon_{\mathrm{p}}^2 = V_{\mathrm{A}}^2$ has eigenvectors corresponding to a nearly transverse electromagnetic wave with only a small longitudinal (i.e., sound wave) component, and the anisotropic root $\upsilon_{\mathrm{p}}^2 = V_{\mathrm{S}}^2 \cos^2\theta$ has eigenvectors corresponding to a sound wave with only a small electromagnetic component ($\tilde{\mathbf{E}} \simeq 0$ and $\tilde{\mathbf{B}} \simeq 0$).

Since the plasma pressure is much smaller than the magnetic field pressure when $V_{\mathrm{S}}^2 \ll V_{\mathrm{A}}^2$, the fluid motion for the slow magnetosonic wave is constrained to be nearly parallel to the static magnetic field. The group velocity and wave energy flow are also nearly parallel to the static magnetic field. The particle motions and

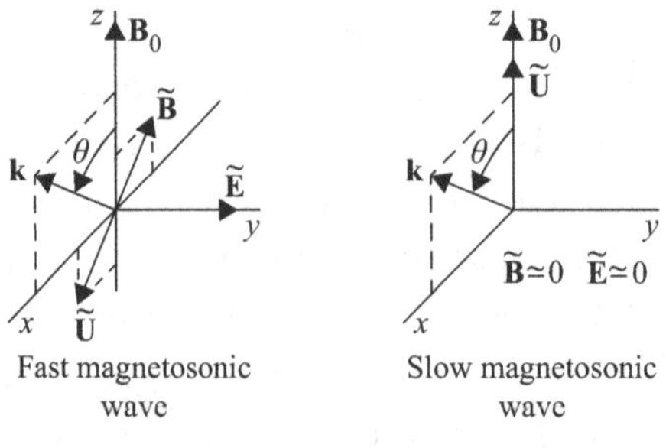

Figure 6.10 The eigenvectors for the fast and slow magnetosonic modes.

wave energy flow for this mode are then similar to the motion of the gas in a parallel array of pipes, with the magnetic field playing the role of the pipes.

6.5.3 MHD Wave Observations

As mentioned earlier, the existence of MHD waves was first predicted theoretically by Alfvén in 1942. It did not take long before these waves were observed in the laboratory, first in experiments with magnetized conducting fluids, such as mercury (Lundquist, 1949), and later in magnetized plasmas (Wilcox et al., 1960). It is now widely recognized that MHD waves are of considerable importance, particularly in space plasmas where the ideal MHD condition is most easily satisfied. We will discuss two such examples.

The first example occurs in the solar wind. When spacecraft-borne magnetic field and plasma measurements became available in the solar wind, it soon became apparent that the solar wind is dominated by large-amplitude, low-frequency fluctuations. By comparing the magnetic field with the plasma flow velocity, Belcher et al. (1969) showed that the fluctuations are caused by Alfvén waves. An example of these fluctuations is shown in Figure 6.11. The top three groups of plots show three components of the solar wind magnetic field, B_N, B_T, and B_R, superposed on suitably scaled components of the solar wind velocity, U_N, U_T, and U_R. A running average has been subtracted from the measurements in order to reveal the fluctuations. The bottom group of plots shows

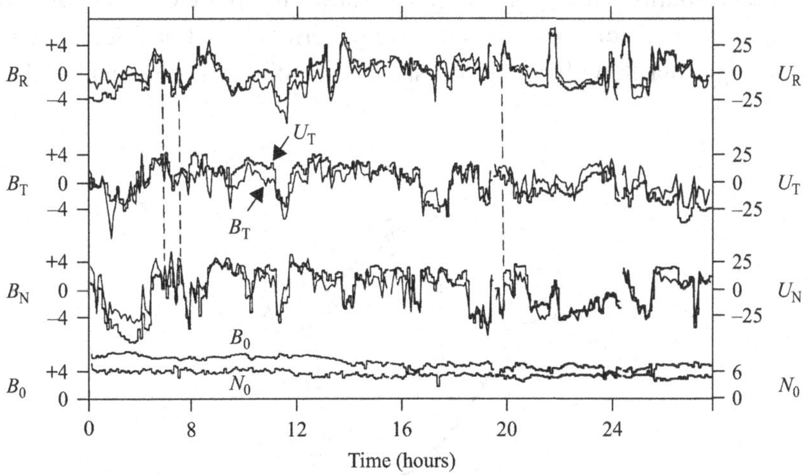

Figure 6.11 Measurements of the solar wind magnetic field, B, and the plasma flow velocity, U, in an orthogonal (N, R, T) coordinate system (from Davis, 1972). The fluctuations are due to transverse Alfvén waves propagating outward from the Sun.

the zero-order magnetic field strength, B_0, and the number density, N_0. As one can see, the corresponding components of the magnetic field and flow velocity fluctuations are closely correlated, as would be expected for the transverse Alfvén mode; see Eqs. (6.5.20) and (6.5.21). The almost total absence of comparable fluctuations in the zero-order magnetic field strength and number density uniquely identifies the mode of propagation as the transverse Alfvén mode. The in-phase nature of the correlation also shows that the waves are propagating outward from the Sun. Turbulent convective motions in the photosphere of the Sun are believed to be the source of these waves. Since the electric field of the transverse mode is always perpendicular to the static magnetic field (see Figure 6.8), this mode is almost totally unaffected by the various collisionless damping processes that can exist in a hot, magnetized plasma (see Chapter 10). Thus, once generated, the waves can propagate great distances with little or no attenuation. In contrast, when hot plasma effects are considered, the fast and slow magnetosonic modes always have a small component of the electric field along the static magnetic field (except for the degenerate case at $\theta = 0$), which leads to damping by the thermal plasma.

The second example is associated with Jupiter's moon Io. In the process of studying radio emissions from Jupiter, Bigg (1964) discovered that Io played a dominant role in controlling radio emissions from Jupiter in the frequency range from about 10 to 40 MHz. To explain the origin of these Io-controlled radio emissions, Goldreich and Lynden-Bell (1969) proposed that Io is a conductor, and that the electromotive force induced by Io's motion through the co-rotating plasma of Jupiter's magnetosphere excites transverse Alfvén waves propagating northward and southward from the moon as shown in Figure 6.12. Because of the finite propagation velocity, the Alfvén wave pattern is swept backward relative to the incident plasma flow, forming a v-shaped standing wave similar to the bow

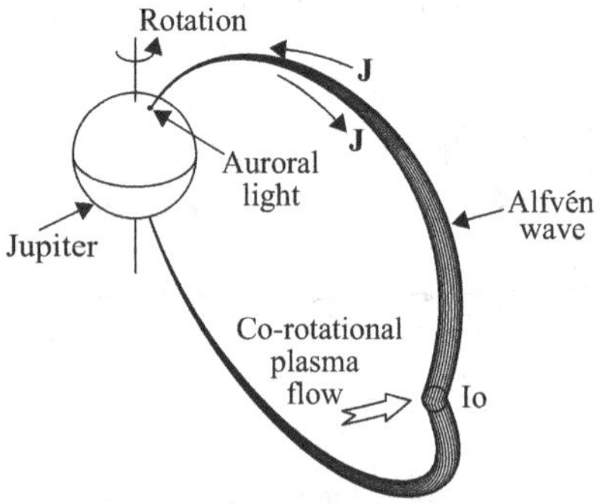

Figure 6.12 A model showing the excitation of transverse Alfvén waves by Jupiter's moon Io.

wave of a ship. The electrical currents associated with these Alfvén waves create two quasi-field-aligned current loops, one in each hemisphere, that link Io to the upper atmosphere of Jupiter. The electrons that carry these currents are believed to be responsible for generating the radio emissions. Magnetic field measurements by the *Voyager 1* spacecraft during a close flyby of Io have since confirmed the existence of these Alfvén waves (Ness et al., 1979). *Voyager 1* also showed that Io has active volcanism. Gases from these volcanos are now known to be a crucial factor in the Alfvén wave excitation by providing an electrically conducting atmosphere around Io, and also via the momentum they transfer to the plasma when the particles are ionized. The Hubble Space Telescope has also provided dramatic confirmation of the existence of these Alfvén waves by detecting auroral light emission from the points where the northward and southward propagating Alfvén waves interact with Jupiter's atmosphere (Connerney et al., 1993).

6.6 Validity of Resistive MHD Equations

As discussed in Section 6.1, the resistive MHD equations are obtained from moments of the kinetic equation. This procedure is rigorous up to a point, beyond which various approximations are required to address the issue of closure, and in developing versions of the model that are applicable for plasma phenomena spanning a specified range of space and time-scales. In this section, we provide an overview and summary of the principal approximations that must hold for the resistive MHD equations to be valid.

Inspection of the steps leading up to the mass continuity equation makes it clear that the mass continuity equation (6.1.31) is an exact equation. The same is true of Eq. (6.1.12), which provides the basis for the momentum equation (6.1.32). However, to obtain Eq. (6.1.32) from (6.1.12), we introduce the assumption of quasi-neutrality, which eliminates the charge density ρ_q from this equation and also eliminates Gauss' law from Maxwell's equations. Furthermore, we also make the assumption that the plasma pressure is a scalar, which requires that the underlying distribution functions for ions as well as electrons are near-Maxwellian. The latter assumption is valid when the system undergoes many collisions over typical macroscopic time-scales and when the typical macroscopic spatial scales are much longer than the collisional mean-free path of ions and electrons.

For the generalized Ohm's law (6.1.23) to reduce to the resistive MHD Ohm's law (6.1.33), the last three terms on the right of Eq. (6.1.23), that is, the Hall current, the electron pressure gradient, and the electron inertia must all be much smaller than the convection term $\mathbf{U} \times \mathbf{B}$ on the left, which is of the same order as the other terms retained in Eq. (6.1.33). As discussed below, all three of these terms are much smaller than the convection term when the ion Larmor radius is much smaller than the typical macroscopic spatial scale, and the characteristic wave frequencies are much smaller than the ion cyclotron frequency.

From the discussion above, we can conclude qualitatively that the resistive MHD model is valid for low-frequency, long-wavelength phenomena in sufficiently collisional non-relativistic plasmas. More precisely, we can formulate mathematical approximations which will delineate the domain of validity of the model. We consider waves of frequency ω and wavenumber k in systems of typical macroscopic spatial scale L. The neglect of the displacement current term and the restriction to non-relativistic dynamics requires the inequalities

$$\omega/k \ll c, \quad C_{\mathrm{e}} \ll c, \quad \text{and} \quad C_{\mathrm{i}} \ll c, \tag{6.6.1}$$

where c is the speed of light, and C_{e} and C_{i} are the electron and ion thermal speeds, respectively (see Eq. (2.1.3)). It can be shown (Problem 6.9) that the quasi-neutrality constraint is valid when

$$k\lambda_{\mathrm{D}} \ll 1. \tag{6.6.2}$$

For the ion and electron distributions to be near-Maxwellian, there must be many collisions in typical time-scales of MHD phenomena, which can be represented for the purpose of this discussion by ω^{-1}. For ions, the typical collisional time τ_{i} is dominated by ion–ion collisions, whereas for electrons the typical collision time τ_{e} is dominated by electron–ion as well as electron–electron collisions (see Chapter 12). The relevant inequalities for the validity of the resistive MHD model are

$$\omega\tau_{\mathrm{e}} \gg 1, \quad \omega\tau_{\mathrm{i}} \gg 1. \tag{6.6.3}$$

To this we add the condition that the mean-free paths for electrons and ions, measured by $C_{\mathrm{e}}\tau_{-}$ and $C_{-}\tau_{-}$ respectively, are much smaller than the macroscopic spatial scale L; that is,

$$C_{\mathrm{e}}\tau_{-} \ll L, \quad C_{-}\tau_{-} \ll L. \tag{6.6.4}$$

It can be shown (Problem 6.10) that the reduction of the generalized Ohm's law to the resistive Ohm's law requires that the ion Larmor radius be much smaller than L and that the relevant wave frequency be much smaller than the ion cyclotron frequency; that is,

$$\rho_{\mathrm{ci}} \ll L, \quad \omega \ll \Omega_{\mathrm{ci}}. \tag{6.6.5}$$

The Hall current and electron pressure gradient terms in the generalized Ohm's law are often referred to as "finite ion-Larmor radius" terms, and their restoration to the resistive MHD Ohm's law is one of the most important extensions of the traditional MHD model. Some of the most important results in contemporary reconnection theory, discussed briefly in Chapter 7, can be attributed to the inclusion of these finite ion-Larmor radius terms.

Problems

6.1. Show for a steady-state flow ($\partial/\partial t = 0$) that the momentum equation with a gravitational force, $\mathbf{F}_{\mathrm{g}} = \rho_{\mathrm{m}}\mathbf{g}$, an isotropic pressure, and no electromagnetic forces ($\mathbf{E} = 0$ and $\mathbf{B} = 0$),

$$\rho_{\mathrm{m}}\left[\frac{\partial \mathbf{U}}{\partial t} + (\mathbf{U}\cdot\nabla)\mathbf{U}\right] = -\nabla P + \rho_{\mathrm{m}}\mathbf{g},$$

can be integrated once along a streamline to obtain Bernoulli's equation,

$$\frac{1}{2}\rho_{\mathrm{m}}U^2 + P + \rho_{\mathrm{m}}gz \simeq \text{constant}.$$

Assume that the gravitational force term can be derived from a potential, $\rho_{\mathrm{m}}\mathbf{g} = -\nabla\Phi_{\mathrm{g}}$, where $\Phi_{\mathrm{g}} = \rho_{\mathrm{m}}gz$ and g is constant.

6.2. If the diffusion coefficient $\bar{\kappa}$ is a constant, show that the scalar diffusion equation

$$\frac{\partial F}{\partial t} = \bar{\kappa}\nabla^2 F$$

has a spherically symmetric solution given by

$$F = \left(\frac{1}{4\pi\bar{\kappa}t}\right)^{3/2} \exp\left[-\frac{r^2}{4\bar{\kappa}t}\right].$$

Make a sketch of $F(r)$ at two times t_1 and t_2.

6.3. Assuming pressure balance, $B^2/(2\mu_0) + P = \text{constant}$, show that the function

$$\mathbf{B} = B_0 \operatorname{erf}\left\{\frac{1}{2}\left(\frac{\mu_0\sigma}{t}\right)^{1/2} y\right\}\hat{\mathbf{x}}$$

is a solution of the static ($\mathbf{U} = 0$) time-dependent equation

$$\frac{\partial \mathbf{B}}{\partial t} = \frac{1}{\mu_0\sigma}\nabla^2\mathbf{B},$$

where

$$\operatorname{erf}(y) = \frac{2}{\sqrt{\pi}}\int_0^y \mathrm{e}^{-v^2}\,\mathrm{d}v.$$

(a) Make a plot of $B_x(y,t)$ as a function of y for a series of times t_1, t_2, and t_3.
(b) Compute the current $J_z(y,t)$ and plot as a function of y for a series of times t_1, t_2, and t_3.
(c) Make a plot of the pressure function, $P(y)$, that would be required to achieve a force-balanced equilibrium for this magnetic field.

6.4. If the magnetic flux is frozen in the plasma and if the first and second adiabatic invariants are conserved, show that

$$\frac{P_\perp}{\rho_{\mathrm{m}} B} = \text{constant and} \quad \frac{P_\parallel P_\perp^2}{\rho_{\mathrm{m}}^5} = \text{constant.}$$

Hint: Note that $P_\perp \sim \rho_{\mathrm{m}} w_\perp$ and that $P_\parallel \sim \rho_{\mathrm{m}} w_\parallel$. The first invariant implies that $w_\perp / B = $ constant and the second invariant implies that $w_\parallel L^2 = $ constant.

6.5. If the magnetic field in the solar wind is "frozen" in the plasma, show that the magnetic field near the solar equatorial plane takes on the shape of an Archimedean spiral

$$\phi = \phi_0 - \frac{\omega_{\mathrm{s}}}{V_{\mathrm{sw}}} \rho,$$

where ϕ is an azimuthal angle viewed from the polar axis, ω_{s} is the angular rotational velocity of the Sun, ρ is the distance from the Sun, and V_{sw} is the solar wind velocity.

6.6. In the derivation of small-amplitude waves in an MHD fluid, we obtained the equation

$$\omega^2 \mathbf{U} = \frac{1}{\mu_0 \rho_{\mathrm{m}0}} \{\mathbf{k} \times (\mathbf{k} \times [\mathbf{U} \times \mathbf{B}_0])\} \times \mathbf{B}_0 + V_{\mathrm{S}}^2 \mathbf{k}(\mathbf{k} \cdot \mathbf{U}).$$

(a) Show that this equation can be represented by the following homogeneous equation:

$$\begin{bmatrix} v_{\mathrm{p}}^2 - V_{\mathrm{A}}^2 - V_{\mathrm{S}}^2 \sin^2\theta & 0 & -V_{\mathrm{S}}^2 \sin\theta\cos\theta \\ 0 & v_{\mathrm{p}}^2 - V_{\mathrm{A}}^2 \cos^2\theta & 0 \\ -V_{\mathrm{S}}^2 \sin\theta\cos\theta & 0 & v_{\mathrm{p}}^2 - V_{\mathrm{S}}^2 \cos^2\theta \end{bmatrix} \begin{bmatrix} U_x \\ U_y \\ U_z \end{bmatrix} = 0,$$

where $V_{\mathrm{A}}^2 = B_0^2/\mu_0 \rho_{\mathrm{m}0}$ and $v_{\mathrm{p}} = \omega/k$.

(b) Show that the determinant of the above matrix is

$$D(k,\omega) = (v_{\mathrm{p}}^2 - V_{\mathrm{A}}^2 \cos^2\theta)[v_{\mathrm{p}}^4 - v_{\mathrm{p}}^2(V_{\mathrm{A}}^2 + V_{\mathrm{S}}^2) + V_{\mathrm{A}}^2 V_{\mathrm{S}}^2 \cos^2\theta] = 0.$$

6.7. If we add viscous effects to the MHD momentum equation by including a term $\nu \nabla^2 \mathbf{U}$, where ν is the dynamic viscosity,

$$\rho_{\mathrm{m}} \frac{\mathrm{d}\mathbf{U}}{\mathrm{d}t} = \mathbf{J} \times \mathbf{B} - \nabla P + \nu \nabla^2 \mathbf{U},$$

and finite conductivity effects via the term $(1/\mu_0\sigma)\nabla^2 B$ in

$$\frac{\partial \mathbf{B}}{\partial t} = \nabla \times (\mathbf{U} \times \mathbf{B}) + \frac{1}{\mu_0 \sigma}\nabla^2 \mathbf{B},$$

show that the dispersion relation for Alfvén waves can be obtained by multiplying both ω^2 and V_S^2 by a factor $(1 + \mathrm{i}k^2/\mu_0\sigma\omega)$ and ω^2 by a factor $(1 + \mathrm{i}\nu k^2/\rho_{m0}\omega)$.

6.8. If the finite conductivity and viscous corrections are small (i.e., $\sigma \to \infty$ and $\nu \to 0$), use the results from Problem 6.7 to show that, for parallel ($\theta = 0$) propagation, the dispersion relation for the transverse Alfvén wave is

$$k \simeq \frac{\omega}{V_A} + \mathrm{i}\frac{\omega^2}{2V_A^3}\left(\frac{1}{\mu_0\sigma} + \frac{\nu}{\rho_{m0}}\right).$$

Note that the viscous and finite conductivity effects increase rapidly with increasing frequency. Give a simple explanation of this frequency dependence.

6.9. (a) Show that in the limit $\epsilon_0 \to 0$ (when the speed of light tends to infinity), the displacement current term $\epsilon_0 \partial \mathbf{E}/\partial t$ in Maxwell's equation as well as the $\epsilon_0 \nabla \cdot \mathbf{E}$ term in Gauss' law can be neglected. This provides a formal justification for the neglect of these terms in the low-frequency limit $\omega/k \ll c$.

(b) Consider isothermal and unmagnetized electron and ion fluids moving under the influence of a steady-state electric field $\mathbf{E}$ that obeys the equation $\nabla \times \mathbf{E} = 0$. Show that when $k\lambda_D \ll 1$, we obtain the inequality $|\epsilon_0 \nabla \cdot \mathbf{E}|/(ne) \ll 1$.

6.10. Obtain the conditions under which terms on the right of the generalized Ohm's law (6.1.23) can be neglected by comparing each of them to the convection term on the left. Assume that the electron pressure is a scalar.

References

Alfvén, H. 1942. Existence of electromagnetic-hydrodynamic waves. *Nature* **150**, 404.

Babcock, H. W. 1961. The topology of the Sun's magnetic field and the 22-year solar cycle. *Astrophys. J.* **133**, 572–587.

Batchelor, G. K. 1967. *An Introduction to Fluid Mechanics*. Cambridge: Cambridge University Press, p. 74.

Belcher, J. W., Davis, L., and Smith, E. J. 1969. Large-amplitude Alfvén waves in the interplanetary medium: Mariner 5. *J. Geophys. Res.* **74**, 2302–2308.

Bigg, E. K. 1964. Influence of the satellite Io on Jupiter's decametric emission. *Nature* **203**, 1008–1010.

Connerney, J. E. P., Baron, R., Satoh, T., and Owen, T. 1993. Images of excited H_3^+ at the foot of the Io flux tube in Jupiter's atmosphere. *Science* **262**, 1035–1038.

Cowling, T. G. 1957. *Magnetohydrodynamics*. New York: Interscience, p. 2.
Davis, L. 1972. The configuration of the interplanetary magnetic field. In *Solar Terrestrial Physics,* Part II, ed. E. R. Dyer. Dordrecht: Reidel, pp. 32–48.
Elsasser, W. M. 1956. Hydromagnetic dynamo theory. *Rev. Mod. Phys.* **28**, 135–163.
Fetter, A. L., and Walecka, J. D. 1980. *Theoretical Mechanics of Particles and Continua*. New York: McGraw-Hill, p. 345.
Goldreich, P., and Lynden-Bell, D. 1969. Io, a Jovian unipolar inductor. *Astrophys. J.* **156**, 59–78.
Lundquist, S. 1949. Experimental demonstration of magneto-hydrodynamic waves. *Nature* **164**, 145–146.
Ness, N. F., Acuna, M., Lepping, R. P., Burlaga, L. F., and Behannon, K. W. 1979. Magnetic field studies at Jupiter by Voyager 1: Preliminary results. *Science* **204**, 982–987.
Wilcox, J. M., Boley, F. I., and DeSilva, A. W. 1960. Experimental study of Alfvén wave propagation. *Phys. Fluids* **3**, 15–19.

Further Reading

Boyd, T. J. M., and Sanderson, J. J. 2003. *The Physics of Plasmas*. Cambridge: Cambridge University Press, Chapters 3 and 4.
Fitzpatrick, R. 2015. *Plasma Physics: An Introduction*. London: CRC Press, Chapter 4.
Freidberg, J. P. 1987. *Ideal Magnetohydrodynamics*. New York: Plenum Press.

7

MHD Equilibria and Stability

This chapter is devoted to the analysis of MHD equilibria and stability. By equilibria, we mean a plasma state that is time-independent. Such states may or may not have equilibrium flows. When the states do not have equilibrium flows, that is, $\mathbf{U} = \mathbf{0}$ in some appropriate frame of reference, the equilibria are called magnetostatic equlibria. When the states have flows that cannot be simply eliminated by a Galilean transformation, the equilbria are called magnetohydrodynamic equilibria. When we introduce small perturbations in a particular equilibrium which is itself time-independent, the time dependence of the perturbations determines the stability of the system. If an equilibrium is unstable, the instability typically grows exponentially in time. The mathematical problem for the stability of magnetostatic equilibria is made tractable due to the formulation of the so-called energy principle. It turns out that when MHD equilibria contain flows that are spatially dependent, the power of the energy principle is weakened significantly, and there has been a general tendency to rely on the normal mode method, for which we provide simple examples.

In nature and in the laboratory, plasmas can be stable according to the equations of ideal MHD. However, even ideally stable plasmas can become unstable in the presence of small departures from idealness, such as a small amount of resistivity. This may appear counter-intuitive upon first glance unless one takes into account the fact that in the presence of even small dissipation the frozen field theorem discussed in Chapter 6 is violated, which enables the plasma to access states of lower potential energy through motions that would be forbidden for ideal plasmas, i.e., by allowing magnetic field lines to slip with respect to the plasma fluid. Such instabilities are called resistive instabilities. These instabilities are part of a general class of phenomena called magnetic reconnection, which is a subject of great interest for space, laboratory, and astrophysical plasmas.

7.1 Magnetostatic Equilibria

Since magnetostatic equilibria, also referred to as static MHD equilibria, are frequently encountered in plasmas, it is useful to examine the specific conditions required to realize them. In a hydrostatic fluid, a static equilibrium is often achieved by balancing the gravitational force with an appropriate pressure gradient force. In a magnetostatic fluid, magnetic forces must also be considered. Since magnetic forces by their very nature are anisotropic, the equilibrium conditions are considerably more complicated than for a hydrostatic fluid.

7.1.1 The Virial Theorem

Before considering the conditions required to achieve a static equilibrium, we first prove an important theorem called the virial theorem. The virial theorem is an exact result that shows it is impossible for an isolated MHD fluid to generate the magnetic forces necessary to maintain a static equilibrium. The importance of this theorem is obvious. An isolated ball of plasma cannot maintain itself in a static equilibrium just due to internally generated forces. To confine a plasma, it is necessary to provide an external force.

The virial theorem can be proved as follows. We start by writing the momentum equation (6.2.8) in index notation:

$$\rho_{\mathrm{m}} \frac{\mathrm{d}U_s}{\mathrm{d}t} = -\sum_r \frac{\partial}{\partial x_r}(P_{rs} + T_{rs}). \tag{7.1.1}$$

If a magnetostatic equilibrium exists ($U_s = 0$), the left-hand side of the above equation must be zero. Next, multiply the above equation by x_s, sum over all three coordinates, and integrate over all space to obtain

$$0 = -\sum_{r,s} \int_V x_s \frac{\partial}{\partial x_r}(P_{rs} + T_{rs})\, \mathrm{d}^3x. \tag{7.1.2}$$

The terms on the right-hand side of this equation can be integrated by parts once to give

$$0 = -\sum_{r,s} \int_S x_s (P_{rs} + T_{rs})\, \mathrm{d}A_r + \sum_{r,s} \int_V \frac{\partial x_s}{\partial x_r}(P_{rs} + T_{rs})\, \mathrm{d}^3x. \tag{7.1.3}$$

Note that the first term involves a surface integral at infinity. For an isolated three-dimensional system, we can assume that the surface integral is zero, the only requirement being that P_{rs} and T_{rs} decrease sufficiently rapidly at infinity (i.e., faster than $1/R^3$). Using the identity $\partial x_s/\partial x_r = \delta_{rs}$, where δ_{rs} is the Kronecker delta, Eq. (7.1.3) simplifies to

$$0 = \int_V \mathrm{Trace}(\overleftrightarrow{\mathbf{P}} + \overleftrightarrow{\mathbf{T}})\,\mathrm{d}^3x, \tag{7.1.4}$$

where the trace is the sum of the diagonal terms in the tensor. However, it can be seen that the traces of $\overleftrightarrow{\mathbf{P}}$ and $\overleftrightarrow{\mathbf{T}}$ are positive definite,

$$\mathrm{Trace}\,\overleftrightarrow{\mathbf{P}} = \sum_{r,s} m_s \int_V (v_r - U_r)^2 f_s(\mathbf{v})\,\mathrm{d}^3v, \tag{7.1.5}$$

$$\mathrm{Trace}\,\overleftrightarrow{\mathbf{T}} = B^2/2\mu_0,$$

so the right-hand side of Eq. (7.1.4) cannot be zero. Therefore, the original assumption, that the system is in magnetostatic equilibrium, must be false. A magnetostatic equilibrium for an isolated MHD system is impossible.

It is important to note the conditions for which the virial theorem applies. The plasma must be finite in three dimensions and completely isolated, with no external forces or flows. The theorem does not apply to systems that are infinite in one or two dimensions, or to systems that contain rigid current-carrying conductors (due to the external mechanical forces on the conductors).

7.1.2 Conditions for a Magnetostatic Equilibrium

If we choose a reference frame in which the fluid is at rest, the conditions for a magnetostatic equilibrium are obtained by simply setting $\partial/\partial t = 0$ and $\mathbf{U} = 0$ in the MHD equations given in Section 6.1.5. The relevant equations are then

$$\mathbf{J} \times \mathbf{B} = \nabla P, \tag{7.1.6}$$

$$\nabla \times \mathbf{B} = \mu_0 \mathbf{J}, \tag{7.1.7}$$

and

$$\nabla \cdot \mathbf{B} = 0, \tag{7.1.8}$$

where the plasma pressure, P, is assumed to be isotropic. By taking the dot product of Eq. (7.1.6) with $\mathbf{J}$ and $\mathbf{B}$, it follows that $\mathbf{J} \cdot \nabla P = 0$ and $\mathbf{B} \cdot \nabla P = 0$. Since $\nabla P = 0$ is perpendicular to a surface of constant pressure, it also follows that the current density $\mathbf{J}$ and the magnetic field $\mathbf{B}$ must lie on surfaces of constant pressure.

Static equilibria can be divided into two classes, depending upon whether the $\mathbf{J} \times \mathbf{B}$ force vanishes or not. If $\mathbf{J} \times \mathbf{B} = 0$, the equilibrium is said to be force free. If $\mathbf{J} \times \mathbf{B} \neq 0$, the equilibrium is said to be force balanced.

7.1.3 Force-Free Equilibria

Force-free MHD equilibria typically occur in very tenuous plasmas where the pressure gradient forces are so small that they can be neglected. Since the condition for a force-free equilibrium is $\mathbf{J}\times\mathbf{B}=0$, it is clear that the force-free condition is satisfied whenever $\mu_0\mathbf{J}=\nabla\times\mathbf{B}=\alpha\mathbf{B}$, where α is a scalar function.

To understand the implications of the equation $\nabla\times\mathbf{B}=\alpha\mathbf{B}$, we must investigate the function α. We start by noting that the divergence of the curl of any vector field is zero, so that

$$\nabla\cdot(\nabla\times\mathbf{B})=\alpha(\nabla\cdot\mathbf{B})+\mathbf{B}\cdot\nabla\alpha=0. \tag{7.1.9}$$

Since $\nabla\cdot\mathbf{B}=0$, this equation simplifies to

$$\mathbf{B}\cdot\nabla\alpha=0. \tag{7.1.10}$$

Since $\nabla\alpha$ is perpendicular to a surface of constant α, the above equation shows that the magnetic field lines must lie on a constant-α surface. Next we show that the constant-α surface cannot be a simple closed surface. To prove this relationship, let C be a closed curve everywhere parallel to the magnetic field. The line integral of $\mathbf{B}$ along this closed path clearly cannot be zero:

$$\int_C \mathbf{B}\cdot \mathrm{d}\boldsymbol{\ell}\neq 0. \tag{7.1.11}$$

Using Stokes' theorem and $\nabla\times\mathbf{B}=\mu_0\mathbf{J}=\alpha\mathbf{B}$, the above integral can be written

$$\int_C \mathbf{B}\cdot \mathrm{d}\boldsymbol{\ell}=\int_S(\nabla\times\mathbf{B})\cdot\mathrm{d}\mathbf{A}=\int_S\alpha\mathbf{B}\cdot\mathrm{d}\mathbf{A}\neq 0, \tag{7.1.12}$$

where S is a simple surface bounded by the curve C. Next, distort the surface S such that it coincides with the constant-α surface. This procedure can only be performed if the surface is simply closed, like a sphere. Otherwise, Stokes' theorem is not valid. Since the function α is constant on this surface, α can be taken outside of the integral, so that

$$\int_C \mathbf{B}\cdot \mathrm{d}\boldsymbol{\ell}=\alpha\int_S\mathbf{B}\cdot\mathrm{d}\mathbf{A}\neq 0. \tag{7.1.13}$$

However, earlier we showed that the magnetic field lines must lie on the constant-α surface, so $\mathbf{B}\cdot\mathrm{d}\mathbf{A}=0$. Thus, the integral on the right-hand side of the above equation is zero, which leads to a contradiction. The constant-α surface cannot be a simply closed surface.

Using a theorem from topology known as the Poincaré–Hopf theorem (Alexandroff and Hopf, 1935), it can be shown that the simplest topological form

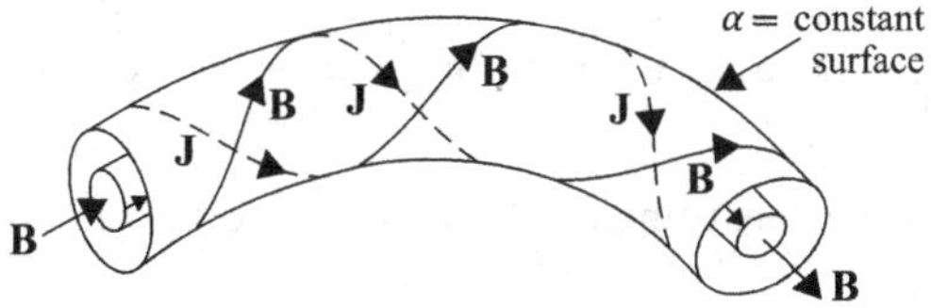

Figure 7.1 The magnetic field and current for a force-free equilibrium form helix-like lines around torus-shaped surfaces of constant α.

for a constant-α surface that satisfies the above restrictions is a torus. The Poincaré–Hopf theorem states that "if a non-vanishing, bounded vector field (**J** and **B**, in our case) lies on a smooth surface (the constant-α surface), then the surface must be toroidal." The current and magnetic field must then form helix-like lines around the torus as shown in Figure 7.1. This result is quite general, and also applies to force-balanced MHD equilibria, since Eq. (7.1.6) implies that $\mathbf{B}\cdot\nabla P = 0$. The surfaces of constant P then play the same role as surfaces of constant α do for force-free equilibria.

Particularly interesting solutions to the static equilibrium equations can be found for two-dimensional geometries with cylindrical symmetry. For such a geometry, it can be shown that the equation $\nabla \times \mathbf{B} = \alpha\mathbf{B}$ is satisfied by a magnetic field $\mathbf{B} = [0, B_\phi, B_z]$ of the form

$$B_\phi = \frac{B_0 k\rho}{1 + k^2\rho^2}, \tag{7.1.14}$$

and

$$B_z = \frac{B_0}{1 + k^2\rho^2}. \tag{7.1.15}$$

In this case, the z component of the current density is given by

$$J_z = \frac{2kB_0/\mu_0}{(1 + k^2\rho^2)^2}, \tag{7.1.16}$$

and the function α is given by

$$\alpha = \frac{\mu_0 J_z}{B_z} = \frac{2k}{1 + k^2\rho^2}. \tag{7.1.17}$$

The magnetic field lines consist of helices that spiral around the central axis with pitch angles that vary from 0 along the central axis to $\pi/2$ at large distances from the central axis. Such a magnetic field configuration is often called a flux rope (see Figure 7.2).

Magnetic field geometries of the type described above have been observed in Earth's magnetosphere and in the solar corona. During disturbed periods large

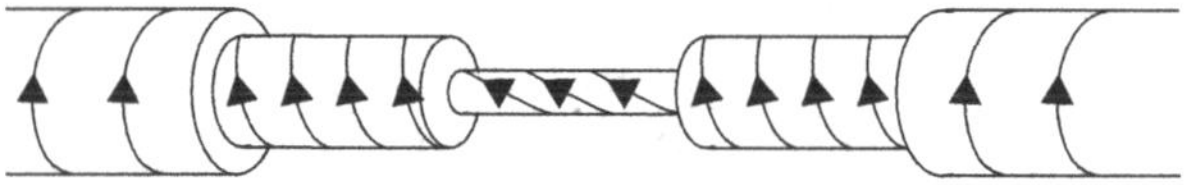

Figure 7.2 A flux rope consisting of helical magnetic field lines nested around a central axis.

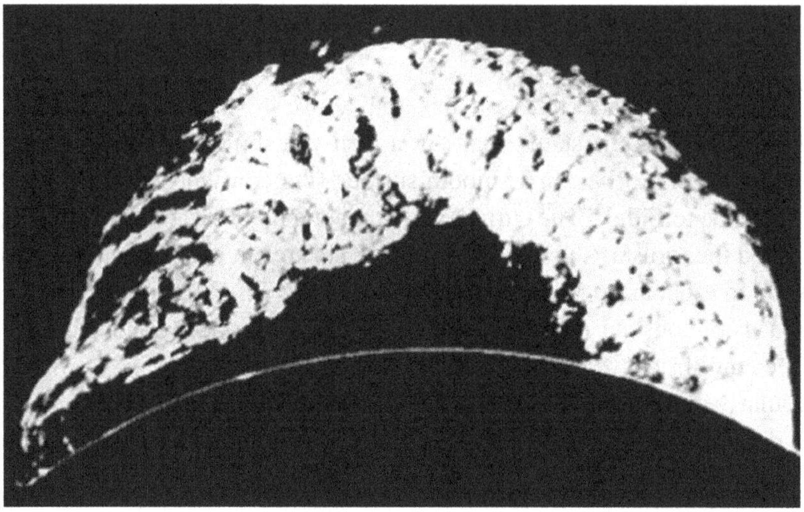

Figure 7.3 The Sun's magnetic field often has helical flux rope-like structures (image courtesy of High Altitude Observatory).

currents often flow along magnetic field lines from one region of the Sun to another. These currents often heat the plasma to such a high temperature that plasma trapped on the magnetic field lines can be detected optically from Earth. The field lines typically consist of an arch with a helical structure, as shown in Figure 7.3.

Another class of force-free solutions with cylindrical symmetry is of the form

$$B_\phi = B_0 J_1(\alpha\rho), \tag{7.1.18}$$

and

$$B_z = B_0 J_0(\alpha\rho), \tag{7.1.19}$$

where α and B_0 are positive constants and J_0 and J_1 are Bessel functions of order 0 and 1. Note that for a cylindrical plasma of radius a, Eq. (7.1.19) leads to a reversal of the axial field at the first zero of $J_0(x)$, i.e., at $\alpha\rho = 2.404$. Configurations of this type are known as reversed field pinches (RFPs) and are among the leading candidates for magnetic confinement of fusion plasmas.

Taylor (1974) proposed that RFPs are generated spontaneously when a turbulent plasma relaxes to a state of minimum potential energy subject to the constancy of a quantity called the magnetic helicity, which is defined by

$$\mathcal{H} = \int_V \mathbf{A} \cdot \mathbf{B}\, \mathrm{d}^3 x, \tag{7.1.20}$$

where $\mathbf{A}$ is the vector potential, i.e., $\mathbf{B} = \nabla \times \mathbf{A}$. Magnetic helicity is an exact invariant for a plasma that obeys the frozen field condition (Eq. (6.3.7)), or equivalently, $\mathbf{E} + \mathbf{U} \times \mathbf{B} = 0$. The invariance of $\mathcal{H}$ can be shown by considering the time derivative of Eq. (7.1.20), which is given by

$$\frac{\mathrm{d}\mathcal{H}}{\mathrm{d}t} = \int_V \left(\frac{\partial \mathbf{A}}{\partial t} \cdot \mathbf{B} + \mathbf{A} \cdot \frac{\partial \mathbf{B}}{\partial t} \right) \mathrm{d}^3 x. \tag{7.1.21}$$

Using Faraday's law (6.1.30), the electric field, $\mathbf{E}$, can be written in terms of a scalar potential, Φ, and the vector potential, $\mathbf{A}$, as

$$\mathbf{E} = -\nabla \Phi - \frac{\partial \mathbf{A}}{\partial t}. \tag{7.1.22}$$

Substituting Eq. (7.1.22) for $\partial \mathbf{A}/\partial t$ and Eq. (6.1.30) for $\partial \mathbf{B}/\partial t$ into Eq. (7.1.21), we obtain

$$\frac{\mathrm{d}\mathcal{H}}{\mathrm{d}t} = -\int_V \mathbf{E} \cdot \mathbf{B}\, \mathrm{d}^3 x - \int_V \mathbf{B} \cdot \nabla \Phi\, \mathrm{d}^3 x - \int_V (\nabla \times \mathbf{E}) \cdot \mathbf{A}\, \mathrm{d}^3 x. \tag{7.1.23}$$

Using the identities

$$\nabla \cdot (\Phi \mathbf{B}) = \mathbf{B} \cdot \nabla \Phi + \Phi (\nabla \cdot \mathbf{B}) = \mathbf{B} \cdot \nabla \Phi, \tag{7.1.24}$$

and

$$\begin{aligned} \nabla \cdot (\mathbf{E} \times \mathbf{A}) &= (\nabla \times \mathbf{E}) \cdot \mathbf{A} - \mathbf{E} \cdot \nabla \times \mathbf{A} \\ &= (\nabla \times \mathbf{E}) \cdot \mathbf{A} - \mathbf{E} \cdot \mathbf{B}, \end{aligned} \tag{7.1.25}$$

Eq. (7.1.23) can be rewritten in the form

$$\frac{\mathrm{d}\mathcal{H}}{\mathrm{d}t} = -2 \int_V \mathbf{E} \cdot \mathbf{B}\, \mathrm{d}^3 x - \int_V \nabla \cdot (\Phi \mathbf{B})\, \mathrm{d}^3 x - \int_V \nabla \cdot (\mathbf{E} \times \mathbf{A})\, \mathrm{d}^3 x. \tag{7.1.26}$$

The second and third volume integrals on the right-hand side of the above equation can then be converted to surface integrals using Gauss' theorem. This gives

$$\frac{\mathrm{d}\mathcal{H}}{\mathrm{d}t} = -2 \int_V \mathbf{E} \cdot \mathbf{B}\, \mathrm{d}^3 x - \int_S \hat{\mathbf{n}} \cdot (\mathbf{B}\Phi + \mathbf{E} \times \mathbf{A})\, \mathrm{d}A, \tag{7.1.27}$$

where $\hat{\mathbf{n}}$ is the unit normal (pointing outward) at the area element dA on the surface S. Assuming that the plasma is bounded by a perfectly conducting wall that is identified with the surface S, the $\mathbf{E}$ and $\mathbf{B}$ fields obey the boundary conditions

$$\hat{\mathbf{n}} \times \mathbf{E} = 0, \qquad \hat{\mathbf{n}} \cdot \mathbf{B} = 0. \tag{7.1.28}$$

From these boundary conditions, it can be seen that the surface integrals on the right-hand side of Eq. (7.1.27) vanish. Also, because in an ideal plasma ($\mathbf{E} + \mathbf{U} \times \mathbf{B} = 0$) we have $\mathbf{E} \cdot \mathbf{B} = 0$, it follows that the volume integral on the right-hand side of Eq. (7.1.27) also vanishes, so $d\mathcal{H}/dt = 0$. Thus, we have shown that the magnetic helicity is a global invariant, i.e.,

$$\mathcal{H} = \text{constant}. \tag{7.1.29}$$

The proof of the invariance of $\mathcal{H}$ given above relies on the assumption that the plasma is ideal, which no plasmas are, in reality. Taylor (1974) conjectured that in a turbulent non-ideal plasma the helicity remains an approximate constant while the potential energy, given by Eq. (6.4.21), decays to a minimum value. It can be shown that these minimum energy equilibria are force free and obey the equations $\nabla \times \mathbf{B} = \alpha \mathbf{B}$ and $P = \text{constant}$, where α is a constant. In a cylindrical geometry, assuming that equilibrium quantities depend only on ρ, one then obtains the Bessel function solutions (7.1.18) and (7.1.19).

Taylor's theory has been successful not only in accounting for the dominant behavior of RFPs, but has also found applications in space and astrophysical plasmas. For further discussion of this interesting subject, see Taylor (1986) and Priest and Forbes (2000).

7.1.4 Force-Balanced Equilibria

If the plasma is not sufficiently tenuous for the force-free condition to apply, then the $\mathbf{J} \times \mathbf{B}$ force must be balanced by the pressure gradient force. Using the momentum equation (6.2.8) with $\mathbf{U} = 0$, the force-balance condition can be written in the form

$$\nabla \left(P + \frac{B^2}{2\mu_0} \right) = \frac{1}{\mu_0} (\mathbf{B} \cdot \nabla) \mathbf{B}, \tag{7.1.30}$$

where the gradient terms are collected on the left-hand side and the non-gradient terms are on the right-hand side. Since there are an infinite number of solutions to this equation, to proceed further we must limit the discussion to certain magnetic field geometries that are of special interest.

Cylindrical Geometries

We start by discussing two-dimensional cylindrical geometries that are characterized by a magnetic field of the form $\mathbf{B} = [0, B_\phi(\rho), B_z(\rho)]$. For such a magnetic field, it can be shown (Problem 7.6) that the radial component of Eq. (7.1.30) becomes

$$\frac{\mathrm{d}}{\mathrm{d}\rho}\left(P + \frac{B_\phi^2}{2\mu_0} + \frac{B_z^2}{2\mu_0}\right) = -\frac{B_\phi^2}{\mu_0 \rho}. \tag{7.1.31}$$

Notice that the radial component of the $(\mathbf{B} \cdot \nabla)\mathbf{B}$ force is entirely controlled by the ϕ component of the magnetic field. This force is caused by the magnetic tension. Because of the curvature of the magnetic field, the magnetic tension force is directed radially inward. If the magnetic field is known, the components of the current density required to produce this field can be calculated from Ampère's law (6.1.28), which in cylindrical coordinates is given by

$$J_\rho = 0, \tag{7.1.32}$$

$$J_\phi = -\frac{1}{\mu_0}\frac{\mathrm{d}B_z}{\mathrm{d}\rho}, \tag{7.1.33}$$

$$J_z = \frac{1}{\mu_0 \rho}\frac{\mathrm{d}}{\mathrm{d}\rho}(\rho B_\phi). \tag{7.1.34}$$

Equation (7.1.31) involves three functions: $B_\phi(\rho)$, $B_z(\rho)$, and $P(\rho)$. Any two of these functions can be specified, whereupon the third can be determined by solving the resulting differential equation, subject to the appropriate boundary conditions. Since considerable freedom exists in our choice of the two "free" functions, Eq. (7.1.31) itself permits an infinite number of possible solutions. In the following, we discuss a few of this very large class of possible solutions.

The Z-pinch. In the Z-pinch, also sometimes called a Bennett pinch, a cylindrically symmetric plasma is confined by a purely azimuthal magnetic field $B_\phi(\rho)$. A sketch of the basic field geometry of a Z-pinch is shown in Figure 7.4. The confining magnetic field B_ϕ is entirely generated by the axial current density J_z according to

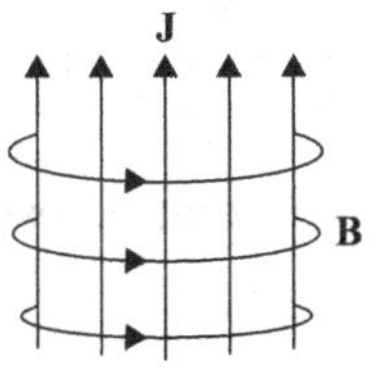

Figure 7.4 The magnetic field of a Z-pinch is produced internally by a current along the axis of a plasma column.

Eq. (7.1.34). Since $B_z = 0$, Eq. (7.1.31) reduces to

$$\frac{\mathrm{d}}{\mathrm{d}\rho}\left(P + \frac{B_\phi^2}{2\mu_0}\right) = -\frac{B_\phi^2}{\mu_0 \rho}. \tag{7.1.35}$$

The above equation permits an infinite number of solutions. Consider a simple case in which the axial current density, J_{z0}, is constant within a plasma column of radius $\rho = a$, and zero outside:

$$\begin{aligned} J_z(\rho) &= J_{z0} \quad \text{for } \rho \le a \text{ and} \\ J_z(\rho) &= 0 \quad \text{for } \rho > a. \end{aligned} \tag{7.1.36}$$

With $J_z(\rho)$ specified by the above equation, Eq. (7.1.34) can be solved for $B_\phi(\rho)$ to yield

$$\begin{aligned} B_\phi(\rho) &= \frac{\mu_0 I \rho}{2\pi a^2} \quad \text{for } \rho \le a \text{ and} \\ B_\phi(\rho) &= \frac{\mu_0 I}{2\pi \rho} \quad \text{for } \rho > a. \end{aligned} \tag{7.1.37}$$

The total current is given by

$$I = \int_0^a J_z \, 2\pi\rho \, \mathrm{d}\rho = \pi a^2 J_{z0}. \tag{7.1.38}$$

Substituting expression (7.1.37) for $B_\phi(\rho)$ into Eq. (7.1.35), and assuming that the plasma pressure at the edge of the column vanishes, i.e., $P(a) = 0$, it can be shown that the pressure is given by

$$P(\rho) = \mu_0 \left(\frac{I}{2\pi a}\right)^2 \left(1 - \frac{\rho^2}{a^2}\right). \tag{7.1.39}$$

If there are sufficient collisions to maintain an isotropic pressure distribution, one can assume that the pressure is given by the ideal gas law, $P = n\kappa T$. To proceed further, it is convenient to define a line number density given by

$$N_\ell = \int_0^a n 2\pi\rho \, \mathrm{d}\rho. \tag{7.1.40}$$

Assuming that the temperature is constant across the plasma column, the above equation can be rewritten as

$$N_\ell \kappa T = \int_0^a n\kappa T 2\pi\rho \, \mathrm{d}\rho = 2\pi \int_0^a P\rho \, \mathrm{d}\rho. \tag{7.1.41}$$

Integrating the above equation by parts once gives

$$N_\ell \kappa T = \left[\pi P\rho^2\right]_0^a - \pi \int_0^a \rho^2 (\mathrm{d}P/\mathrm{d}\rho)\,\mathrm{d}\rho. \tag{7.1.42}$$

If the pressure is assumed to go to zero at the outer boundary of the plasma column, the first term on the right vanishes. Substituting $\mathrm{d}P/\mathrm{d}\rho = -J_z B_\phi$ into the second term then gives

$$N_\ell \kappa T = \pi \int_0^a \rho^2 J_z B_\phi \,\mathrm{d}\rho. \tag{7.1.43}$$

The above equation can then be used to eliminate J_z, which gives

$$N_\ell \kappa T = \frac{\pi}{\mu_0}\int_0^a \rho B_\phi \frac{\mathrm{d}}{\mathrm{d}\rho}(\rho B_\phi)\,\mathrm{d}\rho = \frac{\pi}{2\mu_0}\int_0^a \frac{\mathrm{d}}{\mathrm{d}\rho}(\rho B_\phi)^2\,\mathrm{d}\rho = \frac{\pi}{2\mu_0} a^2 B_\phi^2(a). \tag{7.1.44}$$

To interpret the above equation, it is useful to note that the total current in the column can be evaluated in terms of the magnetic field, $B_\phi(a)$, at the edge of the column:

$$I = \int_0^a J_z 2\pi\rho\,\mathrm{d}\rho = \frac{1}{\mu_0}\int_0^a \frac{1}{\rho}\frac{\mathrm{d}}{\mathrm{d}\rho}(\rho B_\phi) 2\pi\rho\,\mathrm{d}\rho = \frac{2\pi}{\mu_0} a B_\phi(a). \tag{7.1.45}$$

Combining the above expression with Eq. (7.1.44) then gives

$$I^2 = \frac{8\pi N_\ell \kappa T}{\mu_0}. \tag{7.1.46}$$

This equation is called the Bennett relation, after Bennett (1934). Note that the total current is completely independent of the details of the plasma pressure profile.

The Z-pinch can be set up in the laboratory by simply applying a voltage across electrodes at each end of a plasma column. As will be discussed later in Section 7.3.2, the Z-pinch is unstable and often evolves into periodic sausage-like structures.

The ϕ-pinch. The ϕ-pinch is complementary to the Z-pinch in the sense that the only non-vanishing component of the current density is in the ϕ direction, $J_\phi = J_\phi(\rho)$, rather than in the z direction. This current produces a purely axial magnetic field, $B_z = B_z(\rho)$. In practice, this magnetic field must be imposed by an external boundary condition, usually a constant field, B_0, parallel to the z axis, as shown in Figure 7.5. For this magnetic field geometry, Eq. (7.1.31) reduces to

$$\frac{\mathrm{d}}{\mathrm{d}\rho}\left(P + \frac{B_z^2}{2\mu_0}\right) = 0, \tag{7.1.47}$$

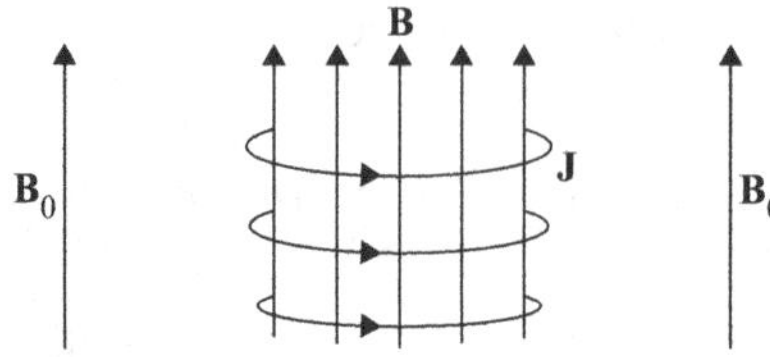

Figure 7.5 The magnetic field of a ϕ-pinch is externally imposed.

which can be integrated directly to give

$$P + \frac{B_z^2}{2\mu_0} = \text{constant}. \tag{7.1.48}$$

The constant on the right-hand side of the above equation can be determined by assuming that the plasma pressure vanishes at the edge of the column, with a constant externally applied axial field B_0 in the vacuum surrounding the plasma. The above equation then becomes

$$P + \frac{B_z^2}{2\mu_0} = \frac{B_0^2}{2\mu_0}. \tag{7.1.49}$$

This equation also applies even when the plasma column does not have cylindrical symmetry. The only requirement is that the magnetic field lines be straight. The $(\mathbf{B} \cdot \nabla)\mathbf{B}$ term on the right of Eq. (7.1.30) then vanishes. A self-consistent model illustrating this type of equilibrium was previously discussed in Section 3.3.4 based on gradient drift and magnetization currents.

The screw pinch. In the screw pinch, both the axial and azimuthal components of the magnetic field are non-zero, that is, $B_\phi \neq 0$ and $B_z \neq 0$. Clearly, the Z-pinch and ϕ-pinch are special cases of the screw pinch. In a screw pinch, the magnetic field lines do not close on themselves, but instead form helical lines around the axis of the cylinder. This is not hard to visualize. Imagine adding a B_z field to the closed B_ϕ lines of a Z-pinch. The result is a set of nested cylindrical surfaces covered by field lines that wrap around the cylinders in helical trajectories, like the flux rope shown in Figure 7.2. The magnetic field lines are also sheared in the radial direction. The amount of shear from one radial position to another depends on the radial dependences of B_ϕ and B_z. We assume that the screw pinch is periodic along z, with a periodicity length $2\pi R$, where R is a constant and a is the radius of the plasma column. The amount of shear is then measured by the radial dependence of the so-called *q-factor*, defined by the equation

$$q(\rho) = \frac{\rho B_z}{R B_\phi}. \tag{7.1.50}$$

The q-factor is also used for the description of a toroidal plasma (see the next section), in which case R and a are the major and minor radii of the plasma torus. In a screw pinch, there are two free functions that need to be specified before a unique equilibrium can be obtained. If we choose $B_\phi(\rho)$ and $B_z(\rho)$ as these two functions, Eq. (7.1.31) can be integrated to determine the plasma pressure $P(\rho)$. The current densities $J_\phi(\rho)$ and $J_z(\rho)$ can then be obtained from Eqs. (7.1.33) and (7.1.34), respectively.

A simple example of a screw pinch equilibrium is one in which a constant B_z field,

$$B_z = B_0 = \text{constant}, \tag{7.1.51}$$

is added to a Z-pinch. Note that the addition of a constant B_z field makes no contribution to the equilibrium equation (7.1.31). If the B_ϕ field is specified by Eq. (7.1.37), one obtains the q-profile

$$\begin{aligned} q(\rho) &= \frac{2\pi B_0 a^2}{\mu_0 I R} \qquad \text{for}\, \rho \le a, \quad \text{and} \\ q(\rho) &= \frac{2\pi B_0 \rho^2}{\mu_0 I R} \qquad \text{for}\, \rho > a. \end{aligned} \tag{7.1.52}$$

Within the plasma column $\rho \le a$, the q-profile is flat, that is, $\mathrm{d}q/\mathrm{d}\rho = 0$. The pressure profile for such an equilibrium is given by Eq. (7.1.39).

Toroidal Geometries

In Section 7.1.3 we pointed out from Poincaré-Hopf's theorem that the magnetic field lines for a magnetostatic equilibrium must lie on the surface of a torus. We now discuss such force-balanced equilibria for an axially symmetric toroidal geometry. In Figure 7.6, we show a cross section of a plasma torus in which all of the equilibrium quantities depend only on the coordinates ρ and z and are independent of the azimuthal angle ϕ. Such equilibria are of great interest for the

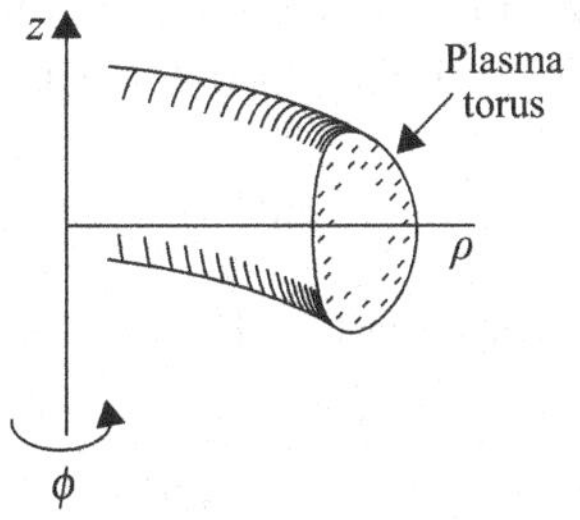

Figure 7.6 The coordinate system used to analyze a toroidal geometry.

tokamak, which is the most widely investigated magnetic confinement device for achieving controlled fusion.

To analyze toroidal geometries, we must develop some techniques for representing toroidal magnetic fields. The assumption of axial symmetry ($\partial/\partial\phi = 0$) implies that $\nabla \cdot \mathbf{B} = 0$ can be written

$$\frac{1}{\rho}\frac{\partial}{\partial \rho}(\rho B_\rho) + \frac{\partial B_z}{\partial z} = 0, \tag{7.1.53}$$

where $\mathbf{B} = [B_\rho, B_\phi, B_z]$. If we represent the magnetic field via a vector potential $\mathbf{A}$, such that $\mathbf{B} = \nabla \times \mathbf{A}$, it follows that $\nabla \cdot \mathbf{B} = 0$ is automatically satisfied, with B_ρ and B_z components given by

$$B_\rho = -\frac{\partial A_\phi}{\partial z} \tag{7.1.54}$$

and

$$B_z = \frac{1}{\rho}\frac{\partial}{\partial \rho}(\rho A_\phi). \tag{7.1.55}$$

The form of these equations motivates the definition of a flux function $\psi(\rho, z) = \rho A_\phi$ such that

$$B_\rho = -\frac{1}{\rho}\frac{\partial \psi}{\partial z} \tag{7.1.56}$$

and

$$B_z = \frac{1}{\rho}\frac{\partial \psi}{\partial \rho}. \tag{7.1.57}$$

Note that with B_ρ and B_z defined by the above two equations, Eq. (7.1.53) is satisfied automatically. Furthermore, it follows that

$$\mathbf{B} \cdot \nabla \psi = B_\rho \frac{\partial \psi}{\partial \rho} + B_z \frac{\partial \psi}{\partial_z} = 0, \tag{7.1.58}$$

which means that magnetic field lines lie on toroidal surfaces of constant ψ. From the condition $\mathbf{B} \cdot \nabla P = 0$ it also follows that if $\nabla P \neq 0$, then

$$P = P(\psi). \tag{7.1.59}$$

Next, we consider the ϕ component of the force-balance equation (7.1.6), which is

$$J_z B_\rho - J_\rho B_z = \frac{1}{\rho}\frac{\partial P}{\partial \phi} = 0. \tag{7.1.60}$$

Using Ampère's law (7.1.7) to express J_z and J_ρ in terms of B_ρ and B_ϕ, the above equation can be rewritten

$$\frac{1}{\rho}\frac{\partial}{\partial \rho}(\rho B_\phi)B_\rho + \frac{\partial B_\phi}{\partial z}B_z = 0. \tag{7.1.61}$$

By defining a new function

$$F = \rho B_\phi, \tag{7.1.62}$$

the above equation can be written in the simple form

$$\frac{\partial F}{\partial \rho}B_\rho + \frac{\partial F}{\partial z}B_z = 0, \tag{7.1.63}$$

or

$$\mathbf{B}\cdot\nabla F = 0. \tag{7.1.64}$$

Since both $\mathbf{B}\cdot\nabla\psi = 0$ and $\mathbf{B}\cdot\nabla F = 0$, it follows that F can be expressed as a function of ψ:

$$F = F(\psi). \tag{7.1.65}$$

Thus, by using two functions, ψ and F, and Eqs. (7.1.56), (7.1.57), and (7.1.62), the magnetic field can be represented in the form

$$\mathbf{B} = \left(\frac{-1}{\rho}\frac{\partial \psi}{\partial z}\right)\hat{\rho} + \left(\frac{1}{\rho}\frac{\partial \psi}{\partial \rho}\right)\hat{\mathbf{z}} + \frac{F(\psi)}{\rho}\hat{\boldsymbol{\phi}}, \tag{7.1.66}$$

or

$$\mathbf{B} = \left(\frac{\nabla\psi}{\rho}\right)\times\hat{\boldsymbol{\phi}} + \frac{F(\psi)}{\rho}\hat{\boldsymbol{\phi}}, \tag{7.1.67}$$

where $\hat{\rho}, \hat{\boldsymbol{\phi}}, \hat{\mathbf{z}}$ are the usual unit vectors in the cylindrical coordinates.

Finally, we consider the ρ component of the force-balance equation (7.1.6), which is given by

$$J_\phi B_z - J_z B\phi = \frac{\partial P}{\partial \rho}. \tag{7.1.68}$$

Following an approach similar to that used for the ϕ component, we proceed by eliminating J_ϕ and J_z using Ampère's law (7.1.7). From the B_ϕ component of

Ampère's law, we can obtain an equation for J_ϕ:

$$\mu_0 J_\phi = \frac{\partial B_\rho}{\partial z} - \frac{\partial B_z}{\partial \rho}$$
$$= -\frac{1}{\rho}\frac{\partial^2 \psi}{\partial z^2} - \frac{1}{\rho}\frac{\partial^2 \psi}{\partial \rho^2} + \frac{1}{\rho^2}\frac{\partial \psi}{\partial \rho}$$
$$= -\frac{1}{\rho}\Delta^* \psi, \tag{7.1.69}$$

where Δ^* is the operator defined by

$$\Delta^* \psi = \rho \frac{\partial}{\partial \rho}\left(\frac{1}{\rho}\frac{\partial \psi}{\partial \rho}\right) + \frac{\partial^2 \psi}{\partial z^2} = \rho^2 \boldsymbol{\nabla} \cdot \left(\frac{\boldsymbol{\nabla} \psi}{\rho^2}\right). \tag{7.1.70}$$

From the z component of Ampère's law (7.1.7), we can obtain an equation for J_z:

$$\mu_0 J_z = -\frac{1}{\rho}\frac{\partial}{\partial \rho}(\rho B_\phi) = -\frac{1}{\rho}\frac{\partial F}{\partial \rho} = -\frac{1}{\rho}\frac{\mathrm{d}F}{\mathrm{d}\psi}\frac{\partial \psi}{\partial \rho}. \tag{7.1.71}$$

By substituting $B_\phi = F(\psi)/\rho$, $\partial P/\partial \rho = (\mathrm{d}P/\mathrm{d}\psi)(\partial \psi/\partial \rho)$, Eq. (7.1.57) for B_z, Eq. (7.1.69) for J_ϕ, and Eq. (7.1.71) for J_z into Eq. (7.1.68), one obtains the equation

$$\Delta^* \psi = -\mu_0 \rho^2 \frac{\mathrm{d}P}{\mathrm{d}\psi} - \frac{1}{2}\frac{\mathrm{d}}{\mathrm{d}\psi}F^2. \tag{7.1.72}$$

This equation is known as the Grad–Shafranov equation and is widely used to calculate magnetostatic equilibria in axisymmetric toroidal systems (Shafranov, 1966). The usual strategy to solve this equation is as follows. First specify two free functions, the plasma pressure $P = P(\psi)$ and the toroidal field function $F = F(\psi)$, and solve Eq. (7.1.72) with specified boundary conditions to determine the flux function $\psi = \psi(\rho, z)$. Once $\psi(\rho, z)$ is known, the three components of the magnetic field can be calculated using Eq. (7.1.66). The pressure profile can then be obtained from the relation $P = P(\psi(\rho, z))$.

In most circumstances of physical interest, the Grad–Shafranov equation is solved numerically. However, the procedure can be illustrated by considering some exact analytical solutions, known as the Solov'ev equilibria (Solov'ev, 1968).

Assume that the two free functions, $F(\psi)$ and $P(\psi)$, are given by

$$\mu_0 P'(\psi) = -C_2 \tag{7.1.73}$$

and

$$F F'(\psi) = C_1, \tag{7.1.74}$$

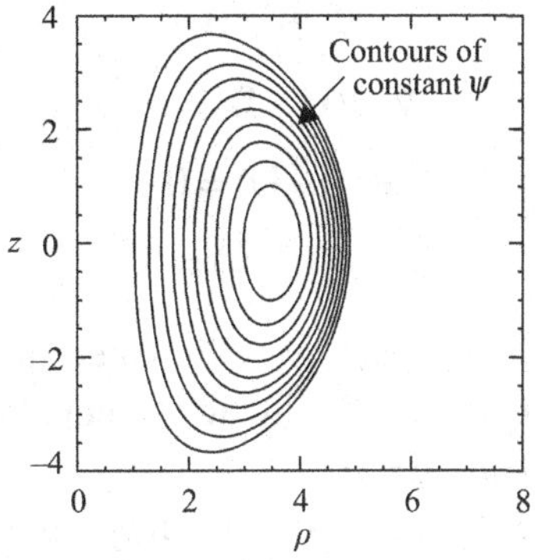

Figure 7.7 A plot of surfaces of constant ψ for the parameters $C_1 = 1$, $C_2 = -8$, $C_3 = -20$, $C_4 = 20$, and $C_5 = 0.2$.

where C_1 and C_2 are constants (Freidberg, 1987). By direct substitution it can be shown that an exact solution of the Grad–Shafranov equation (7.1.72) is given by

$$\psi(\rho, z) = -\frac{C_1}{2}z^2 + \frac{C_2}{8}\rho^4 + C_3 + C_4\rho^2 + C_5(\rho^4 - 4\rho^2 z^2), \tag{7.1.75}$$

where C_1 through C_5 are constants. Since the above equation has five arbitrary constants, the equation clearly represents a large family of solutions. Some of these solutions have surfaces of constant ψ that are nested, as shown in Figure 7.7. Solutions of this type are often used in the analysis of toroidal fusion confinement machines, such as the tokamak. Note that, unlike the cylindrical equilibria discussed above that depend on only one coordinate (ρ), this equilibrium depends on two coordinates (ρ and z).

7.2 Magnetohydrodynamic Equilibria

There are numerous equilibrium configurations of great physical interest in which plasma exhibits spatially inhomogeneous flow velocities, also referred to as sheared flow velocities, which cannot be eliminated by a simple Galilean transformation. It is then necessary to retain the velocity-dependent terms in the MHD equations given in Section 6.1.5. Under steady-state conditions, setting $\partial/\partial t = 0$, and assuming an ideal plasma ($\eta = 0$), we obtain the following equations:

$$\rho_m \mathbf{U} \cdot \nabla \mathbf{U} = \mathbf{J} \times \mathbf{B} - \nabla P, \tag{7.2.1}$$

$$\nabla \times \mathbf{B} = \mu_0 \mathbf{J}, \tag{7.2.2}$$

$$\nabla \cdot \mathbf{B} = 0, \tag{7.2.3}$$

$$\nabla \times (\mathbf{U} \times \mathbf{B}) = 0, \tag{7.2.4}$$

$$\nabla \cdot (\rho_{\mathrm{m}} \mathbf{U}) = 0, \tag{7.2.5}$$

$$\mathbf{U} \cdot \nabla (P/\rho_{\mathrm{m}}^{\gamma}) = 0. \tag{7.2.6}$$

Note that in writing Eq. (7.2.4), we have eliminated the electric field $\mathbf{E}$ between Eqs. (6.1.30) and (6.1.33) to obtain the induction equation (6.3.4), which reduces to (7.2.4) if we set $\partial/\partial t = 0$ and $\eta = 0$.

Clearly, Eqs. (7.2.1)–(7.2.6) are significantly more complicated than the magnetostatic equations (7.1.6)–(7.1.8) to which they reduce if we set $\mathbf{U} = 0$. The presence of a sheared velocity field introduces constraints on the magnetic field $\mathbf{B}$ and the pressure P that are absent for magnetostatic equilbria. Here we consider some simple but nontrivial examples of equilibria with flows that can be treated analytically. One such example is the cylindrical force-balanced equilibrium discussed in Section 7.1.4, but now with a sheared flow. Assuming that the equilibrium depends only on the radial coordinate ρ, the magnetic and velocity fields can be written as $[0, B_\phi(\rho), B_z(\rho)]$ and $[0, U_\phi(\rho), U_z(\rho)]$. Then the radial component of Eq. (7.2.1), which is the only surviving component of the equation, becomes

$$\frac{\partial}{\partial \rho}\left(P + \frac{B_\phi^2}{2\mu_0} + \frac{B_z^2}{2\mu_0}\right) = -\frac{B_\phi^2}{\mu_0 \rho} + \frac{\rho_m U_\phi^2}{\rho}. \tag{7.2.7}$$

It can be checked that all the equations (7.2.2)–(7.2.6) are satisfied automatically under these conditions (Problem 7.7). Note that the essential effect of the sheared velocity field is to introduce a centrifugal force due to the azimuthal velocity $U_\phi(\rho)$, represented by the last term in Eq. (7.2.7), and the axial velocity $U_z(\rho)$ makes no contribution to the radial force-balance condition. The presence of the centrifugal term introduces the mass density ρ_{m} into the MHD equilibrium equation, absent in the magnetostatic equilibrium equation (7.1.31). Rather remarkably, it follows that if we set $U_\phi = 0$, then Eq. (7.2.7) simply reduces to the magnetostatic force-balance equation (7.1.31), even if an arbitrary axial flow velocity $U_z(\rho)$ is included. In other words, all solutions of the magnetostatic force-balanced equation are also solutions of the MHD equilibria with arbitrary axial flow.

In astrophysical objects, for which gravity is an essential ingredient of equilibrium, Eq. (7.2.7) needs to be modified to take into account Newton's gravitational force. If we do so, we obtain

$$\frac{\partial}{\partial \rho}\left(P + \frac{B_\phi^2}{2\mu_0} + \frac{B_z^2}{2\mu_0}\right) = -\frac{B_\phi^2}{\mu_0 \rho} + \frac{\rho_{\mathrm{m}} U_\phi^2}{\rho} - \rho_{\mathrm{m}} g. \tag{7.2.8}$$

Here g is the acceleration due to gravity, and for an annular disk for which $a < \rho < b$, where a and b are constants, g is given by

$$g = \frac{GM}{\rho^2}, \tag{7.2.9}$$

where G is the gravitational constant and M is the mass of the central astrophysical object. An interesting class of equilibria is given by the following radial profiles that identically satisfy equation (7.2.8):

$$\rho_{\mathrm{m}} = \rho_{\mathrm{m}0}\left(\frac{a}{\rho}\right)^{3/2}, \tag{7.2.10}$$

$$P = P_0\left(\frac{a}{\rho}\right)^{5/2}, \tag{7.2.11}$$

$$B_\phi = -\alpha_1\sqrt{\frac{2\mu_0 P_0}{\beta_0(\alpha_1^2+\alpha_2^2)}}\left(\frac{a}{\rho}\right)^{-5/4}, \tag{7.2.12}$$

$$B_z = \alpha_2\sqrt{\frac{2\mu_0 P_0}{\beta_0(\alpha_1^2+\alpha_2^2)}}\left(\frac{a}{\rho}\right)^{-5/4}, \tag{7.2.13}$$

$$U_\phi = U_0\left(\frac{a}{\rho}\right)^{-1/2}, \tag{7.2.14}$$

where

$$\rho_{\mathrm{m}0}U_0^2 = \rho_{\mathrm{m}0}GM - \frac{P_0}{2\beta_0(\alpha_1^2+\alpha_2^2)}[5(1+\beta_0)(\alpha_1^2+\alpha_2^2) - 4\alpha_1^2], \tag{7.2.15}$$

and a is a reference radius. If we define an angular speed $\Omega_\phi = U_\phi/\rho$, we obtain from Eq. (7.2.14) the profile

$$\Omega_\phi = U_0\left(\frac{a}{\rho}\right)^{3/2}. \tag{7.2.16}$$

The radial dependence of the angular velocity is typical of Keplerian orbits of planets. Equations (7.2.10)–(7.2.16) are a useful point of departure for the study of accretion disks in the cylindrical limit.

7.3 Stability of Ideal Magnetostatic Equilibria

In the previous section we showed that a large variety of solutions to the static MHD equilibrium equations can be found. However, the existence of a solution to the equilibrium equations does not prove that the configuration can actually exist in nature, since the equilibrium may be unstable. If the equilibrium is unstable then the system will start to evolve in time, and will either evolve to some other completely different equilibrium configuration or never reach a steady state. In order to understand which equilibria can actually exist in nature it is important to consider the issue of stability. In this section we discuss some of the techniques used to analyze the stability of ideal MHD equilibria.

To develop physical intuition about some of the techniques involved, we start by discussing some simple configurations that can be shown to be unstable using somewhat restrictive assumptions. We then proceed to develop two methods, called (i) the normal mode method, and (ii) the energy method, that can be used to test whether an arbitrary MHD equilibrium is stable or unstable.

7.3.1 MHD Stability: A Mechanical Analogy

It is useful to start the discussion of stability by considering a simple mechanical analog, which is the motion of a particle in a one-dimensional potential energy function, $W(x)$, such as the one shown in Figure 7.8.

For such a system, static equilibrium solutions exist at all points for which the force $F_x = -\partial W/\partial x = 0$. To test the stability at a given equilibrium point, the procedure is to give the particle a small displacement, or perturbation. If the particle returns to the equilibrium point the system is stable. If it does not return it is unstable. From simple physical considerations one can see that the condition for stability is that the equilibrium point must be at a local minimum in the potential energy. To prove this assertion formally, expand the potential energy in a Taylor series around one of the equilibrium points. Since $\partial W/\partial x = 0$ at the equilibrium point, then to lowest order in the displacement, Δx, the change in the potential energy relative to the potential energy at the equilibrium point is given by

$$\delta W = \frac{1}{2}\left(\frac{\partial^2 W}{\partial x^2}\right)\Delta x^2, \tag{7.3.1}$$

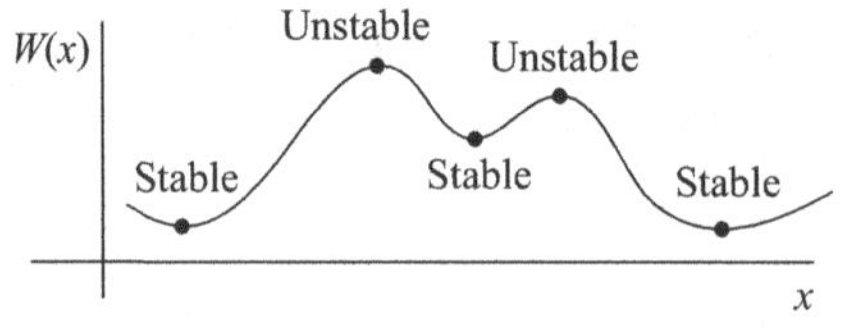

Figure 7.8 The potential energy of a mechanical system has points of stable equilibrium, characterized by $\partial^2 W/\partial x^2 > 0$, and points of unstable equilibrium, characterized by $\partial^2 W/\partial x^2 < 0$.

where $(\partial^2 W/\partial x^2)$ is evaluated at the equilibrium point. For such a potential it can be shown that the equation of motion is given by

$$m\frac{\mathrm{d}^2\Delta x}{\mathrm{d}t^2}+\left(\frac{\partial^2 W}{\partial x^2}\right)\Delta x=0, \tag{7.3.2}$$

which is the simple harmonic oscillator equation. Consider what happens if the particle is given a small displacement at $t=0$. If $\partial^2 W/\partial x^2>0$, the particle executes a simple harmonic oscillation and, if there is even a small amount of damping, the system eventually returns to the equilibrium point. The system is stable. On the other hand, if $\partial^2 W/\partial x^2<0$, the displacement grows exponentially and the particle never returns to the equilibrium point. The system is then unstable.

If the potential energy depends on n coordinates $(x_1,x_2,\ldots,x_n)$, then static equilibria exist at points that simultaneously satisfy the n equations:

$$\frac{\partial W(x_1,x_2,\ldots,x_n)}{\partial x_\alpha}=0, \qquad \alpha=1,2,\ldots,n. \tag{7.3.3}$$

Following the same type of analysis as above, it can be seen that a given equilibrium point is stable if and only if, for all α,

$$\frac{\partial^2 W}{\partial x_\alpha^2}>0, \qquad \alpha=1,2,\ldots,n. \tag{7.3.4}$$

For a system with a very large number of degrees of freedom, such as an MHD fluid, there are so many conditions required for the system to be stable that a completely stable equilibrium is more the exception than the rule.

7.3.2 Interchange Instabilities

Although a wide variety of instabilities can occur in an MHD system, we start by giving a heuristic discussion of a specific class of instabilities called interchange instabilities. Interchange instabilities involve a specific type of perturbation, namely the interchange of plasma between two adjacent magnetic flux tubes. Since the frozen field theorem applies, the magnetic flux must be conserved in this interchange process. From Eq. (6.4.21) for the potential energy, W, one can see that if such a perturbation, δ, is applied to the system, the resulting change in the total energy consists of two terms

$$\delta W=\delta\int_V\frac{P}{\gamma-1}\mathrm{d}^3x+\delta\int_V\frac{B^2}{2\mu_0}\mathrm{d}^3x, \tag{7.3.5}$$

where the first term represents the change in the internal energy, and the second term represents the change in the magnetic energy (see Section 6.4). To simplify

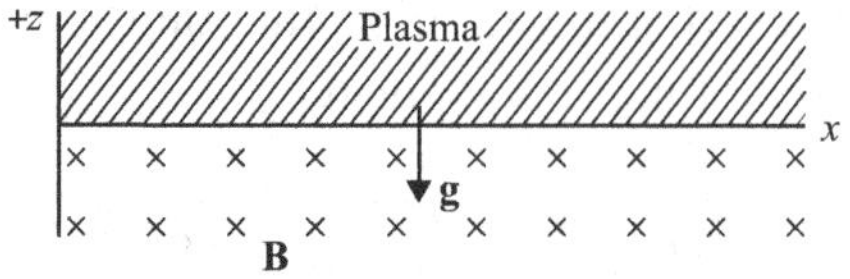

Figure 7.9 A geometry in which an MHD fluid in a gravitational field **g** is supported by a magnetic field **B**.

the analysis further, we assume that the change in the internal energy is zero, so that the change in the potential energy is controlled entirely by the change in the magnetic energy. Although restrictive, this simple model can be used to show that a number of important MHD equilibria are unstable. The disadvantage is that we cannot show that a system is stable, since we have not considered all possible perturbations or the possible destabilizing effect of the internal energy. To determine whether a system is stable in all circumstances we need some more powerful tools, which will be discussed later in this chapter.

As a simple example, consider an MHD fluid that is supported against gravity by a uniform horizontal magnetic field, as shown in Figure 7.9. The static equilibrium condition for this geometry is simply that the gravitation force per unit area acting on the fluid must be balanced by the magnetic pressure on the underside of the fluid,

$$\frac{B^2}{2\mu_0} = \int_0^\infty \rho_m g \, dz. \tag{7.3.6}$$

The magnetic force arises because of a current **J** that must flow along the boundary between the plasma and the magnetic field. By adding a gravitational potential energy term, $\rho_m g z$, to the potential energy Eq. (6.4.21), it is easy to see that the change in the potential energy of the system is given by

$$\delta W = \delta \int_v \frac{B^2}{2\mu_0} d^3x + \delta \int_V \rho_m g z \, d^3x. \tag{7.3.7}$$

Suppose now that a tube of fluid is interchanged with a magnetic flux tube of the same volume as shown in Figure 7.10. Since the magnetic field lines are straight and the volume elements before and after the interchange are the same, this perturbation does not change the magnetic energy. However, it does cause a change in the gravitational potential energy. The change in the potential energy of the system is given by

$$\delta W = \int_V \rho_m g \, \delta z d^3x, \tag{7.3.8}$$

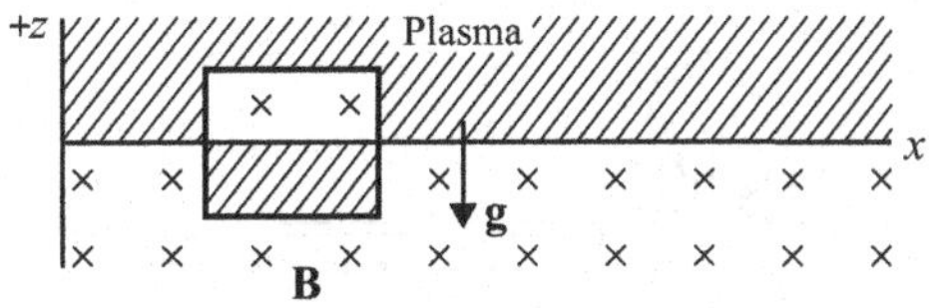

Figure 7.10 An example of an interchange motion that replaces a magnetic flux tube with a tube of fluid from the other side of the boundary.

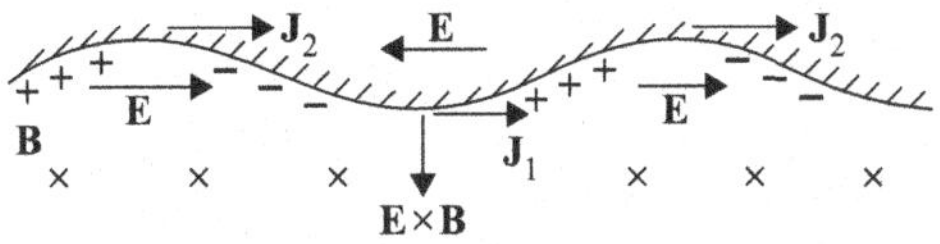

Figure 7.11 When a perturbation develops, electrical currents cause charges to accumulate along the boundary. These charges lead to electric fields and $\mathbf{E} \times \mathbf{B}$ drifts that amplify the original perturbation.

which is negative, since the vertical displacement of the fluid element, δz, is negative. Thus, since the potential energy of the system has decreased, it follows that the system is unstable. This instability, which was first analyzed in detail by Kruskal and Schwarzschild (1954), is the MHD analog of the well-known Rayleigh–Taylor instability in fluid dynamics; see Batchelor (1967).

The physical origin of the Kruskal–Schwarzschild instability can be seen by considering the $\mathbf{J} \times \mathbf{B}$ forces that occur at the boundary between the plasma and the magnetic field. As shown in Figure 7.11, if a ripple-like perturbation develops in the boundary, the current at the low point, $\mathbf{J}_1$, must be larger than the current at the high points, $\mathbf{J}_2$, because more plasma is being supported against the gravitational force. The resulting difference in the boundary current causes regions of positive and negative charge to develop along the boundary, thereby producing oppositely directed $\mathbf{E}$ fields at the high and low points of the ripple. The electric fields are directed in such a way that the $\mathbf{E} \times \mathbf{B}$ drifts reinforce the original disturbance, thereby causing the ripple to grow in amplitude, ultimately leading to a large-scale transport of plasma across the boundary.

Similar types of interchange motions also occur in plasmas that have large inertial forces caused by curvature in the plasma flow. For example, strong radial electric fields often develop in cylindrically symmetric plasma devices. As discussed earlier, these electric fields cause the plasma to rotate rapidly about the axis of the device. The centrifugal force produced by this rotation plays the same role as gravity, and leads to the development of interchange motions on the outer boundary of the plasma. The resulting perturbations look like the flutes on a Greek column, as shown in Figure 7.12. For this reason the instability is sometimes called a "flute" instability.

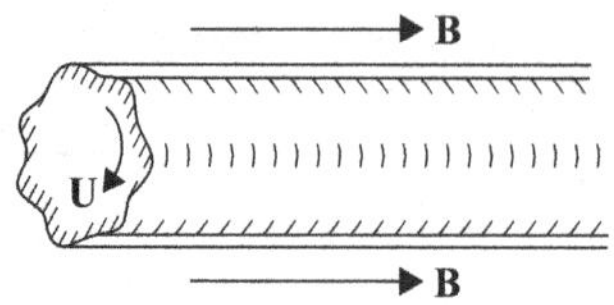

Figure 7.12 In a rapidly rotating plasma column, the centrifugal force can drive an interchange instability, which leads to the development of flute-like ripples on the boundary of the column.

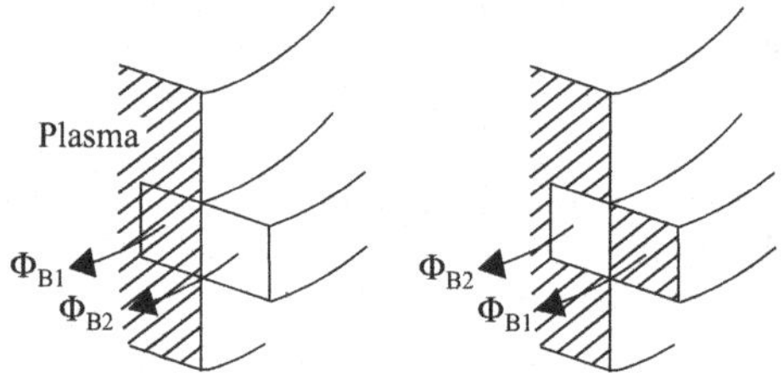

Figure 7.13 An interchange motion consists of interchanging two flux tubes Φ_{B1} and Φ_{B2} at an interface between two plasmas.

Interchange instabilities can also occur in the absence of gravity if the plasma is confined within a curved boundary by a magnetic field. Consider the geometry shown in Figure 7.13. To test the stability of this static equilibrium, we compute the change in the potential energy, δW, when two adjacent flux tubes are interchanged while conserving the magnetic flux, Φ_{B1} and Φ_{B2}. The change in the potential energy is then given by

$$\delta W = \delta \int_V \frac{B^2}{2\mu_0} A \, d\ell = \delta \frac{\Phi_B^2}{2\mu_0} \int_V \frac{d\ell}{A}, \tag{7.3.9}$$

where A is the cross-sectional area of the flux tube and $d\ell$ is the distance along the magnetic field line. The change in magnetic energy of the system due to the interchange is then given by

$$\begin{aligned} \delta W &= \frac{1}{2\mu_0}\left[\Phi_{B2}^2\left(\int_V \frac{d\ell_1}{A_1} - \int_V \frac{d\ell_2}{A_2}\right) + \Phi_{B1}^2\left(\int_V \frac{d\ell_2}{A_2} - \int_V \frac{d\ell_1}{A_1}\right)\right] \\ &= \frac{1}{2\mu_0}(\Phi_{B2}^2 - \Phi_{B1}^2)\left(\int_V \frac{d\ell_1}{A_1} - \int_V \frac{d\ell_2}{A_2}\right) \\ &= \frac{1}{2\mu_0}(\Phi_{B2}^2 - \Phi_{B1}^2)\left(\int_V \left[1 - \frac{A_1}{A_2}\frac{d\ell_2}{d\ell_1}\right]\frac{d\ell_1}{A_1}\right). \end{aligned} \tag{7.3.10}$$

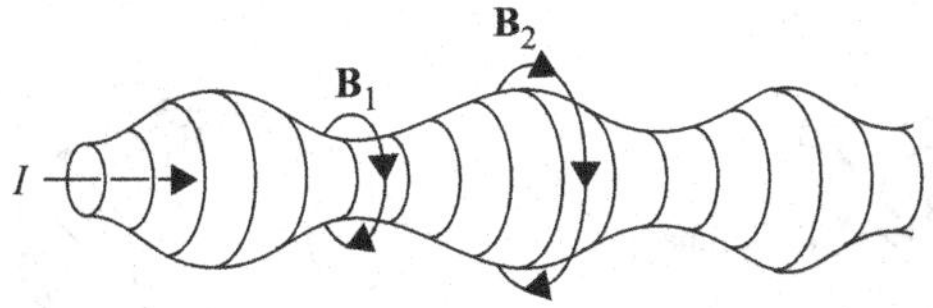

Figure 7.14 The sausage-mode instability in a Z-pinch.

If the fluid is incompressible, which is assumed to be the case, we have $A_1\,\mathrm{d}\ell_1 = A_2\,\mathrm{d}\ell_2$, which gives

$$\delta W = \frac{1}{2\mu_0}\left(\Phi_{\mathrm{B2}}^2 - \Phi_{\mathrm{B1}}^2\right)\left\{\int_V \left[1 - \left(\frac{\mathrm{d}\ell_2}{\mathrm{d}\ell_1}\right)^2\right]\frac{\mathrm{d}\ell_1}{A_1}\right\}. \tag{7.3.11}$$

Since the magnetic field must be stronger outside the fluid (to maintain the pressure balance), $\Phi_{\mathrm{B2}} > \Phi_{\mathrm{B1}}$. The above equation shows that if the field lines are concave towards the fluid ($\mathrm{d}\ell_2 > \mathrm{d}\ell_1$) the system is unstable. For example, the Z-pinch configuration is unstable to interchange motions. This instability manifests itself by the formation of "sausage-shaped" deformations in the outer surface of the plasma column, as shown in Figure 7.14. This sausage-mode instability is essentially geometric in origin. Because the total current, I, that flows along the column must be constant, Ampère's law, $B = \mu_0 I/(2\pi\rho)$, shows that the magnetic field increases in a region of smaller radius, which acts to increase the external magnetic pressure and thereby further shrinks the radius. Once the instability starts, the pinch is rapidly disrupted. Lightning discharges sometimes display this effect (known as bead lightning). A Z-pinch can be stabilized by adding a magnetic field along the z axis. The axial magnetic field produces an internal magnetic field pressure in the region of smaller radius, thereby stabilizing the configuration. Lightning discharges also tend to be stabilized by atmospheric pressure. Therefore, only very intense discharges develop the bead-like structure.

Interchange instabilities also occur in the magnetospheres of Earth and Jupiter. In both cases, relatively low β plasmas (the plasmasphere at Earth and the Io plasma torus at Jupiter) occur deep within the magnetosphere, as shown in Figure 7.15.

Since the magnetic field lines in both cases are concave inward toward the region of higher plasma density, the plasmas tend to be unstable to interchange motions. The interchange motions form flute-like ripples aligned parallel to the magnetic field, as shown in Figure 7.16. Eventually the peaks in the ripple evolve into outward-moving plumes that transport plasma away from the planet, and the valleys evolve into inward-moving plumes that transport lower-density plasma toward the planet. Interchange motions of this type can be stabilized by a variety of effects, one of which is the closure of magnetic field-aligned currents through

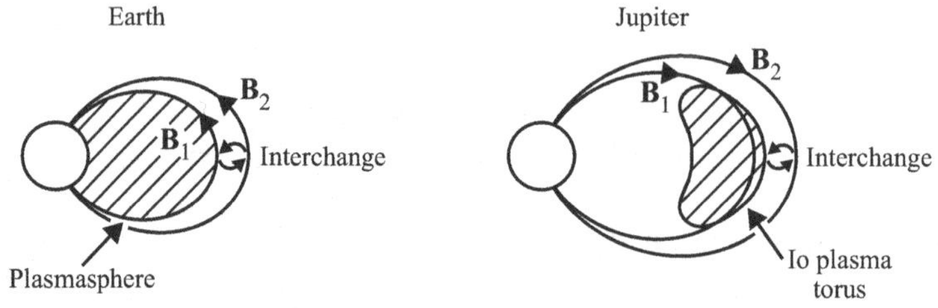

Figure 7.15 The interchange instability is believed to play a major role in the radial transport of plasma in the magnetospheres of Earth and Jupiter.

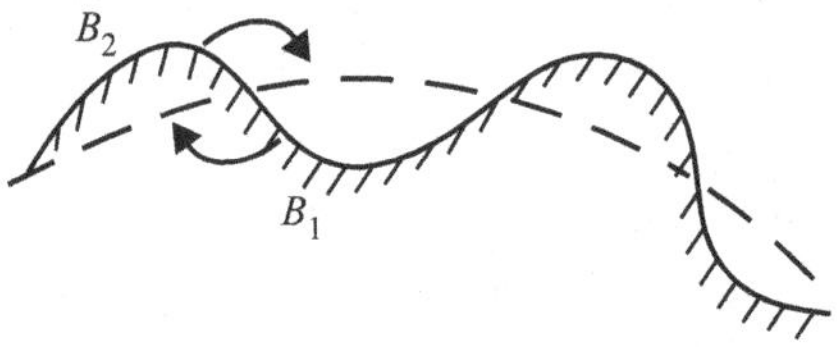

Figure 7.16 The interchange instability tends to form flute-like ripples along the boundary of the plasma.

a resistive layer in the ionosphere. A full discussion of these stabilizing effects is beyond the scope of this book.

Since interchange instabilities of the type described above have their origin in the curvature of magnetic field lines, it is interesting to see what can be done to avoid these instabilities. This issue is particularly important for laboratory magnetic confinement devices, which attempt to confine a hot plasma with an external magnetic field. A simple magnetic mirror field, such as that discussed in Section 3.4, tends to be unstable to interchange motions, since the plasma boundary is concave toward the central axis of the machine. One way of stabilizing such a device against interchange motions is to have the magnetic field lines concave outward away from the plasma. An example of such a configuration is the cusp mirror geometry. In the cusp mirror geometry, the plasma is confined in the low magnetic field region formed by four or more symmetrically arranged conductors with currents alternately in opposite directions, as shown in Figure 7.17. Since the boundaries of the plasma are concave outward away from the plasma, this geometry stabilizes the system against interchange motions.

7.3.3 The Linear Force Operator for Magnetostatic Equilibria

Although the above discussion illustrates some of the basic considerations involved in MHD stability, it is rather specialized. To take into account the effects of plasma pressure, current, and more general perturbations in inhomogeneous plasmas, we need to develop some more powerful tools for analyzing the change

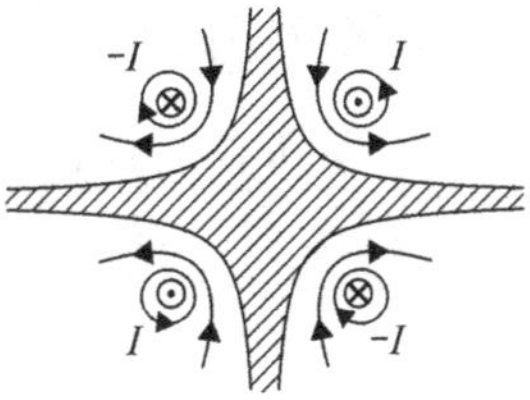

Figure 7.17 The cusp mirror geometry.

in the potential energy. Our immediate objective will be to re-express the forces on the right-hand side of the momentum equation (6.1.32) in the form of a linear operator, called the "linear force operator." The linear force operator can then be used to evaluate the change in the potential energy for any arbitrary perturbation. As in previous analyses involving small perturbations, we will express all of the relevant MHD quantities as the sum of a time-independent zero-order (0) quantity plus a small first-order perturbation. The time-independent zero-order quantities, such as $\mathbf{B}_0$, $\mathbf{J}_0$, and P_0, are assumed to obey the equations for magnetostatic equilibrium, i.e., Eqs. (7.1.6)–(7.1.8). Note that, in contrast to previous perturbation analyses of this type, we cannot assume that the zero-order quantities are homogeneous. Unfortunately, this greatly increases the complexity of the analysis.

Since it is desirable to work with displacements rather than velocities, we introduce the perturbed fluid displacement vector $\boldsymbol{\xi}(\mathbf{r},t)$ by means of the equation

$$\mathbf{U} = \frac{\partial \boldsymbol{\xi}(\mathbf{r},t)}{\partial t}. \tag{7.3.12}$$

Using this relation, it can be shown that the linearized version of the momentum equation (6.1.32) is given by

$$\rho_{\mathrm{m}0}\frac{\partial^2 \boldsymbol{\xi}}{\partial t^2} = \frac{1}{\mu_0}[(\nabla \times \mathbf{B}) \times \mathbf{B}_0 + (\nabla \times \mathbf{B}_0) \times \mathbf{B}] - \nabla P, \tag{7.3.13}$$

where the current density has been eliminated using Ampère's law (6.1.28) and, as usual, we have omitted the subscript 1 on the first-order perturbed quantities. The induction equation (6.3.7) can be integrated over time to give

$$\mathbf{B} = \nabla \times (\boldsymbol{\xi} \times \mathbf{B}_0). \tag{7.3.14}$$

To obtain an equation for the pressure we combine the linearized form of the mass continuity equation (6.1.31),

$$\frac{\partial \rho_{\mathrm{m}}}{\partial t} + \rho_{\mathrm{m}0}\nabla \cdot \mathbf{U} + \mathbf{U} \cdot \nabla \rho_{\mathrm{m}0} = 0, \tag{7.3.15}$$

With the linearized form of the adiabatic equation of state (6.1.34),

$$\frac{\partial P}{\partial t} + \mathbf{U} \cdot \boldsymbol{\nabla} P_0 - \frac{\gamma P_0}{\rho_{m0}} \frac{\partial \rho_m}{\partial t} - \frac{\gamma P_0}{\rho_{m0}} \mathbf{U} \cdot \boldsymbol{\nabla} \rho_{m0} = 0, \tag{7.3.16}$$

to obtain

$$\frac{\partial P}{\partial t} + \mathbf{U} \cdot \boldsymbol{\nabla} P_0 + \gamma P_0 (\boldsymbol{\nabla} \cdot \mathbf{U}) = 0. \tag{7.3.17}$$

Replacing $\mathbf{U}$ by $\partial \boldsymbol{\xi}/\partial t$, and integrating over time gives

$$P = -\boldsymbol{\xi} \cdot \boldsymbol{\nabla} P_0 - \gamma P_0 \boldsymbol{\nabla} \cdot \boldsymbol{\xi}. \tag{7.3.18}$$

Substituting expressions (7.3.14) and (7.3.18) for $\mathbf{B}$ and P, respectively, into Eq. (7.3.13), we obtain the linear differential equation

$$\rho_{m0} \frac{\partial^2 \boldsymbol{\xi}}{\partial t^2} = \mathbf{F}(\boldsymbol{\xi}), \tag{7.3.19}$$

where $\mathbf{F}(\boldsymbol{\xi})$, the linear force operator, is given by

$$\begin{aligned} \mathbf{F}(\boldsymbol{\xi}) = {} & \frac{1}{\mu_0} [(\boldsymbol{\nabla} \times \{\boldsymbol{\nabla} \times (\boldsymbol{\xi} \times \mathbf{B}_0)\}) \times \mathbf{B}_0 + (\boldsymbol{\nabla} \times \mathbf{B}_0) \times \{\boldsymbol{\nabla} \times (\boldsymbol{\xi} \times \mathbf{B}_0)\}] \\ & + \boldsymbol{\nabla}[\boldsymbol{\xi} \cdot \boldsymbol{\nabla} P_0 + \gamma P_0 (\boldsymbol{\nabla} \cdot \boldsymbol{\xi})]. \end{aligned} \tag{7.3.20}$$

In principle, given an arbitrary static equilibrium, the equation of motion (7.3.19) can then be solved to determine all possible plasma displacements $\boldsymbol{\xi}(\mathbf{r}, t)$. In practice, because of the complexity of the linear force operator, such an approach is difficult (even on a computer) for all but the simplest cases. However, the linear force operator has some very nice mathematical properties that greatly simplify the analysis, and lead to two useful and sometimes complementary approaches, called (1) the normal mode method and (2) the energy method. We next describe these two methods.

7.3.4 The Normal Mode Method

In the normal mode method, we assume that the displacement vector can be written as a simple harmonic function of the form

$$\boldsymbol{\xi}_n(\mathbf{r}, t) = \boldsymbol{\xi}_n(\mathbf{r}) \exp(-i\omega_n t), \tag{7.3.21}$$

where the subscript n designates a particular normal mode of frequency ω_n. Since the equation of motion (7.3.19) is linear in $\boldsymbol{\xi}$, we can construct a general solution

$\boldsymbol{\xi}(\mathbf{r},t)$ by a linear superposition, i.e.,

$$\boldsymbol{\xi}(\mathbf{r},t) = \sum_n \boldsymbol{\xi}_n(\mathbf{r})\exp(-\mathrm{i}\omega_n t). \tag{7.3.22}$$

To find the constraints on the normal mode frequencies, we start by substituting expression (7.3.21) into Eq. (7.3.19), which gives

$$-\rho_{\mathrm{m}0}\,\omega_n^2\,\boldsymbol{\xi}_n = \mathbf{F}(\boldsymbol{\xi}_n). \tag{7.3.23}$$

Next, we state the following theorem, the proof of which is assigned as an exercise (Problem 7.5).

Theorem. *The linear force operator* $\mathbf{F}$ *is self-adjoint, i.e., if there are two eigenfunctions* $\boldsymbol{\eta}_1$ and $\boldsymbol{\eta}_2$ *(satisfying standard boundary conditions), then*

$$\int_V \boldsymbol{\eta}_2\cdot\mathbf{F}(\boldsymbol{\eta}_1)\,\mathrm{d}^3x = \int_V \boldsymbol{\eta}_1\cdot\mathbf{F}(\boldsymbol{\eta}_2)\,\mathrm{d}^3x. \tag{7.3.24}$$

This theorem has the following corollaries.

Corollary 1. ω_n^2 *is always real.*

To prove Corollary 1, consider the complex conjugate of Eq. (7.3.23):

$$-\rho_{\mathrm{m}0}\,\omega_n^{2*}\,\boldsymbol{\xi}_n^* = \mathbf{F}(\boldsymbol{\xi}_n^*). \tag{7.3.25}$$

Taking the scalar product of both sides of the above equation with $\boldsymbol{\xi}_n$ and integrating over the entire volume of the system gives

$$-\rho_{\mathrm{m}0}\omega_n^{2*}\int_V \boldsymbol{\xi}_n\cdot\boldsymbol{\xi}_n^*\,\mathrm{d}^3x = \int_V \boldsymbol{\xi}_n\cdot\mathbf{F}(\boldsymbol{\xi}_n^*)\,\mathrm{d}^3x. \tag{7.3.26}$$

Identifying $\boldsymbol{\eta}_1 = \boldsymbol{\xi}_n$ and $\boldsymbol{\eta}_2 = \boldsymbol{\xi}_n^*$, we can then invoke Eq. (7.3.24) and rewrite the above equation as

$$\begin{aligned}-\rho_{\mathrm{m}0}\omega_n^{2*}\int_V \boldsymbol{\xi}_n\cdot\boldsymbol{\xi}_n^*\,\mathrm{d}^3x &= \int_V \boldsymbol{\xi}_n^*\cdot F(\boldsymbol{\xi}_n)\,\mathrm{d}^3x \\ &= -\rho_{\mathrm{m}0}\omega_n^2\int_V \boldsymbol{\xi}_n^*\cdot\boldsymbol{\xi}_n\,\mathrm{d}^3x.\end{aligned} \tag{7.3.27}$$

Since $\int \boldsymbol{\xi}_n\cdot\boldsymbol{\xi}_n^*\,\mathrm{d}^3x = \int |\boldsymbol{\xi}_n|^2\,\mathrm{d}^3x$ is non-zero for any non-trivial eigenfunction, it follows from the above equation that

$$\omega_n^2 = \omega_n^{2*}. \tag{7.3.28}$$

Therefore, ω_n^2 is always real. This result implies that the eigenfrequency ω_n is either purely real or purely imaginary. If ω_n is imaginary, then the system is unstable, since one of the two roots leads to exponential growth in Eq. (7.3.21). Note that, if the self-adjointness of **F** is destroyed, as in an equilibrium with flows, i.e., $\mathbf{U}_0 \neq 0$, then Eq. (7.3.28) does not hold.

Corollary 2. *The eigenmodes of* **F** *are orthonormal, i.e., they obey the condition*

$$\int_V \rho_{\mathrm{m0}}\, \boldsymbol{\xi}_m^* \cdot \boldsymbol{\xi}_n \,\mathrm{d}^3x = \delta_{mn}, \tag{7.3.29}$$

where δ_{mn} is the Kronecker delta function ($\delta_{mn} = 1$ if $m = n$, or 0 if $m \neq n$).

To prove Corollary 2, consider the expression

$$\begin{aligned}\left(\omega_m^{2*} - \omega_n^2\right) \int_V \rho_{\mathrm{m0}}\, \boldsymbol{\xi}_m^* \cdot \boldsymbol{\xi}_n \,\mathrm{d}^3x \\ = -\int_V [\boldsymbol{\xi}_n \cdot \mathbf{F}(\boldsymbol{\xi}_m^*) - \boldsymbol{\xi}_m^* \cdot \mathbf{F}(\boldsymbol{\xi}_n)]\,\mathrm{d}^2x \\ = 0.\end{aligned} \tag{7.3.30}$$

The last step follows from Eq. (7.3.24). If $m \neq n$, Eq. (7.3.30) implies that

$$\int_V \rho_{\mathrm{m0}}\, \boldsymbol{\xi}_m^* \cdot \boldsymbol{\xi}_n \,\mathrm{d}^3x = 0. \tag{7.3.31}$$

If $m = n$, Eq. (7.3.30) is satisfied automatically because of relation (7.3.28). In that case, one can normalize the linear eigenfunction $\boldsymbol{\xi}$ so that

$$\int_V \rho_{\mathrm{m0}}\, \boldsymbol{\xi}_m^* \cdot \boldsymbol{\xi}_m \,\mathrm{d}^3x = 1. \tag{7.3.32}$$

Combining Eqs. (7.3.31) and (7.3.32), we obtain Eq. (7.3.29).

As mentioned above, an important consequence of Corollary 1 is that, for an ideal MHD magnetostatic equilibrium, the eigenfrequency is either purely oscillatory (i.e., ω_n is purely real) or purely growing (i.e., ω_n is purely imaginary). This makes the marginal frequency condition, $\omega^2 = 0$, an important case to study, since it separates stable solutions from unstable solutions. Thus, Corollary 1 is of great help in determining the stability of the system, since one need not search the entire complex ω^2 plane but only the real ω^2 axis for solutions, with the origin $\omega^2 = 0$ separating unstable solutions ($\omega^2 < 0$) from stable ones ($\omega^2 > 0$). Schematically, a possible frequency spectrum of the ideal MHD force operator is shown in Figure 7.18. The origin 0 is the marginal point. To the left of the origin are discrete and unstable frequencies. To the right of the origin are stable frequencies.

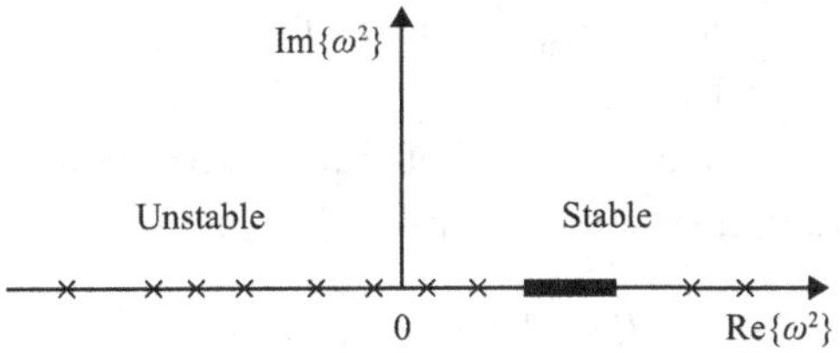

Figure 7.18 The frequency spectrum of the ideal MHD force operator.

It is possible to obtain not only discrete stable frequencies, but also a continuous band of stable frequencies, indicated by the dark shaded band.

The normal mode method is a brute force method for determining the eigenmodes and eigenfrequencies of the system using Eq. (7.3.23). Once these are known, the general solution (7.3.22) can be constructed by superposition. Next, we discuss the energy principle, which is a more subtle and even more powerful method of testing the stability of a magnetostatic equilibrium.

7.3.5 The Energy Principle

To analyze the stability of a static MHD equilibrium using the energy principle, we essentially follow the same procedure discussed in Section 7.3.1 for a particle moving in a one-dimensional potential. However, the analysis is more complicated, since an MHD fluid has an infinite number of degrees of freedom. The first step is to compute the change in the potential energy, δW, for a small arbitrary perturbation, $\boldsymbol{\xi}$, relative to some equilibrium state. To obtain the change in the potential energy, δW, we start by showing that

$$\frac{\mathrm{d}}{\mathrm{d}t}\int_V \left[\frac{1}{2}\rho_{\mathrm{m}0}\left|\frac{\partial \boldsymbol{\xi}}{\partial t}\right|^2 - \frac{1}{2}\boldsymbol{\xi}^* \cdot \mathbf{F}(\boldsymbol{\xi})\right]\mathrm{d}^3x = 0, \tag{7.3.33}$$

where $\mathbf{F}(\boldsymbol{\xi})$ is the linear force operator defined by Eq. (7.3.20). To prove the above result, consider the left-hand side, which can be written

$$\frac{1}{2}\int_V \left[\rho_{\mathrm{m}0}\frac{\partial \boldsymbol{\xi}^*}{\partial t}\cdot\frac{\partial^2 \boldsymbol{\xi}}{\partial t^2} + \rho_{\mathrm{m}0}\frac{\partial \boldsymbol{\xi}}{\partial t}\cdot\frac{\partial^2 \boldsymbol{\xi}^*}{\partial t^2} - \frac{\partial \boldsymbol{\xi}^*}{\partial t}\cdot\mathbf{F}(\boldsymbol{\xi}) - \boldsymbol{\xi}^*\cdot\mathbf{F}\left(\frac{\partial \boldsymbol{\xi}}{\partial t}\right)\right]\mathrm{d}^3x$$

$$= \frac{1}{2}\int_V \left[\frac{\partial \boldsymbol{\xi}^*}{\partial t}\cdot\left\{\rho_{\mathrm{m}0}\frac{\partial^2 \boldsymbol{\xi}}{\partial t^2} - \mathbf{F}(\boldsymbol{\xi})\right\} + \frac{\partial \boldsymbol{\xi}}{\partial t}\cdot\left\{\rho_{\mathrm{m}0}\frac{\partial^2 \boldsymbol{\xi}^*}{\partial t^2} - \mathbf{F}(\boldsymbol{\xi}^*)\right\}\right]\mathrm{d}^3x, \tag{7.3.34}$$

where in the last step we have used Eq. (7.3.24) to write

$$\int_V \boldsymbol{\xi}^*\cdot\mathbf{F}\left(\frac{\partial \boldsymbol{\xi}}{\partial t}\right)\mathrm{d}^3x = \int_V \frac{\partial \boldsymbol{\xi}}{\partial t}\cdot\mathbf{F}(\boldsymbol{\xi}^*)\,\mathrm{d}^3x. \tag{7.3.35}$$

It is obvious now that, upon using the linear force equation (7.3.19), expression (7.3.34) vanishes identically.

By simple inspection, one can see that Eq. (7.3.33) is an energy conservation equation for the perturbed system. It implies that

$$\delta K + \delta W = C = \text{constant}, \tag{7.3.36}$$

where

$$\delta K = \frac{1}{2}\int_V \rho_{m0}\left|\frac{\partial \boldsymbol{\xi}}{\partial t}\right|^2 \mathrm{d}^3x \tag{7.3.37}$$

is the perturbed kinetic energy and

$$\delta W = -\frac{1}{2}\int_V \boldsymbol{\xi}^* \cdot \mathbf{F}(\boldsymbol{\xi})\,\mathrm{d}^3x \tag{7.3.38}$$

is the perturbed potential energy. Note that this equation differs from Eq. (6.4.21) for the total energy, W, because it represents the energy change, δW, relative to the equilibrium state, due to the linearized small-amplitude perturbation.

We now prove that $\delta W \geq 0$ is a necessary and sufficient condition for stability. If a fluid displacement is unstable, the kinetic energy δK will increase in time without bound. (It will increase exponentially for a linear instability.) But while δK increases, δW will have to decrease and become negative in order to keep the energy C given by Eq. (7.3.36) constant. In contrast, if $\delta W \geq 0, \delta K$ must be bounded from above according to the relation $\delta K \leq C - \delta W$. In this case, there can be no unbounded growth of kinetic energy and the plasma is stable. Hence, $\delta W \geq 0$ is a sufficient condition for stability.

To prove that $\delta W \geq 0$ is also a necessary condition for stability, let us consider an initial displacement such that $\boldsymbol{\xi}(\mathbf{r},0) \neq 0$ but $\partial\boldsymbol{\xi}(\mathbf{r},0)/\partial t = 0$. Furthermore, we require that this initial displacement makes δW negative. It follows that

$$\begin{aligned} C(t=0) &= \delta W(t=0) + \delta K(t=0) \\ &= \delta W(t=0) < 0. \end{aligned} \tag{7.3.39}$$

Let us define the quantity

$$I(t) = \frac{1}{2}\int_V \rho_{m0}\,|\boldsymbol{\xi}|^2\,\mathrm{d}^3x. \tag{7.3.40}$$

The first derivative of I is given by

$$\frac{\mathrm{d}I}{\mathrm{d}t} = \frac{1}{2}\int_V \rho_{m0}\left(\boldsymbol{\xi}^* \cdot \frac{\partial \boldsymbol{\xi}}{\partial t} + \boldsymbol{\xi} \cdot \frac{\partial \boldsymbol{\xi}^*}{\partial t}\right)\mathrm{d}^3x, \tag{7.3.41}$$

and the second derivative is given by

$$\frac{d^2 I}{dt^2} = \frac{1}{2}\int_V \rho_{m0}\left(2\left|\frac{\partial \boldsymbol{\xi}}{\partial t}\right|^2 + \boldsymbol{\xi}^* \cdot \frac{\partial^2 \boldsymbol{\xi}}{\partial t^2} + \boldsymbol{\xi} \cdot \frac{\partial^2 \boldsymbol{\xi}^*}{\partial t^2}\right) d^3x$$
$$= \frac{1}{2}\int_V \left[2\rho_{m0}\left|\frac{\partial \boldsymbol{\xi}}{\partial t}\right|^2 + \boldsymbol{\xi}^* \cdot \mathbf{F}(\boldsymbol{\xi}) + \boldsymbol{\xi} \cdot \mathbf{F}(\boldsymbol{\xi}^*)\right] d^3x. \qquad (7.3.42)$$

Finally, using Eqs. (7.3.24), (7.3.37), and (7.3.38), we obtain

$$\frac{d^2 I}{dt^2} = 2(\delta K - \delta W)$$
$$= 2(2\delta K - C), \qquad (7.3.43)$$

where the last step follows from Eq. (7.3.36). Since δK is positive definite at all times and C is negative by virtue of Eq. (7.3.39), we can write

$$\frac{d^2 I}{dt^2} > -2C > 0. \qquad (7.3.44)$$

The above equation implies that I, and hence $\boldsymbol{\xi}$, increases without bound if there exists any initial perturbation that makes δW negative. To ensure stability, $\delta W \geq 0$ must hold for all permissible plasma displacements. This proves the necessity of the stability condition $\delta W \geq 0$.

In addition to the strength of the energy principle, $\delta W \geq 0$, as a criterion for ideal MHD stability, there is yet another property of δW that makes it a very useful computational tool. It can be shown that δW is variational (Freidberg, 1987). By this we mean the following: Consider a trial function $\boldsymbol{\xi}$ that obeys the boundary conditions that a true plasma displacement of the system must obey, although $\boldsymbol{\xi}$ itself may not be a true displacement. If this trial function makes δW negative, then the variational property of δW guarantees that there exists a true displacement $\boldsymbol{\xi}$ that makes δW even more negative.

The variational property of δW enables us to settle rigorously the stability or instability of any given magnetostatic equilibrium without knowledge of the true eigenfunctions of the system. All we need to do is to have some intuition regarding a trial function obeying the appropriate boundary conditions that can make δW negative, and that is quite enough to demonstrate that the equilibrium is unstable, because a true eigenfunction will make δW even more negative. We state this property without proof. See Freidberg (1987) for a further discussion of this topic.

7.3.6 A More Useful Form for δW

Equation (7.3.38) for the perturbed potential is often not in a convenient form for calculations. A more useful form can be obtained by substituting Eq. (7.3.20) for the linear force operator, $\mathbf{F}(\boldsymbol{\xi})$, into Eq. (7.3.38) and simplifying the resulting equation. After carrying out this substitution, Eq. (7.3.38) becomes

$$\delta W = -\frac{1}{2}\int_V \boldsymbol{\xi}^* \cdot \left\{ \frac{1}{\mu_0}[(\nabla \times \{\nabla \times (\boldsymbol{\xi} \times \mathbf{B}_0)\}) \times \mathbf{B}_0 + (\nabla \times \mathbf{B}_0) \right.$$
$$\left. \times \{\nabla \times (\boldsymbol{\xi} \times \mathbf{B}_0)\}] + \nabla[\boldsymbol{\xi} \cdot \nabla P_0 + \gamma P_0(\nabla \cdot \boldsymbol{\xi})] \right\} \mathrm{d}^3 x. \tag{7.3.45}$$

To proceed further, we assume that the plasma is surrounded by a perfectly conducting wall. In this case, the plasma displacement $\boldsymbol{\xi}$ as well as the perturbed magnetic field $\mathbf{B}_1$, given by Eq. (7.3.14), obey the boundary conditions

$$\hat{\mathbf{n}} \cdot \boldsymbol{\xi} = 0 \tag{7.3.46}$$

and

$$\hat{\mathbf{n}} \cdot \mathbf{B}_1 = \hat{\mathbf{n}} \cdot [\nabla \times (\boldsymbol{\xi} \times \mathbf{B}_0)] = 0. \tag{7.3.47}$$

We next consider the terms in the integrand on the right of Eq. (7.3.45). Using the vector identity $\nabla \cdot (\mathbf{F} \times \mathbf{G}) = \mathbf{G} \cdot (\nabla \times \mathbf{F}) - \mathbf{F} \cdot (\nabla \times \mathbf{G})$, the first term in the integrand can be written

$$\begin{aligned}
&\frac{1}{\mu_0}\boldsymbol{\xi}^* \cdot \{(\nabla \times \mathbf{B}_1) \times \mathbf{B}_0\} \\
&= \frac{1}{\mu_0}(\mathbf{B}_0 \times \boldsymbol{\xi}^*) \cdot (\nabla \times \mathbf{B}_1) \\
&= -\frac{1}{\mu_0}\nabla \cdot \{(\mathbf{B}_0 \times \boldsymbol{\xi}^*) \times \mathbf{B}_1\} + \frac{1}{\mu_0}\mathbf{B}_1 \cdot \nabla \times (\mathbf{B}_0 \times \boldsymbol{\xi}^*) \\
&= \frac{1}{\mu_0}\nabla \cdot \{(\boldsymbol{\xi}^* \times \mathbf{B}_0) \times \mathbf{B}_1\} - \frac{1}{\mu_0}|\nabla \times (\boldsymbol{\xi} \times \mathbf{B}_0)|^2.
\end{aligned} \tag{7.3.48}$$

The second term in the integrand of Eq. (7.3.45) can be rewritten as

$$\frac{1}{\mu_0}\boldsymbol{\xi}^* \cdot (\nabla \times \mathbf{B}_0) \times \{\nabla \times (\boldsymbol{\xi} \times \mathbf{B}_0)\} = \boldsymbol{\xi}^* \cdot \mathbf{J}_0 \times \{\nabla \times (\boldsymbol{\xi} \times \mathbf{B}_0)\}. \tag{7.3.49}$$

The third term needs no reprocessing at this point, so we move on to the fourth term, which can be written

$$\begin{aligned}
\boldsymbol{\xi}^* \cdot \nabla\{\gamma P_0(\nabla \cdot \boldsymbol{\xi})\} &= \nabla \cdot \{\gamma P_0\, \boldsymbol{\xi}^*(\nabla \cdot \boldsymbol{\xi})\} - \gamma P_0(\nabla \cdot \boldsymbol{\xi}^*)(\nabla \cdot \boldsymbol{\xi}) \\
&= \nabla \cdot \{\gamma P_0\, \boldsymbol{\xi}^*(\nabla \cdot \boldsymbol{\xi})\} - \gamma P_0|\nabla \cdot \boldsymbol{\xi}|^2.
\end{aligned} \tag{7.3.50}$$

Substituting Eqs. (7.3.48), (7.3.49), and (7.3.50) into expression (7.3.45) for δW, and using Gauss' theorem to convert the divergence terms in Eqs. (7.3.48) and

(7.3.50) to surface terms, we obtain

$$\delta W = \frac{1}{2}\int_V \left[\frac{1}{\mu_0}|\nabla\times(\xi\times\mathbf{B}_0)|^2 + \gamma P_0|\nabla\cdot\xi|^2 \right.$$
$$\left. - \xi^*\cdot\mathbf{J}_0\times\{\nabla\times(\xi\times\mathbf{B}_0)\} - \xi^*\cdot\nabla(\xi\cdot\nabla P_0)\right]\mathrm{d}^3x$$
$$-\frac{1}{2}\int_s \hat{\mathbf{n}}\cdot\left[\frac{1}{\mu_0}\{(\xi^*\times\mathbf{B}_0)\times\mathbf{B}_1\} + \gamma P_0\xi^*(\nabla\cdot\xi)\right]\mathrm{d}A, \tag{7.3.51}$$

where $\hat{\mathbf{n}}$ is the unit normal on the surface S bounding the plasma. When S is perfectly conducting, the surface integral vanishes because of boundary conditions (7.3.46) and (7.3.47), and we are left with

$$\delta W = \frac{1}{2}\int_V \left[\frac{1}{\mu_0}|\nabla\times(\xi\times\mathbf{B}_0)|^2 + \gamma P_0|\nabla\cdot\xi|^2 \right.$$
$$\left. - \xi^*\cdot\mathbf{J}_0\times\{\nabla\times(\xi\times\mathbf{B}_0)\} - \xi^*\cdot\nabla(\xi\cdot\nabla P_0)\right]\mathrm{d}^3x. \tag{7.3.52}$$

The first two terms on the right-hand side of the above equation are positive and stabilizing, while the third and fourth terms can be destabilizing. The first term represents the energy required to bend field lines. The second term represents the energy required to compress a plasma with non-zero equilibrium pressure. The third term, which depends explicitly on the equilibrium current density, $\mathbf{J}_0$, can potentially drive instabilities of the "kink" type. The fourth term, which depends explicitly on the equilibrium pressure gradient, can potentially drive instabilities of the "ballooning" or "interchange" type. For an ideal MHD plasma, an instability always arranges its eigenfunction in such a way as to minimize the stabilizing contributions to δW (the first and second terms). For example, in an infinite cylinder or a torus, the marginally stable ($\omega_n = 0$) eigenfunctions are incompressible and obey the condition $\nabla\cdot\xi = 0$, which reduces the second stabilizing term in δW to zero.

The current density and the pressure gradient in any given magnetostatic equilibrium are potential sources of free energy for ideal MHD instabilities. In most cases of physical interest, δW needs to be computed numerically to determine the stability for various perturbations, such as kink, ballooning, and interchange motions. In the next subsection, we consider an analytically tractable example.

7.3.7 The Rayleigh–Taylor Instability

We next discuss the well-known instability known as the Rayleigh–Taylor instability, which is similar to the Kruskal–Schwarzschild instability discussed

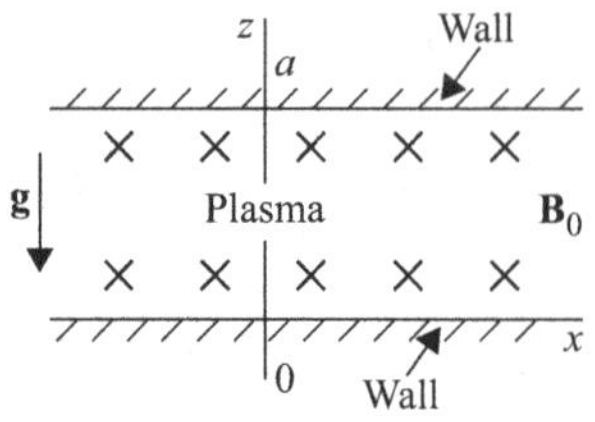

Figure 7.19 An ideal MHD fluid between two perfectly conducting walls (at $z = 0$ and $z = a$) with a downward applied gravitational force, $\rho_m g$. The magnetic field, indicated by the $\times$ symbols, is directed into the paper.

earlier, except instead of an abrupt boundary the vertical distribution of plasma density is continuous. To illustrate the techniques involved, we will carry out the analysis using both the normal mode method and the energy method. The basic geometry is shown in Figure 7.19. In this model we assume that the plasma is bounded by perfectly conducting walls at $z = 0$ and $z = a$, that the gravitational acceleration, $\mathbf{g}$, is directed downward in the $-z$ direction, and that the magnetic field, $\mathbf{B}_0$, is in the $+y$ direction (i.e., into the paper). To provide for a simple vertical variation of the fluid density, we assume that the equilibrium mass density profile is given by an exponential,

$$\rho_{m0}(z) = \rho_0 \exp(-z/H_S), \tag{7.3.53}$$

where ρ_0 is the mass density at $z = 0$ and H_S is a constant that is called the scale height in the ionospheric literature. If $H_S > 0$, the equilibrium density decreases with increasing height, and if $H_S < 0$, the equilibrium density increases with increasing height. Because of the downward gravitational force, if $H_S < 0$ we expect the equilibrium to be unstable, since this represents a situation in which a heavy fluid lies on top of a light fluid.

In the presence of gravity, the momentum equation (6.4.2) must be modified to include the gravitational force, $\rho_m \mathbf{g}$, so that

$$\rho_m \frac{d\mathbf{U}}{dt} = \frac{1}{\mu_0}(\mathbf{B} \cdot \nabla)\mathbf{B} - \nabla\left(\frac{B^2}{2\mu_0} + P\right) + \rho_m \mathbf{g}. \tag{7.3.54}$$

In magnetostatic equilibrium, i.e., $U_0 = 0$, the above equation with $\mathbf{B} = B_0\hat{\mathbf{y}}$ reduces to

$$\frac{\partial}{\partial z}\left(\frac{B_0^2}{2\mu_0} + P_0\right) + \rho_{m0} g = 0. \tag{7.3.55}$$

Linearizing this equation with respect to small perturbations around the zero-order equilibrium gives

$$\rho_{m0}\frac{\partial \mathbf{U}}{\partial t} = \frac{1}{\mu_0}(\mathbf{B}_0 \cdot \nabla)\mathbf{B} + \frac{1}{\mu_0}(\mathbf{B} \cdot \nabla)\mathbf{B}_0 - \nabla\left(\frac{\mathbf{B}_0 \cdot \mathbf{B}}{\mu_0} + P\right) + \rho_m \mathbf{g}, \tag{7.3.56}$$

where, as usual, we have suppressed the subscript 1 on the first-order terms. Next consider a plane wave perturbation of the form $\exp(\mathrm{i}kx - \mathrm{i}\omega t)$. Note that in considering such a form, we have assumed that the perturbations have no spatial variation along y, and hence, the equilibrium magnetic field suffers no spatial modulation along y. Also, because the equilibrium depends explicitly on z, we do not assume plane wave solutions along z, but solve for the z dependence of the solutions from the linearized equations and boundary conditions. Under these conditions, the first two terms on the right of Eq. (7.3.56) vanish identically. It is convenient to eliminate the third term by taking the curl of both sides of Eq. (7.3.56), which gives

$$\nabla \times \left(\rho_{m0}\frac{\partial \mathbf{U}}{\partial t}\right) = \nabla \times (\rho_m \mathbf{g}). \tag{7.3.57}$$

The y component of the above equation is given by

$$\omega\left[\mathrm{i}k\rho_{m0}\widetilde{U}_z - \frac{\partial}{\partial z}(\rho_{m0}\widetilde{U}_x)\right] = kg\tilde{\rho}_m, \tag{7.3.58}$$

where we have used $U_z = \widetilde{U}_z \exp(\mathrm{i}kx - \mathrm{i}\omega t)$.

Next we assume, for simplicity, that the perturbed velocity is incompressible, i.e.,

$$\nabla \cdot \mathbf{U} = 0. \tag{7.3.59}$$

For the assumed plane wave perturbation, the incompressibility condition then becomes

$$\mathrm{i}k\widetilde{U}_x + \frac{\partial \widetilde{U}_z}{\partial z} = 0. \tag{7.3.60}$$

Since $\nabla \cdot \mathbf{U} = 0$, the linearized mass continuity equation (7.3.15) simplifies to

$$\frac{\partial \rho_m}{\partial t} + \mathbf{U} \cdot \nabla \rho_{m0} = 0, \tag{7.3.61}$$

which for the assumed plane wave perturbation yields

$$-\mathrm{i}\omega\tilde{\rho}_m + \widetilde{U}_z\frac{\partial \rho_{m0}}{\partial z} = 0. \tag{7.3.62}$$

It is straightforward to eliminate $\widetilde{U}_x$ and $\tilde{\rho}_{m0}$ from Eq. (7.3.58) by using Eqs. (7.3.60) and (7.3.62), which gives

$$\frac{1}{\rho_{m0}}\frac{\partial}{\partial z}\left(\rho_{m0}\frac{\partial \widetilde{U}_z}{\partial z}\right) - k^2\left(1 - \frac{g}{H_S\omega^2}\right)\widetilde{U}_z = 0, \tag{7.3.63}$$

where $1/H_S = -(1/\rho_{m0})\partial\rho_{m0}/\partial z$.

The above equation is a second-order differential equation for $\widetilde{U}_z$, to be solved using the boundary conditions $\widetilde{U}_z = 0$ at $z = 0$ and $z = a$. To solve Eq. (7.3.63) in closed form, we make the substitution

$$\widetilde{U}_z(z) = f(z)\exp\left(\frac{z}{2H_S}\right), \tag{7.3.64}$$

which gives

$$\frac{d^2 f}{dz^2} + \alpha^2 f = 0, \tag{7.3.65}$$

where

$$\alpha^2 = k^2\left(\frac{g}{H_S\omega^2} - 1\right) - \frac{1}{4H_S^2}. \tag{7.3.66}$$

Since $f(0) = f(a) = 0$, we obtain the eigenfunctions

$$f_n(z) = f_0 \sin\frac{n\pi z}{a}, \qquad n = 1, 2, \ldots, \tag{7.3.67}$$

where f_0 is a constant, and the eigenvalue relation

$$\frac{n^2\pi^2}{a^2} = k^2\left(\frac{g}{H_S\omega_n^2} - 1\right) - \frac{1}{4H_S^2}. \tag{7.3.68}$$

The above equation can be solved for the frequencies of the normal modes which are given by

$$\omega_n^2 = \left(\frac{g}{H_S}\right)\frac{4k^2a^2H_S^2}{a^2 + 4H_S^2(k^2a^2 + n^2\pi^2)}. \tag{7.3.69}$$

Note that:

1. If $H_S > 0$, the frequency, ω_n, of the nth mode is purely real and the system is stable. On the other hand, if $H_S < 0$, the frequency is purely imaginary and the system is unstable.

2. The largest growth rate, $\sqrt{g/H_S}$, occurs in the limit $k \to \infty$ and the smallest growth rate occurs in the limit $k \to 0$.
3. For fixed k, the growth rate decreases with increasing mode number n.

The Rayleigh–Taylor instability can also be analyzed by means of the energy method. Note that in this case, we need to add the gravity term $\rho_m \mathbf{g}$ to the linear force operator $\mathbf{F}(\boldsymbol{\xi})$, Eq. (7.3.20). Assuming incompressibility, i.e., $\nabla \cdot \boldsymbol{\xi} = 0$, we can write

$$\begin{aligned}\rho_m \mathbf{g} &= -\nabla \cdot (\rho_{m0} \boldsymbol{\xi})\mathbf{g} \\ &= \xi_z (\partial \rho_{m0}/\partial z) g \hat{\mathbf{z}}.\end{aligned} \tag{7.3.70}$$

The contribution of the term $\rho_m \mathbf{g}$ to δW is given by $(-1/2)\boldsymbol{\xi}^* \cdot \rho_m \mathbf{g} = -(1/2)|\xi_z|^2 g \, \partial \rho_{m0}/\partial z$. While this contribution needs to be added to the integrand in Eq. (7.3.52), note that the second and third terms in the integrand vanish because $\nabla \cdot \boldsymbol{\xi} = 0$ and $\mathbf{J}_0 = 0$, respectively. Hence, we obtain

$$\delta W = \frac{1}{2} \int_V \left[\frac{1}{\mu_0} |\nabla \times (\boldsymbol{\xi} \times \mathbf{B}_0)|^2 - \boldsymbol{\xi}^* \cdot \nabla(\boldsymbol{\xi} \cdot \nabla P_0) - |\xi_z|^2 g \, \partial \rho_{m0}/\partial z \right] d^3x. \tag{7.3.71}$$

Note that the second term in the integrand on the right-hand side of the above equation can be written as $\nabla \cdot [\boldsymbol{\xi}^* (\boldsymbol{\xi} \cdot \nabla P_0)]$, which can be transformed by Gauss' law to a surface term that vanishes upon using the boundary condition (7.3.46). This leaves the third term in the integrand as the only possible destabilizing term. Hence a sufficient condition for stability is

$$\frac{\partial \rho_{m0}}{\partial z} < 0 \quad \text{or} \quad H_S > 0, \tag{7.3.72}$$

which is consistent with the results determined from the normal mode method. A good example of the Rayleigh–Taylor instability occurs in Earth's ionosphere near the magnetic equator, where the magnetic field is nearly horizontal. Under certain

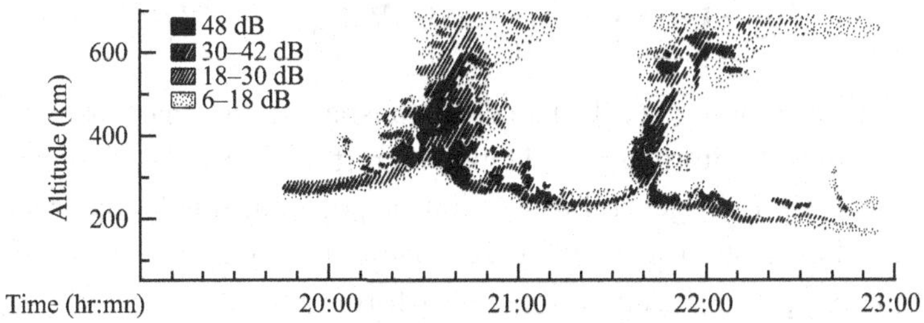

Figure 7.20 A radar profile of Earth's ionosphere that shows buoyant rising plumes of plasma caused by the Rayleigh–Taylor instability (from Basu and Kelley, 1979).

conditions, "bubbles" of low-density plasma from near the base of the ionosphere rise upwards into the ionosphere, causing plume-like disturbances, such as that shown in Figure 7.20, that can be detected by ground-based radars (Basu and Kelley, 1979).

7.4 Stability of Ideal Magnetohydrodynamic Equilibria

In the presence of spatially dependent equilibrium flow, $\mathbf{U}_0$, which cannot be eliminated by a Galilean transformation, the problem of linear stability of MHD equilibria becomes much more complicated than that for magnetostatic equilibria. Generalizing the procedure followed in Section 7.3.3 and taking into account the effect of equilibrium flow, we introduce again the perturbed fluid displacement vector $\boldsymbol{\xi}(\mathbf{r},t)$. It can be shown, after significant algebra, that the linearized equation of motion can be written in the form

$$\rho_{m0}\frac{\partial^2\boldsymbol{\xi}}{\partial t^2}+2\rho_{m0}\mathbf{U}_0\cdot\boldsymbol{\nabla}\frac{\partial\boldsymbol{\xi}}{\partial t}=\mathbf{G}(\boldsymbol{\xi}), \tag{7.4.1}$$

where $\mathbf{G}(\boldsymbol{\xi})$ is the generalized linear force operator (operating on $\boldsymbol{\xi}$), given by

$$\mathbf{G}(\boldsymbol{\xi})=\mathbf{F}(\boldsymbol{\xi})+\boldsymbol{\nabla}\cdot(\rho_{m0}\boldsymbol{\xi}\mathbf{U}_0\cdot\boldsymbol{\nabla}\mathbf{U}_0-\rho_{m0}\mathbf{U}_0\mathbf{U}_0\cdot\boldsymbol{\nabla}\boldsymbol{\xi}); \tag{7.4.2}$$

see Frieman and Rotenberg (1960) and Goedbloed et al. (2010). Here $\mathbf{F}(\boldsymbol{\xi})$ is the force operator in the absence of equilibrium flows, and is given by Eq. (7.3.20). Using the equilibrium equation (7.2.5), it can be shown further that Eq. (7.4.1) can be rewritten

$$\rho_{m0}\left(\frac{\partial}{\partial t}+\mathbf{U}_0\cdot\boldsymbol{\nabla}\right)^2\boldsymbol{\xi}=\mathbf{F}(\boldsymbol{\xi})+\boldsymbol{\nabla}\cdot(\rho_{m0}\boldsymbol{\xi}\mathbf{U}_0\cdot\boldsymbol{\nabla}\mathbf{U}_0), \tag{7.4.3}$$

where

$$\left(\frac{\partial}{\partial t}+\mathbf{U}_0\cdot\boldsymbol{\nabla}\right)^2\boldsymbol{\xi}\equiv\left(\frac{\partial}{\partial t}+\mathbf{U}_0\cdot\boldsymbol{\nabla}\right)\left(\frac{\partial}{\partial t}+\mathbf{U}_0\cdot\boldsymbol{\nabla}\right)\boldsymbol{\xi}. \tag{7.4.4}$$

The left-hand side of Eq. (7.4.3) makes transparent that part of the effect of equilibrium flows to introduce a Doppler shift through the operator $\mathbf{U}_0\cdot\boldsymbol{\nabla}$. However, since $\mathbf{U}_0$ is in general a function of position, this Doppler shift varies from one spatial location to another. A wave with a Doppler-shifted frequency at a particular spatial location then transforms to a wave with a different Doppler-shifted frequency at another spatial location. An initial perturbation on this system can exhibit multi-frequency effects even in the absence of nonlinear terms in Eq. (7.4.1).

Let us consider normal modes of the form $\boldsymbol{\xi}(\mathbf{r},t) = \hat{\boldsymbol{\xi}}(\mathbf{r})\exp(-\mathrm{i}\omega t)$ in Eq. (7.4.1), which gives

$$\rho_{\mathrm{m}0}\omega^2\hat{\boldsymbol{\xi}} + 2\mathrm{i}\rho_{\mathrm{m}0}\omega\mathbf{U}_0\cdot\boldsymbol{\nabla}\hat{\boldsymbol{\xi}} + \mathbf{G}\hat{\boldsymbol{\xi}} = 0. \tag{7.4.5}$$

Even though the generalized force operator, $\mathbf{G}$, as well as the operator $\mathrm{i}\rho_{\mathrm{m}0}\mathbf{U}_0\cdot\boldsymbol{\nabla}$ are both self-adjoint, the combination of the two in the form of Eq. (7.4.5) is not and produces complex frequencies ω rather than the purely real or imaginary frequencies that are obtained for static equilibria for which $\mathbf{U}_0 = 0$. In the latter case, a linear perturbation is either purely oscillatory or growing. However, in the presence of flows, an equilibrium can be "overstable," which is actually a form of instability by which the system exhibits oscillatory behavior as well as an amplitude growing with time.

The complexities introduced by the presence of equilibrium flows into the stability problem deprive us of powerful results like the energy principle for magnetostatic plasmas. As demonstrated by Frieman and Rotenberg (1960), one can obtain a sufficient condition for stability analogous to the energy principle, but such a condition is not necessary for stability. Therefore, brute-force normal mode analyses become essential in order to obtain quantitative results on stability. We illustrate this approach in the next section by considering an important instability for astrophysical plasmas—the magnetorotational instability.

7.4.1 The Magnetorotational Instability

The magnetorotational instability (MRI) was first discovered by Velikhov (1959) and Chandrasekhar (1960), but it was independently rediscovered by Balbus and Hawley (1991) who made its application to accretion disks what it is today—a problem of great interest in plasma physics that has literally launched an entirely new area of research. Before we provide a mathematical treatment of the instability, we provide some background on why the instability is important.

Astrophysical objects such as black holes cannot be seen directly, but their presence in the universe is usually detected by their effect on surrounding objects. Accretion disks are formed by diffused matter in orbital motion around a massive central object such as a black hole. While the large kinetic or gravitational energy of accreting matter in such disks can decay away readily by radiation, accounting for their luminosity in the night sky, it requires significant viscosity to enable the accretion process itself. Without such viscosity, the law of conservation of angular momentum would tend to preserve the orbits of the objects spinning around the central object as it tends to do for planets orbiting around the Sun, and there would be very little accretion. Simple estimates of the molecular viscosity in such systems produce numbers for accretion rates that are much too small. As is often the case in such situations, one turns to the possible effects of instability and turbulence to enhance the observed dissipation. But what type of instability can an accretion disk support? Is it hydrodynamic in origin? Or is it magnetohydrodynamic, requiring the intervention of a magnetic field? In what follows, we will give a heuristic analysis of the instability beginning from the MHD equations, but will also consider the limit in which the magnetic field tends to zero. It will turn out that this limit is quite subtle.

We will give here a highly simplified treatment of the so-called magnetorotational instability (MRI) using the ideal MHD equations (Balbus and Hawley, 1991). Consider a disk, parameterized in cylindrical coordinates (ρ, ϕ, z), with a uniform vertical magnetic field $\mathbf{B}_0 = B_0\hat{\mathbf{z}}$ and Keplerian azimuthal flow $U_{\phi 0} = \rho\Omega_0$, where the subscript 0 denotes equilibrium quantities. We assume that all components of the linear perturbation $\hat{\xi}$ in Eq. (7.4.5) have no dependence on ρ and ϕ, and are proportional to $\exp(\mathrm{i}kz - \mathrm{i}\omega t)$. After some algebra (Problem 7.9), we obtain the dispersion relation

$$(v_\rho^2 - V_\mathrm{A}^2)[v_\rho^4 - v_\rho^2(V_\mathrm{A}^2 + V_\mathrm{S}^2) + V_\mathrm{A}^2 V_\mathrm{S}^2] - \left(\frac{1}{k^2}\right)\left[\kappa^2 v_\rho^4 - v_\rho^2\left(\kappa^2 V_\mathrm{S}^2 + V_\mathrm{A}^2\frac{\partial\Omega_0^2}{\partial\ln\rho}\right) + V_\mathrm{A}^2 V_\mathrm{S}^2\frac{\partial\Omega_0^2}{\partial\ln\rho}\right] = 0, \tag{7.4.6}$$

where κ, defined by the relation

$$\kappa^2 = \frac{1}{\rho^3}\frac{\mathrm{d}(\rho^4\Omega_0^2)}{\mathrm{d}\rho}, \tag{7.4.7}$$

is called the epicyclic frequency, and other quantities have the same definitions as in Section 6.5. Note that κ^2 is a measure of how much the angular momentum per unit mass, represented by $\Omega_0\rho^2$, deviates from a constant.

In the case without rotation, Eq. (7.4.7) reduces to Eq. (6.5.16) with $\cos\theta = 1$. As discussed in Chapter 6, the dispersion equation factors into the shear Alfvén mode, the slow magnetosonic mode, and the fast magnetosonic mode. Without any sources of free energy such as the plasma current density, pressure, or flows, there are no instabilities in the system. However, something remarkable happens when flows are introduced, and Eq. (7.4.6) applies. Let us consider the case $v_p^2 = 0$, whereby Eq. (7.4.6) reduces to the condition

$$k^2 V_\mathrm{A}^2 + \frac{\partial\Omega_0^2}{\partial\ln\rho} = 0. \tag{7.4.8}$$

It turns out that Eq. (7.4.8) is precisely the transition point separating roots that are stable ($\omega^2 > 0$) from roots that are unstable ($\omega^2 < 0$). One can show that if k is small enough there is always instability, unless

$$\frac{\partial \ln \Omega_0^2}{\partial \ln \rho} > 0. \tag{7.4.9}$$

To see this in a relatively simple way as compared to the complicated dispersion relation (7.4.6), we consider the incompressible limit, $\gamma \to \infty$, when $V_S \to \infty$. This is conveniently done by dividing the dispersion relation (7.4.6) by V_S^2, and then taking the limit $V_S \to \infty$. We obtain (Problem 7.10)

$$v_\rho^4 - (v_\rho^2/k^2)(\kappa^2 + 2k^2 V_A^2) + (V_A^2/k^2)\left(k^2 V_A^2 + \frac{\partial \Omega_0^2}{\partial \ln \rho}\right) = 0. \tag{7.4.10}$$

If $k \to 0$, that is, when the wavelength of the mode tends to extremely large values, we obtain the stability condition (7.4.9). This is easily violated for accretion disk profiles, such as the Keplerian profile specified by Eq. (7.2.16). In more physically realistic cases with a finite system size, there is an upper bound to the largest wavelength, but a weak magnetic field can break the stability condition

$$\kappa^2 V_A^2 + \frac{\partial \Omega_0^2}{\partial \ln \rho} > 0. \tag{7.4.11}$$

From Eq. (7.4.6), it can be shown that the mode with the largest growth rate is one for which

$$\omega_{\max} = \frac{1}{2}\left|\frac{\partial \Omega_0}{\partial \ln \rho}\right|. \tag{7.4.12}$$

It is interesting to note that while this maximum growth rate is independent of the magnetic field, the MRI actually needs a weak magnetic field to be unstable. The source of free energy is the velocity shear, not the magnetic field. What the magnetic field does is to provide a way for the instability to access this source of free energy, but the instability is intrinsically a MHD instability, and not a hydrodynamic instability.

7.5 Resistive Instabilities

As discussed at the beginning of this chapter, a plasma configuration that is ideally stable may become unstable in the presence of a small but non-zero resistivity. Under these conditions, the frozen field theorem, proved in Section 6.3.1, is violated. Recall that the frozen field theorem is a direct consequence of Ohm's law in ideal MHD,

$$\mathbf{E} + \mathbf{U} \times \mathbf{B} = 0, \tag{7.5.1}$$

which can be satisfied by the simple choice

$$\mathbf{U} = \frac{\mathbf{E}\times\mathbf{B}}{B^2}. \tag{7.5.2}$$

The usual interpretation of Eq. (7.5.2) is that it represents the velocity with which field lines move. Since electrons and ions also move with the same velocity (independent of their charge), the velocity of a field line, under this interpretation, is identical to the plasma velocity. This gives us a definite way to track the identity of a field line during ideal plasma motion: plasma elements that lie on it remain so for all times, enabling us to keep track of the identity of the field line for all times. Note that as $\mathbf{B}$ tends to zero, $\mathbf{U}$ tends to infinity, which is unphysical, pointing to the vulnerability of this definition at null points of the magnetic field (or at x-type neutral lines, as mentioned in Section 3.8.3), where even small departures from ideal behavior can have dramatic consequences.

In the presence of a small but finite resistivity, it is not possible to define a field-line velocity in the manner of Eq. (7.5.2). Under such circumstances, we cannot give field lines identities by labeling them with fluid elements and having the lines move with the fluid. The field lines themselves can "break" and "reconnect," producing new magnetic configurations of lower energy. These lower-energy states may not be accessible if field lines are constrained during their motion by the frozen field theorem, as they do not allow magnetic field lines to reconfigure their initial topological preparation. This process of access to lower-energy states is often facilitated by resistive instabilities, to which we now turn.

A classic example of a magnetic field configuration that can support resistive instabilities is the so-called planar or Harris current sheet (Harris, 1961), shown in Figure 7.21, for which the magnetic field is given by

$$\mathbf{B}_0 = \mathbf{B}_0 \tanh(y/a)\hat{\mathbf{x}} + B_T\hat{\mathbf{z}}, \qquad \mathbf{B}_0 > 0, \qquad B_T > 0, \tag{7.5.3}$$

where $\mathbf{B}_0$ and B_T are positive constants, and a is the width of a smooth current density distribution in the z direction, which can be calculated by taking the curl of

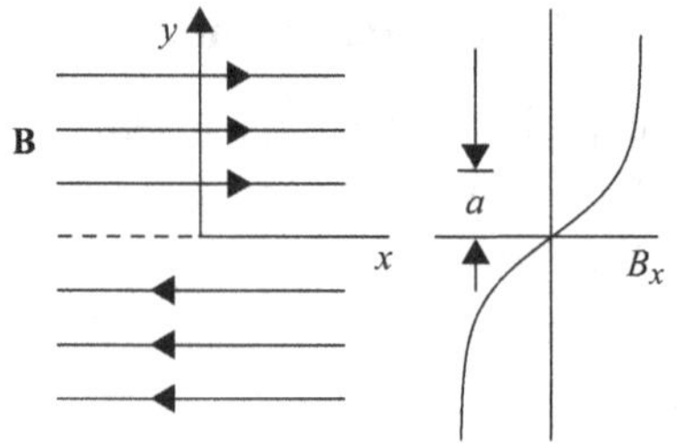

Figure 7.21 The magnetic field geometry across a planar current sheet.

the above equation. Here the z component of the magnetic field is often referred to as the "guide field" component, which may or may not be small depending on the application one has in mind. Note that at the $y = 0$ surface, $B_{0x} = 0$. For $y > 0$, the magnetic field in the x, y plane points in the $+x$ direction, while for $y < 0$, it points in the $-x$ direction. The neutral surface $y = 0$, represented by the dashed line in Figure 7.21, has a special geometric significance because it separates two regions of oppositely directed magnetic field. It turns out that the Harris current sheet is ideally completely stable, but unstable in the presence of resistivity. To analyze this instability, which is called the tearing instability, we begin with the linearized incompressible resistive MHD equations:

$$\rho_0 \frac{\partial \mathbf{U}_1}{\partial t} = \mathbf{J}_0 \times \mathbf{B}_1 + \mathbf{J}_1 \times \mathbf{B}_0 - \boldsymbol{\nabla} P_1, \tag{7.5.4}$$

$$\frac{\partial \mathbf{B}_1}{\partial t} = \boldsymbol{\nabla} \times (\mathbf{U}_1 \times \mathbf{B}_0) + \frac{\eta}{\mu_0} \boldsymbol{\nabla}^2 \mathbf{B}_1, \tag{7.5.5}$$

$$\boldsymbol{\nabla} \cdot \mathbf{B}_1 = 0, \tag{7.5.6}$$

$$\boldsymbol{\nabla} \cdot \mathbf{U}_1 = 0, \tag{7.5.7}$$

where $\mathbf{J}_0 = \boldsymbol{\nabla} \times \mathbf{B}_0 / \mu_0$ and $\mathbf{J}_1 = \boldsymbol{\nabla} \times \mathbf{B}_1 / \mu_0$. Note that the incompressibility condition (7.5.7) can be obtained formally from Eq. (7.3.17) by taking the formal limit $\gamma \to \infty$. Extension of the present calculation to include the effects of plasma compressibility is technically more demanding, but not essential for the discussion of the classical tearing instability (Furth et al., 1963).

Using Eqs. (7.5.6) and (7.5.7), we may represent the linearized perturbations $\mathbf{B}_1$ and $\mathbf{U}_1$ in the form

$$\mathbf{B}_1 = \hat{\mathbf{z}} \times \boldsymbol{\nabla} \psi_1(x, y, t) \tag{7.5.8}$$

and

$$\mathbf{U}_1 = \hat{\mathbf{z}} \times \boldsymbol{\nabla} \phi_1(x, y, t), \tag{7.5.9}$$

where $\psi_1(x, y, t)$ and $\phi_1(x, y, t)$ are scalar functions. This approach has the advantage that two vector functions, which have six scalar components, can be represented in terms of two scalar functions, and the constraints imposed by Eqs. (7.5.6) and (7.5.7) are included in a natural way. Note that for simplicity we have assumed that the linear perturbations are two-dimensional, and we do not have any components nor any spatial variation along z. We write them in the normal mode form

$$\psi_1(x, y, t) = \tilde{\psi}_1(y) \exp(ikx + \gamma t) \tag{7.5.10}$$

and

$$\phi_1(x,y,t) = \tilde{\phi}_1(y)\exp(\mathrm{i}kx + \gamma t), \tag{7.5.11}$$

where γ is the growth rate of the instability. Substituting these forms in the x component of Eq. (7.5.5), we obtain

$$\gamma\tilde{\psi}_1 - \mathrm{i}kb\tilde{\phi}_1 = \frac{\eta}{\mu_0}\left(\frac{\partial^2}{\partial y^2} - k^2\right)\tilde{\psi}_1, \tag{7.5.12}$$

where $b(y/a) = b_0\tanh(y/a)$. Taking the curl of Eq. (7.5.4) and then its scalar product with $\hat{\mathbf{z}}$, it follows that

$$\mu_0\rho_{\mathrm{m}0}\gamma\left(\frac{\partial^2}{\partial y^2} - k^2\right)\tilde{\phi}_1 = \mathrm{i}kb\left(\frac{\partial^2}{\partial y^2} - k^2\right)\tilde{\psi}_1 - \mathrm{i}k\frac{b''}{a^2}\tilde{\psi}_1, \tag{7.5.13}$$

where the prime on the function b represents the derivative of b with respect to its argument. We now introduce dimensionless versions of the independent and dependent variables introduced above. We redefine a new independent variable y as the old variable y/a, that is, $y/a \to y$, and similarly $ka \to k$, $\gamma\tau_{\mathrm{A}} \to \gamma$ where $\tau_{\mathrm{A}} = a/V_{\mathrm{A}}$ and $V_{\mathrm{A}} = b_0/\sqrt{\mu_0\rho_{\mathrm{m}0}}$, $b \to b/b_0$, $k\tilde{\psi}_1/b_0 \to \tilde{\psi}_1$, and $\mathrm{i}k^2\tilde{\phi}_1/\gamma \to \tilde{\phi}_1$. Furthermore, we define the dimensionless Lundquist number $S = \tau_R/\tau_A$.

In terms of these dimensionless quantities, Eqs. (7.5.12) and (7.5.13) can be written as

$$\gamma\left(\tilde{\psi}_1 - b\tilde{\phi}_1\right) = S^{-1}\left(\frac{\partial^2}{\partial y^2} - k^2\right)\tilde{\psi}_1 \tag{7.5.14}$$

and

$$\gamma^2\left(\frac{\partial^2}{\partial y^2} - k^2\right)\tilde{\phi}_1 = -b\left(\frac{\partial^2}{\partial y^2} - k^2\right)\tilde{\psi}_1 + b''\tilde{\psi}_1. \tag{7.5.15}$$

Resistive instabilities, when they exist, typically have growth rates that are smaller than ideal instabilities that usually grow on a time-scale of the order of the Alfvén time-scale τ_A, but much faster than the time-scale of resistive diffusion τ_R. In dimensional variables, this implies that $\gamma\tau_A \ll 1$ and $\gamma\tau_R \gg 1$, which translates in dimensionless variables to $\gamma \ll 1 \ll S\gamma$. Under these conditions, Eqs. (7.5.14) and (7.5.15) reduce, respectively, to

$$\tilde{\psi}_1 - b\tilde{\phi}_1 = 0 \tag{7.5.16}$$

and

$$\left(\frac{\partial^2}{\partial y^2} - k^2\right)\psi_1'' - \frac{b''}{b}\tilde{\psi}_1 = 0. \tag{7.5.17}$$

Note that Eqs. (7.5.16) and (7.5.17) are completely independent of resistivity and inertia. They are thus obtainable from the ideal MHD equations, which neglect resistivity, and furthermore, without including any time dependence. The spatial region over which these equations hold is called the "ideal region" (or the "exterior region" in the language of boundary-layer theory). These equations do not hold everywhere, as can be clearly anticipated by inspection of Eq. (7.5.16), which predicts that $\phi_1 \to \infty$ as $b \to 0$. In other words, the perturbed velocity $\mathbf{U}_1$ tends to infinity as $y \to 0$, which is unphysical. It will be necessary to introduce the effects of inertia and resistivity in a narrow region, called the resistive layer (or the "interior region"), near $y = 0$ to resolve this unphysical singularity in the perturbed velocity. Since a mathematically rigorous analysis using the tools of boundary-layer theory is outside the scope of this text, we will use heuristic arguments here to obtain essential results.

We note that the second-order differential equation (7.5.17) for $\tilde{\psi}_i$ can be solved by imposing suitable boundary conditions on y, such as $\tilde{\psi}_1 \to 0$ as $y \to \pm\infty$ (or equivalently, by prescribing $\tilde{\psi}_1$ at physical boundaries). Exploiting the linearity of the equation, the two branches of the solution for $\tilde{\psi}_1(y)$, one extending from $-\infty$ to zero and the other from zero to $+\infty$, can and should be made to match as $y \to 0$ at a common value $\tilde{\psi}_1(0)$. While this would ensure continuity of $\tilde{\psi}_1(0)$ required by Maxwell's equations, in general the derivative of $\tilde{\psi}_1(y)$ will not be continuous as $|y| \to 0$, that is, as y tends to zero through positive and negative values of y. This discontinuity in the logarithmic derivative of $\tilde{\psi}_1$ yields the so-called tearing stability parameter, Δ' defined as

$$\Delta' = \frac{1}{\tilde{\psi}_1(0)}\left[\left(\frac{\partial\tilde{\psi}_1}{\partial y}\right)_{y\to 0+} - \left(\frac{\partial\tilde{\psi}_1}{\partial y}\right)_{y\to 0-}\right]. \tag{7.5.18}$$

Further analysis, discussed below, indicates that tearing modes are unstable if $\Delta' > 0$. It can be shown that for the Harris equilibrium (7.5.3), one can solve for $\tilde{\psi}_1(y)$ exactly and obtain

$$\Delta' = \frac{1}{k^2}(1 - k^2), \tag{7.5.19}$$

which implies that tearing modes are unstable for $k < 1$ (or equivalently, $ka < 1$ in dimensional variables). Note that we can then determine the stability of tearing modes by simply solving the ideal MHD (or external region) equations without even solving the equations for the resistive layer (or the interior region). However, in order to determine the growth rate of the instability, we need to solve the resistive layer equations.

The resistive layer is a narrow layer localized in the vicinity of the $y = 0$ surface. In this narrow layer, for which $|y| \ll 1$, the spatial derivative $\partial/\partial y \gg 1$. Under these conditions, we can approximate the magnetic field $b(y) \cong b_0 y$ and assume that within the layer $\partial\tilde{\psi}_1/\partial y \gg k\tilde{\psi}_1$. Under these conditions, Eq. (7.5.14) becomes

$$\gamma(\tilde{\psi}_1 - y\tilde{\phi}_1) \cong S^{-1}\frac{\partial^2\tilde{\psi}_1}{\partial y^2}. \tag{7.5.20}$$

Similarly, Eq. (7.5.15) reduces to

$$\gamma^2\frac{\partial^2\tilde{\phi}_1}{\partial y^2} \cong -y\frac{\partial^2\tilde{\psi}_1}{\partial y^2}. \tag{7.5.21}$$

Combining Eqs. (7.5.20) and (7.5.21), we obtain

$$\gamma\frac{\partial^2\tilde{\phi}_1}{\partial y^2} \cong -yS\,[\tilde{\psi}_1(0) - y\tilde{\phi}_1], \tag{7.5.22}$$

where we have imposed the "constant-psi" approximation, $\tilde{\psi}_1 \cong \tilde{\psi}_1(0)$, within the interior region. We define a new dependent variable $\Phi_1 = -\delta^{1/4}\tilde{\phi}_1/\tilde{\psi}_1(0)$ and a new independent variable $Y = \delta^{-1/4}y$ where $\delta = \gamma/S$ to obtain the equation

$$\frac{\partial^2\Phi_1}{\partial Y^2} - Y^2\Phi_1 = Y. \tag{7.5.23}$$

The new independent variable Y is a "stretched coordinate," typical of boundary-layer analyses, whereby Y is a dimensionless coordinate that becomes of order unity when the real space coordinate y becomes of the order $\delta^{1/4} = (\gamma/S)^{1/4}$, which represents the typical thickness of the resistive layer (or the interior region). This layer becomes narrower as the resistivity $\eta \to 0$ or $S \to \infty$, and depends on the growth rate γ which also tends to smaller values as $\eta \to 0$ (verified later). The narrowness of the boundary layer is a qualitative justification for the "constant-psi" approximation used earlier, which assumes that resistive diffusion across this very narrow layer is enough to flatten out the perturbed flux $\tilde{\psi}_1$.

We can also rewrite Eq. (7.5.21) in new variables by defining $\tilde{\psi}_1(y)/\tilde{\psi}_1(0) = \Psi_1(Y)$, and obtain

$$\frac{\gamma^2}{\delta^{1/2}}\frac{\partial^2\Phi_1}{\partial Y^2} = Y\frac{\partial^2\Psi_1}{\partial Y^2}. \tag{7.5.24}$$

The interior and exterior region solutions should be matched by means of the parameter Δ', which has been calculated previously using the exterior region solution, and now needs to be calculated from the interior region solution. The matching condition is

$$\int_{-\infty}^{\infty}\frac{\partial^2\Psi_1}{\partial Y^2}\frac{1}{\delta^{1/4}}\mathrm{d}Y = \Delta', \tag{7.5.25}$$

with the left-hand side calculated using the interior region solution. Equivalently, we may write, using Eq. (7.5.24),

$$\frac{\gamma^2}{\delta^{3/4}} \int_{-\infty}^{\infty} \frac{1}{Y} \frac{\partial^2 \Psi_1}{\partial Y^2} \mathrm{d}Y = \Delta'. \tag{7.5.26}$$

The integral on the left-hand side of the above equation can be rewritten using Eq. (7.5.23), whereupon we obtain

$$\frac{\gamma^2}{\delta^{3/4}} \int_{-\infty}^{\infty} (1 + Y\Phi_1) \mathrm{d}Y = \Delta', \tag{7.5.27}$$

which can be inverted to give

$$\gamma = \Delta'^{4/5} S^{-3/5} \left[\int_{-\infty}^{\infty} (1 + Y\Phi_1) \mathrm{d}Y \right]^{-4/5}. \tag{7.5.28}$$

We note that Eq. (7.5.28) for the growth rate γ obeys the inequality $\gamma \ll 1 \ll S\gamma$, consistent with our earlier assumption. In other words, the tearing instability does grow much faster than that of simple resistive diffusion, and much less rapidly than that of ideal Alfvénic motion.

In addition to the classical linear tearing instability, which has a growth rate proportional to $S^{-3/5}$, there is yet another faster tearing mode for which the growth rate is proportional to $S^{-1/3}$. The latter is obtained when the dimensionless wave number k is much less than unity, which makes the tearing stability parameter Δ', defined by Eq. (7.5.27), much greater than unity. Under these conditions, the constant-psi approximation breaks down, calling for a different asymptotic treatment of the interior layer equations than given above. For more details, we refer the reader to the monograph by White (2006).

While understanding linear tearing modes is important, we should point out that the time-scale over which tearing mode growth occurs according to the dictates of linear theory is typically very short. As discussed in Section 3.10, a perturbation of the flux function of the type Ψ_1 will open up magnetic islands of width w, given by Eq. (3.10.23). It turns out that linear tearing mode theory breaks down as soon as the width of a magnetic island becomes of the order of the resistive (or interior) layer width, which is a very narrow width for systems characterized by high values of the Lundquist number. For most situations of physical interest, by the time a magnetic island produced by a linear tearing mode becomes detectable by even sophisticated experimental diagnostics, it is already in the nonlinear regime.

In the nonlinear regime, both types of linear tearing modes slow down from exponential growth in time to very slow algebraic growth in time. The classical theory for the nonlinear evolution of the slow tearing mode was developed by Rutherford (1973), and has found extensive application in fusion plasma physics (White, 2006). However, the fast tearing mode has a qualitatively different dynamical behavior, anticipated by the nonlinear Sweet–Parker model discussed in the next section.

7.6 Magnetic Reconnection

In Section 3.8.3, we discussed particle motion in a basic x-type magnetic field configuration and the possibility of plasma transport across a boundary between regions of different magnetic field topologies. This type of plasma transport in two dimensions is called magnetic reconnection or field line merging (Vasyliunas, 1975). In what follows, we give a brief discussion of some of the important features and models of this important phenomenon. A classic example of a magnetic field configuration that can support reconnection is the previously discussed Harris current sheet, shown in Figure 7.21, for which the magnetic field is given by

$$\mathbf{B} = B_0 \tanh\left(\frac{y}{a}\right)\hat{\mathbf{x}} + B_{\mathrm{T}}\hat{\mathbf{z}}, \tag{7.6.1}$$

where $B_0 > 0$ and $B_{\mathrm{T}} > 0$.

In addition to the proper magnetic geometry, magnetic reconnection requires some form of dissipation that violates the frozen field theorem, as discussed in Section 6.3.1. If plasma is transported across a boundary between regions with different magnetic field topologies, as required in magnetic reconnection, it must be that the plasma is not frozen to the magnetic field, but slips with respect to the field. The frozen field theorem, which is deduced from the condition $\mathbf{E}+\mathbf{U}\times\mathbf{B}=0$, requires that $E_{||}=0$. It follows that a necessary condition for magnetic reconnection to occur is that the component of the electric field parallel to the magnetic field, denoted by $E_{||}$, be non-zero. In a resistive plasma with Ohm's law (6.1.33), this implies that

$$E_{||} = \eta J_{||} \neq 0. \tag{7.6.2}$$

When $E_{||} \neq 0$, the magnetic flux threading any closed curve moving with the plasma is not constant, violating Statement 1 of Section 6.3.1. Instead of flux conservation, we can now have flux annihilation.

Although necessary, the existence of a non-zero, parallel electric field, $E_{||}$, is not sufficient to ensure that magnetic reconnection will occur. The change of magnetic flux in a closed loop due to resistive diffusion can produce a non-zero $E_{||}$, but diffusion does not necessarily imply reconnection. One way to distinguish diffusion from reconnection is by considering characteristic time-scales. In the absence of resistivity, the characteristic time-scale of field line motion is the so-called Alfvén time-scale,

$$\tau_A = \frac{L}{V_A}, \tag{7.6.3}$$

where L is on the order of the system size and $V_A = B_0/\sqrt{\mu_0\rho_{m0}}$ is the characteristic Alfvén speed corresponding to the component B_x. The presence of resistivity introduces another time-scale that can be obtained by balancing the term on the left of Eq. (6.3.4) with the second term on the right. We then obtain the resistive diffusion time-scale,

$$\tau_R = \mu_0\,\sigma L^2 = \mu_0\,\eta^{-1}L^2. \tag{7.6.4}$$

It is convenient at this point to define a dimensionless number, the Lundquist number, by the ratio

$$S = \frac{\tau_R}{\tau_A} = \mu_0\,\sigma\,V_A L = \mu_0\eta^{-1}V_A L. \tag{7.6.5}$$

The Lundquist number is very similar to the magnetic Reynolds number, defined earlier by Eq. (6.3.6), except that U is replaced by V_A. The Lundquist number for hot fusion plasmas, solar coronal plasmas, or magnetotail plasmas can lie in the range 10^8–10^{14}, so there is a very large separation between τ_A and τ_R in such systems. What is interesting about magnetic reconnection in a resistive plasma is that its characteristic time-scale, τ_r, lies between τ_A and τ_R, i.e., $\tau_A \ll \tau_r \ll \tau_R$. This allows flux annihilation to occur much faster when field lines reconnect than when they diffuse in a high-S plasma. The geometry of magnetic field lines plays a crucial role in determining whether such rapid flux annihilation can occur. In the geometry represented in Figure 7.21, one intuitively expects dramatic changes to occur in a small region $\Delta \ll d$ near $y = 0$. One expects a characteristic reconnection time $\tau_r = \mu_0\eta^{-1}\Delta^2$, which is much shorter than the resistive diffusion time τ_R, defined by Eq. (7.6.4). This is borne out by the Sweet–Parker model of magnetic reconnection, which is discussed next.

7.6.1 The Sweet–Parker Model

Parker (1957) and Sweet (1958) developed a simple model of magnetic reconnection that has been a cornerstone of the literature for more than five decades and has been verified by computer simulations. This model is now known as the Sweet–Parker model. The problem they addressed is the following: Consider the field configuration given by Eq. (7.6.1) in which the component B_x tends to $\pm B_0\hat{\mathbf{x}}$ as $y \to \pm\infty$. If these oppositely directed fields are convected inward toward each

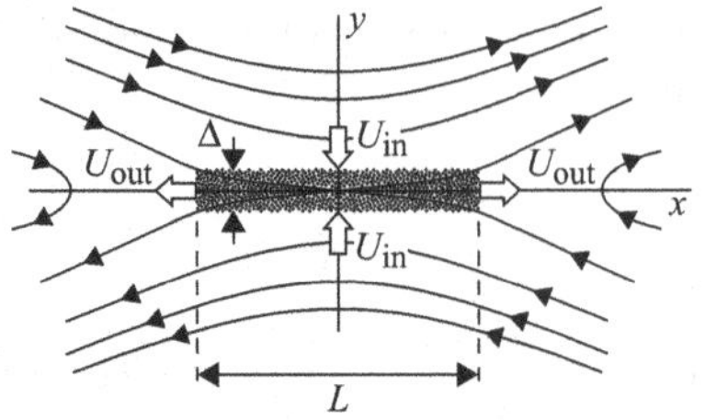

Figure 7.22 The magnetic field geometry used to evaluate reconnection near an x-type neutral line.

other, what is the speed U_{in} (along y) at which the oppositely directed fields merge and annihilate each other? We assume that the plasma is incompressible. It can then be shown that the field B_T plays no role in the dynamics.

As shown in Figure 7.22, the inward flow U_{in} induces a steady outward flow U_{out} in a narrow channel of width Δ (along x). In steady state, the momentum balance condition along x gives

$$\rho_m U_x \frac{\partial U_x}{\partial x} = -\frac{\partial P}{\partial x}. \tag{7.6.6}$$

Note that variation of the magnetic field along x is neglected. In an incompressible plasma, the density ρ_m is constant, and we can integrate Eq. (7.6.6) from $x = 0$ to a value of x sufficiently large as to lie just outside the reconnection region. Let this asymptotic distance be $x = L/2$. We then obtain

$$\rho_m \frac{U_{out}^2}{2} = \delta P, \tag{7.6.7}$$

where U_{out} is the value of U_x in the asymptotic region, $U_x = 0$ at $x = 0$, and δP is the pressure differential of the fluid between $x = 0$ and $L/2$. This pressure differential is essentially set up by the equilibrium condition, $P + B^2/2\mu_0$ = constant, across the width of the sheet, with P largest at the neutral line where $B = 0$. Hence, we obtain $\delta P = B_0^2/2\mu_0$ which, when substituted in Eq. (7.6.7), gives

$$U_{out} = \left(\frac{B_0}{\mu_0 \rho_m}\right)^{1/2} = V_A. \tag{7.6.8}$$

By mass continuity of the incompressible fluid in steady state, we obtain

$$U_{in} L = U_{out} \Delta = V_A \Delta. \tag{7.6.9}$$

During reconnection, a thin and intense current sheet develops over a narrow width and represents the main site for Ohmic dissipation. In steady state, the resistive dissipation rate in the reconnection layer of width Δ must balance the energy inflow, i.e.,

$$\eta J^2 \Delta = U_{\text{in}} \frac{B_0^2}{2\mu_0}, \tag{7.6.10}$$

where

$$J = \frac{B_0}{\mu_0 \Delta}. \tag{7.6.11}$$

From the above equations, we obtain

$$U_{\text{in}} = \frac{2\eta}{\mu_0 \Delta}. \tag{7.6.12}$$

Eliminating U_{in} between Eqs. (7.6.9) and (7.6.12), it follows that

$$\Delta = \left(\frac{2\eta L}{V_{\text{A}} \mu_0} \right)^{1/2} = \frac{\sqrt{2} L}{S^{1/2}}. \tag{7.6.13}$$

Substituting expression (7.6.13) for Δ into Eq. (7.6.9), we obtain

$$U_{\text{in}} = \frac{\sqrt{2} V_{\text{A}}}{S^{1/2}}. \tag{7.6.14}$$

Note that for $S \gg 1, \Delta \ll 2L$, as anticipated earlier. Hence, the time-scale of reconnection is shorter than the time-scale of diffusion. Note that this separation of time-scales depends crucially on the geometry of the magnetic field lines. If the magnetic field did not point in opposite directions (across the neutral line) but in the same direction, then the parameter Δ in Eqs. (7.6.10) and (7.6.11) would be replaced by L, with the consequence that $U_{\text{in}} = \sqrt{2} V_{\text{A}}/S$, which is indistinguishable from simple diffusion. Thus, the reconnection velocity U_{in} is $S^{1/2}$ times larger than the diffusion velocity. Note, however, this large U_{in} is still smaller than the Alfvén speed V_{A} by a factor of $S^{1/2}$.

7.6.2 Fast Reconnection

Neither the linear tearing instability (both slow and fast), discussed in Section 7.5, nor the nonlinear Sweet–Parker model, discussed in Section 7.6.1, are examples of fast reconnection. This is because many systems of great interest in laboratory, astrophysical, and space plasmas are characterized by Lundquist numbers S that are very high (varying from 10^4 to numbers in excess of 10^{12} by several orders of magnitude). Under these conditions the time-scale of growth of a linear tearing mode or the time needed to reconnect a significant amount of magnetic flux in the Sweet–Parker scenario becomes extremely long, and does not agree with those that are typically observed. For example, in the solar corona, for which the typical Lundquist number is estimated to be 10^{12} or higher, the reconnection rate predicted by the Sweet–Parker model, as measured by U_{in}, is one millionth of the Alfvén speed V_A or lower. If this is a measure of the rate at which magnetic energy is released by the process of merging of magnetic field lines during a solar flare, it would take approximately 2–3 years for a solar flare to release stored magnetic energy in the Sun, when the observed time-scale is of the order of 20 minutes or so. The key limitation of the Sweet–Parker model is the strong and inverse dependency of the reconnection rate on the Lundquist number, proportional to $S^{-1/2}$. Although much faster than the rate of simple resistive diffusion, which occurs at a characteristic rate of S^{-1}, the reconnection rate predicted by the Sweet–Parker model is simply not fast enough.

One of the exciting advances in reconnection theory over the last few years has been the discovery that there is a pathway to much faster reconnection even within the framework of the resistive MHD model. It is beyond the scope of this book to provide a detailed quantitative description of these new results, so we will provide a qualitative overview. It turns out that the thin and intense current layer (also often referred to as the "current sheet") embedded in a Sweet–Parker reconnection layer, which becomes increasingly intense as S increases, is itself unstable with respect to a very rapidly growing linear tearing instability (Loureiro et al., 2007). This linear instability can be shown to be a special case of the standard tearing instability, if one takes into account the fact that the width of an equilibrium current layer, assumed to be a in Eq. (7.5.3) and independent of S, is actually a width that is strongly dependent on S for the Sweet–Parker model, and given by Eq. (7.6.13) (Bhattacharjee et al., 2009). As S becomes larger and exceeds a critical threshold, $S = S_c \sim 10^4$, the Sweet–Parker layer breaks up due to a virulent tearing instability, referred to as the "plasmoid instability."

What happens to this linear plasmoid instability when it gets to the nonlinear regime? If the instability were to fizzle out nonlinearly or lead to a Sweet–Parker regime, it would not be all that interesting from the viewpoint of fast reconnection. It actually turns out that the instability continues to persist in the nonlinear regime, producing a copious number of plasmoids of various sizes that are convected along the reconnection layer, either coalescing with each other to form larger plasmoids or convected out of the system in the horizontal direction if the ends are open. During this process, the original and transient Sweet–Parker current sheet thins down drastically (Figure 7.23).

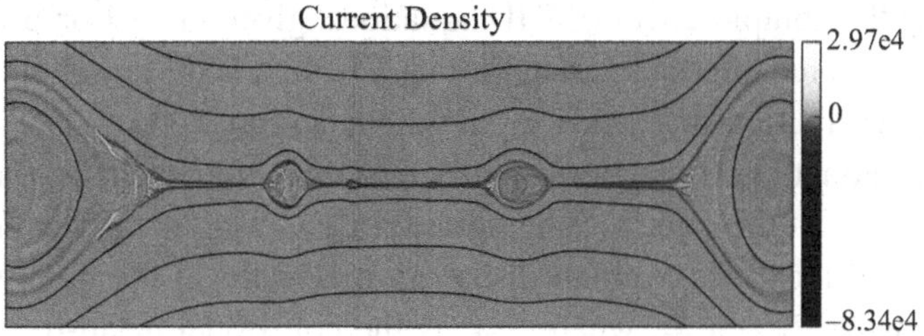

Figure 7.23 Plasmoid instability of a Sweet–Parker current sheet that breaks up into a large number of plasmoids, and the current sheet becomes thinner. Small plasmoids coalesce to form larger plasmoids that have longer lives, unless they are lost by convection out of the two ends (figure courtesy of Yi-Min Huang).

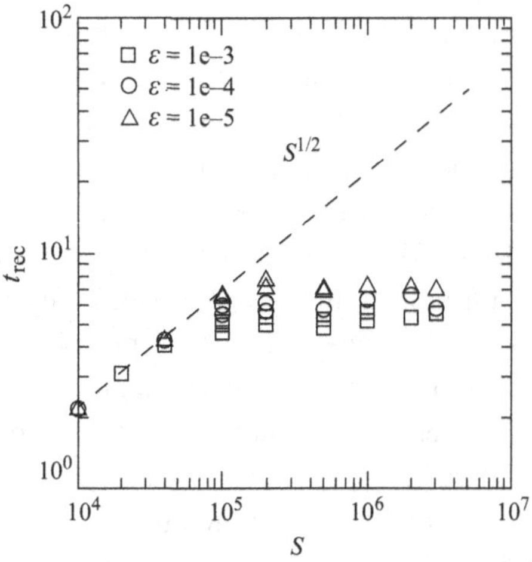

Figure 7.24 This plot shows the time needed to reconnect approximately 25% of the magnetic flux with the initial condition for Figure 7.23. Note that after crossing a threshold of approximately $S = 10^4$, the time to reconnect becomes insensitive to the Lundquist number S. Below this threshold the time for reconnection scales according to the prediction of the Sweet–Parker model. The various points in the plot correspond to various choices for the initial level of (white) noise imposed in the simulations, which are based on the nonlinear MHD equations. Small plasmoids coalesce to form larger plasmoids that have longer lives, unless they are lost by convection out of the two ends (Huang and Bhattacharjee, 2010).

It turns out rather remarkably that in the nonlinear regime the plasmoid instability leads to a new nonlinear regime in which the reconnection rate becomes nearly independent of the Lundquist number S. This is shown in the plot given in Figure 7.24 from Huang and Bhattacharjee (2010). The independence of the reconnection rate with respect to the mechanism (in this case, the plasma resistivity) that breaks field lines produces a regime of fast reconnection.

In deriving the complete set of MHD equations, given in Section 6.1.5, we have explicitly stated three important assumptions. The first is the assumption of a local Maxwellian distribution that leads to an adiabatic equation of state of the form (6.1.25). This equation of state essentially describes the adiabatic evolution of a plasma and neglects the effects of heat transport, as well as viscous heating. The assumption of a local Maxwellian distribution is valid in the highly collisional regime, i.e., when $L \gg \lambda_{\mathrm{mfp}}$, where λ_{mfp} is the collisional mean-free path of the system. The second, and less serious, assumption is the neglect of the displacement current term in Ampère's law (6.1.28) that is valid in a non-relativistic plasma as long as the phase velocity of the waves is much smaller than the speed of light. Finally, the third assumption is that of quasi-neutrality, or zero charge density, valid when $L \gg \lambda_{\mathrm{D}}$.

In addition to the three assumptions mentioned above, we had forewarned the reader that the conditions under which the generalized Ohm's law (6.1.23) reduces to the resistive Ohm's law (6.1.24) are quite stringent. Of the three neglected terms on the right-hand side of Eq. (6.1.23), the neglect of the last term is easiest to justify: it is valid to do so when $m_{\mathrm{e}} \ll m_{\mathrm{i}}$ and $\rho_{\mathrm{c}} \ll L$, where ρ_{c} is the ion cyclotron radius Eq. (3.1.4). However, the neglect of the second and third terms, which are called finite Larmor radius terms, is valid only when $\rho_{\mathrm{c}} \ll L$. It turns out that the neglect of these terms is much more delicate than neglect of the electron inertia term. This is especially so in magnetic reconnection problems, where the reconnection layer width Δ can often be comparable to ρ_{c} in higher density plasmas. In such cases, $\rho_{\mathrm{c}} \gtrsim \Delta$ even though $\rho_{\mathrm{c}} \ll L$, which requires that the physics of the reconnection layer be treated by adding the finite Larmor radius terms to the resistive Ohm's law, which now becomes

$$\mathbf{E} + \mathbf{U} \times \mathbf{B} = \eta \mathbf{J} + \frac{\mathbf{J} \times \mathbf{B}}{ne} - \frac{\boldsymbol{\nabla} \cdot \overleftrightarrow{\mathbf{P}}_{\mathrm{e}}}{ne}. \qquad (7.6.15)$$

In weakly collisional plasmas, the resistivity η is very low and the Lundquist number S is very high. In such cases it is easy to obtain situations where the parameter Δ, defined by Eq. (7.6.13), becomes sufficiently small that it is important to replace the resistive Ohm's law by the generalized Ohm's law (7.6.15). In what follows, we will describe qualitatively some of the changes brought about in the reconnection dynamics by the presence of the Hall current term $\mathbf{J} \times \mathbf{B}/ne$. For simplicity, we will consider the case in which the out-of-plane component B_{T} is zero and neglect the effect of the electron pressure tensor $\overleftrightarrow{\mathbf{P}}_{\mathrm{e}}$. Using the equation $\mathbf{J} = ne(\mathbf{U}_{\mathrm{i}} - \mathbf{U}_{\mathrm{e}})$ and the approximation $\mathbf{U} \simeq \mathbf{U}_{\mathrm{i}}$, we can combine the term $\mathbf{U} \times \mathbf{B}$ on the left and the Hall current term on the right to rewrite Eq. (7.6.15) in the form

$$\mathbf{E} + \mathbf{U}_{\mathrm{e}} \times \mathbf{B} \simeq \eta \mathbf{J}. \qquad (7.6.16)$$

In magnetic reconnection, the effect of the resistivity η is felt most strongly in a small region Δ near the neutral surface $y = 0$, also known as the diffusion region. Away from the diffusion region, the effect of η can be neglected whereby Eq. (7.6.16) reduces to

$$\mathbf{E} + \mathbf{U}_e \times \mathbf{B} = 0. \tag{7.6.17}$$

The above equation implies that in the presence of the Hall current, the electron fluid, which moves with the velocity $\mathbf{U}_e$, is frozen to the magnetic field, but the ion fluid, which moves with the velocity $\mathbf{U}_i$, is not. This is one of the principal signatures of Hall MHD reconnection; the ions and electrons move as separate fluids.

Because the electron fluid is frozen to the magnetic field $\mathbf{B}$, it follows the magnetic field all the way to the diffusion region and ends up carrying most of the current in this region. The ions, which are not tied to the magnetic field, flow in a channel much broader than the electron channel. Computer simulations show that, under such conditions, the merging flow U_{in}, which should be identified with the ion merging flow, can exceed the Sweet–Parker result U_{in} significantly and depends rather weakly on the Lundquist number S. For more details, the reader is referred to the monograph by Biskamp (2000).

Problems

7.1. To find solutions of a force-free MHD equilibrium, we frequently look for solutions of the form $\nabla \times \mathbf{B} = \alpha \mathbf{B}$, where α is a constant.

(a) In rectangular coordinates, show that $\mathbf{B} = (0, B_0 \sin \alpha x, B_0 \cos \alpha x)$ is a solution.

(b) In cylindrical coordinates, show that $\mathbf{B} = [B_\rho, B_\phi, B_z] = [0, B_0 J_1(\alpha\rho), B_0 J_0(\alpha\rho)]$ is a solution where J_0 and J_1 are the zeroth- and first-order Bessel functions. Hint: You will need to research the properties of Bessel functions to do this problem. Make a sketch of the magnetic field lines in the region $0 < \alpha\rho < 2.4$ (i.e., inside the first zero of $J_0(\alpha\rho)$).

7.2. Solutions for force-free MHD equilibria can also be found for which α is not a constant.

(a) In cylindrical coordinates, show that

$$\mathbf{B} = [B_\rho, B_\phi, B_z] = \left[0, \frac{B_0 k\rho}{1 + k^2\rho^2}, \frac{B_0}{1 + k^2\rho^2}\right]$$

is a solution where k is a constant. In this case show that

$$J_z = \frac{2kB_0/\mu_0}{(1 + k^2\rho^2)^2}.$$

(b) Make a sketch of B_ϕ, B_z, and J_z as a function of ρ.

7.3. Show that a time-dependent solution of the force-free ($\nabla \times \mathbf{B} = \alpha\mathbf{B}$) static equilibrium equation

$$\frac{\partial \mathbf{B}}{\partial t} = \frac{1}{\mu_0 \sigma} \nabla^2 \mathbf{B}$$

is given by

$$\mathbf{B} = \mathbf{B}_0 \, \mathrm{e}^{-t/\tau},$$

where $\mathbf{B}_0$ is the solution of the vector Helmholtz equation

$$\nabla^2 \mathbf{B}_0 + \alpha^2 \mathbf{B}_0 = 0$$

and $\tau = (\mu_0 \sigma)/\alpha^2$, with α = constant.

Note: Solutions of the vector Helmholtz equation can be found in many electricity and magnetism books.

7.4. For the "Bennett pinch" type of pressure-balanced equilibrium, the solution for $n(\rho)$ depends on the conductivity relation. Assuming the thermal conductivity is so large that the temperature is independent of radius, analyze the following cases:

(a) Lorentz type conductivity. Using $J_z = \frac{e^2 n}{m\nu} E_z$ show that $n(\rho) = n_0/[1 + (\rho/R)^2]^2$, where R is a constant.

(b) Fully ionized gas. Using $J_z = \sigma E$, where σ is a constant, show that $n(\rho) = n_0(1 - \rho^2/R^2)$.

7.5. Show that the linear force operator

$$\mathbf{F}(\boldsymbol{\xi}) = \frac{1}{\mu_0}[(\nabla \times \{\nabla \times (\boldsymbol{\xi} \times \mathbf{B}_0)\}) \times \mathbf{B}_0 + (\nabla \times \mathbf{B}_0) \times \{\nabla \times (\boldsymbol{\xi} \times \mathbf{B}_0)\}]$$
$$+ \nabla[\boldsymbol{\xi} \cdot \nabla P_0 + \gamma P_0(\nabla \cdot \boldsymbol{\xi})]$$

is self-adjoint, i.e., if there are two eigenfunctions η_1 and η_2 satisfying standard boundary conditions, then

$$\int_V \eta_2 \cdot \mathbf{F}(\eta_1)\,\mathrm{d}^3x = \int_V \eta_1 \cdot \mathbf{F}(\eta_2)\,\mathrm{d}^3x.$$

7.6. For a force-balanced MHD equilibrium in a cylindrical geometry with $\mathbf{B} = [0, B_\phi(\rho), B_z(\rho)]$ the radial component of the pressure balance condition $\mathbf{J} \times \mathbf{B} = \nabla P$ can be written

$$\frac{\partial}{\partial \rho}\left(P + \frac{B_\phi^2}{2\mu_0} + \frac{B_z^2}{2\mu_0}\right) = [(\mathbf{B} \cdot \nabla)\mathbf{B}]_\rho.$$

Show that $[(\mathbf{B}\cdot\nabla)\mathbf{B}]_\rho = -B_\phi^2/\rho$.
Hint: Use the identity $\nabla(\mathbf{F}\cdot\mathbf{G}) = (\mathbf{F}\cdot\nabla)\mathbf{G} + (\mathbf{G}\cdot\nabla)\mathbf{F} + \mathbf{F}\times(\nabla\times\mathbf{G}) + \mathbf{G}\times(\nabla\times\mathbf{G})$.

7.7. Assuming that the magnetohydrodynamic equilibrium profiles depend only on the radial coordinate ρ, show from Eqs. (7.2.1)–(7.2.6) that the magnetic and velocity fields can be written in the form $[0, B_\phi(\rho), B_z(\rho)]$ and $[0, U_\phi(\rho), U_z(\rho)]$ and that the equilibrium condition is given by Eq. (7.2.7). Check that under these conditions all of Eqs. (7.2.1)–(7.2.6) are satisfied identically.

7.8. Show by direct substitution that the equilibrium profiles, given by Eqs. (7.2.10)–(7.2.15), satisfy Eq. (7.2.7).

7.9. Derive the dispersion relation for the magnetorotational instability (7.4.6) from the linearized equation of motion (7.4.5).

7.10. (a) Derive the dispersion relation (7.4.10) from the more general dispersion relation (7.4.6) for the MRI in the incompressible limit.
(b) Derive the MRI stability condition (7.4.11) from the dispersion relation (7.4.10).
(c) Obtain the largest growth rate for the instability, and show that it is independent of the magnitude of the magnetic field.

7.11. For the Harris sheet, Eq. (7.5.17) permits an exact solution. Obtain this solution, and demonstrate that the tearing stability parameter is given by Eq. (7.5.10).

References

Alexandroff, P., and Hopf, H. 1935. *Topologie*. Berlin: Springer Verlag.

Balbus, S., and Hawley, J. 1991. A powerful local shear instability in weakly magnetized disks: 1. Linear analysis. *Astrophys. J.* **376**, 214–222.

Basu, S., and Kelley, M. C. 1979. A review of recent observations of equatorial scintillations and their relationship to current theories of F region irregularity generation. *Radio Sci.* **14**, 471–485.

Batchelor, G. K. 1967. *An Introduction to Fluid Mechanics*. Cambridge: Cambridge University Press, p. 74.

Bennett, W. H. 1934. Magnetically self-focussing streams. *Phys. Rev.* **45**, 890–897.

Bhattacharjee, A., Huang, Y-M., Yang, H, and Rogers, B. 2009. Fast reconnection in high-Lundquist-number plasmas due to the plasmoid instability. *Phys. of Plasmas* **16**, 112102.

Biskamp, D. 2000. *Magnetic Reconnection in Plasmas*. Cambridge: Cambridge University Press.

Chandrasekhar, S. 1960. The stability of non-dissipative Couette flow in hydromagnetics. *Proc. Natl Acad. Sci.* **46**, 253–257.

Freidberg, J. P. 1987. *Ideal Magnetohydrodynamics*. New York: Plenum Press.

Frieman, D., and Rotenberg, M. 1960. On hydromagnetic stability of stationary equilibria. *Rev. Mod. Phys.* **32**, 898–902.

Furth, H. P., Killeen, J., and Rosenbluth, M. N. 1963. Finite-resistivity instabilities of a sheet pinch. *Phys. Fluids* **6**, 459–484.

Goedbloed, J., Keppens, R., and Poedts, S. 2010. *Advanced Magnetohydrodynamics: With Applications to Laboratory and Astrophysical Plasmas*. Cambridge: Cambridge University Press.

Harris, E. G. 1961. Plasma instabilities associated with anisotropic velocity distributions. *J. Nucl. Energy Part C*. **2**, 138–145.

Huang, Y.-M., and Bhattacharjee, A. 2010. Scaling laws of resistive magnetohydrodynamic reconnection in the high-Lundquist-number, plasmoid-unstable regime. *Phys. of Plasmas*. **17**, 062104.

Kruskal, M. D., and Schwarzschild, M. 1954. Some instabilities of a completely ionized plasma. *Proc. R. Soc. London, Ser. A* **223**, 348–360.

Loureiro, N. F., Schekochihin, A. A., and Cowley, S. C. 2007. Instability of current sheets and formation of plasmoid chains. *Phys. of Plasmas* **14**, 100703.

Parker, E. N. 1957. Sweet's mechanism for merging magnetic fields in conducting fluids. *J. Geophys. Res.* **62**, 509–520.

Priest, E., and Forbes, T. 2000. *Magnetic Reconnection: MHD Theory and Applications*. Cambridge: Cambridge University Press, p. 261.

Rutherford, P. H. 1973. Nonlinear growth of the tearing mode. *Phys. Fluids* **16**, 1903–1908.

Shafranov, V. D. 1966. Plasma equilibrium in a magnetic field. In *Reviews of Plasma Physics*, Vol. 2, ed. M. A. Leontovich. : New York: Consultants Bureau, p. 103.

Solov'ev, L. S. 1968. The theory of hydrodynamic stability of toroidal plasma configurations. *Sov. Phys. JETP* **26**, 400–407.

Sweet, P. A. 1958. The production of high energy particles in solar flares, *Nuovo Cimento Suppl. Ser. X* **8**, 188–196.

Taylor, J. B. 1974. Relaxation of toroidal plasma and generation of reversed magnetic fields. *Phys. Rev. Lett.* **33**, 1139–1141.

Taylor, J. B. 1986. Relaxation and magnetic reconnection in plasmas. *Rev. Mod. Phys.* **53**, 741–763.

Vasyliunas, V. M. 1975. Theoretical models of magnetic field line merging, 1. *Rev. Geophys. Space Phys.* **13** (1), 303–336.

Velikhov, E. P. 1959. Stability of an ideally conducting liquid flowing between cylinders rotating in a magnetic field. *Sov. Phys. JETP* **36**, 995–998.

White, R. B. 2006. *The Theory of Toroidally Confined Plasma*, Second Edition. London: Imperial College Press.

Further Reading

Biskamp, D. 2000. *Magnetic Reconnection in Plasmas*. Cambridge: Cambridge University Press.

Chandrasekhar, S. 1981. *Hydrodynamic and Hydromagnetic Stability*. Mineola, NY: Dover. Originally published in 1961 by Oxford University Press.

Cowling, T. G. 1957. *Magnetohydrodynamics*. New York: Wiley Interscience.

Fitzpatrick, R. 2015. *Plasma Physics: An Introduction*. London: CRC Press.

Goedbloed, J., Keppens, R., and Poedts, S. 2010. *Advanced Magnetohydrodynamics With Applications to Laboratory and Astrophysical Plasmas*. Cambridge: Cambridge University Press.

Kulsrud, R. M. 2005. *Plasma Physics for Astrophysics*. Princeton, NJ: Princeton University Press.

White, R. B. 2006. *The Theory of Toroidally Confined Plasma*, Revised Second Edition. London: Imperial College Press.

8

Discontinuities and Shock Waves

Most fluids, including plasmas, have a remarkable tendency to form discontinuities. The formation of discontinuities can usually be traced to a nonlinear effect called wave steepening. For example, consider the propagation of a sound wave in an ordinary gas. As we discussed earlier, the propagation speed of a sound wave is given by $V_S^2 = \mathrm{d}P/\mathrm{d}\rho_m$. For an adiabatic equation of state, P/ρ_m^γ = constant, it can be shown that the propagation speed increases as the pressure increases (V_S is proportional to P^α, where $\alpha = (\gamma - 1)/2\gamma$ is a positive constant). Thus, if a pressure pulse is introduced into the gas, the trailing edge of the pulse, which has a higher pressure, tends to catch up with the leading edge, as shown in Figure 8.1. This amplitude dependence causes the pulse to steepen, until it becomes a nearly discontinuous change in pressure. Discontinuities of this type are called shock waves. Shock waves propagate at a speed that is intermediate between the sound speeds upstream and downstream of the shock. The thickness of a shock is ultimately controlled by some microscopic length-scale. In an ordinary gas, the relevant length-scale is the mean-free path. In a collisionless plasma, which has an infinite mean-free path, other length-scales, such as the ion cyclotron radius, play a role in determining the thickness.

To investigate the types of discontinuities that can exist in an ideal MHD fluid, we start by assuming that a planar discontinuity of zero thickness separates two otherwise uniform fluids. Conservation of mass, momentum, and energy, and

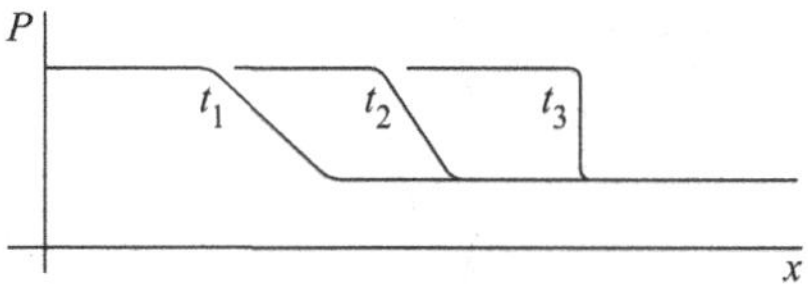

Figure 8.1 Pulse shapes at a sequence of times t_1, t_2, and t_3 that illustrate wave steepening.

the boundary conditions for the electric and magnetic fields are then used to put constraints on the fluid parameters and fields across the discontinuity. These constraints are called jump conditions. The derivation of the MHD jump conditions is discussed below.

8.1 The MHD Jump Conditions

To derive the jump conditions for an MHD fluid, we start with the laws of conservation of mass, momentum, and energy, given by Eqs. (6.4.1), (6.4.2), and (6.4.14), which after some minor rearrangements can be written in the following forms.

1. Conservation of mass:

$$\frac{\partial \rho_{\mathrm{m}}}{\partial t} + \nabla \cdot (\rho_{\mathrm{m}} \mathbf{U}) = 0. \tag{8.1.1}$$

2. Conservation of momentum:

$$\frac{\partial}{\partial t}(\rho_{\mathrm{m}} \mathbf{U}) + \nabla \cdot \left[\rho_{\mathrm{m}} \mathbf{U}\mathbf{U} + \left(P + \frac{B^2}{2\mu_0}\right)\overleftrightarrow{\mathbf{1}} - \frac{\mathbf{B}\mathbf{B}}{\mu_0}\right] = 0. \tag{8.1.2}$$

3. Conservation of energy:

$$\frac{\partial}{\partial t}\left(\frac{1}{2}\rho_{\mathrm{m}} U^2 + \xi + \frac{B^2}{2\mu_0}\right) + \nabla \cdot \left(\frac{1}{2}\rho_{\mathrm{m}} U^2 \mathbf{U} + h\mathbf{U} + \frac{1}{\mu_0}\mathbf{E} \times \mathbf{B}\right) = 0, \tag{8.1.3}$$

where ξ is the internal energy and h is the enthalpy. Since we want to relate the conditions on opposite sides of the discontinuity, consider a pill-box-shaped volume V containing the discontinuity, as shown in Figure 8.2. The discontinuity surface divides the volume V into two volumes V_1 and V_2, each with outer surfaces A_1 and A_2 as shown. The normal to the discontinuity surface, $\hat{\mathbf{n}}$, is assumed to be directed from side 1 to side 2, and the cross-sectional area of the discontinuity surface is A'.

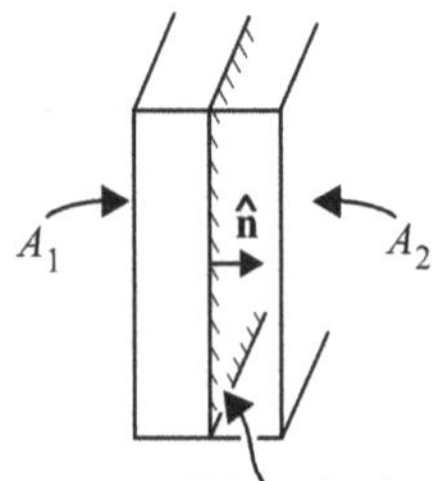

Figure 8.2 The pill-box-shaped surface used to derive the MHD jump conditions.

To derive the first jump condition we integrate the mass conservation equation over the volume V:

$$\int_V \left[\frac{\partial \rho_m}{\partial t} + \nabla \cdot (\rho_m \mathbf{U})\right] d^3x = 0. \tag{8.1.4}$$

Using the divergence theorem, the integral involving the divergence can be converted to surface integrals over A_1 and A_2, and the integral involving the time derivative can be split into two volume integrals over V_1 and V_2:

$$\int_{V_1} \frac{\partial \rho_m}{\partial t} d^3x + \int_{V_2} \frac{\partial \rho_m}{\partial t} d^3x + \int_{A_1} (\rho_m \mathbf{U}) \cdot d\mathbf{A} + \int_{A_2} (\rho_m \mathbf{U}) \cdot d\mathbf{A} = 0. \tag{8.1.5}$$

Next, let the volumes V_1 and V_2 shrink to zero in such a way that the area A' remains constant. The first and second terms (the volume integrals) then go to zero. Since $A_1 \to A'$ and $A_2 \to A'$, the third and fourth terms (the surface integrals) can be rewritten as

$$\int_{A'} (\rho_{m1} \mathbf{U}_1 \cdot (-\hat{\mathbf{n}}))\, dA' + \int_{A'} (\rho_{m2} \mathbf{U}_2 \cdot \hat{\mathbf{n}})\, dA' = 0 \tag{8.1.6}$$

or

$$\int_{A'} (\rho_{m1} \mathbf{U}_1 \cdot \hat{\mathbf{n}})\, dA' = \int_{A'} (\rho_{m2} \mathbf{U}_2 \cdot \hat{\mathbf{n}})\, dA', \tag{8.1.7}$$

where $\hat{\mathbf{n}}$ is a unit vector normal to the surface of the discontinuity. Since the above result is true for arbitrary A', the integrands must be equal, which gives the first jump condition, *conservation of mass flux*

$$\rho_{m1} \mathbf{U}_1 \cdot \hat{\mathbf{n}} = \rho_{m2} \mathbf{U}_2 \cdot \hat{\mathbf{n}}. \tag{8.1.8}$$

Jump conditions are often represented using the symbol $\{\}$, which denotes the difference of the enclosed quantity between the two sides of the discontinuity. Using the jump symbol, the above equation can be written

$$\{\rho_m \mathbf{U} \cdot \hat{\mathbf{n}}\} = 0. \tag{8.1.9}$$

Note that the above jump condition can be obtained by simply replacing the divergence operator, $\nabla \cdot ()$, in the mass continuity equation with the unit normal, and then setting the jump in this quantity to zero, i.e., $\{\hat{\mathbf{n}} \cdot ()\} = 0$. Following this procedure, it is straightforward to obtain the second jump condition, which is *conservation of momentum flux*

$$\left\{\rho_m \mathbf{U}(\mathbf{U} \cdot \hat{\mathbf{n}}) + \left(P + \frac{B^2}{2\mu_0}\right)\hat{\mathbf{n}} - \frac{\mathbf{B}}{\mu_0}(\mathbf{B} \cdot \hat{\mathbf{n}})\right\} = 0, \tag{8.1.10}$$

and the third jump condition, which is *conservation of energy flux*

$$\left\{\left(\frac{1}{2}\rho_{\mathrm{m}}U^2 + h\right)(\mathbf{U}\cdot\hat{\mathbf{n}}) + \frac{1}{\mu_0}(\mathbf{E}\times\mathbf{B})\cdot\hat{\mathbf{n}}\right\} = 0. \tag{8.1.11}$$

In addition to the above jump conditions, Maxwell's equations require that the tangential component of the electric field and the normal component of the magnetic field must be continuous across the discontinuity,

$$\{\mathbf{E}_{\mathrm{t}}\} = 0 \tag{8.1.12}$$

and

$$\{\mathbf{B}\cdot\hat{\mathbf{n}}\} = 0. \tag{8.1.13}$$

The above set of five jump conditions is sufficient to determine the possible discontinuities that can exist in an MHD fluid.

The jump conditions can be put into a more useful form by introducing the subscripts n and t for the normal and tangential components relative to the discontinuity surface and eliminating the electric field using $\mathbf{E} = -\mathbf{U}\times\mathbf{B}$. After some minor manipulations it can be shown that the jump conditions are given by

1. $$\{\rho_{\mathrm{m}}U_{\mathrm{n}}\} = 0, \tag{8.1.14}$$

2a. $$\left\{\rho_{\mathrm{m}}U_{\mathrm{n}}^2 + P + \frac{B_{\mathrm{t}}^2}{2\mu_0}\right\} = 0, \tag{8.1.15}$$

2b. $$\left\{\rho_{\mathrm{m}}U_{\mathrm{n}}\mathbf{U}_{\mathrm{t}} - B_{\mathrm{n}}\frac{\mathbf{B}_{\mathrm{t}}}{\mu_0}\right\} = 0, \tag{8.1.16}$$

3. $$\left\{\left(\frac{1}{2}\rho_{\mathrm{m}}U^2 + h + \frac{B^2}{\mu_0}\right)U_{\mathrm{n}} - (\mathbf{U}\cdot\mathbf{B})\frac{B_{\mathrm{n}}}{\mu_0}\right\} = 0, \tag{8.1.17}$$

4. $$\{\mathbf{U}_{\mathrm{n}}\times\mathbf{B}_{\mathrm{t}} + \mathbf{U}_{\mathrm{t}}\times\mathbf{B}_{\mathrm{n}}\} = 0, \tag{8.1.18}$$

5. $$\{B_{\mathrm{n}}\} = 0. \tag{8.1.19}$$

Before proceeding further, it is helpful to note certain simple algebraic rules concerning the jump symbol, $\{\}$. First, as can be easily verified, the jump symbol is distributive, i.e., $\{a+b\} = \{a\} + \{b\}$. Thus, $\{B^2\} = \{B_{\mathrm{n}}^2 + B_{\mathrm{t}}^2\} = \{B_{\mathrm{n}}^2\} + \{B_{\mathrm{t}}^2\}$. Since jump condition 5 above implies that $\{B_{\mathrm{n}}^2\} = 0$, it follows that $\{B^2\} = \{B_{\mathrm{t}}^2\}$. This result was used in the derivation of jump condition 2b above. Second, whenever some quantity, such as $(\rho_{\mathrm{m}}U_{\mathrm{n}})$, is known to be conserved, i.e., $\{\rho_{\mathrm{m}}U_{\mathrm{n}}\} = 0$, this quantity can be factored out of any term in which it occurs. Thus, in jump condition 2b, we can write $\{\rho_{\mathrm{m}}U_{\mathrm{n}}\mathbf{U}_{\mathrm{t}}\} = (\rho_{\mathrm{m}}U_{\mathrm{n}})\{\mathbf{U}_{\mathrm{t}}\}$. Since the conserved quantity is the same on both sides of the discontinuity, it is not necessary to specify on which side of the

Table 8.1 *Classification of MHD discontinuities*

	$U_n = 0$	$U_n \neq 0$
$\{\rho_m\} = 0$	trivial	rotational discontinuity
$\{\rho_m\} \neq 0$	contact discontinuity	shock wave

discontinuity the quantity is to be evaluated. However, if the conserved quantity consists of a product of two or more terms, these terms must all be evaluated on the same side of the discontinuity.

8.2 Classification of Discontinuities

To analyze the types of discontinuities that can occur in an MHD fluid, it is convenient to categorize them according to whether U_n and $\{\rho_m\}$ are zero or non-zero. This classification scheme was first introduced by Landau and Lifshitz (1960). Of the four possible combinations, the case $U_n = 0$ and $\{\rho_m\} = 0$ is trivial, since no change occurs across the discontinuity. The remaining three cases (see Table 8.1) are called a contact discontinuity ($U_n = 0, \{\rho_m\} \neq 0$), a rotational discontinuity ($U_n \neq 0, \{\rho_m\} = 0$), and a shock wave ($U_n \neq 0, \{\rho_m\} \neq 0$). Each of these cases will be analyzed separately. Sometimes it is also useful to consider further subcategories depending on whether B_n is zero or non-zero.

8.2.1 Contact Discontinuities

Contact discontinuities are characterized by no fluid flow across the discontinuity, $U_n = 0$, and a non-zero density jump, $\{\rho_m\} \neq 0$. If $B_n \neq 0$, it is easily verified from the jump conditions that $\{\mathbf{U}_t\} = 0, \{\mathbf{B}\} = 0$, and $\{P\} = 0$. Thus, only the density can change across the discontinuity. Because the pressure cannot change, the density change must be compensated by a corresponding change in temperature such that $n_1 \kappa T_1 = n_2 \kappa T_2$. This type of discontinuity is called a *contact discontinuity.* Contact discontinuities are often encountered in explosive processes, such as chemical explosions, where a new source of hot gas is suddenly injected into the system. Although contact discontinuities are common in ordinary gases, such discontinuities (with $B_n \neq 0$) are rarely observed in plasmas, since the plasma can readily diffuse along the magnetic field between the two regions.

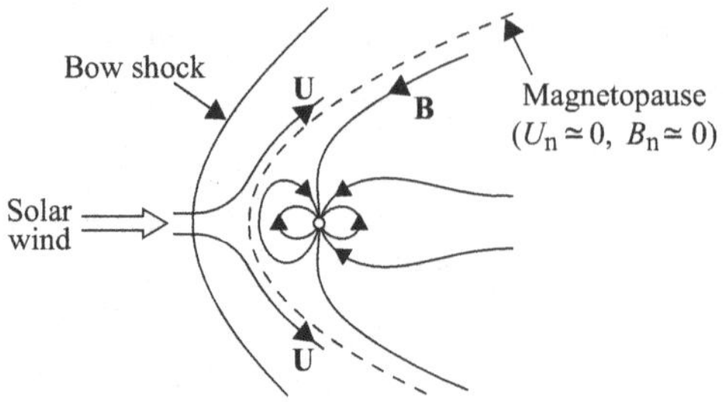

Figure 8.3 The interaction of the solar wind with a magnetized planet produces two types of discontinuities, the *bow shock*, where the supersonic solar wind becomes subsonic, and the *magnetopause*, where the planetary magnetic field pressure balances the solar wind pressure.

A special case of a contact discontinuity occurs when $B_n = 0$. If $B_n = 0$, it is easy to show from the jump conditions that $\{\mathbf{U}_t\} \neq 0, \{\mathbf{B}_t\} \neq 0$, and

$$\left\{ P + \frac{B^2}{2\mu_0} \right\} = 0. \tag{8.2.1}$$

This special case of a contact discontinuity is called a *tangential discontinuity*. For a tangential discontinuity, the fluid velocity and magnetic field are parallel to the surface of the discontinuity but change in magnitude or direction. The above equation shows that the sum of the plasma and magnetic field pressures must be constant across the discontinuity.

Since plasmas cannot easily diffuse across magnetic field lines, tangential discontinuities are commonly observed in plasmas. A good example occurs in association with planetary magnetospheres. Planetary magnetospheres typically have a well-defined boundary between the solar wind and the planetary magnetic field. This boundary is called the magnetopause (Parks, 1991). If the reconnection rate between the planetary and solar wind magnetic fields is small, then the normal components of the flow velocity and the magnetic field at the magnetopause are also small ($U_n \simeq 0$ and $B_n \simeq 0$). The magnetosphere is then said to be closed, since the solar wind plasma and magnetic field do not penetrate into the magnetosphere; see Figure 8.3. Under these conditions the magnetopause is to a good approximation a tangential discontinuity. The plasma and magnetic field pressures on the two sides of the magnetopause must then satisfy the pressure balance condition given by Eq. (8.2.1). Tangential discontinuities are also sometimes observed separating regions with different pressures and magnetic fields in the solar wind (Smith, 1973). The heliopause, which separates the solar wind plasma from the interstellar plasma (Axford, 1996), is also a tangential discontinuity.

8.2.2 Rotational Discontinuities

Rotational discontinuities are characterized by a fluid flow across the discontinuity, $U_n \neq 0$, and no change in the density, $\{\rho_m\} = 0$. Since the density is constant it immediately follows that the enthalpy and pressure are unchanged across the discontinuity, $\{h\} = 0$ and $\{P\} = 0$. From jump conditions 1 and 2a, it is also easily shown that $\{U_n\} = 0$ and $\{B_t^2\} = 0$.

From jump conditions 2b and 4, one can see that $\{\mathbf{U}_t\}$ and $\{\mathbf{B}_t\}$ must simultaneously satisfy

$$(\rho_m U_n)\{\mathbf{U}_t\} = \frac{B_n}{\mu_0}\{\mathbf{B}_t\} \tag{8.2.2}$$

and

$$\mathbf{U}_n \times \{\mathbf{B}_t\} + \{\mathbf{U}_t\} \times \mathbf{B}_n = 0. \tag{8.2.3}$$

Eliminating $\{\mathbf{U}_t\}$ from the above equations, we obtain

$$\mathbf{U}_n \times \{\mathbf{B}_t\} = B_n \frac{\mathbf{B}_n \times \{\mathbf{B}_t\}}{\mu_0(\rho_m U_n)}, \tag{8.2.4}$$

which is satisfied identically if

$$U_n^2 = \frac{B_n^2}{\mu_0 \rho_m}. \tag{8.2.5}$$

This type of discontinuity is called a *rotational discontinuity* since the magnetic field $\mathbf{B}_t$ remains constant in magnitude but rotates in the plane of the discontinuity. The field geometry viewed normal to the discontinuity is as shown in Figure 8.4. The speed at which the discontinuity propagates relative to the rest frame of the fluid is $U_n = V_A \cos\theta$, where θ is the angle between the magnetic field and the normal, and $V_A = B/\sqrt{\mu_0 \rho_m}$ is the Alfvén speed.

It is easily recognized that for small amplitudes a rotational discontinuity is simply a transverse Alfvén wave. In the linearized treatment, $\mathbf{U}_1$ and $\mathbf{B}_1$ correspond

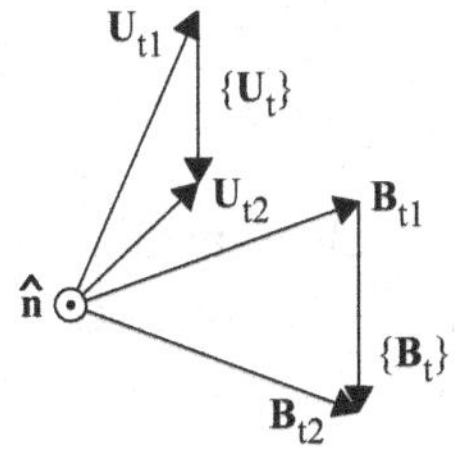

Figure 8.4 The field geometry of a rotational discontinuity as viewed normal to the plane of the discontinuity.

to $\{\mathbf{U}_\mathrm{t}\}$ and $\{\mathbf{B}_\mathrm{t}\}$, the wave number $\mathbf{k}$ is parallel to the normal, $B_0 \cos\theta$ corresponds to B_n, and $V_\mathrm{A} \cos\theta$ corresponds to U_n.

If the reconnection rate between the solar wind magnetic field and the planetary magnetic field is substantial, as tends to occur when the solar wind magnetic field is directed opposite to the planetary magnetic field, then the solar wind can penetrate significantly into the magnetosphere ($U_\mathrm{n} \neq 0$ and $B_\mathrm{n} \neq 0$). The magnetosphere is then said to be open. Under these conditions, the magnetopause becomes a rotational discontinuity (Sonnerup et al., 1981). Rotational discontinuities are also frequently observed in the solar wind (Smith, 1973).

8.3 Shock Waves

Shock waves are characterized by a fluid flow across the discontinuity, $U_\mathrm{n} \neq 0$, and a non-zero density jump, $\{\rho_\mathrm{m}\} \neq 0$. Of the three basic types of discontinuities, shock waves are by far the most difficult to analyze, since all of the jump conditions must be used in the analysis. To simplify the analysis, it is useful to make a specific choice for the coordinate system, called the de Hoffmann–Teller frame of reference, after de Hoffmann and Teller (1950), who first pointed out the advantages of using this coordinate system.

8.3.1 The de Hoffmann–Teller Frame

In the de Hoffmann–Teller frame of reference, the fluid velocity $\mathbf{U}$ and the magnetic field $\mathbf{B}$ are parallel both upstream and downstream of the shock, and lie in a plane perpendicular to the discontinuity surface. To prove that such a coordinate system can be found, start with jump conditions 2b and 4. After recognizing that $(\rho_\mathrm{m} U_\mathrm{n})$ and B_n are conserved across the discontinuity, these two equations can be written

$$(\rho_\mathrm{m} U_\mathrm{n})\{\mathbf{U}_\mathrm{t}\} - B_n \left\{\frac{\mathbf{B}_\mathrm{t}}{\mu_0}\right\} = 0, \tag{8.3.1}$$

and

$$(\rho_\mathrm{m} U_\mathrm{n}) \left\{\frac{\mathbf{B}_\mathrm{t}}{\rho_\mathrm{m}}\right\} - B_\mathrm{n} \{\mathbf{U}_\mathrm{t}\} = 0. \tag{8.3.2}$$

From the above two equations, it is clear that the vectors $\{\mathbf{B}_\mathrm{t}/\rho_\mathrm{m}\}$, $\{\mathbf{U}_\mathrm{t}\}$, and $\{\mathbf{B}_\mathrm{t}\}$ are all parallel. Solving the second equation for $\{\mathbf{U}_\mathrm{t}\}$ and substituting it into the first equation then gives the equation

$$(\rho_\mathrm{m} U_\mathrm{n})^2 \left\{\frac{\mathbf{B}_\mathrm{t}}{\rho_\mathrm{m}}\right\} - B_\mathrm{n}^2 \left\{\frac{\mathbf{B}_\mathrm{t}}{\mu_0}\right\} = 0, \tag{8.3.3}$$

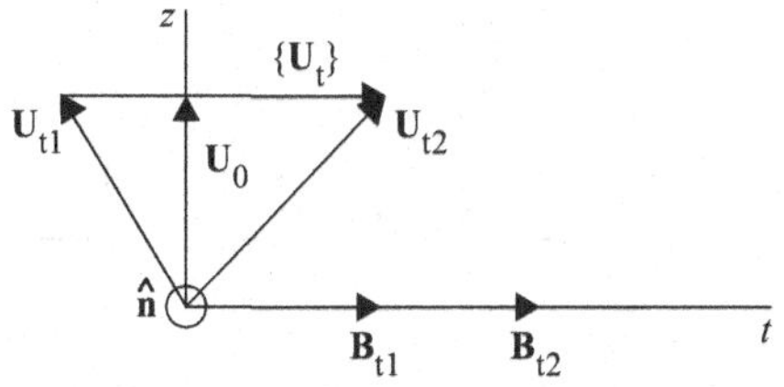

Figure 8.5 The field geometry of a shock wave viewed normal to the plane of the discontinuity. The normal, $\hat{\mathbf{n}}$, is directed upward out of the paper.

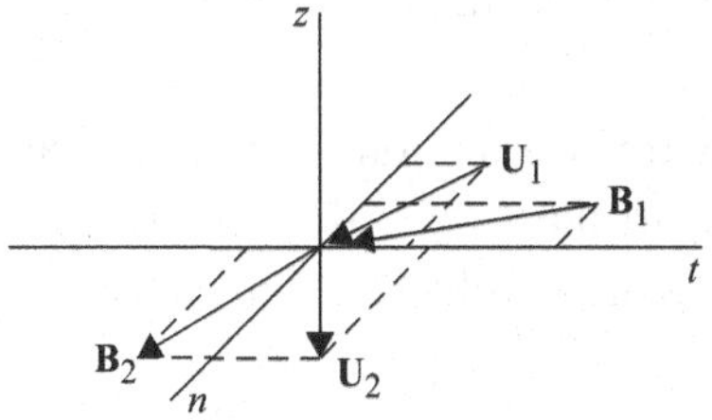

Figure 8.6 By performing a transformation along the z axis, a coordinate system can be found where the flow velocity $\mathbf{U}$ and the magnetic field $\mathbf{B}$, upstream and downstream of the shock, are coplanar.

which can be rewritten in the form

$$\mathbf{B}_{t1}\left(\frac{(\rho_m U_n)^2}{\rho_{m1}} - \frac{B_n^2}{\mu_0}\right) = \mathbf{B}_{t2}\left(\frac{(\rho_m U_n)^2}{\rho_{m2}} - \frac{B_n^2}{\mu_0}\right). \tag{8.3.4}$$

If $\rho_{m1} \neq \rho_{m2}$, the above equations show that $\mathbf{B}_{t1}$ and $\mathbf{B}_{t2}$ are parallel. From Eq. (8.3.2) and the definition of $\{\mathbf{B}_t\}$, it then follows that $\mathbf{B}_{t1}, \mathbf{B}_{t2}, \{\mathbf{B}_t\}$, and $\{\mathbf{U}_t\}$ are all parallel to each other and lie in the plane of the shocks. The field geometry viewed normal to the discontinuity surface is then as shown in Figure 8.5, where for notational convenience we have selected the tangential t axis to be parallel to $\mathbf{B}_{t1}$ and $\mathbf{B}_{t2}$. The z axis completes the right-hand (n,t,z) coordinate system. Unless otherwise specified, side 1 is assumed to be the upstream side (i.e., U_{n1} is positive).

From the above geometry, it is clear that the fluid velocities $\mathbf{U}_{t1}$ and $\mathbf{U}_{t2}$ can be made parallel to $\mathbf{B}_{t1}$ and $\mathbf{B}_{t2}$ by transforming to a coordinate system moving with a velocity $\mathbf{U}_0$ parallel to the z axis. Thus, without loss of generality, a coordinate system can be found in which $\mathbf{U}$ and $\mathbf{B}$ lie in the n,t plane. The field geometry in this coordinate system is shown in Figure 8.6.

It is easily verified that the electric field $\mathbf{E} = -\mathbf{U} \times \mathbf{B}$ in this new coordinate system is in the z direction and is given by

$$E_z = -(U_{n1}B_{t1} - U_{t1}B_n) = -(U_{n2}B_{t2} - U_{t2}B_n). \tag{8.3.5}$$

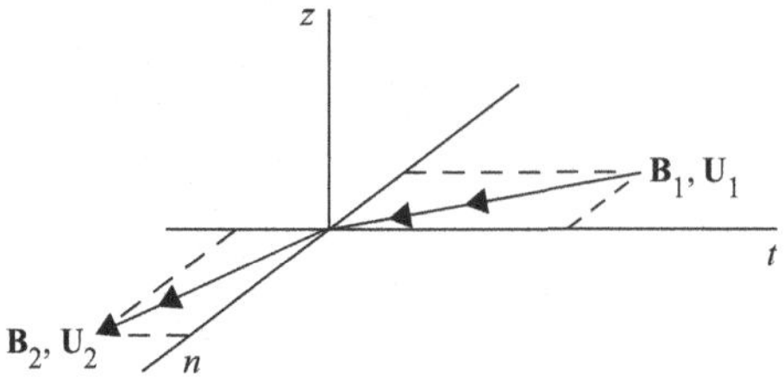

Figure 8.7 By performing a second transformation, this time along the t axis, a coordinate system can be found where the flow velocity **U** and the magnetic field **B** are parallel both upstream and downstream of the shock. This coordinate system is called the de Hoffmann–Teller frame.

By performing a second transformation to a coordinate system moving with a constant velocity parallel to the t axis, this electric field can be transformed to zero. The appropriate translation velocity is given by

$$U_{t0} = U_{t1} - U_{n1}\left(\frac{B_{t1}}{B_n}\right). \tag{8.3.6}$$

This is the de Hoffmann–Teller frame. Since $\mathbf{E} = -\mathbf{U}\times\mathbf{B} = 0$, it follows that in this frame of reference **U** and **B** are parallel. The field geometry is then as shown in Figure 8.7. Thus, by performing two coordinate transformations, we have reduced an inherently three-dimensional problem to two dimensions (i.e., the n,t plane), thereby greatly simplifying the analysis.

8.3.2 The Shock Propagation Speed

The next step in the analysis is to determine the normal component of upstream flow velocity, U_n, as a function of the upstream parameters, thereby giving the propagation speed of the shock relative to the upstream fluid. At this point it is convenient to introduce the speed of sound, which as before is defined by

$$V_S^2 = \frac{\gamma P}{\rho_m}. \tag{8.3.7}$$

In terms of the shock coordinates (n,t,z) described earlier, the jump conditions given by Eqs. (8.1.14)–(8.1.19) can then be written

1. $$\rho_{m1}U_{n1} = \rho_{m2}U_{n2}, \tag{8.3.8}$$

2a. $$\rho_{m1}U_{n1}^2 + \rho_{m1}\frac{V_{S1}^2}{\gamma} + \frac{B_{t1}^2}{2\mu_0} = \rho_{m2}U_{n2}^2 + \rho_{m2}\frac{V_{S2}^2}{\gamma} + \frac{B_{t2}^2}{2\mu_0}, \tag{8.3.9}$$

2b.
$$\rho_{m1} U_{n1} U_{t1} - B_{n1} \frac{B_{t1}}{\mu_0} = \rho_{m2} U_{n2} U_{t2} - B_{n2} \frac{B_{t2}}{\mu_0}, \tag{8.3.10}$$

3.
$$\rho_{m1} U_{n1} \left(\frac{V_{S1}^2}{\gamma - 1} + \frac{U_{t1}^2 + U_{n1}^2}{2} \right) + \frac{B_{t1}}{\mu_0} (B_{t1} U_{n1} - B_n U_{t1})$$
$$= \rho_{m2} U_{n2} + \left(\frac{V_{S2}^2}{\gamma - 1} + \frac{U_{t2}^2 + U_{n2}^2}{2} \right) + \frac{B_{t2}}{\mu_0} (B_{t2} U_{n2} - B_n U_{t2}), \tag{8.3.11}$$

4a.
$$U_{n1} B_{t1} - U_{t1} B_{n1} = U_{n2} B_{t2} - U_{t2} B_{n2}, \tag{8.3.12}$$

4b.
$$U_{n2} B_{t2} - U_{t2} B_{n2} = 0 \quad \text{(i.e., the } E_z = 0 \text{ condition)}, \tag{8.3.13}$$

5.
$$B_{n1} = B_{n2}, \tag{8.3.14}$$

where we have used $h = \gamma P/(\gamma - 1)$ for the enthalpy. These equations are known as the *Rankine–Hugoniot* equations, after a similar set of well-known equations in fluid mechanics (Currie, 1974).

The procedure used to solve the above system of equations is to specify all of the upstream parameters except **U** and then to solve for U_{n1} as a function of the upstream parameters. In the above jump conditions, there are twelve unknowns ($\rho_{m1}, \rho_{m2}, U_{n1}, U_{n2}, U_{t1}, U_{t2}, V_{S1}, V_{S2}, B_{t1}, B_{t2}, B_{n1}$, and B_{n2}) of which four upstream parameters ($\rho_{m1}, V_{S1}, B_{t1}$, and B_{n1}) are specified. The jump conditions provide seven equations for $12 - 4 = 8$ unknown quantities. Since the number of unknowns exceeds the number of equations by one, we must specify one more quantity in order to solve the system of equations. This indeterminacy is due to the fact that no information has been provided about the strength of the shock. Ultimately, this information involves the boundary conditions on the fluid flow. To make the problem tractable we introduce a new parameter,

$$r = \frac{\rho_{m2}}{\rho_{m1}}, \tag{8.3.15}$$

called the shock strength, which is the ratio of the downstream to the upstream plasma densities. With this parameter specified, enough equations are now available to determine completely all of the unknown parameters.

Before proceeding, to provide some further simplification, it is useful to introduce two dimensionless quantities:

$$M_A = \frac{U_n}{V_{An}} = \frac{U_n \sqrt{\mu_0 \rho_m}}{B_n} \tag{8.3.16}$$

and

$$M_{\rm S} = \frac{U_{\rm n}}{V_{\rm S}} = U_{\rm n}\sqrt{\frac{\rho_{\rm m}}{\gamma P}}. \tag{8.3.17}$$

The quantity $M_{\rm A}$ is called the Alfvén Mach number and is the ratio of the normal component of the flow velocity to the normal component of the Alfvén velocity. The quantity $M_{\rm S}$ is called the sonic Mach number and is the ratio of the normal component of the flow speed to the sound speed. It is also convenient to define an angle θ between the magnetic field and the shock normal. This angle is given by

$$\tan\theta = B_{\rm t}/B_{\rm n}. \tag{8.3.18}$$

The next step is to eliminate all quantities on side 2 of the jump conditions except the density $\rho_{\rm m2}$, which will be combined as a ratio with $\rho_{\rm m1}$ to introduce the shock strength r.

First, we solve for $B_{\rm t2}$ and $U_{\rm t2}$ from jump conditions 2b and 4b. After substituting $B_{\rm n}$ for $B_{\rm n1}$ and $B_{\rm n2}$, these two jump conditions become

$$\rho_{\rm m2}U_{\rm n2}U_{\rm t2} - \frac{B_{\rm n}B_{\rm t2}}{\mu_0} = \rho_{\rm m1}U_{\rm n1}U_{\rm t1} - \frac{B_{\rm n}B_{\rm t1}}{\mu_0} \tag{8.3.19}$$

and

$$U_{\rm n2}B_{\rm t2} - B_{\rm n}U_{\rm t2} = 0. \tag{8.3.20}$$

Using the second equation above to eliminate $U_{\rm t2}$ from the first equation and solving for $B_{\rm t2}$ gives

$$B_{\rm t2} = \frac{\left(\rho_{\rm m1}U_{\rm n1}U_{\rm t1} - \dfrac{B_{\rm t1}B_{\rm n}}{\mu_0}\right)}{\left(\rho_{\rm m2}U_{\rm n2}^2 - \dfrac{B_{\rm n}^2}{\mu_0}\right)}B_{\rm n}. \tag{8.3.21}$$

Next, use jump conditions 4a and 4b ($B_{\rm n}U_{\rm t1} = U_{\rm n1}B_{\rm t1}$) to eliminate $U_{\rm t1}$, which gives

$$B_{\rm t2} = \frac{\left(\rho_{\rm m1}U_{\rm n1}^2 - \dfrac{B_{\rm n}^2}{\mu_0}\right)}{\left(\rho_{\rm m2}U_{\rm n2}^2 - \dfrac{B_{\rm n}^2}{\mu_0}\right)}B_{\rm t1}. \tag{8.3.22}$$

By dividing the numerator and denominator by $B_{\rm n}^2/\mu_0$ and noting that $r = \rho_{\rm m2}/\rho_{\rm m1}$, the above equation can be written in terms of the Alfvén Mach number as follows:

$$B_{\rm t2} = \frac{M_{\rm A1}^2 - 1}{(M_{\rm A1}^2/r) - 1}B_{\rm t1}. \tag{8.3.23}$$

Next, the tangential velocity U_{t2} can be determined by substituting B_{t2} from Eq. (8.3.21) into Eq. (8.3.20) to give

$$U_{t2} = U_{n2} \frac{\left(\rho_{m1} U_{n1} U_{t1} - \dfrac{B_n B_{t1}}{\mu_0}\right)}{\left(\rho_{m2} U_{n2}^2 - \dfrac{B_n^2}{\mu_0}\right)}, \tag{8.3.24}$$

which can be reorganized in the form

$$U_{t2} = U_{t1} + \frac{\left(\dfrac{B_n^2 U_{t1}}{\mu_0} - \dfrac{U_{n2} B_n B_{t1}}{\mu_0}\right)}{\left(\rho_{m2} U_{n2}^2 - \dfrac{B_n^2}{\mu_0}\right)} = U_{t1} + \frac{\left[\dfrac{U_{t1}}{U_{n1}} - \left(\dfrac{U_{n2}}{U_{n1}}\right)\left(\dfrac{B_{t1}}{B_n}\right)\right]}{\left(\dfrac{M_{A1}^2}{r} - 1\right)} U_{n1}. \tag{8.3.25}$$

Using jump conditions 1 and 4a and introducing $\tan\theta_1 = B_{t1}/B_n$, the above equation can be written in terms of the Alfvén Mach number as follows:

$$U_{t2} = U_{t1} + \frac{(r-1)\tan\theta_1}{(M_{A1}^2 - r)} U_{n1}. \tag{8.3.26}$$

Having obtained equations for B_{t2} and U_{t2} in terms of quantities on side 1 of the shock, these equations can then be used to evaluate various quantities in the jump conditions. In jump condition 2a the quantity

$$\frac{B_{t2}^2 - B_{t1}^2}{2\mu_0} = \left[\left(\frac{M_{A1}^2 - 1}{\dfrac{M_{A1}^2}{r} - 1}\right)^2 - 1\right] \frac{B_{t1}^2}{2\mu_0} \tag{8.3.27}$$

simplifies to

$$\frac{B_{t2}^2 - B_{t1}^2}{2\mu_0} = \frac{(r-1)[(r+1)M_{A1}^2 - 2r]M_{A1}^2}{(M_{A1}^2 - r)^2} \left(\frac{B_{t1}^2}{2\mu_0}\right). \tag{8.3.28}$$

In jump condition 3, after factoring out $(\rho_{m1} U_{n1}) = (\rho_{m2} U_{n2})$, the quantity

$$\frac{B_{t2}^2}{\mu_0 \rho_{m2}} - \frac{B_{t1}^2}{\mu_0 \rho_{m1}} = \left[\frac{1}{r}\left(\frac{M_{A1}^2 - 1}{\dfrac{M_{A1}^2}{r} - 1}\right)^2 - 1\right] \frac{B_{t1}^2}{\mu_0 \rho_{m1}} \tag{8.3.29}$$

simplifies to

$$\frac{B_{t2}^2}{\mu_0 \rho_{m2}} - \frac{B_{t1}^2}{\mu_0 \rho_{m1}} = \frac{(r-1)(M_{A1}^4 - r)}{(M_{A1}^2 - r)^2} \left(\frac{B_{t1}^2}{\mu_0 \rho_{m1}}\right). \tag{8.3.30}$$

Similarly, the quantity

$$\frac{1}{2}(U_{t2}^2 - U_{t1}^2)$$

$$= \frac{1}{2}(U_{t2}^2 - U_{t1}^2)\frac{M_{A1}^2 B_n^2}{U_{n1}^2 \mu_0 \rho_{m1}} = \frac{1}{2}(U_{n2}^2 B_{t2}^2 - U_{n1}^2 B_{t1}^2)\frac{M_{A1}^2}{U_{n1}^2 \mu_0 \rho_{m1}}$$

$$= \frac{1}{2}\left(\frac{B_{t2}^2}{r^2} - B_{t1}^2\right)\frac{M_{A1}^2}{\mu_0 \rho_{m1}} = \frac{1}{2}\left[\left(\frac{M_{A1}^2 - 1}{M_{A1}^2 - r}\right)^2 - 1\right]\frac{M_{A1}^2 B_{t1}^2}{\mu_0 \rho_{m1}} \tag{8.3.31}$$

simplifies to

$$\frac{1}{2}(U_{t2}^2 - U_{t1}^2) = \frac{(r-1)}{2}\left[\frac{2M_{A1}^2 - (1+r)}{(M_{A1}^2 - r)^2}\right]\frac{M_{A1}^2 B_{t1}^2}{\mu_0 \rho_{m1}}, \tag{8.3.32}$$

and the quantity

$$\frac{B_n}{\mu_0 \rho_{m1} U_{n1}}(B_{t2}U_{t2} - B_{t1}U_{t1}) = \frac{B_{t1}}{\mu_0 \rho_{m1} U_{t1}}(B_{t2}U_{t2} - B_{t1}U_{t1})$$

$$= \frac{B_{t1}}{\mu_0 \rho_{m1}}\left[B_{t2}\left(\frac{U_{t2}}{U_{t1}}\right) - B_{t1}\right]$$

$$= \left[\frac{1}{r}\left(\frac{M_{A1}^2 - 1}{\frac{M_{A1}^2}{r} - 1}\right)^2 - 1\right]\frac{B_{t1}^2}{\mu_0 \rho_{m1}} \tag{8.3.33}$$

simplifies to

$$\frac{B_n}{\mu_0 \rho_{m1} U_{n1}}(B_{t2}U_{t2} - B_{t1}U_{t1}) = \frac{(r-1)(M_{A1}^4 - r)}{(M_{A1}^2 - r)^2}\frac{B_{t1}^2}{\mu_0 \rho_{m1}}. \tag{8.3.34}$$

These equations are then used to eliminate the corresponding quantities from the jump conditions in which they appear. Substituting Eq. (8.3.28) into jump condition 2a gives

$$(\rho_{m2}U_{n2}^2 - \rho_{m1}U_{n1}^2) + \left(\rho_{m2}\frac{V_{S2}^2}{\gamma} - \rho_{m1}\frac{V_{S1}^2}{\gamma}\right) + \left(\frac{B_{t1}^2}{2\mu_0}\right)$$

$$\times \frac{M_{A1}^2(r-1)[(r+1)M_{A1}^2 - 2r]}{(M_{A1}^2 - r)^2} = 0, \tag{8.3.35}$$

which, after dividing by $B_n^2/2\mu_0$ and introducing M_A^2 and M_S^2 from Eqs. (8.3.16) and (8.3.17), simplifies to

$$M_{A2}^2 - M_{A1}^2 + \frac{1}{\gamma}\left(\frac{M_{A2}^2}{M_{S2}^2} - \frac{M_{A1}^2}{M_{S1}^2}\right) + \frac{M_{A1}^2(r-1)[(r+1)M_{A1}^2 - 2r]\tan^2\theta_1}{2(M_{A1}^2 - r)^2} = 0, \tag{8.3.36}$$

where $\tan^2\theta_1$ has been substituted for B_{t1}^2/B_n^2.

Next, dividing jump condition 3 on both sides by $\rho_{m1}U_{n1} = \rho_{m2}U_{n2}$, we obtain

$$\frac{1}{2}(U_{n2}^2 - U_{n1}^2) + \frac{1}{\gamma-1}(V_{S2}^2 - V_{S1}^2) + \frac{1}{2}(U_{t2}^2 - U_{t1}^2) + \left(\frac{B_{t2}^2}{\mu_0\rho_{m2}} - \frac{B_{t1}^2}{\mu_0\rho_{m1}}\right) - \frac{B_n}{\mu_0\rho_{m1}U_{n1}}(B_{t2}U_{t2} - B_{t1}U_{t1}) = 0, \tag{8.3.37}$$

which, after dividing by $B_n^2/\mu_0\rho_{m2}$ and introducing M_A^2 and M_S^2 from Eqs. (8.3.16) and (8.3.17), and substituting Eqs. (8.3.30), (8.3.32), and (8.3.34), simplifies to

$$\frac{1}{2}(M_{A2}^2 - rM_{A1}^2) + \frac{1}{\gamma-1}\left(\frac{M_{A2}^2}{M_{S2}^2} - \frac{M_{A1}^2}{M_{S1}^2}\right) - \left(\frac{r-1}{\gamma-1}\right)\frac{M_{A1}^2}{M_{S1}^2} + r\frac{(r-1)}{2}\frac{(2M_{A1}^2 - r - 1)}{(M_{A1}^2 - r)^2}M_{A1}^2\tan^2\theta_1 = 0, \tag{8.3.38}$$

where $\tan^2\theta_1$ has again been substituted for B_{t1}^2/B_n^2. Finally, the desired equation for M_{A1} is obtained by eliminating the term $[(M_{A2}^2/M_{S2}^2) - (M_{A1}^2/M_{S1}^2)]$ from Eqs. (8.3.36) and (8.3.38), and by eliminating the M_{A2}^2/M_{A1}^2 term from the resulting equation using

$$\frac{M_{A2}^2}{M_{A1}^2} = \frac{U_{n2}^2}{(B_n^2/\mu_0\rho_{m2})}\frac{(B_n^2/\mu_0\rho_{m1})}{U_{n1}^2} = \left(\frac{\rho_{m1}}{\rho_{m2}}\right)^2\left(\frac{\rho_{m2}}{\rho_{m1}}\right) = \frac{1}{r}. \tag{8.3.39}$$

The result, after some lengthy but straightforward algebra, is

$$(M_{A1}^2 - r)^2\left[M_{S1}^2 - \frac{2r}{r+1-\gamma(r-1)}\right] - r\tan^2\theta_1 M_{S1}^2\left[\frac{2r-\gamma(r-1)}{r+1-\gamma(r-1)}M_{A1}^2 - r\right] = 0. \tag{8.3.40}$$

At this point it is convenient to convert the upstream Mach numbers back to the upstream flow velocity using the relations $M_{A1} = U_{n1}/(V_{A1}\cos\theta_1)$ and $M_{S1} = U_{n1}/V_{S1}$. After making these substitutions and simplifying, the above equation becomes

$$(U_{n1}^2 - rV_{A1}^2\cos^2\theta_1)^2\left[U_{n1}^2 - \frac{2r\,V_{S1}^2}{r+1-\gamma(r-1)}\right]$$
$$- r\sin^2\theta_1 U_{n1}^2 V_{A1}^2\left[\frac{2r-\gamma(r-1)}{r+1-\gamma(r-1)}U_{n1}^2 - rV_{A1}^2\cos^2\theta_1\right] = 0. \qquad (8.3.41)$$

This equation is called the *shock adiabatic* (Anderson, 1963). It gives the propagation speed of the shock as a function of the shock strength and the upstream parameters. The shock adiabatic is closely analogous to the dispersion relation obtained from linear wave analysis, except the propagation speed now depends on the wave amplitude. As can be seen from the above equation, the shock adiabatic is a cubic equation in U_{n1}^2. Thus, three pairs of roots exist, corresponding to three pairs of oppositely propagating shocks. In order of increasing speeds, these shocks are called the slow, intermediate, and fast shocks. Since the general solution is a complicated function of the various upstream parameters, it is useful to consider certain limiting cases.

8.3.3 The Weak Shock Limit

To gain insight into the general form of the solutions of the shock adiabatic, we start by considering the weak shock limit, $r = 1$. This case is important because each root of the shock adiabatic can be traced back to a root at $r = 1$. Setting $r = 1$ in Eq. (8.3.41) gives

$$\left(U_{n1}^2 - V_{A1}^2\cos^2\theta_1\right)^2\left(U_{n1}^2 - V_{S1}^2\right) - \sin^2\theta_1 U_{n1}^2 V_{A1}^2\left(U_{n1}^2 - V_{A1}^2\cos^2\theta_1\right) = 0, \qquad (8.3.42)$$

which simplifies to

$$\left(U_{n1}^2 - V_{A1}^2\cos^2\theta_1\right)\left[U_{n1}^4 - U_{n1}^2\left(V_{A1}^2 + V_{S1}^2\right) + V_{S1}^2 V_{A1}^2\cos^2\theta_1\right] = 0. \qquad (8.3.43)$$

The above equation is seen to be identical to the dispersion relation, Eq. (6.5.16), for small-amplitude waves in an MHD fluid. The shock speed, U_{n1}, plays the same role as the phase speed, v_p, of an MHD wave. This correspondence is not surprising, since for a small-amplitude discontinuity the field quantities can be decomposed into an infinite series of plane waves all propagating at the same speed. Explicit solutions for the propagation speeds of the slow (s), intermediate

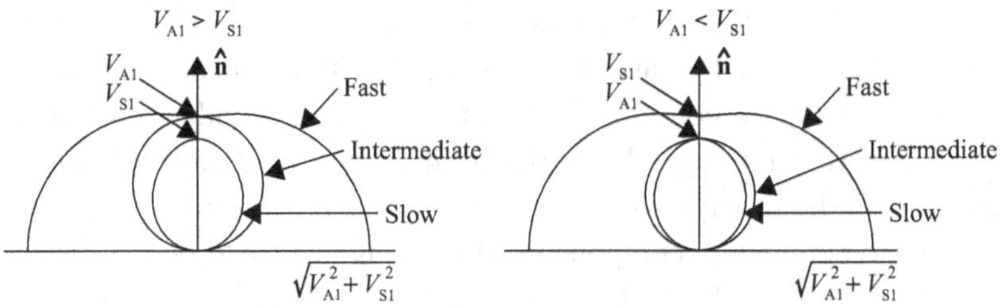

Figure 8.8 Polar plots of the propagation speeds of the slow, intermediate, and fast shocks in the weak shock limit for the two cases $V_{A1} > V_{S1}$ and $V_{A1} < V_{S1}$.

(i), and fast (f) shocks are given below for $r = 1$:

$$U_{n1}^2(s) = \frac{1}{2}\left(V_{A1}^2 + V_{S1}^2\right) - \frac{1}{2}\left[\left(V_{A1}^2 - V_{S1}^2\right)^2 + 4V_{A1}^2 V_{S1}^2 \sin^2\theta_1\right]^{1/2}, \quad (8.3.44)$$

$$U_{n1}^2(i) = V_{A1}^2 \cos^2\theta_1, \quad (8.3.45)$$

$$U_{n1}^2(f) = \frac{1}{2}\left(V_{A1}^2 + V_{S1}^2\right) + \frac{1}{2}\left[\left(V_{A1}^2 - V_{S1}^2\right)^2 + 4V_{A1}^2 V_{S1}^2 \sin^2\theta_1\right]^{1/2}. \quad (8.3.46)$$

Polar plots of the propagation speeds of these shocks are shown as a function of the angle θ between the magnetic field and the shock normal in Figure 8.8. These plots show the shape of the shock fronts that would be produced by a small impulsive disturbance at the origin. As with MHD waves, two cases must be considered: $V_{A1} > V_{S1}$ and $V_{A1} < V_{S1}$. The slow and fast shocks correspond to the slow and fast MHD waves, and the intermediate shock corresponds to the transverse Alfvén wave. The field geometries for the three basic types of small-amplitude shocks can be determined by referring to the eigenvector diagrams shown in Figures 6.8 and 6.10. In the small amplitude $r = 1$ limit, the intermediate shock degenerates to a rotational discontinuity.

8.3.4 Parallel and Perpendicular Shocks

From simple inspection of the shock adiabatic Eq. (8.3.41) it is evident that the equation is greatly simplified whenever either $\theta_1 = 0$ or $\theta_1 = \pi/2$. The case $\theta_1 = 0$ is called a parallel shock, and the case $\theta_1 = \pi/2$ is called a perpendicular shock. For a parallel shock ($\theta_1 = 0$), it can be shown that the shock adiabatic has two solutions:

$$U_{n1}^2 = \frac{2rV_{S1}^2}{r + 1 - \gamma(r-1)} \quad (8.3.47)$$

and

$$U_{n1}^2 = rV_{A1}^2. \quad (8.3.48)$$

In the limit $r \to 1$, the first root corresponds to the slow shock if $V_{A1} > V_{S1}$, and to the fast shock if $V_{A1} < V_{S1}$, and the second root corresponds to the intermediate shock. It is interesting to note that the first root has no dependence on the Alfvén speed, and therefore no dependence on the magnetic field strength. The reason that the magnetic field plays no role in this shock can be seen from Eq. (8.3.23), which relates the tangential components of the magnetic field on the two sides of the shock. For a parallel shock B_{t1} is zero. Provided that $M_{A1}^2 \neq r$, which is always the case for the first root, it follows that B_{t2} is zero. Thus, the magnetic field is parallel to the normal on both sides of the shock. Since the normal component of the magnetic field is continuous across the shock ($B_{n1} = B_{n2}$), it follows that the magnetic field is completely unaffected by the shock. The reason the magnetic field plays no role is that for a parallel shock the fluid motion is purely compressional, with no velocity component perpendicular to the magnetic field and hence no $\mathbf{U} \times \mathbf{B}$ force. This shock is indistinguishable from a shock in an ordinary gas, and forms because of the nonlinear wave steepening effect illustrated in Figure 8.1. Since γ is always greater than one, it is obvious from inspection of Eq. (8.3.47) that the speed of this shock increases as r increases, and goes to infinity as the denominator goes to zero at $r = r_m$, where

$$r_m = \frac{\gamma + 1}{\gamma - 1}. \tag{8.3.49}$$

No solution exists for $r > r_m$. The quantity r_m is the maximum attainable shock strength, and therefore represents an upper limit to the density compression ratio, $r = \rho_{m2}/\rho_{m1}$. For $\gamma = 5/3$ it can be shown that $r_m = 4$. To show the asymptotic dependence more explicitly as $r \to r_m$, it is useful to rewrite the $r + 1 - \gamma(r - 1)$ term in the denominator of Eq. (8.3.47) as $(\gamma - 1)(r_m - r)$. The shock propagation speed is then given by

$$U_{n1}^2 = \frac{2r V_{S1}^2}{(\gamma - 1)(r_m - r)}, \tag{8.3.50}$$

which explicitly shows that as $r \to r_m$ the speed varies as $1/(r_m - r)^{1/2}$.

In contrast to the first root, the magnetic field plays an essential role in the second root, $U_{n1}^2 = rV_{A1}^2$, which corresponds to the intermediate shock. The reason that the magnetic field plays a role in the propagation of this shock, even though B_{t1} is zero, is that for this root the denominator on the right-hand side of Eq. (8.3.23) is zero ($M_{A1}^2 = r$). This leads to an indeterminancy in B_{t2}, even though B_{t1} is zero. For this shock the magnetic field can be parallel to the normal on the upstream side of the shock, but still have a tangential component on the downstream side. Because of the abrupt appearance of a tangential component to the magnetic field, this shock is sometimes called a *switch-on shock.*

For a perpendicular shock ($\theta = \pi/2$), it can be shown that the shock adiabatic (Eq. (8.3.41)) has only one solution:

$$U_{\rm n1}^2 = \frac{2r[V_{\rm S1}^2 + (r - \gamma(r-1)/2)V_{\rm A1}^2]}{(\gamma - 1)(r_{\rm m} - r)}. \qquad (8.3.51)$$

In the weak shock limit ($r = 1$), this solution corresponds to the fast magnetosonic mode. Note that the propagation speed of this shock depends on both the sound speed and the Alfvén speed.

The reason for this mixed dependence is that the plasma and the magnetic field are both compressed as the plasma flows across the shock. Since there is no field line stretching for a perpendicular shock ($\ell_1 = \ell_2$), the frozen field theorem shows that the magnetic field ratio varies in direct proportion to the density compression ratio, $B_2/B_1 = \rho_{\rm m2}/\rho_{\rm m1}$. This same conclusion can be obtained from Eq. (8.3.23) (note that for $\theta_1 = \pi/2$, $M_{\rm A1} = U_{\rm n1}/V_{\rm A1}\cos\theta_1 \to \infty$). Since both the plasma pressure and the magnetic field pressure increase across the shock, it is easy to see why the shock propagation speed depends on both the sound speed and the Alfvén speed. Note also that the Alfvén speed increases as the magnetic field strength increases. This amplitude dependence causes the magnetic field to steepen in a way that is analogous to the steepening of a pressure pulse in an ordinary gas.

From inspection of Eq. (8.3.51), one can see that the propagation speed of a perpendicular shock goes to infinity as the denominator goes to zero, just as is the case in a parallel shock. A perpendicular shock has exactly the same upper limit to the shock strength, $r_{\rm m}$, and asymptotic dependence, $1/(r_{\rm m} - r)^{1/2}$, as a parallel shock. Shortly, we will show that this upper limit applies for any orientation of the upstream magnetic field.

8.3.5 The Strong Shock Limit

Since both parallel and perpendicular shocks have an upper limit to the shock strength, it is useful to explore the general solution to the shock adiabatic as r approaches $r_{\rm m}$. For this purpose, we rewrite the $r + 1 - \gamma(r-1)$ term as $(\gamma - 1)(r_{\rm m} - r)$, and then evaluate the limiting form as $r \to r_{\rm m}$. Expanding the shock adiabatic equation (8.3.41) in powers of $U_{\rm n1}$ gives the following equation:

$$U_{\rm n1}^6 - U_{\rm n1}^4\left[\frac{2rV_{\rm S1}^2 + 2r\sin^2\theta_1 V_{\rm A1}^2(r - \gamma(r-1)/2)}{(\gamma-1)(r_{\rm m} - r)} + 2rV_{\rm A1}^2\cos^2\theta_1\right]$$

$$+ U_{\rm n1}^2 r^2 V_{\rm A1}^2\cos^2\theta_1\left[V_{\rm A1}^2 + \frac{4V_{\rm S1}^2}{(\gamma-1)(r_{\rm m}-r)}\right] - \frac{2r^3V_{\rm A1}^4V_{\rm S1}^2\cos^4\theta_1}{(\gamma-1)(r_{\rm m}-r)} = 0. \qquad (8.3.52)$$

As $r \to r_{\rm m}$, it is clear that one of the roots for $U_{\rm n1}^2$ must approach infinity as $1/(r_{\rm m} - r)$. This root is controlled by the $U_{\rm n1}^6$ and $U_{\rm n1}^4$ terms in the above equation and is given by

$$U_{\rm n1}^2 = \frac{2r[V_{\rm S1}^2 + \sin^2\theta_1 V_{\rm A1}^2(r - \gamma(r-1)/2)]}{(\gamma - 1)(r_{\rm m} - r)}. \tag{8.3.53}$$

It is interesting to note that even though the above equation is rigorously valid only for a strong shock, it still reduces exactly to Eq. (8.3.50) for a parallel shock and exactly to Eq. (8.3.51) for a perpendicular shock. Note that the shock speed goes to infinity as $1/(r_{\rm m} - r)^{1/2}$ for all orientations of the upstream magnetic field, $0 \le \theta \le \pi/2$.

The remaining two roots for $U_{\rm n1}^2$ can be investigated by multiplying Eq. (8.3.52) by $(r_{\rm m} - r)$ and taking the limit as r approaches $r_{\rm m}$, which gives the equation

$$U_{\rm n1}^4[2V_{\rm S1}^2 + \sin^2\theta_1 V_{\rm A1}^2(r_m - 1)] - U_{\rm n1}^2 4r_{\rm m}\cos^2\theta_1\, V_{\rm A1}^2 V_{\rm S1}^2$$
$$+ 2r_{\rm m}^2\, V_{\rm A1}^4 V_{\rm S1}^2 \cos^4\theta_1 = 0. \tag{8.3.54}$$

The above equation is a quadratic in $U_{\rm n1}^2$ and in the limit $r = r_{\rm m}$ has a discriminant

$$-8r_{\rm m}^2(r_{\rm m} - 1)\cos^4\theta_1 \sin^2\theta_1\, V_{\rm A1}^6 V_{\rm S1}^2. \tag{8.3.55}$$

Since $r_{\rm m}$ is always greater than one, the discriminant is always negative. Thus, the remaining two roots, which correspond to the slow and intermediate shocks, merge and become complex (i.e., non-propagating) at some intermediate shock strength, $r_{\rm c}$, that is less than $r_{\rm m}$. It follows then that $r_{\rm m}$ represents an absolute upper limit to the shock strength for all three types of shocks.

8.3.6 Arbitrary Shock Strengths and Magnetic Field Orientations

From the above discussion, we can arrive at the following qualitative understanding of the solutions of the shock adiabatic for arbitrary shock strengths and magnetic field orientations:

(1) The shock adiabatic has three roots, called the slow, intermediate, and fast shocks.
(2) In the weak shock limit, $r \simeq 1$, the propagation speeds of the three shocks correspond to the propagation speeds of the slow magnetosonic wave, $U_{\rm n}({\rm s})$, the transverse Alfvén wave, $U_{\rm n}({\rm i})$, and the fast magnetosonic wave, $U_{\rm n}({\rm f})$.
(3) As the shock strength increases, the speeds of the slow and intermediate shocks merge at an intermediate value of the shock strength, $r_{\rm c}$. For any larger value

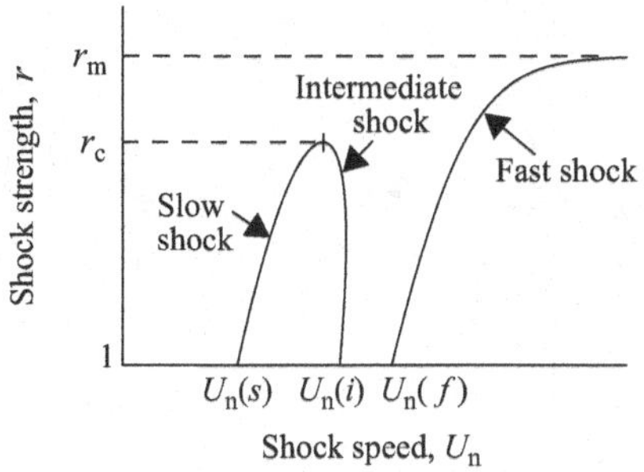

Figure 8.9 The shock propagation velocity, U_n, as a function of the shock strength, r, for a constant angle, θ, between the upstream magnetic field direction and the shock normal.

of r, these shocks no longer exist. Thus, there is an upper limit, r_c, to the strengths of the slow and intermediate shocks.

(4) For shock strengths greater than r_c, only one solution exists (the fast shock), and the speed of this shock asymptotically approaches infinity as $1/(r_m - r)^{1/2}$, where r_m is given by Eq. (8.3.49). No solutions exist for shock strengths greater than r_m.

From these results we can conclude that a plot of the shock speed, U_n, as a function of the shock strength, r, must have the form shown in Figure 8.9. The exact shape of the plot depends on the specific choices for the upstream parameters V_{A1}, V_{S1}, and θ_1, and can only be determined by carrying out a full numerical solution of Eq. (8.3.41). The plot in Figure 8.9 was made using $V_{A1} = 4, V_{S1} = 1$, and $\theta_1 = 60°$.

An important distinction can be made between fast and slow shocks. From Eq. (8.3.23) it can be shown that $B_{t2} > B_{t1}$ for fast shocks and $B_{t2} < B_{t1}$ for slow shocks. Thus, the magnetic field is rotated away from the normal for a fast shock, $\theta_2 > \theta_1$, and toward the normal, $\theta_2 < \theta_1$, for a slow shock. The corresponding geometries are illustrated in Figure 8.10. The change in the direction of the flow velocity at the shock is caused by the $\mathbf{J} \times \mathbf{B}$ force acting on the fluid at the discontinuity. The current in the discontinuity surface flows upward (out of the paper) in Figure 8.10 for a fast shock, and downward (into the paper) for a slow shock.

8.3.7 Entropy and Reversibility

Inspection of Eq. (8.3.41) clearly shows that for each of the three roots of the shock adiabatic there are two solutions, U_{n1} and $-U_{n1}$. This reversibility arises because the Rankine–Hugoniot equations (8.3.8)–(8.3.14) are time reversible. Since our convention is always to label the upstream side 1 and the downstream side 2, it follows that, upon reversing the direction of propagation, the shock strength

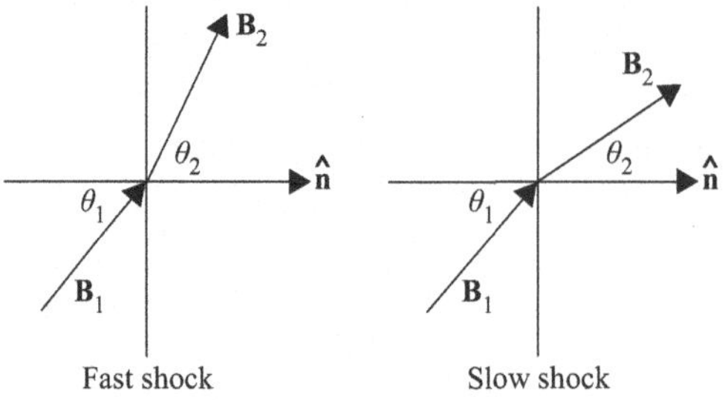

Figure 8.10 Fast and slow shocks can be distinguished by the deflection of the downstream magnetic field, which is away from the normal for a fast shock, and toward the normal for a slow shock.

changes from r to $1/r$. It can be shown that the shock adiabatic only has solutions for r values in the range $(1/r_{\mathrm{m}}) < r < r_{\mathrm{m}}$. This includes values less than one. A value of r less than one implies that the density downstream of the shock is less than the density upstream of the shock. However, such non-compressive shocks never occur in nature. The reason for this is that such shocks violate the second law of thermodynamics. To understand the thermodynamic limitations placed on solutions of the Rankine–Hugoniot equations, we must return to a discussion of the compression that occurs as the plasma flows across the shock. Since the kinetic energy decreases as the fluid crosses the shock, it is clear that some of the flow energy is converted to some other form during the compression process, either internal energy (i.e., heat) or magnetic energy. Although the compression in the shock transition region is abrupt, so that there is little time for heat to flow, the adiabatic equation of state (6.1.25) cannot be used. The reason for this is that the fluid is not in a *succession of equilibrium states*, which is a basic assumption in the derivation of the adiabatic equation of state. Such abrupt non-equilibrium processes are always accompanied by dissipation of energy and are irreversible (Fermi, 1956). In a shock the energy dissipation occurs during the sudden compression, which irreversibly converts some of the incoming flow energy into heat. Thus, instead of being determined by the adiabatic equation of state, the downstream parameters are determined by the Rankine–Hugoniot equations. That the adiabatic equation of state is not used may seem surprising, since the energy equation (6.4.14) was derived using the adiabatic equation of state. However, the use of the adiabatic equation of state in the derivation of the energy equation was just a convenient way of deriving the internal energy equation (6.4.15) and enthalpy equation (6.4.16). These quantities are state variables and are independent of the process used to go from the initial state to the final state.

To show that the compression that takes place at a shock is an irreversible process, we next compute the change in the entropy as the fluid flows across the shock. The second law of thermodynamics states that the entropy must either increase or be constant and, if it increases, the process is irreversible. Following the usual procedure used in thermodynamics, the change in the entropy, dS, is computed by integrating dS = dQ/T for some reversible process between the initial and final states. For an ideal gas it can be shown that the change in the specific entropy, $\Delta s = s_2 - s_1$, for an abrupt adiabatic (i.e., no heat loss) compression from some initial pressure and density $(P_1, \rho_{\mathrm{m1}})$ to a final pressure and density $(P_2, \rho_{\mathrm{m2}})$ is given by

$$\Delta s = c_{\mathrm{V}} \ln\left[\frac{P_2}{P_1}\left(\frac{\rho_{\mathrm{m1}}}{\rho_{\mathrm{m2}}}\right)^{\gamma}\right], \tag{8.3.56}$$

where c_{V} is the specific heat capacity per unit volume, and γ is the ratio of the heat capacity at constant pressure to the heat capacity at constant volume (Fetter and Walecka, 1980). To evaluate the entropy change using the above equation, the first step is to compute the pressure ratio, P_2/P_1, across the shock as a function of the shock strength, $r = \rho_{\mathrm{m2}}/\rho_{\mathrm{m1}}$. The pressure ratio is most easily obtained from Eq. (8.3.36), which, after substituting for M_{A} and M_{S} using Eqs. (8.3.16) and (8.3.17), gives

$$\frac{P_2}{P_1} = 1 + \gamma\frac{U_{\mathrm{n1}}^2}{V_{\mathrm{S1}}^2}\frac{(r-1)}{r}\left[1 - \frac{rV_{\mathrm{A1}}^2[(r+1)U_{\mathrm{n1}}^2 - 2rV_{\mathrm{A1}}^2\cos^2\theta_1]}{2(U_{\mathrm{n1}}^2 - rV_{\mathrm{A1}}^2\cos^2\theta_1)^2}\sin^2\theta_1\right]. \tag{8.3.57}$$

To compute the pressure ratio we must solve the shock adiabatic equation (8.3.41) for the shock speed U_{n1} and then substitute the shock speed into the above equation. Once the pressure ratio is known, the entropy change can be computed from Eq. (8.3.56), noting that $\rho_{\mathrm{m2}}/\rho_{\mathrm{m1}} = r$. Usually this computation must be done numerically. However, in certain cases it is possible to carry out the calculation analytically. For example, in the strong shock limit, where the shock speed varies as $U_{\mathrm{n1}} \propto 1/(r_{\mathrm{m}} - r)^{1/2}$, the second term in the brackets becomes negligibly small as $r \to r_{\mathrm{m}}$. The pressure ratio is then determined by the term in front of the brackets, which goes to plus infinity as $1/(r_{\mathrm{m}} - r)$. Thus, for a strong compressible shock the entropy change is clearly positive, which shows that the shock can exist, and that the process is irreversible. In the weak shock limit, it is evident from Eq. (8.3.57) that the pressure ratio goes to one as $r \to 1$. The entropy change then goes to zero. Thus, in the weak shock limit the shock propagation is reversible. This is to be expected, since as was shown earlier, small-amplitude MHD waves propagate with essentially no energy dissipation. That the entropy is positive for all values of r greater than one, and negative for all values of r less than one, is difficult to prove in general but is easy to demonstrate for specific cases. For example, for a parallel shock, $\theta_1 = 0$, the shock speed given by Eq. (8.3.50) can be substituted directly into Eq. (8.3.57) to give the following equation for the pressure ratio as a

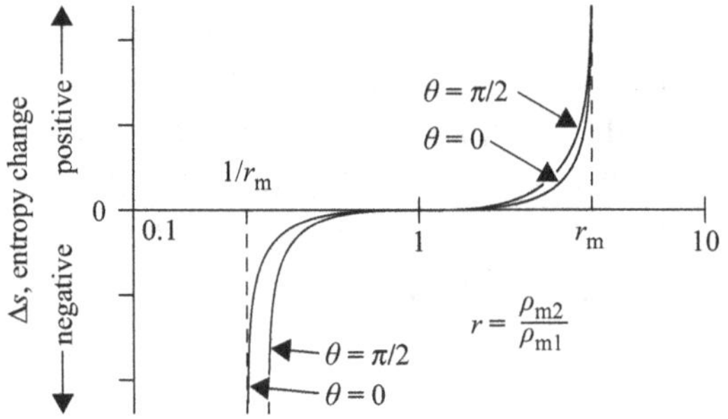

Figure 8.11 A plot of the entropy change, Δs, for a fast shock as a function of the shock strength, $r = \rho_{m2}/\rho_{m1}$, for a specific choice of upstream parameters ($V_{A1} = 40\,\text{km s}^{-1}$, $V_{S1} = 30\,\text{km s}^{-1}$, $\theta = 0$, and $\gamma = 5/3$).

function of the shock strength:

$$\frac{P_2}{P_1} = \frac{r r_m - 1}{r_m - r}, \tag{8.3.58}$$

where we have eliminated γ using $\gamma = (r_m + 1)/(r_m - 1)$, which is derivable from Eq. (8.3.49). A plot of the corresponding entropy change, Δs, computed from the above equation for $\theta_1 = 0$, is shown in Figure 8.11 as a function of the shock strength, r. Also shown is the entropy change computed from Eqs. (8.3.51), (8.3.56), and (8.3.57) for $\theta_1 = \pi/2$. As can be seen in both cases the entropy change goes to zero at $r = 1$, and is positive for $r > 1$ and negative for $r < 1$. This means that these shocks can only occur if the shock strength, $r = \rho_{m2}/\rho_{m1}$, is greater than one (i.e., only if the fluid is compressed).

8.3.8 Nonlinear Evolution and Stability

The Rankine–Hugoniot equations implicitly assume that the flow through the shock is in steady state, i.e., all of the time-dependent terms in the MHD equations are zero. However, one must ask whether such a steady-state solution can actually evolve from some arbitrary initial condition. As discussed in the beginning of this chapter, for a discontinuity to exist, the initial disturbance must undergo the nonlinear process known as wave steepening. If an arbitrary initial disturbance does not undergo wave steepening then a shock cannot form, even though a formal solution to the jump conditions may exist. The question of whether an arbitrary initial disturbance evolves into a shock is difficult to answer in general, since it involves the full nonlinear solution of the time-dependent MHD equations. However, on a qualitative level, both the slow and fast magnetosonic modes have the proper nonlinear characteristics to lead to wave steepening. That this is the case can be seen from a simple inspection of Eqs. (6.5.5), (6.5.14), (6.5.17), and (6.5.19), which shows that the phase velocities of both the slow and fast magnetosonic modes increase as the pressure and/or the magnetic field strength increase (i.e., the larger amplitude part of the disturbance tends to overtake the smaller amplitude part of the disturbance). Thus, slow and fast shocks arise from steepening of the slow and fast magnetosonic modes, a conclusion that can be verified by numerically solving the time-dependent MHD equations. On the other hand, the transverse Alfvén mode, which is linked to the intermediate shock, does not tend to steepen, since the phase velocity, which is determined by the zero-order background magnetic field, is independent of the magnetic field perturbation, which is perpendicular to the zero-order magnetic field (see Figure 6.8). As mentioned earlier, for small amplitudes and $\theta = 0$, the intermediate shock degenerates to a rotational discontinuity. Thus, based on these simple considerations, there are good reasons to believe that a transverse Alfvén wave cannot evolve into an intermediate shock.

An alternative approach to answering the question of whether a shock can exist is to test the stability of the shock by introducing a small-amplitude perturbation in the shock boundary, and then carrying out a time-dependent linear analysis to determine whether the resulting perturbation grows in amplitude. If the perturbation grows, the steady-state solution is unstable and cannot occur in nature, since if a shock somehow formed, it would quickly break up into some other type of non-steady flow. This type of stability analysis is difficult, and will not be pursued here. However, the issues involved have been studied in considerable detail by several authors (Kennel et al., 1989; Kantrowitz and Petschek, 1966; Anderson, 1963), and the conclusions are that the slow and fast shocks are stable, but the intermediate shock is not. Thus, of the three possible solutions of the shock adiabatic, only the fast and slow shocks are believed to actually occur in nature.

8.3.9 Observations of MHD Shocks

MHD shocks have been observed in a large variety of geophysical, laboratory, and astrophysical plasmas. Shocks are known to be produced by supernova explosions, by the outflow of hot gases from certain types of stars, by solar flares, and in the solar wind upstream of planetary magnetospheres. Of these, the shocks that form in the solar wind upstream of planetary magnetospheres have received considerable study and are worthy of some brief comments. This type of shock is often called the bow shock, in analogy with the bow wave of a ship. An example of a bow shock observed upstream of Earth's magnetosphere is shown in Figure 8.12. In this case the angle between the shock normal and the magnetic field, $\theta_1 = 76°$, is quite large, and the upstream flow speed, $U_1 = 294\ \mathrm{km\,s^{-1}}$, is much greater than either the Alfvén speed or the sound speed, which are $V_{A1} = 37.8\ \mathrm{km\,s^{-1}}$ and $V_{S1} =$

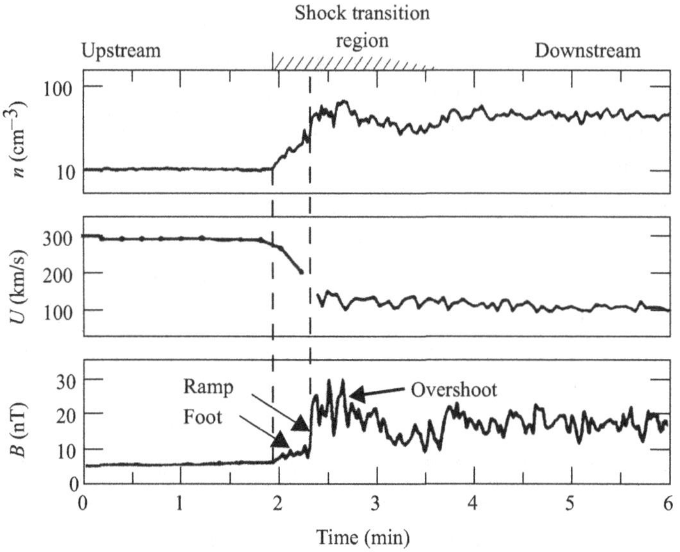

Figure 8.12 An example of a bow shock crossing observed at a geocentric radial distance of 15.4 R_E upstream of Earth (from Scudder et al., 1986).

58.6 km s^{-1}. Therefore, this is a strong, nearly perpendicular fast shock. Because the solar wind velocity, typically about 400 km s^{-1}, is usually much greater than either the Alfvén speed or the sound speed, planetary bow shocks are almost always strong fast shocks. The abrupt increases in the number density, n, and the magnetic field strength, B, and the decrease in the flow velocity, U, at the shock are clearly evident. Also, note that the ratio of the upstream number density to the downstream number density (i.e., $n_2/n_1 = \rho_{m2}/\rho_{m1} = r$) is almost exactly the theoretical limit for a strong shock, i.e., $r_m = 4$.

As one can see from Figure 8.12, the shock transition region has considerable structure extending over a period of nearly two minutes. This time span corresponds to a thickness of several hundred kilometers (Scudder et al., 1986). When standing bow shocks were first discovered in the solar wind, one of the main questions was what controlled the thickness of the shock, which is much smaller than the mean-free path. Typical mean-free paths in the solar wind are a substantial fraction of an astronomical unit (i.e., ~ 10^8 km). It is now known that various characteristic length-scales of the plasma, such as the electron and ion cyclotron radii, play a crucial role in determining the thickness and structure of the shock. Since the cyclotron radius of the ions is much larger than the cyclotron radius of the electrons, when the incoming plasma encounters the main jump in the magnetic field, labeled "ramp" in Figure 8.12, the electrons and ions respond quite differently. Because of their small cyclotron radius relative to the field gradients, the electrons are essentially carried along with the fluid flow (i.e., at the $\mathbf{E}\times\mathbf{B}$ drift velocity) while conserving the first adiabatic invariant. Because of their much larger cyclotron radius, the ions are strongly deflected by the $\mathbf{v}\times\mathbf{B}$ force and move on complicated non-adiabatic trajectories that are quite different from that of the electrons. In some cases the ions can even be reflected back into the upstream region, where they are accelerated by the solar wind electric field and carried back into the shock again by the $\mathbf{E}\times\mathbf{B}$ drift; see Section 8.4. Such reflected ions are responsible for the region labeled "foot" that lies ahead of the main ramp in the magnetic field; see Figure 8.12. The resulting differential motion between the ions and the electrons leads to a charge separation and currents that are self-consistently responsible for the electric and magnetic fields in the shock transition region. As the plasma flows through this system of quasi-static electric and magnetic fields, the distribution functions are drastically altered, especially for the ions, resulting in a hot, highly non-thermal plasma made up of ring-like and shell-like particle distributions. These non-thermal particle distributions are unstable and rapidly lead to the generation of plasma wave turbulence via various plasma instabilities, some of which are discussed in the next two chapters. This turbulent region, sometimes called the "foreshock," then acts to scatter the particles and drive the distribution functions toward thermal equilibrium, effectively playing the role of collisions. The net result is the conversion of the directed kinetic energy of the incoming plasma into random thermal energy, which is very similar to what occurs in a collision dominated shock.

To illustrate the types of plasma waves that occur in a collisionless shock, an electric field spectrogram obtained at Jupiter's bow shock is shown in Figure 8.13. The narrow-band emissions at about 6 kHz are electron plasma oscillations excited by a beam of electrons that escapes into the region upstream of the shock. The intense broadband electric field noise starting at about 30 seconds and extending into the downstream region is plasma wave turbulence excited by unstable particle distributions in the shock transition region. It is this turbulence that plays a key role in producing the irreversible dissipation that converts the incoming flow energy into random thermal energy.

The thickness of the shock and the detailed substructure within the shock depend on many factors, the most important being the angle θ_1 between the upstream magnetic field and the shock normal, and the two upstream Mach numbers, M_{A1} and M_{S1}. The transition region of a quasi-perpendicular shock is usually thin and well defined. The transition region of a quasi-parallel shock is usually more complicated, and often appears thick and poorly defined in the magnetic field data, but less so in the plasma data (see Scudder et al., 1984). The physical parameters that control the thickness are mainly the cyclotron radius of the ions and a characteristic length called the electron inertial length, which is the spatial

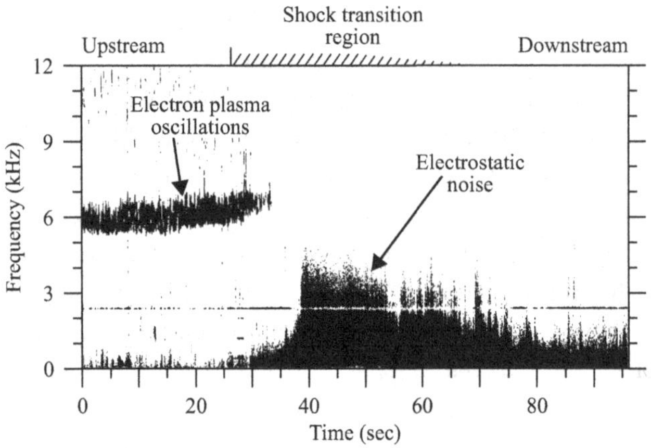

Figure 8.13 An electric field spectrogram showing the plasma waves observed by the *Voyager 1* spacecraft during a crossing of Jupiter's bow shock (Gurnett and Scarf, 1983).

scale over which the electrons can respond to the electrostatic fields in the shock. For a more detailed discussion of the structure of the shock transition layer, see Kivelson and Russell (1995), Baumjohann and Treumann (1997), and Balogh and Treumann (2013).

8.4 Charged Particle Acceleration by MHD Shocks

As spacecraft ventured beyond the protective shell of Earth's atmosphere in the late 1950s, it was soon discovered that intense bursts of high-energy particles, both electrons and ions, were emitted from explosive events on the Sun called "solar flares." The energy of the particles produced by these solar outbursts was found to extend over a very wide range, from a few keV to many tens of MeV. Such events often lasted several days and had such high intensities, especially for the ions, that they posed a potential health risk for astronauts. Subsequent investigations showed that solar flares were associated with dense clouds of plasma ejected outward from the Sun at speeds often exceeding 1000 km/s. These ejections are now called "coronal mass ejections," and are usually preceded by a shock wave that accelerates charged particles to very high energies.

An example of an interplanetary shock and the associated energetic particles is shown in Figure 8.14. These measurements were made by the *Voyager 1* spacecraft at a radial distance of 1.63 AU from the Sun a few days after a solar flare. The top panel shows the solar wind velocity and the bottom panel shows the flux of energetic ions (mostly protons) in the energy range from 53 to 58 keV. The interplanetary shock can be clearly identified by the abrupt jump in the solar wind velocity late in day 337, 1977, as indicated by the vertical dashed line. It is evident

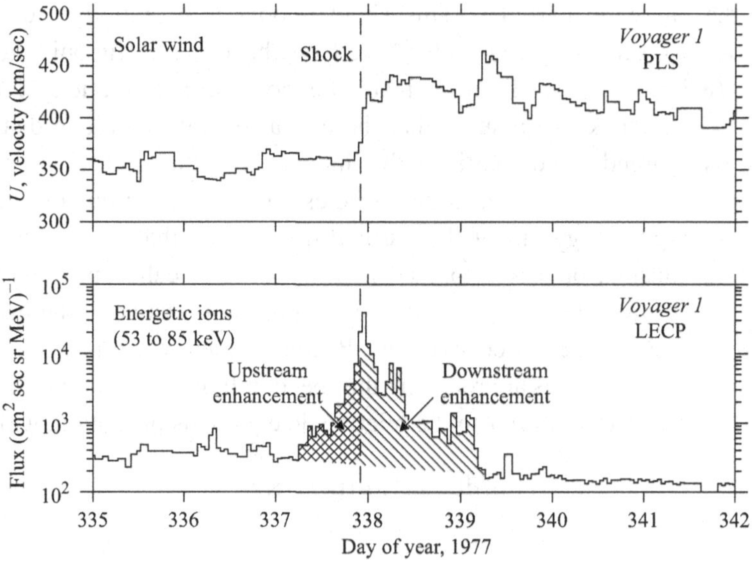

Figure 8.14 The top panel shows the solar wind velocity, U, as measured by the *Voyager 1* plasma instrument (PLS) on the *Voyager 1* spacecraft. The bottom panel shows the intensity of ions (mainly protons) with energies from 53 to 85 keV as measured by the *Voyager 1* low-energy charged particle instrument (adapted from Decker et al., 1981).

that the shock coincides almost exactly with a sharp peak in the energetic ion fluxes, with the enhanced intensities starting several days before the arrival of the shock, and ending a little more than one day later. Although only a single energy channel is shown, the enhanced ion intensities extend over a broad range of energies, from less than 30 keV, the lowest energy measured, to well above 4 MeV, the highest energy measured. The fluxes decrease rapidly with increasing energy, roughly following a power-law energy spectrum, with a spectral index of roughly −2.5 to −3.0. From this and many other similar examples it is clear that interplanetary shocks are very effective at accelerating ions to very high energies, to several MeV or more. As we discuss below, there are two basic mechanisms by which shocks can accelerate charged particles to high energies.

8.4.1 Shock Drift Acceleration

To illustrate one of the mechanisms by which collisionless shocks can accelerate charged particles, we adopt a somewhat simplified model for the shock and for the motion of energetic particles in the vicinity of the shock. We assume that the shock is a planar discontinuity in an otherwise homogeneous MHD plasma and that the thickness of the discontinuity is small compared to the cyclotron radii of the energetic particles. This shock model is essentially the same as that discussed earlier in this chapter based on the Rankin–Hugoniot relations, i.e., Eqs. (8.3.8)–(8.3.14). To investigate the possible acceleration of charged particles by the shock, we assume that there is a separate energetic distribution of positively charged "seed" particles that interact with the electric and magnetic fields of the shock. In practice, these particles could have various origins. They could be the high-energy tail of the thermal distribution that makes up the bulk of the MHD plasma, or newly ionized neutral particles called pickup ions that have gained substantial energies via their initial motion relative to the plasma, or pre-existing particles such as cosmic rays. Because in many applications particles are accelerated to velocities approaching the speed of light, to maintain a high level of generality it is necessary to use the relativistic equations of motion, which are

$$\mathbf{dp}/\mathbf{dt} = q[\mathbf{E} + (\mathbf{p}/m) \times \mathbf{B}] \tag{8.4.1}$$

and

$$\mathbf{dr}/\mathbf{dt} = \mathbf{v} = \mathbf{p}/m, \tag{8.4.2}$$

where $\mathbf{p}$ is the momentum, $m = \gamma m_0$ is the relativistic mass, $\gamma = [1 + (p/m_0 c^2)]^{1/2}$ is the Lorentz factor, and m_0 is the rest mass. In this system the total energy of the particle is given by $T = \gamma m_0 c^2$, and the kinetic energy is given by $w = (\gamma - 1) m_0 c^2$. We also assume that the number of seed particles is so small that their currents and charges do not significantly alter the solution to the MHD jump conditions. Of course, for a sufficiently energetic shock, this simplifying assumption may not be true, in which case the interaction of the seed particles must be incorporated into the overall model of the shock. For a discussion of the physics of such high-energy shocks, see Zank et al. (1993).

Although some analytic simplifications can be achieved by taking advantage of cyclic coordinates and the constancy of their associated conjugate momenta (Whipple et al., 1986), a complete beginning-to-end analytic analysis of their motion is usually not practical. The simplest approach is to use a computer to integrate the particle trajectories using Eqs. (8.4.1) and (8.4.2). Since by assumption the electric and magnetic fields are constant on each side of the shock, the particle motions are quite simple, consisting of full or partial cyclotron orbits plus $\mathbf{E} \times \mathbf{B}$ drifts. For simplicity we assume there is no cross-shock electrostatic potential, although such potentials can be included in more advanced models. Two approaches can be used to compute the particle trajectories: (1) the analysis can be done in the shock frame of reference, in which case the $\mathbf{E} \times \mathbf{B}$ drift must be taken into account in the integration; or (2) the velocity vectors and positions can be transformed to the *de Hoffmann–Teller* frame, in which case the motion is a simple cyclotron motion with no $\mathbf{E} \times \mathbf{B}$ drift. Of course, after the computation in the *de Hoffman–Teller* frame the velocity vectors and positions must be transformed back to the shock frame of reference. In either case, if the particle crosses the shock then the electric and magnetic fields must be changed according to the solutions of the *Rankin–Hugoniot* equations on the each side of the shock. In practice, these computations can be easily done for large numbers of seed particles and the results can then be used to study the average energy gain and angular distribution of the accelerated particles.

In general, when a particle interacts with a shock it is either transmitted through the shock or reflected from the shock. Representative (x,y) particle trajectories that illustrate these two cases are shown in Figure 8.15. In both cases an identical quasi-perpendicular shock is assumed to be present in the $x = 0$ plane with the plasma flowing from left to right with a normal component of velocity U_{n1} on the upstream side of the shock. The upstream magnetic field $\mathbf{B}_1$ is assumed to be oriented in the (x,z) plane at an angle $\theta_{Bn1} = 85°$ with respect to the x axis and the shock is assumed to have a compression ratio of $r = \rho_{m2}/\rho_{m1} = 3$. In both cases the seed particle is assumed to have a positive charge with an injection speed corresponding to a 1 MeV proton in the rest frame of the upstream plasma. In panel (a) the initial pitch angle of the particle was taken to be $\alpha_1 = 38.8°$. As

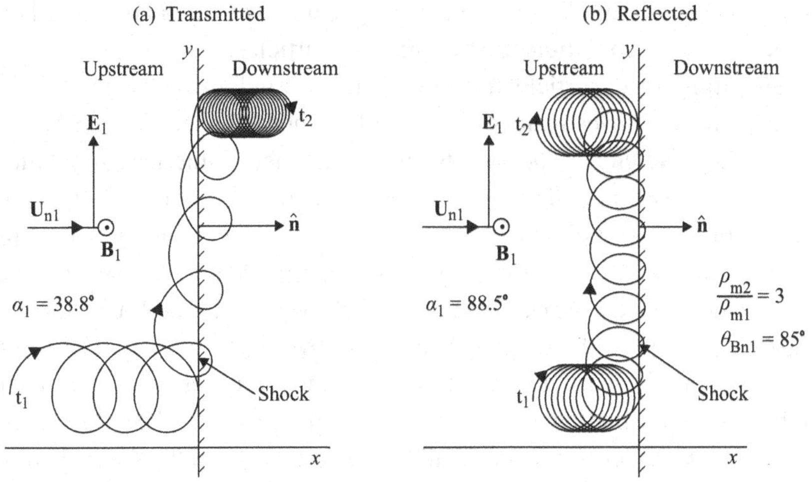

Figure 8.15 The x,y trajectories of two energetic, positively charged particles injected at different pitch angles relative to the magnetic field upstream of identical shocks. In panel (a) a particle with a pitch angle of $\alpha_1 = 38.8°$ is transmitted through the shock. In panel (b) a particle with a larger pitch angle of $\alpha_1 = 88.5°$ is reflected by the shock. Note that in both cases the particles gain energy by drifting in the direction of the motional electric field, $\mathbf{E} = -\mathbf{U} \times \mathbf{B}$, as they gyrate repeatedly through the shock (computer simulation provided by R. Decker).

can be seen, after injection the particle is carried into the shock by the upstream $\mathbf{E} \times \mathbf{B}$ drift. At the shock it then undergoes a series of cycloid-like gyrations through the shock and is eventually carried downstream by the downstream $\mathbf{E} \times \mathbf{B}$ drift. If the initial pitch angle is increased to $\alpha_1 = 88.8°$, after a somewhat similar series of gyrations through the shock, the particle is reflected back into the upstream plasma, as shown in panel (b). Although not immediately apparent in the (x, y) plot, the reflection is caused by the reversal of the component of the particle velocity along the magnetic field, as in a magnetic mirror field. This reversal causes the guiding center of the particle to move back along the magnetic field in the opposite direction from which it arrived. Note that this reflection process depends on the fact that the magnetic field has an oblique angle with respect to the shock front, i.e., a purely perpendicular shock cannot reflect particles back upstream (although it can if there is a cross-shock electrostatic potential).

It is apparent from inspection of Figure 8.15 that in both cases the interaction with the shock causes the particle to "drift" in the direction of the $\mathbf{E} = \mathbf{U} \times \mathbf{B}$ electric field, thereby gaining energy due to the work done by the electric field force. This type of particle acceleration is called "shock drift acceleration." It is easily verified that the energy gain, $w = q\mathbf{E}\Delta y$, is also positive for a negatively charged particle. Because the cyclotron radius of an ion is much greater than that of an electron, ions gain much more energy than electrons. The details of exactly how much energy is gained depend in a complicated way on the parameters of the shock and the initial conditions for the injected particles.

By computing seed particle trajectories for a random distribution of initial conditions, various investigators have studied how the energy gain depends on the shock parameters. The results from one such numerical simulation, by Decker et al. (1988), are shown in Figure 8.16. In this simulation, an isotropic distribution of seed particles with constant speeds and random cyclotron phases was introduced ahead of a strong $r = 4$ quasi-perpendicular shock, and the fractional energy gain, $\Delta w/w$, for each particle was calculated as a function of the angle θ_{Bn1} between the magnetic field and the shock normal. The top panel shows the results for particles reflected from the shock, and the bottom panel shows the results for particles transmitted through the shock. The black dots give the average fractional energy gain for all of the particles, and the vertical error bars give the standard deviation of the resulting distribution. The smooth solid line is the fractional energy gain computed assuming conservation of the first adiabatic invariant; see Section 3.8.2. As can be seen, this line closely follows the black dots. Note from the vertical dashed line in Figure 8.16 that there is a maximum magnetic field angle, $\theta_{Bn1}(\text{max})$, above which particles are not reflected by the shock. For $\theta_{Bn1} > \theta_{Bn1}(\text{max})$ it can be shown that in the *de Hoffman–Teller* frame all of the particles are in the loss cone, and are therefore transmitted through the shock. Also,

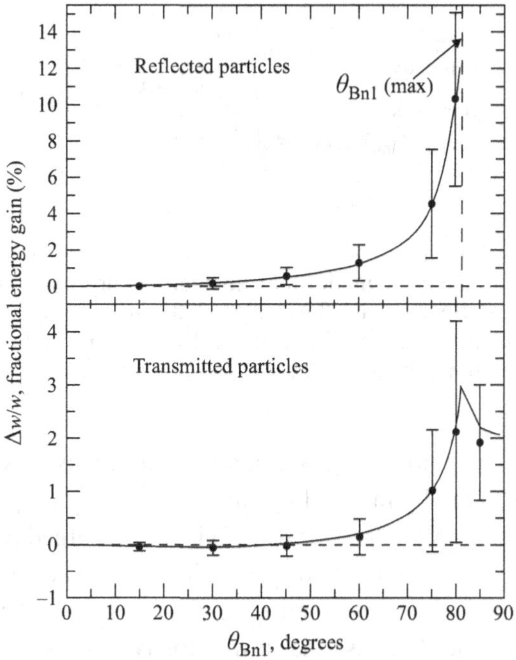

Figure 8.16 The fractional energy gain, $\Delta w/w$, of particles of speed $v = \sqrt{10}U_{n1}$ reflected (top panel) and transmitted (bottom panel) from a strong ($r = 4$) shock as a function of the angle, θ_{Bn1}, between the magnetic field and the shock normal. The black dots indicate the average energy gain, and the vertical bars indicate plus and minus one standard deviation. The smooth solid line assumes conservation of the first adiabatic invariant. The vertical dashed line is the maximum θ_{Bn1} that can be reflected from the shock (adapted from Decker et al., 1988).

note that the average fractional energy gain is always positive for both reflected and transmitted particles, and in both cases near 90° the energy gain increases rapidly with increasing θ_{Bn1}. Thus, shock drift acceleration is much more effective for quasi-perpendicular shocks than for quasi-parallel shocks. However, in both cases the overall fractional energy gain is relatively modest, only a few percent. Thus, although shock drift acceleration can cause charge particle acceleration of shocks, it cannot explain the very large energy gains that are often observed.

8.4.2 Diffusive Shock Acceleration

One of the enduring puzzles in astrophysics is how cosmic rays are accelerated to very high energies, greater than 10^{20} eV. One of the most widely accepted mechanisms for explaining the acceleration of cosmic rays is that they originate from supernova shocks. Since shock drift acceleration during a single transit through a shock cannot explain the very large energy gains often observed, it is

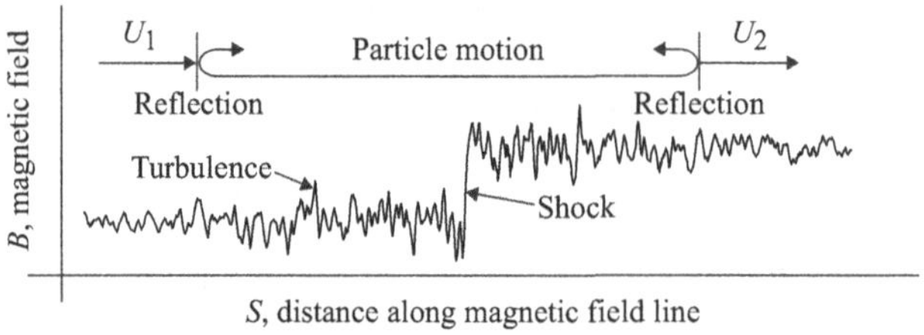

Figure 8.17 An illustration of Fermi acceleration of a charged particle by reflection from turbulent magnetic field fluctuations upstream and downstream of an MHD shock. Because the flow velocity downstream of the shock is less than the flow velocity upstream of the shock, $U_2 < U_1$, the particle usually gains energy in the reflection process, especially for large θ_{Bn} angles.

clear that some other contributing acceleration mechanism is needed. A ready answer is provided by the Fermi mechanism. As originally proposed to explain the acceleration of cosmic rays, Fermi (1949) envisioned charged particles trapped on a magnetic field line embedded in two approaching plasma clouds, with the energy being gained by repeated magnetic mirror reflections from the two approaching clouds, as in Figure 3.36. Because the flow velocity always decreases at a shock, one can see that essentially the same mechanism applies to particles bouncing back and forth along a magnetic field line between reflection points on either side of the shock; see Figure 8.17. In general, energy is gained by the upstream reflection and lost in the downstream reflection. However, because the upstream flow velocity is always greater than the downstream flow velocity, $U_{n2} < U_{n1}$, on average there is always a net energy gain. The only missing ingredient is a mechanism for reflecting the particles. This missing ingredient is provided by the fact that there is always a certain level of turbulent magnetic field fluctuations both upstream and downstream of the shock, and such fluctuations can cause the particles to be reflected. In fact, the fluctuations are believed to be driven by instabilities caused by particles accelerated by the shock, as in the electron and ion foreshocks discussed in the previous section. The usual assumption is that the reflections are caused by pitch-angle scattering from relatively small wave fields, the accumulative effect of which can be substantial even if the fluctuations are small. The magnetic field of Alfvén waves is particularly effective in producing pitch-angle scattering, since the wave magnetic field typically has a component perpendicular to the background magnetic field. For this reason pitch-angle scattering from turbulent Alfvén waves is often regarded as the primary mechanism for reflecting the particles. Such turbulent scattering is inherently statistical and is typically described by a pitch-angle diffusion process.

Because of the strong role of diffusion in the Fermi shock acceleration mechanism, this mechanism has come to be called *diffusive shock acceleration*

(Axford et al., 1977; Fisk, 1971; Bell, 1978; Blanford and Ostriker, 1978) and is thought to be the primary mechanism by which supernova shocks accelerate cosmic rays. It is also believed to be an important particle acceleration process for interplanetary shocks. In this model the number of times the particle crosses the shock becomes a random variable. At each shock crossing, energy is gained by both the Fermi reflection process and the shock drift acceleration process, the exact ratio of which depends on the type of shock and other details. Although many particles only cross the shock once, or only a few times, there are some that cross many times and these are the particles that are accelerated to the highest energies. As the amount of energy gain per crossing tends to be proportional to the initial energy of the particle, the resulting energy spectrum typically develops into a steeply decreasing power-law energy distribution. Although supernova shocks are thought to play the dominant role in the acceleration of cosmic rays via the diffusive shock acceleration mechanism, the current consensus is that this mechanism is only able to account for cosmic ray energies up to about 10^{14} eV per nucleon (Lagage and Cesarsky, 1983). At this point there is a distinct increase in the slope, or "knee," in the cosmic ray energy spectrum. Above this knee, diffusive shock acceleration is no longer effective because the cyclotron radius becomes larger than the typical radius of a supernova shock. How cosmic rays above this knee are accelerated to the highest observed energies, $\sim 10^{20}$ eV, remains one of the great puzzles of modern astrophysics.

Problems

8.1. By eliminating $[(M_{A2}^2/M_{S2}^2)-(M_{A1}^2/M_{S1}^2)]$ from Eqs. (8.3.36) and (8.3.38) and eliminating M_{A2}^2 using Eq. (8.3.39), show that the Alfvén Mach number, M_{A1}, and the sonic Mach number, M_{S1}, must satisfy

$$(M_{A1}^2 - r)^2 \left[M_{S1}^2 - \frac{2r}{r+1-\gamma(r-1)} \right]$$

$$- r\tan^2\theta_1 M_{S1}^2 \left[\frac{2r-\gamma(r-1)}{r+1-\gamma(r-1)} M_{A1}^2 - r \right] = 0.$$

8.2. By making the substitutions $M_{A1} = U_{n1}/(V_{A1}\cos\theta_1)$ and $M_{S1} = U_{n1}/V_{S1}$, show that the above equation simplifies to

$$(U_{n1}^2 - rV_{A1}^2\cos^2\theta_1)^2 \left[U_{n1}^2 - \frac{2rV_{S1}^2}{r+1-\gamma(r-1)} \right]$$

$$- r\sin^2\theta_1 U_{n1}^2 V_{A1}^2 \left[\frac{2r-\gamma(r-1)}{r+1-\gamma(r-1)} U_{n1}^2 - rV_{A1}^2\cos^2\theta_1 \right] = 0.$$

8.3. Show that there is only one solution to the shock adiabatic in the limit $r \to r_{\rm m}$. Hint: After using the root given by Eq. (8.3.52), show that in the limit $r \to r_{\rm m}$ the remaining two roots to the shock adiabatic are solutions to the equation

$$U_{\rm n}^4\left[2V_{\rm S1}^2 + \sin^2\theta_1\, V_{\rm A1}^2(r_{\rm m}-1)\right] - U_{\rm n1}^2\, 4r_{\rm m}\cos^2\theta_1 V_{\rm A1}^2 V_{\rm S1}^2 + 2r_{\rm m}^2 V_{\rm A1}^4 V_{\rm S1}^2\cos^4\theta_1 = 0,$$

and that this quadratic equation has a discriminant

$$-8r_{\rm m}^2(r_{\rm m}-1)\cos^4\theta_1\sin^2\theta_1 V_{\rm A1}^6 V_{\rm S1}^2,$$

which is always negative for $r_{\rm m} > 1$.

8.4. By numerically solving the shock adiabatic equation (8.3.41), show that the three roots for $U_{\rm n1}$ have the qualitative form given in Figure 8.9. Make a plot of U_{n1} versus θ for each of the three roots. For representative parameters use $V_{\rm A1} = 40\,{\rm km\,s^{-1}}$, $V_{\rm S1} = 30\,{\rm km\,s^{-1}}$, and $\theta = 45°$.

8.5. By substituting $M_{\rm S} = U_{\rm n}\sqrt{\rho_{\rm m}/\gamma P}$ for $M_{\rm S2}$ and $M_{\rm S1}$ into Eq. (8.3.36) and substituting $M_{\rm A1}^2 = \mu_0\rho_{\rm m1}U_{\rm n1}^2/B_{\rm n}^2$, show that the downstream to upstream pressure ratio is given by

$$\frac{P_2}{P_1} = 1 + \gamma\frac{U_{\rm n1}^2}{V_{\rm S1}^2}\frac{(r-1)}{r}\left[1 - \frac{rV_{\rm A1}^2[(r+1)U_{\rm n1}^2 - 2rV_{\rm A1}^2\cos^2\theta_1]}{2(U_{\rm n1}^2 - rV_{\rm A1}^2\cos^2\theta_1)^2}\sin^2\theta_1\right].$$

8.6. Show that for a parallel ($\theta_1 = 0$) fast mode shock the downstream to upstream pressure ratio simplifies to

$$\frac{P_2}{P_1} = \frac{r\,r_{\rm m} - 1}{r_{\rm m} - r},$$

where $\gamma = (r_{\rm m}+1)/(r_{\rm m}-1)$ has been substituted for γ. Plot P_2/P_1 as a function of r for $r_{\rm m} = 4$.

References

Anderson, J. E. 1963. *Magnetohydrodynamic Shock Waves*. Cambridge, MA: MIT Press, p. 21.

Axford, W. I., Lear, E., and Skadron, G. 1977. The acceleration of cosmic rays by shock waves. *Proc. 15th Int. Cosmic Ray Conf., Plovdiv*, Sofia: Bulgarian Academy of Sciences, vol. 11, pp. 132–137.

Axford, W. I. 1996. The heliosphere. In *The Heliosphere in the Local Interstellar Medium*, ed. R. von Steiger, R. Lallement, and M. A. Lee. Dordrecht: Kluwer Academic Publishers, pp. 9–14.

Baumjohann, W., and Treumann, R. A. 1997. *Basic Space Plasma Physics*. London: World Scientific Publishing, Chapter 8.

Bell, A. R. 1978. The acceleration of cosmic rays in shock fronts, II. *Mon. Not. R. Astron. Soc.* **182**, 443–455.

Blanford, R. D. and Ostriker, J. P. 1978. Particle acceleration by astrophysical shocks. *Astrophys. J.* **221**, L29–32.

Currie, I. G. 1974. *Fundamental Mechanics of Fluids*. New York: McGraw-Hill, p. 337.

de Hoffmann, F., and Teller, E. 1950. Magneto-hydrodynamic shocks. *Phys. Rev.* **80**, 691–703.

Decker, R. B., Pesses, M. E., and Krimigis, S. M., 1981. Shock-associated low-energy ion enhancements observed by Voyagers 1 and 2. *J. Geophys. Res.* **86**, 8819–8831.

Decker, R. B. 1988. Computer modeling of test particle acceleration at oblique shocks. *Space Sci. Rev.* **48**, 195–262.

Fermi, E., 1949. On the origin of cosmic rays, *Phys. Rev.* **75**, 1169–1174.

Fermi, E., 1956. *Thermodynamics*. New York: Dover Publications.

Fetter, A. L., and Walecka, J. D. 1980. *Theoretical Mechanics of Particles and Continua*. New York: McGraw-Hill, p. 345.

Fisk, L. A. 1971. Increase in the low-energy cosmic ray intensity at the front of propagating interplanetary shock waves. *J. Geophys. Res.* **76**, 1662–1672.

Gurnett, D. A., and Scarf, F. L. 1983. Plasma waves in the Jovian magnetosphere. In *Physics of the Jovian Magnetosphere*, ed. A. J. Dessler. Cambridge: Cambridge University Press, p. 292.

Kantrowitz, A., and Petschek, H. E. 1966. MHD characteristics and shock waves. In *Plasma Physics in Theory and Application*, ed. W. B. Kunkel. New York: McGraw-Hill, pp. 148–206.

Kennel, C. F., Blandford, R. D., and Coppi, P. 1989. MHD intermediate shock discontinuities. Part 1. Rankine–Hugoniot conditions. *J. Plasma Phys.* **43** (2), 299–319.

Kivelson, M. G., and Russell, C. T. 1995. *Introduction to Space Physics*. Cambridge: Cambridge University Press, Chapter 5.

Lagage, P. O., and Cesarsky, C. J. 1983. The maximum energy of cosmic rays accelerated by supernova shocks. *Astron. Astrophys.* **125**, 249–257.

Landau, L. D., and Lifshitz, E. M. 1960. *Electrodynamics of Continuous Media*. New York: Pergamon Press, p. 224.

Parks, G. K. 1991. *Physics of Space Plasmas: An Introduction*. New York: Addison-Wesley, p. 7.

Scudder, J. D., Burlaga, L. F., and Greenspan, E. W. 1984. Scale lengths at quasi-parallel shocks. *J. Geophys. Res.* **89**, 7545–7550.

Scudder, J. D., Mangeney, A., Lacombe, C., Harvey, C. C., Aggson, T. L., Anderson, R. R., Gosling, J. T., Paschmann, G., and Russell, C. T. 1986. The resolved layer of a collisionless, high β, supercritical, quasi-perpendicular shock wave: 1. Rankine–Hugoniot geometry, currents and stationarity. *J. Geophys. Res.* **91**, 11019–11052.

Smith, E. J. 1973. Identification of interplanetary tangential and rotational discontinuities. *J. Geophys. Res.* **78**, 2054–2063.

Sonnerup, B. U. O., Paschmann, G., Papamastorakis, I., Sckopke, N., Haerendel, G., Bame, S. J., Asbridge, J. R., Gosling, J. T., and Russell, C. T. 1981. Evidence for magnetic field reconnection at the Earth's magnetopause. *J. Geophys. Res.* **86**, 10049–10067.

Whipple, E. C., Northrop, T. G., and Birmingham, T. J. 1986. Adiabatic theory in regions of strong field gradients. *J. Geophys. Res.* **91**, 3149–4156.

Zank, G. P., Webb, G. M., and Donohue, D. J. 1993. Particle injection and the structure of energetic-particle-modified shocks. *Astrophys. J.* **406**, 67–91.

Further Reading

Anderson, J. E. 1963. *Magnetohydrodynamic Shock Waves*. Cambridge, MA: MIT Press, Chapter 2.

Balogh, A., and Treumann, R. 2013. *Physics of Collisionless Shocks*. New York: Springer.

Blandford, R., and Eichler, D. 1987. Particle acceleration at astrophysical shocks: A theory of cosmic ray origin. *Phys. Rep.* **154**, 1–75.

Boyd, T. J. M., and Sanderson, J. J. 2003. *The Physics of Plasmas*. Cambridge: Cambridge University Press, Chapter 5.

Drury, L. O. C. 1983. An introduction to the theory of diffusive shock acceleration of energetic particles in tenuous plasmas. *Rep. Prog. Phys.* **46**, 973–1027.

Landau, L. D., and Lifshitz, E. M. 1960. *Electrodynamics of Continuous Media*. New York: Pergamon Press, Chapter 8.

Tidman, D. A., and Krall, N. A. 1971. *Shock Waves in Collisionless Plasmas*. New York: Wiley, Chapter 1.

Zank, G. P. 2014. *Transport Processes in Space Physics and Astrophysics, Lecture Notes in Physics*, vol. 877, New York: Springer.

9

Electrostatic Waves in a Hot Unmagnetized Plasma

In this chapter we investigate the propagation of small-amplitude waves in a hot unmagnetized plasma. Because of the shortcomings of the moment equations, the approach used is to solve the Vlasov equation directly using a linearization procedure similar to that used in the analysis of waves in cold plasmas and MHD. Although both electromagnetic and electrostatic solutions exist, the discussion in this chapter is limited to solutions that are purely electrostatic, i.e., the electric field is derivable from the gradient of a potential, $\mathbf{E} = -\nabla\Phi$. Electromagnetic solutions are discussed in the next chapter.

From Faraday's law it is easily verified that electrostatic waves have no magnetic component. This greatly simplifies the Vlasov equation by eliminating the $\mathbf{v} \times \mathbf{B}$ force. For electrostatic waves, it is usually easier to solve for the potential rather than for the electric field. Therefore, in the following analysis, the electric field is replaced by $\mathbf{E} = -\nabla\Phi$ and the potential is calculated from Poisson's equation, $\nabla^2\Phi = -\rho_q/\epsilon_0$.

9.1 The Vlasov Approach

In an initial attempt to analyze the problem, we assume that normal modes of the form $\exp(-i\omega t)$ exist and represent them by using Fourier transforms, following the same basic procedure used in Chapter 4. This is the approach used by Vlasov (1945), who first considered this problem. As we will see, the Vlasov approach encounters a mathematical difficulty that can only be resolved by reformulating the linearized problem not as a normal mode problem using Fourier transform, but as an initial value problem using the technique of Laplace transforms.

As we have done before, we start by only considering the motion of the electrons. The ions are considered to be immobile, with the same zero-order number density, n_0, as the electrons. The effects of ion motions will be considered later. The equations that must be solved then consist of Vlasov's equation (with $q = -e, \mathbf{E} = -\nabla\Phi$, and $\mathbf{B} = 0$)

$$\frac{\partial f}{\partial t} + \mathbf{v} \cdot \nabla f + \frac{e}{m} \nabla\Phi \cdot \nabla_{\mathbf{v}} f = 0 \tag{9.1.1}$$

and Poisson's equation

$$\nabla^2 \Phi = -\frac{\rho_q}{\epsilon_0} = -\frac{e}{\epsilon_0}\left[n_0 - \int_{-\infty}^{\infty} f \, \mathrm{d}^3 v\right]. \tag{9.1.2}$$

Following the usual linearization procedure, we assume that the electron velocity distribution function, $f(\mathbf{v})$, consists of a constant uniform zero-order distribution, $f_0(\mathbf{v})$, plus a small first-order perturbation, $f_1(\mathbf{v})$:

$$f(\mathbf{v}) = f_0(\mathbf{v}) + f_1(\mathbf{v}). \tag{9.1.3}$$

Similarly, we write $\Phi = \Phi_1$. Linearizing Eqs. (9.1.1) and (9.1.2) and noting that $\int f_0 \mathrm{d}^3 v = n_0$, we obtain for the first-order equations

$$\frac{\partial f_1}{\partial t} + \mathbf{v} \cdot \nabla f_1 + \frac{e}{m} \nabla\Phi_1 \cdot \nabla_{\mathbf{v}} f_0 = 0 \tag{9.1.4}$$

and

$$\nabla^2 \Phi_1 = \frac{e}{\epsilon_0} \int_{-\infty}^{\infty} f_1(\mathbf{v}) \, \mathrm{d}^3 v. \tag{9.1.5}$$

Next, we attempt to solve these equations using the usual Fourier transform approach. Since no preferred direction exists, the z axis can be aligned parallel to $\mathbf{k}$. After dropping the subscript 1 on the first-order terms and making the usual operator substitutions ($\partial/\partial t \to -\mathrm{i}\omega$ and $\nabla \to \mathrm{i}\mathbf{k}$), it is easy to see that the Fourier-transformed equations are

$$-\mathrm{i}\omega \tilde{f} + \mathrm{i}k v_z \tilde{f} + \mathrm{i}\frac{e}{m} k \tilde{\Phi} \frac{\partial f_0}{\partial v_z} = 0 \tag{9.1.6}$$

and

$$k^2 \tilde{\Phi} = -\frac{e}{\epsilon_0} \int_{-\infty}^{\infty} \tilde{f}(\mathbf{v}) \, \mathrm{d}^3 v. \tag{9.1.7}$$

Solving for $\tilde{f}$ from Eq. (9.1.6),

$$\tilde{f} = \frac{-1}{(k v_z - \omega)} \frac{e}{m} k \tilde{\Phi} \frac{\partial f_0}{\partial v_z}, \tag{9.1.8}$$

and substituting into Eq. (9.1.7) gives

$$k^2\tilde{\Phi} = \frac{e^2}{\epsilon_0 m} k\tilde{\Phi} \int_{-\infty}^{\infty} \frac{(\partial f_0/\partial v_z)}{(kv_z - \omega)} \mathrm{d}^3 v. \tag{9.1.9}$$

Rearranging the terms and factoring out $\tilde{\Phi}$ then gives the homogeneous equation

$$\left[1 - \frac{e^2}{\epsilon_0 m k^2} \int_{-\infty}^{\infty} \frac{\partial f_0/\partial v_z}{(v_z - \omega/k)} \mathrm{d}^3 v\right] \tilde{\Phi} = 0. \tag{9.1.10}$$

For the potential $\tilde{\Phi}$ to have a non-trivial solution, the term in the brackets must be zero, which gives the dispersion relation

$$D(k,\omega) = 1 - \frac{e^2}{\epsilon_0 m k^2} \int_{-\infty}^{\infty} \frac{\partial f_0/\partial v_z}{(v_z - \omega/k)} \mathrm{d}^3 v = 0. \tag{9.1.11}$$

At this point, it is useful to define a new normalized one-dimensional distribution function:

$$F_0(v_z) = \frac{1}{n_0} \int_{-\infty}^{\infty} f_0(\mathbf{v}) \mathrm{d}v_x \mathrm{d}v_y. \tag{9.1.12}$$

The function $F_0(v_z)$ is sometimes called the reduced distribution function. Note that by dividing by n_0 the reduced distribution function is normalized such that $\int F_0(v_z)\mathrm{d}v_z = 1$. Recognizing that $\omega_\mathrm{p}^2 = n_0 e^2/(\epsilon_0 m)$, the dispersion relation equation (9.1.11) can be written

$$D(k,\omega) = 1 - \frac{\omega_\mathrm{p}^2}{k^2} \int_{-\infty}^{\infty} \frac{\partial F_0/\partial v_z}{(v_z - \omega/k)} \mathrm{d}v_z = 0. \tag{9.1.13}$$

In some applications it is useful to put the dispersion relation into a slightly different form by integrating by parts once, which gives

$$\int_{-\infty}^{\infty} \frac{\partial F_0/\partial v_z}{(v_z - \omega/k)} \mathrm{d}v_z = \left[\frac{F_0}{(v_z - \omega/k)}\right]_{-\infty}^{\infty} + \int_{-\infty}^{\infty} \frac{F_0}{(v_z - \omega/k)^2} \mathrm{d}v_z. \tag{9.1.14}$$

Since F_0 goes to zero at $v_z = \pm\infty$, the first term in the integration by parts vanishes, so the dispersion relation becomes

$$D(k,\omega) = 1 - \frac{\omega_\mathrm{p}^2}{k^2} \int_{-\infty}^{\infty} \frac{F_0}{(v_z - \omega/k)^2} \mathrm{d}v_z = 0. \tag{9.1.15}$$

Both of the above forms of the dispersion relation suffer from a serious problem. Because the denominator goes to zero at $v_z = \omega/k$, the integrals do not converge unless F_0 and $\partial F_0/\partial v_z$ are zero at $v_z = \omega/k$. Physically, the dispersion relation exists only if there are no particles moving with a velocity equal to the phase velocity of the wave. For the moment, we restrict our consideration to distribution functions for which F_0 and $\partial F_0/\partial v_z$ are zero at $v_z = \omega/k$. In the next section we will return to the issue of how to deal with a distribution function that is not zero at $v_z = \omega/k$.

Since the charge only enters into the dispersion equation as a squared, e^2, term in $\omega_{\rm p}^2$, it is easy to see that ion motions can be incorporated by replacing $\omega_{\rm p}$ and F_0 by $\omega_{{\rm p}s}$ and F_{s0}, respectively, and summing over all species. For an arbitrary number of species, the general form of the dispersion relation is then

$$Ð(k,\omega) = 1 - \sum_s \frac{\omega_{{\rm p}s}^2}{k^2} \int_{-\infty}^{\infty} \frac{\partial F_{s0}/\partial v_z}{(v_z - \omega/k)} \, {\rm d}v_z = 0, \tag{9.1.16}$$

or equivalently,

$$Ð(k,\omega) = 1 - \sum_s \frac{\omega_{{\rm p}s}^2}{k^2} \int_{-\infty}^{\infty} \frac{F_{s0}}{(v_z - \omega/k)^2} \, {\rm d}v_z = 0. \tag{9.1.17}$$

These forms will be useful later when we consider ion effects.

9.1.1 The Bohm–Gross Dispersion Relation

To get back to familiar ground (Langmuir oscillations), we next evaluate the dispersion relation assuming that the phase velocity is much greater than the electron thermal velocity, i.e., $(\omega/k)^2 \gg \langle v_z^2 \rangle$. Then the integrand in Eq. (9.1.15) can be expanded in powers of the small quantity kv_z/ω:

$$\frac{1}{(1 - kv_z/\omega)^2} = 1 + 2\left(\frac{kv_z}{\omega}\right) + 3\left(\frac{kv_z}{\omega}\right)^2 + \cdots \tag{9.1.18}$$

This expansion is sometimes called the high phase velocity expansion since $\omega/k \gg v_z$. Using this expansion, the dispersion relation can be written

$$Ð(k,\omega) = 1 - \frac{\omega_{\rm p}^2}{\omega^2} \int_{-\infty}^{\infty} \left[1 + 2\left(\frac{kv_z}{\omega}\right) + 3\left(\frac{kv_z}{\omega}\right)^2 + \cdots\right] F_{\rm e0} \, {\rm d}v_z. \tag{9.1.19}$$

If we assume that there is no zero-order current, the first moment is zero, $\int v_z F_{\rm e0} \, {\rm d}v_z = 0$. Retaining terms up to the second moment, $\langle v_z^2 \rangle = \int v_z^2 F_{\rm e0} \, {\rm d}v_z$, then gives

$$Ð(k,\omega) = 1 - \frac{\omega_{\rm p}^2}{\omega^2}\left(1 + 3\frac{k^2}{\omega^2}\langle v_z^2 \rangle\right) = 0, \tag{9.1.20}$$

or

$$\omega^2 = \omega_p^2\left(1 + 3\frac{k^2\langle v_z^2\rangle}{\omega^2}\right). \tag{9.1.21}$$

This equation is most conveniently solved by using the method of successive approximations. Since $(\omega/k)^2 \gg \langle v_z^2\rangle$, the first approximation is

$$\omega^2 = \omega_p^2. \tag{9.1.22}$$

Substituting $\omega^2 = \omega_p^2$ into the right-hand side of Eq. (9.1.21) then gives the second approximation:

$$\omega^2 = \omega_p^2 + 3\langle v_z^2\rangle k^2. \tag{9.1.23}$$

If the velocity distribution function is a Maxwellian, it is easy to show that $\langle v_z^2\rangle = \kappa T_e/m_e$. The dispersion relation then becomes

$$\omega^2 = \omega_p^2 + 3\left(\frac{\kappa T_e}{m_e}\right)k^2. \tag{9.1.24}$$

Note that if the temperature is zero, we recover the cold plasma result, $\omega^2 = \omega_p^2$, which is a pure oscillation at the electron plasma frequency.

Since $C_e^2 = \kappa T_e/m_e$, the dispersion relation given by Eq. (9.1.24) is seen to be identical to the Bohm–Gross dispersion relation derived in Chapter 5 using the adiabatic equation of state

$$\omega^2 = \omega_p^2 + \gamma_e C_e^2 k^2, \tag{9.1.25}$$

provided we use $\gamma_e = 3$. The fact that $\gamma_e = 3$ implies that the electrons move with one degree of freedom (i.e., $f = 1$ in Eq. (5.4.43)), a result that is not obvious from the moment equation approach.

9.1.2 Cold Beam Instabilities

Next, we consider an electron velocity distribution function consisting of two or more beams of cold electrons moving parallel to the z axis. Letting n_j and V_j be the number density and velocity of the jth beam, the zero-order distribution function can be represented by a product of delta functions:

$$f_{e0}(\mathbf{v}) = \sum_j n_j \delta(v_x)\delta(v_y)\delta(v_z - V_j). \tag{9.1.26}$$

Using Eq. (9.1.12), the reduced distribution function is given by

$$F_{e0}(v_z) = \frac{1}{n_0}\int_{-\infty}^{\infty} f_{e0}(\mathbf{v})\, \mathrm{d}v_x\, \mathrm{d}v_y = \sum_j \frac{n_j}{n_0}\delta(v_z - V_j), \tag{9.1.27}$$

where $n_0 = \sum_j n_j$. Since $F_{e0}(v_z)$ is zero except when $v_z = V_j$, Eq. (9.1.15) can be used and yields

$$Đ(k,\omega) = 1 - \frac{\omega_p^2}{k^2} \sum_j \frac{n_j}{n_0} \int_{-\infty}^{\infty} \frac{\delta(v_z - V_j)}{(v_z - \omega/k)^2} dv_z = 0. \tag{9.1.28}$$

Evaluating the integral and defining $\omega_{pj}^2 = n_j e^2/(\epsilon_0 m_e)$ as the plasma frequency of the jth beam gives the dispersion relation

$$Đ(k,\omega) = 1 - \sum_j \frac{\omega_{pj}^2}{k^2} \frac{1}{(V_j - \omega/k)^2} = 0. \tag{9.1.29}$$

Note that since the charge enters into the dispersion relation as e^2, it does not matter if the beam particles are positively or negatively charged. Therefore, Eq. (9.1.29) can also be used to analyze electron-ion plasmas. The only requirement is that the total zero-order charge density be zero (i.e., electrical neutrality).

The roots of the dispersion relation equation (9.1.29) are most conveniently analyzed by writing the equation as

$$H(\omega/k) = \sum_j \frac{\omega_{pj}^2}{(V_j - \omega/k)^2} = k^2, \tag{9.1.30}$$

which for two beams has the form shown in Figure 9.1. As one can see, each beam introduces two roots. If $k > k_c$, the roots are all real. These roots correspond to electrostatic oscillations that move along the beam at phase velocities near the beam velocities. If $k < k_c$, the roots that were located between V_1 and V_2 become imaginary.

Since H is real, these roots always occur as complex conjugate pairs, $\omega_- = \omega_r - i\gamma$ and $\omega_+ = \omega_r + i\gamma$, where γ is assumed to be positive. The root ω_+ corresponds

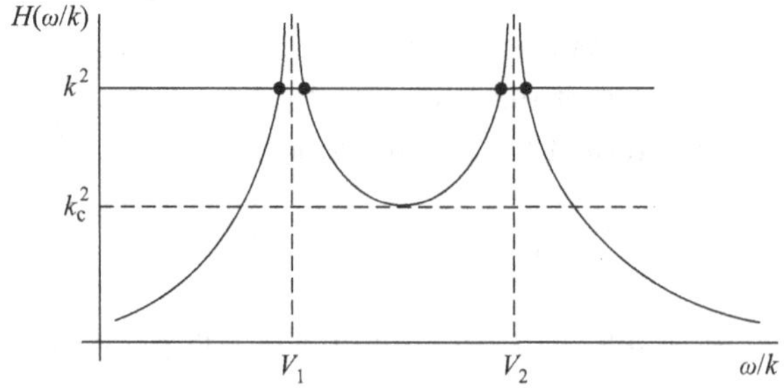

Figure 9.1 A plot of $H(\omega/k)$ as a function of the phase velocity ω/k for two equal density beams of velocities V_1 and V_2.

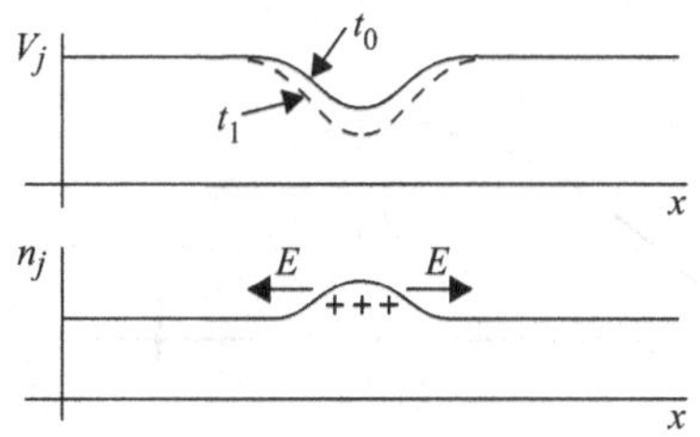

Figure 9.2 An illustration showing that a perturbation at time t_0 causes a charge imbalance to develop that leads to an even larger perturbation (dashed line) at a later time t_1.

to a wave that is growing exponentially in time:

$$E = E_0\,\mathrm{e}^{-\mathrm{i}\omega_r t}\,\mathrm{e}^{\gamma t}. \tag{9.1.31}$$

The quantity γ is called the growth rate. The above analysis shows that for sufficiently long wavelengths, $k < k_c$, a cold multi-beam plasma is always unstable. This instability is called the *two-stream instability*, since the interaction involves two or more streams of particles.

The physical basis for the two-stream instability is as follows. For a steady-state beam streaming through a fixed background of neutralizing particles, conservation of particles implies that the product of the number density of the beam and the velocity of the beam must be constant:

$$n_j V_j = \text{constant}. \tag{9.1.32}$$

If some perturbation causes a reduction in the velocity at time t_0, the number density must increase, as illustrated in Figure 9.2. Assuming that the beam consists of positive charges (the same basic argument holds for negative charges), the resulting increase in the number density of the beam causes an excess of positive charge. This positive charge produces an electric field that is directed away from the region of positive charge, which in turn causes a further decrease in the beam velocity, thereby causing the initial perturbation to grow in amplitude, as at time t_1. It is this feedback, via the electric field, that causes the two-stream instability.

Two Oppositely Directed Beams of Equal Density

For the special case of two beams of equal density ($n_1 = n_2 = n_0$) and equal but opposite velocities ($V_1 = -V_2 = V$), it is possible to obtain a simple analytical expression for the growth rate. The dispersion relation in this case is

$$D(k,\omega) = 1 - \frac{\omega_p^2}{(\omega - kV)^2} - \frac{\omega_p^2}{(\omega + kV)^2} = 0, \tag{9.1.33}$$

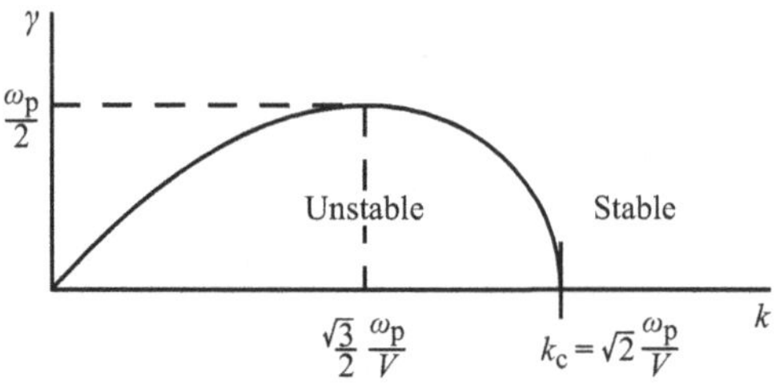

Figure 9.3 A plot of the growth rate, γ, of the unstable solution as a function of the wave number, k, for two oppositely directed beams of equal density.

where $\omega_{\rm p}^2 = n_0 e^2/(\epsilon_0 m)$. The above equation, which is quadratic in ω^2, has four roots given by

$$\omega^2 = \omega_{\rm p}^2 + k^2V^2 \pm (\omega_{\rm p}^4 + 4k^2V^2\omega_{\rm p}^2)^{1/2}. \tag{9.1.34}$$

It is easy to show that if $k > \sqrt{2}\omega_{\rm p}/V = k_{\rm c}$, then ω is real and all of the modes are purely oscillatory. However, if $k < k_{\rm c}$, two of the oscillatory modes are replaced by an exponentially growing mode and an exponentially decaying mode. Writing $\omega = \pm {\rm i}\gamma$ for these two roots, the growth rate of the unstable mode is given by

$$\gamma = \left[(\omega_{\rm p}^4 + 4k^2V^2\omega_{\rm p}^2)^{1/2} - (\omega_{\rm p}^2 + k^2V^2)\right]^{1/2}. \tag{9.1.35}$$

The maximum growth rate is given by $\gamma_{\rm max} = \omega_{\rm p}/2$ and occurs at $k_{\rm max} = \sqrt{3/8}\,k_{\rm c}$. A plot of the growth rate as a function of wave number is shown in Figure 9.3. Note that in the unstable region, the fields grow exponentially in time with no oscillations ($\omega = 0$). The absence of an oscillatory component is a consequence of the choice of the reference frame which, in this case, is such that the two electron beams have equal and opposite velocities (i.e., $V_1 = -V_2$). In a frame of reference moving at a velocity V', the fields would oscillate at a frequency $\omega' = kV'$, where kV' is the Doppler shift.

The Weak-Beam Approximation

In various applications, one sometimes encounters situations where a very low-density electron beam of plasma frequency $\omega_{\rm b}$ streams through a background of electrons initially at rest with a plasma frequency $\omega_{\rm p}$. The dispersion relation in this case is easily shown to be

$$D(k,\omega) = 1 - \frac{\omega_{\rm p}^2}{\omega^2} - \frac{\omega_{\rm b}^2}{(kV-\omega)^2} = 0, \tag{9.1.36}$$

which can be rewritten in the form

$$\left(1-\frac{\omega_{\mathrm{p}}^2}{\omega^2}\right)(kV-\omega)^2=\omega_{\mathrm{b}}^2. \tag{9.1.37}$$

If the term on the right is small, one can assume that the solution is essentially the same as with no beam (i.e., $\omega=\omega_{\mathrm{p}}$ and $k=\omega_{\mathrm{p}}/V$) plus a small frequency shift $\omega=\omega_{\mathrm{p}}+\delta\omega$. Substituting these approximations into the above equation and expanding to first order in $\delta\omega/\omega_{\mathrm{p}}$ gives

$$\left[1-\frac{\omega_{\mathrm{p}}^2}{\omega_{\mathrm{p}}^2}\left(1-2\frac{\delta\omega}{\omega_{\mathrm{p}}}\right)\right](\delta\omega)^2=\omega_{\mathrm{b}}^2. \tag{9.1.38}$$

This equation simplifies to $\delta\omega^3=(1/2)\omega_{\mathrm{b}}^2\omega_{\mathrm{p}}$, which has three roots:

$$\delta\omega=\left(\frac{1}{2}\right)^{1/3}\omega_{\mathrm{b}}^{2/3}\omega_{\mathrm{p}}^{1/3}\quad\text{and}\quad\delta\omega=\left(\frac{1}{2}\right)^{1/3}\omega_{\mathrm{b}}^{2/3}\omega_{\mathrm{p}}^{1/3}\left[-\frac{1}{2}\pm\mathrm{i}\frac{\sqrt{3}}{2}\right]. \tag{9.1.39}$$

The root with the positive imaginary part corresponds to an exponentially growing oscillation. It is easy to show that the frequency ω_{r} and growth rate γ of the unstable solution are given by

$$\omega_{\mathrm{r}}=\omega_{\mathrm{p}}\left[1-\frac{1}{2}\left(\frac{1}{2}\right)^{1/3}\left(\frac{n_{\mathrm{b}}}{n_0}\right)^{1/3}\right]\quad\text{and}\quad\gamma=\frac{\sqrt{3}}{2}\left(\frac{1}{2}\right)^{1/3}\left(\frac{n_{\mathrm{b}}}{n_0}\right)^{1/3}\omega_{\mathrm{p}}. \tag{9.1.40}$$

This beam-driven mode is shifted downward in frequency from the plasma frequency and grows at a rate proportional to the beam density to the one-third power, $\gamma\propto n_{\mathrm{b}}^{1/3}$. Note that as the beam density goes to zero, this mode reduces to an undamped oscillation at the plasma frequency.

The Buneman Instability

Another important instability, first investigated by Buneman (1959), consists of a cold electron beam streaming through a single species of positive ions initially at rest. Using Eq. (9.1.29), the dispersion relation is given by

$$D(k,\omega)=1-\frac{\omega_{\mathrm{pi}}^2}{\omega^2}-\frac{\omega_{\mathrm{pe}}^2}{(kV-\omega)^2}=0, \tag{9.1.41}$$

where ω_{pi} and ω_{pe} are the ion and electron plasma frequencies and $n_{\mathrm{i}}=n_{\mathrm{e}}=n_0$.

Note that since the ions are initially assumed to be at rest, the beam velocity associated with the ion term is zero. If one assumes that $\omega_{\mathrm{pi}}^2\ll\omega^2$ (which must be confirmed later), the above equation can be written in the approximate form,

$$kV-\omega=\pm\omega_{\mathrm{pe}}\left[1+\frac{\omega_{\mathrm{pi}}^2}{2\omega^2}\right], \tag{9.1.42}$$

where the $(1-\omega_{\rm pi}^2/\omega^2)^{1/2}$ term has been expanded to first order in the small quantity $\omega_{\rm pi}^2/\omega^2$. The minus sign gives a purely real solution for the frequency. For the plus sign, one can show by direct substitution that the frequency is complex and is given by

$$\omega = (\omega_{\rm pe}\omega_{\rm pi}^2 \cos\theta)^{1/3} {\rm e}^{{\rm i}\theta}, \tag{9.1.43}$$

where θ is a parameter determined by k; see Problem 9.3. Since this solution is complex, the corresponding mode is unstable whenever $\theta > 0$. By substituting $\omega = \omega_{\rm r} + {\rm i}\gamma$ into the above equation and separating the real and imaginary parts, it can be shown that the maximum growth rate occurs at $\theta = \pi/3$. The growth rate and frequency at $\theta = \pi/3$ are given by

$$\gamma = \frac{\sqrt{3}}{2}\left(\frac{m_{\rm e}}{2m_{\rm i}}\right)^{1/3} \omega_{\rm pe} \tag{9.1.44}$$

and

$$\omega_{\rm r} = \frac{1}{2}\left(\frac{m_{\rm e}}{2m_{\rm i}}\right)^{1/3} \omega_{\rm pe}. \tag{9.1.45}$$

This very fast growing instability is called the Buneman mode and occurs whenever a cold electron beam streams through a background of cold ions. Later, when we consider beams with a finite temperature, we will show that the drift velocity must exceed the electron thermal speed before the instability can occur. Also, we will show that the finite temperature introduces another instability called the ion acoustic instability.

9.2 The Landau Approach

As pointed out in Section 9.1, if the velocity distribution function is non-zero at a velocity equal to the phase velocity of the wave, the dispersion relation does not exist because of the singularity at $v_z = \omega/k$. This difficulty is inherent in the Vlasov approach, which does not properly treat the initial transient phase of a disturbance and, more importantly, assumes that a normal mode of the form $\exp(-{\rm i}\omega t)$ exists for all times. In a classic paper, Landau (1946) resolved this difficulty by pointing out that it is necessary to consider the long-time response to a disturbance imposed at $t = 0$. This type of problem is called an initial-value problem and is usually solved by using Laplace transforms. Since the solution involves some subtle points regarding the use of Laplace transforms, we start by giving a brief review of Laplace transforms.

9.2.1 Laplace Transforms: A Brief Review

The Laplace transform of a function $f(t)$ is defined by

$$\tilde{f}(p) = \int_0^{\infty} f(t)\mathrm{e}^{-pt}\,\mathrm{d}t, \tag{9.2.1}$$

where p is a complex number, $p = \gamma - \mathrm{i}\omega$. (Here we have defined the imaginary part of p as $-\mathrm{i}\omega$ with ω real in order to maintain consistency with the Fourier analysis notation.) Since the integral diverges if the real part of p is negative, denoted by $\mathrm{Re}\{p\} < 0$, the transform, $\tilde{f}(p)$, exists only if $\mathrm{Re}\{p\} > 0$ and only for functions that grow less rapidly than an exponential. The inverse of a Laplace transform is given by the integral

$$f(t) = \frac{1}{2\pi\mathrm{i}} \int_{\sigma-\mathrm{i}\infty}^{\sigma+\mathrm{i}\infty} \tilde{f}(p)\,\mathrm{e}^{pt}\,\mathrm{d}p, \tag{9.2.2}$$

where the integration path is along a line $\mathrm{Re}\{p\} = \sigma$ that is located to the right of all the poles of $\tilde{f}(p)$, which are points where the function $\tilde{f}(p)$ goes to infinity. The required integration path is shown in Figure 9.4, where the "×" symbols indicate the poles of $\tilde{f}(p)$.

By integrating Eq. (9.2.1) by parts once, a useful relation can be established for the Laplace transform of the first-derivative of a function:

$$\begin{aligned}\tilde{f}'(p) &= \int_0^{\infty} f'(t)\mathrm{e}^{-pt}\,\mathrm{d}t \\ &= [f(t)\mathrm{e}^{-pt}]_0^{\infty} + p\int_0^{\infty} f(t)\mathrm{e}^{-pt}\,\mathrm{d}t \\ &= p\tilde{f}(p) - f(0).\end{aligned} \tag{9.2.3}$$

Note that the Laplace transform of the derivative of a function involves the initial value of the function, $f(0)$, evaluated at $t = 0$. By repeated use of the above

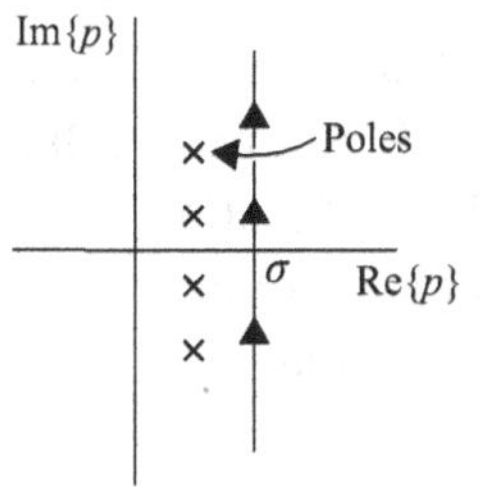

Figure 9.4 The integration contour for the inverse Laplace transform must be along a line $\sigma = \mathrm{Re}\{p\}$ that passes to the right of all poles of $\tilde{f}(p)$.

equation, one can obtain the Laplace transform of higher-order derivatives. For example, the Laplace transform of the second derivative of a function is given by

$$\tilde{f}''(p) = p[p\tilde{f}(p) - f(0)] - f'(0). \tag{9.2.4}$$

Laplace transforms can be used to obtain solutions of differential equations in much the same way as Fourier transforms. However, Laplace transforms have the advantage that the initial conditions are automatically included in the solution.

To illustrate the use of Laplace transforms for solving a differential equation, we solve the problem of a damped harmonic oscillator subject to a specific set of initial conditions at $t = 0$. The differential equation for a damped harmonic oscillator is given by

$$f'' + 2\gamma_0 f' + \omega_0^2 f = 0, \tag{9.2.5}$$

where γ_0 is the damping constant and ω_0 is the characteristic frequency. The Laplace transform of the above equation is

$$\tilde{f}'' + 2\gamma_0 \tilde{f}' + \omega_0^2 \tilde{f} = 0. \tag{9.2.6}$$

Using Eqs. (9.2.3) and (9.2.4), we obtain

$$p^2\tilde{f} - pf(0) - f'(0) + 2\gamma_0 p\tilde{f} - 2\gamma_0 f(0) + \omega_0^2 \tilde{f} = 0. \tag{9.2.7}$$

This purely algebraic equation can then be solved for $\tilde{f}$, which gives

$$\tilde{f}(p) = \frac{f'(0) + (p + 2\gamma_0)f(0)}{p^2 + 2\gamma_0 p + \omega_0^2}. \tag{9.2.8}$$

To obtain $f(t)$, we must evaluate the inverse Laplace transform of this function. One way to obtain the inverse Laplace transform is by simply looking up the inverse in a table. However, this cannot always be done, and for our purposes it is instructive to compute the inverse directly by evaluating the integral in Eq. (9.2.2). Complex integrals of this type are usually evaluated using the Cauchy residue theorem (see Arfken, 1970). This theorem allows one to compute the contour integral of a function $g(z)$ that is analytic except for a finite number of poles. An analytic function is a function that has a continuous unique derivative in some region R of the complex plane. Any function that can be expanded as a power series is an analytic function. If the function contains a first-order pole z_0 of the form $g(z) = G(z)/(z - z_0)$ within a contour C, then by shrinking the contour to a small circle around z_0 one can show that

$$\int_C g(z)\,\mathrm{d}z = \int_C \frac{G(z)}{z - z_0}\,\mathrm{d}z = 2\pi\mathrm{i}\,\mathrm{Res}[g(z), z_0] = 2\pi\mathrm{i}\,G(z_0), \tag{9.2.9}$$

where the quantity $\mathrm{Res}[g(z), z_0]$ is called the residue of the function $g(z)$ evaluated at z_0. For a first-order pole, the residue is simply the function $G(z)$. If the function contains several first-order poles, then the integral is simply $2\pi\mathrm{i}$ times the sum of the residues within the contour. For a pole of order n, such that $g(z) = G(z)/(z - z_0)^n$, the residue is given by

$$\mathrm{Res}[g(z), z_0] = \frac{1}{(n-1)!} \frac{\mathrm{d}^{n-1}}{\mathrm{d}z^{n-1}} G(z)\Bigg|_{z=z_0}. \tag{9.2.10}$$

To evaluate the inverse Laplace transform using the residue theorem, the line of integration, $\mathrm{Re}\{p\} = \sigma$, must be distorted into a closed contour. To ensure that the additional integration path produces no contribution to the integral, the line of integration must be closed at infinity in the left half of the complex p-plane, $\mathrm{Re}\{p\} < 0$, without intersecting any poles, as shown in Figure 9.5. The integral diverges if the contour is closed at infinity in the right half-plane. Since the Laplace transform is only defined for $\mathrm{Re}\{p\} > 0$, the function $\tilde{f}(p)$ must be extended into the left half of the complex p-plane by analytic continuation (see Flanigan, 1983). In the case of functions involving the ratio of simple polynomials, the analytic continuation is the same function, with unchanged coefficients for the polynomials. Later we will encounter a case where the analytic continuation must be treated more delicately.

To use the residue theorem, we must determine the location of the poles of $\tilde{f}(p)$. As can be seen by inspection of Eq. (9.2.8), the poles are located at the points where the denominator is zero, which are at

$$p_{\pm} = -\gamma_0 \pm \mathrm{i}\sqrt{\omega_0^2 - \gamma_0^2}. \tag{9.2.11}$$

The inverse Laplace transform is then given by

$$f(t) = \frac{1}{2\pi\mathrm{i}} \int_C \left[\frac{f'(0) + (p + 2\gamma_0) f(0)}{(p - p_+)(p - p_-)}\right] \mathrm{e}^{pt}\, \mathrm{d}p, \tag{9.2.12}$$

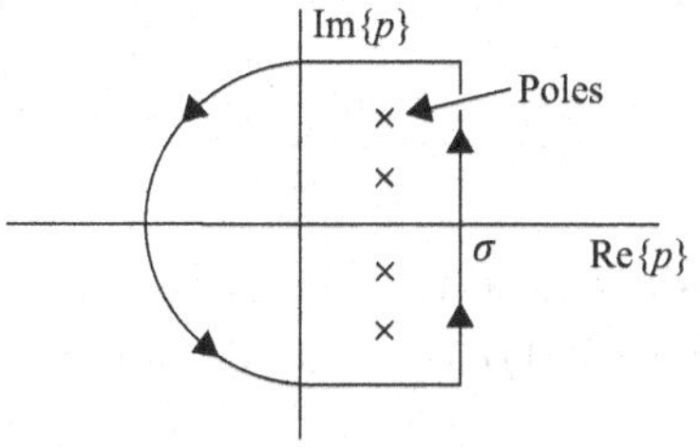

Figure 9.5 To assure convergence, the integral in the inverse Laplace transform must be closed at infinity in the left half of the complex p-plane without intersecting any poles.

where the contour C now passes around the poles at p_+ and p_-. Evaluating the appropriate residues, it can be shown that

$$f(t) = \frac{f'(0) + (p_- + 2\gamma_0)f(0)}{p_- - p_+}\,\mathrm{e}^{p_- t} + \frac{f'(0) + (p_+ + 2\gamma_0)f(0)}{p_+ - p_-}\,\mathrm{e}^{p_+ t}, \tag{9.2.13}$$

which after substituting for p_+ and p_- simplifies to

$$f(t) = \frac{f'(0) + \gamma_0 f(0)}{\omega}\,\mathrm{e}^{-\gamma_0 t}\sin\omega t + f(0)\,\mathrm{e}^{-\gamma_0 t}\cos\omega t, \tag{9.2.14}$$

where $\omega = \sqrt{\omega_0^2 - \gamma_0^2}$. Note that the roots of the denominator of $\tilde{f}(p)$ completely determine the oscillation frequency and damping.

9.2.2 The Dispersion Relation for a Plasma of Hot Electrons and Immobile Ions

As an initial application of the Landau method, we start by considering the small-amplitude waves that can occur in a plasma of hot electrons and immobile ions. The relevant equations are (9.1.4) and (9.1.5). To solve these equations for an arbitrary initial condition, we perform a Fourier transform in space (with the **k** vector along the z axis) and a Laplace transform in time. After dropping the subscript 1 on the first-order terms, the transforms of these equations are

$$p\tilde{f} - f(0) + \mathrm{i}kv_z\tilde{f} + \mathrm{i}\frac{e}{m}\,k\tilde{\Phi}\,\frac{\partial f_0}{\partial v_z} = 0 \tag{9.2.15}$$

and

$$k^2\tilde{\Phi} = -\frac{e}{\epsilon_0}\int_{-\infty}^{\infty}\tilde{f}\,\mathrm{d}^3v, \tag{9.2.16}$$

where $f(0)$ is the first-order distribution function evaluated at $t = 0$. It should be noted that even though it is evaluated at $t = 0$, $f(0)$ is not a constant but depends on position and velocity coordinates. Solving Eq. (9.2.15) for $\tilde{f}$ gives

$$\tilde{f} = \frac{-\mathrm{i}(e/m)\,k\tilde{\Phi}\,(\partial f_0/\partial v_z) + f(0)}{p + \mathrm{i}kv_z}. \tag{9.2.17}$$

Substituting this expression into Eq. (9.2.16) and introducing the reduced distribution functions, $F_0(v_z) = (1/n_0)\int f_0\mathrm{d}v_x\mathrm{d}v_y$ and $F(0) = (1/n_0)\int f(0)\mathrm{d}v_x\mathrm{d}v_y$, we obtain the equation

$$k^2\tilde{\Phi} = -\frac{en_0}{\epsilon_0}\int_{-\infty}^{\infty}\frac{F(0)}{p + \mathrm{i}kv_z}\,\mathrm{d}v_z + \mathrm{i}\omega_\mathrm{p}^2\,k\tilde{\Phi}\int_{-\infty}^{\infty}\frac{\partial F_0/\partial v_z}{p + \mathrm{i}kv_z}\,\mathrm{d}v_z, \tag{9.2.18}$$

where $F(0)$ is the reduced first-order distribution function evaluated at $t = 0$. The above equation can then be solved for $\tilde{\Phi}$ and written in the form

$$\tilde{\Phi}(k,p) = \frac{N(k,p)}{D(k,p)}, \tag{9.2.19}$$

where the numerator is given by

$$N(k,p) = \mathrm{i}\,\frac{en_0}{\epsilon_0 k^3}\int_{-\infty}^{\infty}\frac{F(0)}{v_z - \mathrm{i}p/k}\,\mathrm{d}v_z \tag{9.2.20}$$

and the denominator is given by

$$D(k,p) = 1 - \frac{\omega_\mathrm{p}^2}{k^2}\int_{-\infty}^{\infty}\frac{\partial F_0/\partial v_z}{v_z - \mathrm{i}p/k}\,\mathrm{d}v_z. \tag{9.2.21}$$

Note that the denominator can be obtained by simply changing ω to $\mathrm{i}p$ in the dispersion relation obtained from the Fourier analysis approach, i.e., Eq. (9.1.13).

To determine the temporal behavior of the electrostatic potential we must carry out the inverse Laplace transform

$$\Phi(k,t) = \frac{1}{2\pi\mathrm{i}}\int_{\sigma-\mathrm{i}\infty}^{\sigma+\mathrm{i}\infty}\tilde{\Phi}(k,p)\,\mathrm{e}^{pt}\,\mathrm{d}p, \tag{9.2.22}$$

where the integration is along the line $\mathrm{Re}\{p\} = \sigma$. As discussed earlier, this line must be to the right of all poles of $\tilde{\Phi}(k, p)$.

Following the procedure discussed in Section 9.2.1, we evaluate the inverse Laplace transform by closing the contour of integration in the left half-plane, $\mathrm{Re}\{p\} < 0$, and distorting the integration contour so that the integral can be evaluated by integrating around the poles of $\tilde{\Phi}(k,p)$ using the residue theorem. These steps are illustrated in Figure 9.6. To use the residue theorem, there are two requirements. First, the function $\tilde{\Phi}(k,p)$ must be analytic except for an isolated number of poles. This means that we cannot consider velocity distribution functions that have discontinuities, such as step functions and various other pathological functions. To meet this requirement, we restrict the analysis to distribution functions that are analytic everywhere, that is, can be represented by a power series $F_0(v_z) = \sum_n a_n v_z^n$. Second, the function $\tilde{\Phi}(k,p)$ must be analytically continued from the right half of the complex p-plane into the left half-plane. Unfortunately, this analytic continuation is not straightforward, since the integrals in $N(k,p)$ and $D(k,p)$ have a discontinuity along the line $\gamma = \mathrm{Re}\{p\} = 0$, which separates the right-hand and left-hand halves of the complex p-plane. This difficulty is closely related to the problem that was encountered in the Vlasov approach, namely the singularity at $v_z = \omega/k$.

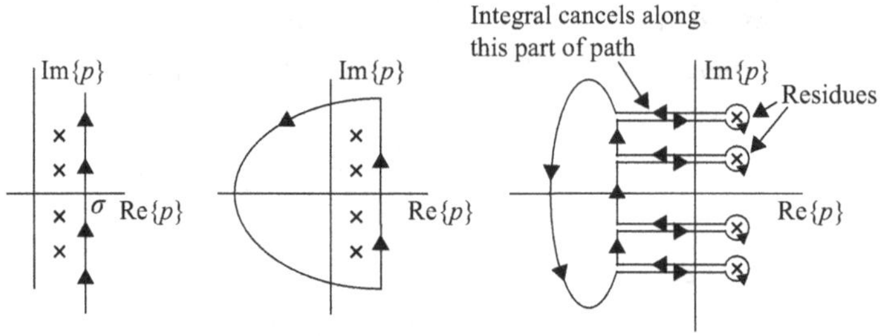

Figure 9.6 The final stage in evaluating the inverse Laplace transform involves distorting the integration contour around the poles and arranging for the integral to cancel around the rest of the contour. The resulting integrals around the poles are called the residues.

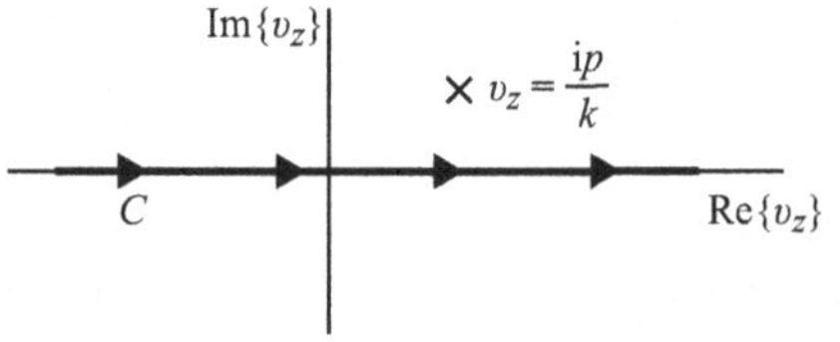

Figure 9.7 For Re{p} positive the integration contour in $Ɖ(k,p)$ is along the Re{v_z} axis.

The resolution of the difficulty at $v_z = \omega/k$ was provided by Landau, who carried out a careful analytic continuation of the functions $N(k,p)$ and $Ɖ(k,p)$. The procedure is essentially the same for the numerator and denominator, so we only discuss the denominator. The requirement for an analytic continuation is that the function must vary smoothly across the boundary between the two regions. From the Laplace transform analysis given above, Eq. (9.2.21) is a valid representation of $Ɖ(k,p)$ in the right half-plane, Re{p} > 0. Since the sign of k determines the sign of ip/k, we must consider two cases: k positive and k negative. First, we consider the case where k is positive. Even though the integral in Eq. (9.2.21) is along the real velocity axis, it is useful to visualize the integration path in the complex v_z plane, as shown in Figure 9.7. For Re{p} > 0 and $k > 0$, the pole at $v_z = ip/k$ lies above the real v_z axis. As the real part of p changes from positive to negative, the pole moves downward across the real v_z axis. To avoid a discontinuity when the pole crosses the real v_z axis, Landau simply distorted the integration contour so that it always stays below the pole, as shown in Figure 9.8. It is then a simple matter to define a function $Ɖ(k,p)$ that is valid for any value of Re{p}:

$$Ɖ(k,p) = 1 - \frac{\omega_p^2}{k^2} \int_C \frac{\partial F_0/\partial v_z}{v_z - ip/k} \, dv_z, \tag{9.2.23}$$

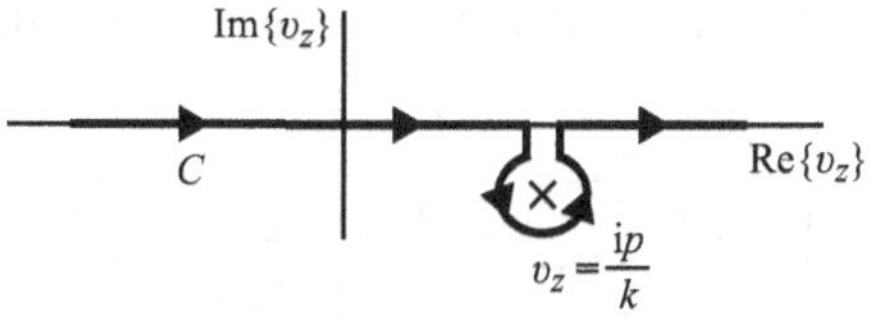

Figure 9.8 To continue $\mathit{Đ}(k,p)$ analytically into the left half of the complex p-plane, the integration contour must be distorted such that it passes below the pole at $v_z = \mathrm{i}p/k$. This diagram is for $k > 0$.

where the contour C passes below the pole at $v_z = \mathrm{i}p/k$. Note that the only difference in the two cases is the shape of the integration contour. For $\mathrm{Re}\{p\} > 0$, the integration contour extends along the real v_z axis from minus infinity to plus infinity. For $\mathrm{Re}\{p\} < 0$ the integration contour includes an additional loop around the pole at $v_z = \mathrm{i}p/k$.

Another way to view the above analytical continuation is to define two functions $Đ_+(k,p)$ and $Đ_-(k,p)$ that are valid for $\mathrm{Re}\{p\} > 0$ and $\mathrm{Re}\{p\} < 0$. These functions are

$$Đ_+(k,p) = 1 - \frac{\omega_\mathrm{p}^2}{k^2}\int_{-\infty}^{\infty} \frac{\partial F_0/\partial v_z}{v_z - \mathrm{i}p/k}\,\mathrm{d}v_z, \quad \mathrm{Re}\{p\} > 0, \tag{9.2.24}$$

and

$$Đ_-(k,p) = 1 - \frac{\omega_\mathrm{p}^2}{k^2}\int_{-\infty}^{\infty} \frac{\partial F_0/\partial v_z}{v_z - \mathrm{i}p/k}\,\mathrm{d}v_z - 2\pi\mathrm{i}\frac{k}{|k|}\frac{\omega_\mathrm{p}^2}{k^2}\left.\frac{\partial F_0}{\partial v_z}\right|_{v_z=\mathrm{i}p/k}, \quad \mathrm{Re}\{p\} < 0. \tag{9.2.25}$$

The $2\pi\mathrm{i}$ term in the above equation comes from evaluating the residue around the pole at $v_z = \mathrm{i}p/k$. This term corrects for the discontinuity as the pole crosses the real v_z axis. Note that the integrals in both of the above two equations are along the real v_z axis from $-\infty$ to $+\infty$. The $k/|k|$ term in the residue takes care of the case where k is negative, which has the effect of reversing the sense (hence sign) of the contour around the pole. An essentially identical procedure is used to obtain the analytic continuation of $N(k,p)$.

Having established the analytic continuation of $N_+(k,p)$ and $Đ_+(k,p)$, the temporal variation of the electrostatic potential can be determined by computing the inverse Laplace transform

$$\Phi(k,t) = \frac{1}{2\pi\mathrm{i}}\int_{\sigma-\mathrm{i}\infty}^{\sigma+\mathrm{i}\infty} \tilde{\Phi}(k,p)\,\mathrm{e}^{pt}\,\mathrm{d}t. \tag{9.2.26}$$

The inverse transform is easily evaluated using the residue theorem, which gives a solution of the form

$$\Phi(k,t) = \sum_i e^{p_i t}\,\mathrm{Res}[\tilde{\Phi}(k,p), p_i], \tag{9.2.27}$$

where p_i is the ith pole of $\tilde{\Phi}(k,p)$. Since $F(0)$ was assumed to be an analytic function of v_z, the numerator of $\tilde{\Phi}(k,p)$ does not have any singularities. Hence, the only poles of $\tilde{\Phi}(k,p)$ are those for which $Ɖ(k,p) = 0$.

Because the p_i are in general complex, the time behavior of the potential (hence the electric field) is in general oscillatory, with exponential damping (growth) if the real part of p_i is negative (positive). As t becomes large, the pole with the largest value for $\mathrm{Re}\{p_i\}$ dominates the oscillatory behavior of the electric field.

Since the roots of $Ɖ(k,p) = 0$ give the frequency and growth rate of the electrostatic potential, the equation $Ɖ(k,p) = 0$ plays the same role as the dispersion relation in the Fourier analysis approach. However, the roots $p_i = \gamma_i - i\omega_i$ are now complex.

9.2.3 The Cauchy Velocity Distribution Function

As a simple example that illustrates the above solution, consider a plasma consisting of electrons with a velocity distribution function given by

$$F_0(v_z) = \frac{C}{\pi}\frac{1}{C^2 + v_z^2} \tag{9.2.28}$$

and an equal density of immobile ions. This distribution function, which is called the Cauchy distribution function, has the advantage of being relatively easy to analyze and has a shape similar to a Maxwellian. The easiest way to evaluate the dispersion relation is by integrating Eq. (9.2.23) by parts once, which gives

$$Ɖ(k,p) = 1 - \frac{\omega_p^2}{k^2}\int_C \frac{F_0(v_z)}{(v_z - ip/k)^2}\,dv_z = 0. \tag{9.2.29}$$

After substituting $F_0(v_z)$ from Eq. (9.2.28) into the above equation, the dispersion relation can be written

$$Ɖ(k,p) = 1 - \frac{\omega_p^2}{k^2}\frac{C}{\pi}\int_C \frac{dv_z}{(v_z - iC)(v_z + iC)(v_z - ip/k)^2} = 0. \tag{9.2.30}$$

The integrand in the above equation has three poles, at $v_z = iC, -iC$, and ip/k. For the purpose of computing the integral, we start by assuming that $\mathrm{Re}\{p\}$ is positive. Two cases must be considered, $k > 0$ and $k < 0$. For $k > 0$, the poles are located as shown in Figure 9.9. The integral is easily evaluated using the residue

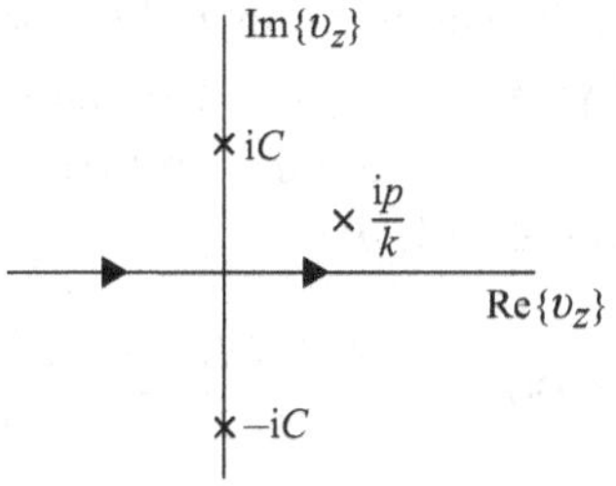

Figure 9.9 The location of the poles in the integral for $Đ(k, p)$ for the Cauchy distribution function. This diagram assumes that $\text{Re}\{p\}$ is positive and that $k > 0$.

theorem. In this case it is convenient to close the contour in the lower half of the complex v_z plane for $k > 0$, and in the upper half of the complex v_z plane for $k < 0$. This procedure minimizes the number of residues that must be considered. After evaluating the residues, the dispersion relation can be written as a single equation that is valid for both $k > 0$ and $k < 0$ (see Problem 9.6):

$$Đ(k, p) = 1 + \frac{\omega_p^2}{(p + C|k|)^2} = 0. \tag{9.2.31}$$

At this point the dispersion relation is only valid in the right half of the complex p-plane, since we have assumed that $\text{Re}\{p\} > 0$. However, the analytic continuation to the lower half-plane is trivial, since the function is a polynomial in p and is therefore continuous throughout the complex p-plane. Note that since a direct analytical continuation can be made, it is not necessary to consider the alternative definitions for $Đ(k, p)$ given by Eq. (9.2.23), or Eqs. (9.2.24) and (9.2.25).

The roots to the above dispersion relation are easily shown to be

$$p + C|k| = \pm i\omega_p. \tag{9.2.32}$$

Since $p = \gamma - i\omega$, the growth rate and frequency are $\gamma = -C|k|$ and $\omega = \pm\omega_p$. The temporal dependence of the electrostatic potential is then given by

$$\tilde{\Phi}(k, t) = \tilde{\Phi}_0\, e^{\pm i\omega_p t}\, e^{-C|k|t}, \tag{9.2.33}$$

where $\tilde{\Phi}_0$ represents the contribution obtained from the initial value term. Note that the growth rate, $\gamma = -C|k|$, is negative. This means that the fields are exponentially damped. This damping is called Landau damping, after Landau (1946), who first identified this collisionless damping process.

9.2.4 The Weak Growth Rate Approximation

To gain further insight into the nature of Landau damping, we next develop an approximate solution of the dispersion relation that is valid whenever the magnitude of the growth rate is small compared to the frequency, $|\gamma| \ll |\omega|$. This approximation is obtained by expanding the dispersion relation in a Taylor series around the point $p = -\mathrm{i}\omega$, where $p = \gamma - \mathrm{i}\omega$. To first order in the small parameter γ, the Taylor series expansion of $Ð(k, p)$ is

$$Ð(k,p) = Ð(k,-\mathrm{i}\omega) + \mathrm{i}\left(\frac{\partial Ð(k,-\mathrm{i}\omega)}{\partial \omega}\right)\gamma = 0. \tag{9.2.34}$$

The first term, $Ð(k, -\mathrm{i}\omega)$, includes both real and imaginary parts, both of which must be evaluated in the limit $\gamma \to 0$. Thus, for the first term we write

$$Ð(k,-\mathrm{i}\omega) = Ð_{\mathrm{r}}(k,-\mathrm{i}\omega) + \mathrm{i}Ð_{\mathrm{i}}(k,-\mathrm{i}\omega). \tag{9.2.35}$$

The second term, $(\partial Ð/\partial\omega)\gamma$, also includes both real and imaginary parts. However, we only keep the real part, since for small $|\gamma|$ the imaginary part gives a negligible correction to the real part of Eq. (9.2.34). Combining the above two equations and setting the real and imaginary parts to zero then gives two equations,

$$Ð_{\mathrm{r}} = 0 \tag{9.2.36}$$

and

$$\gamma = \frac{-Ð_{\mathrm{i}}}{\partial Ð_{\mathrm{r}}/\partial\omega}. \tag{9.2.37}$$

Since we are taking the limit as γ approaches zero, to evaluate $Ð_{\mathrm{i}}$ and $Ð_{\mathrm{r}}$ we must carefully consider the integral in $Ð(k, -\mathrm{i}\omega)$, which after substituting $p = \gamma - \mathrm{i}\omega$ can be written

$$Ð(k,-\mathrm{i}\omega) = 1 - \lim_{\gamma\to 0}\frac{\omega_{\mathrm{p}}^2}{k^2}\int_C \frac{\partial F_0/\partial v_z}{v_z - (\omega/k + \mathrm{i}\gamma/k)}\,\mathrm{d}v_z. \tag{9.2.38}$$

To evaluate this integral we use the Plemelj relation, which is given by

$$\lim_{\epsilon\to 0}\int_{-\infty}^{\infty}\frac{f(x)}{x-(x_0 \pm \mathrm{i}\epsilon)}\,\mathrm{d}x = P\int_{-\infty}^{\infty}\frac{f(x)}{x-x_0}\,\mathrm{d}x \pm \mathrm{i}\pi f(x_0), \tag{9.2.39}$$

where ϵ is a small positive quantity and P refers to the principal value integral, which is defined by

$$P\int_{-\infty}^{\infty}\cdots\mathrm{d}x = \lim_{\delta\to 0}\left[\int_{-\infty}^{x_0-\delta}\cdots\mathrm{d}x + \int_{x_0+\delta}^{\infty}\cdots\mathrm{d}x\right]. \tag{9.2.40}$$

Note that the usual $2\pi\mathrm{i}$ term in the residue around the pole is replaced by $\mathrm{i}\pi$. The change from $2\pi\mathrm{i}$ to $\mathrm{i}\pi$ occurs because in the limit $\epsilon \to 0$ the integration contour only

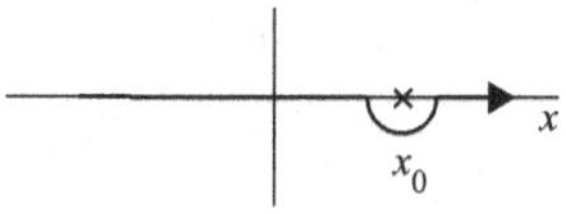

Figure 9.10 In the limit $\epsilon \to 0$ the integration contour in the Plemelj relation goes half-way around the pole at x_0.

goes half-way around the pole, as shown in Figure 9.10. After using the Plemelj relation, $Ð_r$ and $Ð_i$ become

$$Ð_r = 1 - \frac{\omega_p^2}{k^2} P \int_{-\infty}^{\infty} \frac{\partial F_0/\partial v_z}{v_z - \omega/k} \, dv_z \tag{9.2.41}$$

and

$$Ð_i = -\pi \frac{k}{|k|} \frac{\omega_p^2}{k^2} \frac{\partial F_0}{\partial v_z}\bigg|_{v_z = \omega/k} . \tag{9.2.42}$$

The $k/|k|$ term in $Ð_i$ takes into account the change in sign that occurs when $k > 0$ and $k < 0$. As in the Cauchy distribution, the sign of k determines whether the pole approaches the real v_z axis from above ($k > 0$) or from below ($k < 0$), which in turn determines the $\pm$ sign in the Plemelj relation. The real part of $Ð$ is essentially the same as was obtained from the Fourier analysis approach, except the principal value integral now provides a mathematical procedure for avoiding the divergence in the integral at $v_z = \omega/k$. Substituting Eq. (9.2.42) into Eq. (9.2.37), we obtain the following expression for the growth rate:

$$\gamma = \pi \frac{k}{|k|} \frac{\omega_p^2}{k^2 \partial Ð_r/\partial \omega} \frac{\partial F_0}{\partial v_z}\bigg|_{v_z = \omega/k} . \tag{9.2.43}$$

This result is called the weak growth rate approximation and is valid whenever $|\gamma| \ll |\omega|$. The weak growth rate approximation shows that the growth rate is proportional to the slope of the reduced distribution evaluated at the phase velocity of the wave (see Figure 9.11).

For Langmuir waves, we can carry out an explicit evaluation of the $\partial Ð_r/\partial \omega$ term in Eq. (9.2.43). From the real part of the dispersion relation, it is easy to show, using the same procedure as in the Fourier analysis approach, that in the high phase velocity limit the real part of the dispersion relation is given by

$$Ð_r = 1 - \frac{\omega_p^2}{\omega^2}\left(1 + 3\frac{k^2}{\omega^2}\langle v_z^2 \rangle\right). \tag{9.2.44}$$

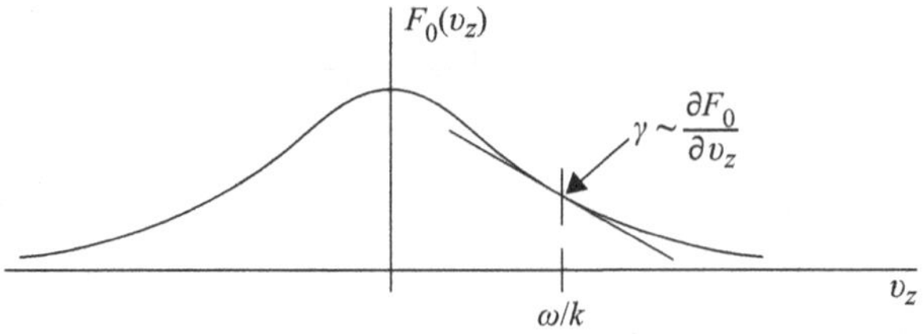

Figure 9.11 In the weak growth rate approximation, the growth rate, γ, is proportional to the slope of the distribution function evaluated at the phase velocity, ω/k, of the wave.

To lowest order $k, D_r = 1 - \omega_p^2/\omega^2$, so $\partial D_r/\partial\omega = 2\omega_p^2/\omega^3$, which, when evaluated at $\omega = \omega_p$, gives

$$\gamma = \frac{\pi}{2}\frac{k}{|k|}\frac{\omega_p^3}{k^2}\left.\frac{\partial F_0}{\partial v_z}\right|_{v_z = \omega/k}. \quad (9.2.45)$$

If the reduced distribution function has a single maximum at $v_z = 0$, such as for a Maxwellian or a Cauchy distribution, one can see from the above equation that the growth rate is always negative (i.e., indicating damping), independent of the direction of propagation. Note that when k is positive, ω/k is positive and $\partial F_0/\partial v_z$ is negative, which makes γ negative. When k is negative, ω/k is negative, and $\partial F_0/\partial v_z$ is positive, which again makes γ negative.

For a Maxwellian distribution, it is easy to show that the frequency and growth rate are given, respectively, by

$$\omega^2 = \omega_p^2 + 3\left(\frac{\kappa T}{m}\right)k^2 \quad (9.2.46)$$

and

$$\gamma = -\sqrt{\frac{\pi}{8}}\frac{\omega_p}{|k\lambda_D|^3}\exp\left[-\frac{1}{2(k\lambda_D)^2} - \frac{3}{2}\right], \quad (9.2.47)$$

where λ_D is the Debye length. The frequency is identical to the result obtained in the Fourier analysis approach (i.e., the Bohm–Gross dispersion relation). The growth rate is always negative, which indicates damping, very similar to the situation with the Cauchy distribution. Note that the $\exp[-3/2]$ factor comes from the second term on the right side of Eq. (9.2.46). Because of the strong exponential dependence of the Maxwellian, it is important to keep this higher-order term. As can be seen, the damping is a strong function of $k\lambda_D$. For long wavelengths, much greater than the Debye length, the damping is very weak. As the wavelength decreases, the damping increases rapidly, becoming very strong as the wavelength approaches the Debye length. Oscillatory solutions do not exist for $k\lambda_D \gtrsim 1$.

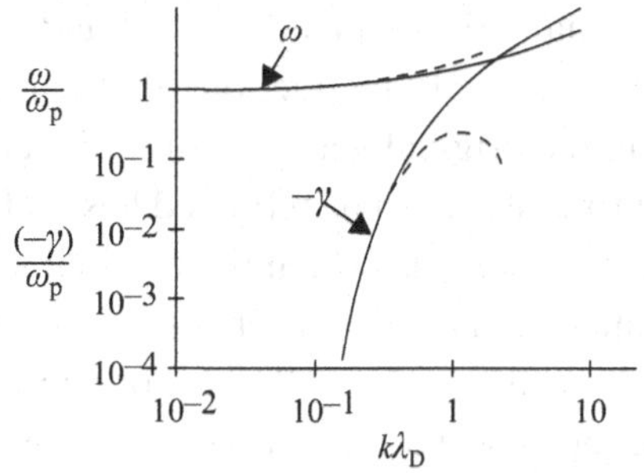

Figure 9.12 A comparison of an approximate analytical solution to the Langmuir wave dispersion relation (dashed lines) with the results of an exact numerical computation (solid lines).

Physically, the damping becomes large as the phase velocity of the wave moves into the region where the slope of the distribution function is large. A plot of the frequency, ω, and damping rate, $-\gamma$, is shown in Figure 9.12. The dashed lines are from Eqs. (9.2.46) and (9.2.47), and the solid lines are from an exact numerical solution of the dispersion relation. As can be seen, Eqs. (9.2.46) and (9.2.47) give a good fit for $k\lambda_D \ll 1$, but fail as $k\lambda_D \to 1$.

9.2.5 The Physical Origin of Landau Damping

The theoretical discovery by Landau (1946) of a wave damping mechanism that does not depend on collisions was not only surprising, but left many unanswered questions regarding the physical processes involved. The surprise lies in the fact that the Vlasov equation does not have any mechanism for irreversible interactions (for example, collisions), and yet Landau's analysis predicts damping. That the damping is real, and not just an artifact of the mathematical techniques used, was shown in a laboratory experiment by Malmberg and Wharton (1966). From our previous analyses, which showed that the growth rate is proportional to the slope of the zero-order distribution function at the phase velocity of the wave, it is clear that Landau damping is somehow closely associated with particles moving at velocities at or near the phase velocity, the so-called resonant particles. The widely accepted explanation of this paradox is that initially, at $t = 0$, the electrons are bunched in configuration space in such a way that they produce the initial spatial electrostatic potential $\Phi(t = 0)$, and that after $t = 0$ the resonant electrons stream away at slightly different velocities, as given by the initial velocity distribution function $F_0(v_z)$. Because of this velocity spread, the original bunch of electrons gradually spreads in configuration space, thereby causing the initial waveform to decay in amplitude. Because the electrons get out of phase relative to the original waveform, this process is often called "phase mixing." Although phase mixing provides a qualitative explanation of Landau damping, because the Vlasov equation has no irreversible interactions that can randomize the phases, the information regarding the original waveform is still "encoded" in the electron phases. That the phase information can be retrieved has, in fact, been demonstrated in carefully controlled experiments in which the original waveform reappears at a later time, yielding so-called "long delayed echoes" (Gould et al., 1967). However, if there is any irreversible scattering by collisions or some other random process, such as interactions with other waves, the encoded information is ultimately lost, leading to an increase in entropy.

Since the equations leading to Landau damping were linearized, it is clear that Landau damping is an inherently linear process. However, there is also a nonlinear aspect to Landau damping, and it is important to recognize the difference between the two processes. That nonlinear effects eventually become involved is easily demonstrated. From the derivation of Eq. (9.1.4), one can see that the key requirement in the linearization process is that

$$|\partial f/\partial v_z| \ll |\partial f_0/\partial v_z|. \tag{9.2.48}$$

Since the zero-order term, $\partial f_0/\partial v_z$, is independent of time, we focus our attention on the first-order term, $\partial f/\partial v_z$. From Eq. (9.2.17), one can see that there is a term of the form

$$\frac{\partial \tilde{f}}{\partial v_z} = -\frac{\mathrm{i}kf(0)}{(p+\mathrm{i}kv_z)^2}. \tag{9.2.49}$$

Using a table of inverse Laplace transforms, it can be shown that the corresponding temporal dependence is given by

$$\frac{\partial f}{\partial v_z} = -\mathrm{i}kf(0)t\mathrm{e}^{-\mathrm{i}kv_z t}. \tag{9.2.50}$$

Since the magnitude of this term increases linearly with time, it is clear that the inequality in the above equation is eventually violated, no matter how small the initial perturbation, $f(0)$.

From the above discussion it is clear that there are two time-scales that must be considered in Landau damping, an early phase in which linear analysis holds, and a later phase in which nonlinear effects dominate. To understand these linear and nonlinear aspects of Landau damping, it is useful to transform into a frame of reference that is moving at the phase velocity of the wave, ω/k. The advantage of this frame of reference is that for a sufficiently small growth rate the electrostatic potential of the wave is nearly time-stationary. The total energy of the particles, $W=(1/2)mv_z^2+q\Phi$, is then conserved, at least to a first approximation. The particle

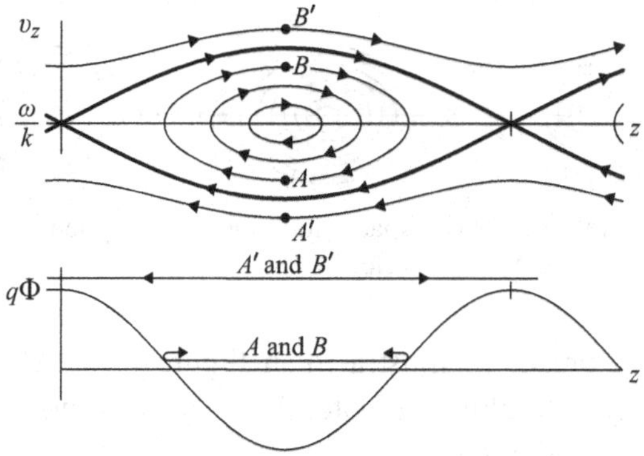

Figure 9.13 Phase-space trajectories in a frame of reference (z, v_z) moving at the phase velocity for a sinusoidal electrostatic potential $\Phi(z) = \Phi_0 \cos kz$.

velocities are then given by

$$v_z = \pm\sqrt{\frac{2}{m}(W - q\Phi)}, \tag{9.2.51}$$

where the (+) sign is for particles moving in the $+z$ direction, and the (−) sign is for particles moving in the $-z$ direction. For an electrostatic potential of the form $\Phi(z) = \Phi_0 \cos kz$, the trajectories in (z, v_z) phase space are as shown in Figure 9.13. The slightly darker lines in the phase-space plot that separate trapped particles from untrapped particles are called separatrices. The total energy of particles on the separatrices is given by $W_0 = q\Phi_0$, where Φ_0 is the amplitude of the electrostatic potential. Particles with total energies $W < W_0$ are trapped and particles with total energies $W > W_0$ are untrapped. The velocity distribution of particles on these trajectories is determined by the zero-order distribution function, $F_0(v_z)$. Once the initial particle distribution function is determined, trapped particles, such as A and B, follow the ellipse-shaped trajectories inside the separatrices, and untrapped particles, such as A' and B', stream freely along the trajectories outside the separatrices. The oscillation frequency of the trapped particles depends on how deep the particles are in the potential well. Particles trapped near the bottom of the potential well, where the potential is parabolic, oscillate at a frequency called the bounce frequency, which is given by

$$\omega_b = k\sqrt{\frac{q\Phi_0}{m}}. \tag{9.2.52}$$

Due to the increasing width of the potential well at higher energies, the oscillation frequency decreases with increasing total energy, ultimately going to zero for particles at the top of the well. Since the oscillation frequency decreases with increasing energy, as the particles move along the ellipse-shaped trajectories, they

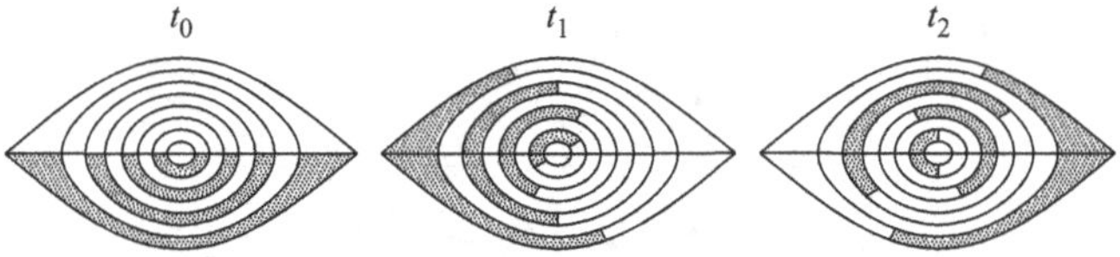

Figure 9.14 The relative phase-space locations of trapped particles at three successive phases of the bounce cycle.

gradually get out of phase, as illustrated in Figure 9.14. Although the trapped particle motions may be highly correlated initially, they eventually become well mixed throughout the trapping region.

The main difference between the phase mixing of the trapped and untrapped particles is that the average velocity of the trapped particles must move at the phase velocity of the wave, whereas the non-trapped particles are able to stream freely relative to the phase of the wave. Although both contribute to the overall phase mixing, as we discuss below, the nonlinearly trapped particles produce effects on the long-term evolution of the wave amplitude that deviate considerably from the exponential damping that is characteristic of the early linear phase of Landau damping.

For sufficiently small wave amplitudes, Eq. (9.2.52) shows that the bounce period of the trapped particles is very long, much longer than the oscillation period of the electrostatic wave. Therefore, during the initial phase, particle trapping effects are not important, since the particles have not had time to complete even a small fraction of a bounce cycle on the time-scale of the exponential decay. It is during this initial phase that Landau's linear analysis applies. The duration of the linear phase is bounded by

$$0 < t \ll \omega_{\mathrm{b}}^{-1} = \frac{1}{k}\sqrt{\frac{m}{q\Phi_0}}. \tag{9.2.53}$$

The easiest way to develop a quantitative understanding of the mechanism responsible for damping of the wave in both the linear and nonlinear phases is to investigate the kinetic energy of the particles interacting with the wave. Since the total energy of the system is conserved, if the kinetic energy of the particles increases, then the electrostatic energy of the wave must decrease. Whether the kinetic energy of the particles increases or decreases in response to the presence of the wave depends on the initial distribution of particle velocities, as characterized by the zero-order particle distribution function, $F_0(v_z)$. If the slope of the zero-order distribution function, $\partial F_0(v_z)/\partial v_z$, is negative (at $v_z \approx \omega/k$), the number of particles initially moving slower than the phase velocity (for example, particles A and A' in Figure 9.13) is greater than the number of particles initially moving faster than the phase velocity (for example, particles B and B' in Figure 9.13). As time advances, these particles proceed to move in the direction of the arrows. From the direction of motion it is evident that the kinetic energy in the plasma rest frame, which is proportional to $(\omega/k + v_z)^2$, increases for particles A and A' and decreases for particles B and B'. Since there are more particles of type A and A' than there are of type B and B', the total kinetic energy contributed by these particles must increase. Note that the increase of the total kinetic energy in the plasma rest frame is a consequence of the quadratic dependence of the kinetic energy on v_z, i.e., $(\omega/k + v_z)^2$. Also, note that since the particles have only moved a small fraction of a bounce cycle, the issue of whether the particles are trapped or not is irrelevant during the initial linear phase. It can be shown (see, for instance, Nicholson (1983)) that the rate of change of the kinetic energy is given by

$$\frac{\mathrm{d}}{\mathrm{d}t}\left\langle \frac{1}{2} n_0 m \left(\frac{\omega}{k} + v_z\right)^2 \right\rangle = -\pi \left(\frac{\epsilon_0 E_0^2}{2}\right) \frac{\omega_\mathrm{p}^3}{k^2} F_0'\left(\frac{\omega}{k}\right), \tag{9.2.54}$$

where E_0 is the electric field amplitude and the angle brackets indicate an average over many cycles. As one can see, if the derivative of the zero-order distribution function, $F_0'(\omega/k)$, is negative, the kinetic energy of the particles increases. The wave energy must then decrease, which implies damping. The above equation gives a damping rate that is in exact agreement with the damping rate obtained using the weak growth rate approximation, i.e., Eq. (9.2.45).

As the linear phase of the Landau damping proceeds, a point is eventually reached where inequality (9.2.48) is no longer satisfied. The bounce motion of the trapped particles then becomes important (O'Neil, 1965). Since the trapped particle motions involve turning points, the dynamical motions are inherently nonlinear. What happens during the nonlinear phase is difficult to determine analytically and is usually investigated by numerically solving the fully nonlinear equations. From such numerical simulations, it is possible to give a relatively simple interpretation of the physical processes involved. During the initial linear phase, the wave gives energy to the resonant particles, both trapped and untrapped. As time progresses into the nonlinear phase, the trapped particles start to give some of the energy back to the wave. However, not all of the initial energy can be returned, because some has been lost to the untrapped particles. Also, the trapped particles start to phase mix, which further diminishes the energy return. The result is that the wave amplitude decreases less rapidly than predicted by the linear damping rate and starts to level off, as shown in Figure 9.15, or may even increase (Armstrong, 1967). Eventually, after many bounces, the trapped particles become thoroughly phase mixed. At this point, the wave damping ceases. Although the trapped particles in this final state may appear to be randomly distributed, they are not. In principle, the information imprinted by the initial state is still contained in

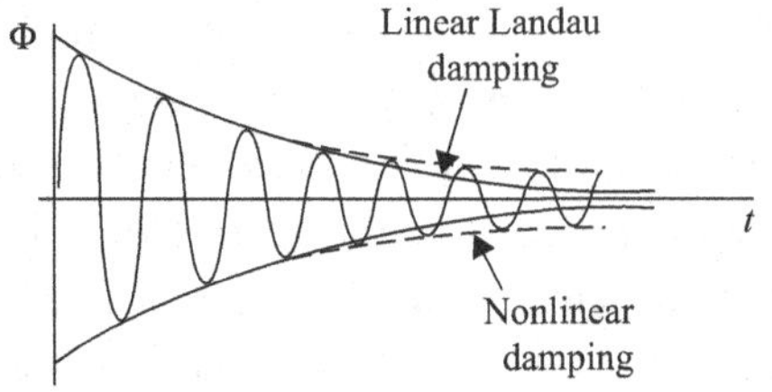

Figure 9.15 The nonlinear effects of particle trapping tend to increase the wave amplitude relative to the predictions of linear Landau damping.

their relative phases. However, as in the linear phase, if there is even a small rate of irreversible interactions, such as due to collisions, the phase-mixed distribution is eventually converted to a completely random distribution.

9.3 The Plasma Dispersion Function

Next we discuss the dispersion relation for a Maxwellian distribution function, which in normalized one-dimensional form is given by

$$F_0(v_z) = \left(\frac{m}{2\pi\kappa T}\right)^{1/2} \exp\left[-\frac{mv_z^2}{2\kappa T}\right]. \tag{9.3.1}$$

Unfortunately, for a Maxwellian distribution function, the integral in the dispersion relation equation (9.2.23) cannot be evaluated using the residue theorem. The reason is that when the integration contour is closed at infinity the integral diverges. Nevertheless, approximations can be carried out in certain limits, and it is useful to develop the tools necessary to carry out these approximations, since we will have use for these tools later. If we make a change of variables to

$$z = \sqrt{\frac{m}{2\kappa T}}v_z \quad \text{and} \quad \zeta = \sqrt{\frac{m}{2\kappa T}}\left(\frac{\mathrm{i}p}{k}\right), \tag{9.3.2}$$

it is easy to show that the dispersion relation equation (9.2.23) can be rewritten in the form

$$D(k,p) = 1 + \frac{1}{(k\lambda_\mathrm{D})^2}\frac{1}{\sqrt{\pi}}\int_C \frac{z\,\mathrm{e}^{-z^2}}{z-\zeta}\,\mathrm{d}z = 0, \tag{9.3.3}$$

where we have made use of the fact that $\sqrt{\kappa T/m} = \omega_\mathrm{p}\lambda_\mathrm{D}$. The integral in the above expression can be rewritten in a standard form by the algebraic reorganization

$$\begin{aligned}\frac{1}{\sqrt{\pi}}\int_C \frac{z\,\mathrm{e}^{-z^2}}{z-\zeta}\,\mathrm{d}z &= \frac{1}{\sqrt{\pi}}\int_C \left(1 + \frac{\zeta}{z-\zeta}\right)\mathrm{e}^{-z^2}\,\mathrm{d}z \\ &= 1 + \zeta Z(\zeta),\end{aligned} \tag{9.3.4}$$

where we have used $\int \exp[-z^2]\,\mathrm{d}z = \sqrt{\pi}$, and the function

$$Z(\zeta) = \frac{1}{\sqrt{\pi}} \int_C \frac{\mathrm{e}^{-z^2}}{z-\zeta}\,\mathrm{d}z \tag{9.3.5}$$

is known as the plasma dispersion function (Fried and Conte, 1961). The contour C is understood to be along the real z axis, passing under the pole at $z = \zeta$. Using the plasma dispersion function, the dispersion relation equation (9.3.3) can then be rewritten in the following form:

$$Ð(k, p) = 1 + \frac{1}{(k\lambda_{\mathrm{D}})^2}[1 + \zeta Z(\zeta)] = 0. \tag{9.3.6}$$

If we define $\zeta = x + \mathrm{i}y$, one can see from Eq. (9.3.2) and the definition $p = \gamma - \mathrm{i}\omega$ that the dimensionless quantities x and y are proportional to the frequency and growth rate

$$x = \sqrt{\frac{m}{2\kappa T}}\,\frac{\omega}{k} \quad \text{and} \quad y = \sqrt{\frac{m}{2\kappa T}}\,\frac{\gamma}{k}. \tag{9.3.7}$$

To proceed further, it is necessary to develop appropriate power series expansions for $Z(\zeta)$. Two expansions in particular are useful. The first is the large-argument expansion ($|\zeta| \gg 1$), valid for high phase velocities, and the second is the small-argument expansion ($|\zeta| \ll 1$), valid for low phase velocities.

We start by discussing the large-argument expansion ($|\zeta| \gg 1$). For many applications the pole in the integration contour is located very close to the real z axis (i.e., $y \ll x$). This is normally the case for the large-argument expansion, since at high phase velocities the magnitude of the growth rate, $|\gamma|$, is usually very small. When the pole is located very close to the real z axis, the Plemelj relation can be used to evaluate the plasma dispersion function. Using the Plemelj relation, it can then be shown that the plasma dispersion function becomes

$$Z(\zeta) = \mathrm{i}\frac{k}{|k|}\sqrt{\pi}\mathrm{e}^{-\zeta^2} + \frac{1}{\sqrt{\pi}} P \int_{-\infty}^{\infty} \frac{\mathrm{e}^{-z^2}}{z-\zeta}\,\mathrm{d}z, \tag{9.3.8}$$

where the $k/|k|$ factor takes into account the sign change that occurs at $k = 0$. To evaluate the integral, one proceeds by expanding the $1/(z-\zeta)$ term as a series in inverse powers of ζ:

$$\frac{1}{z-\zeta} = \frac{-1}{\zeta(1 - z/\zeta)} = -\frac{1}{\zeta}\left[1 + \left(\frac{z}{\zeta}\right) + \left(\frac{z}{\zeta}\right)^2 + \cdots\right]. \tag{9.3.9}$$

Substituting this series into the integral, and integrating term by term, gives the following power series:

$$Z(\zeta) = \mathrm{i}\frac{k}{|k|}\sqrt{\pi}\mathrm{e}^{-\zeta^2} - \left[\frac{1}{\zeta} + \frac{1}{2\zeta^3} + \frac{3}{4\zeta^5} + \cdots\right]. \tag{9.3.10}$$

For the small-argument expansion ($|\zeta| \ll 1$), an alternative representation must be used for $Z(\zeta)$, since the above expansion does not converge if $\zeta/z \ll 1$. By simple substitution, it is straightforward to show that the plasma dispersion function satisfies the linear first-order differential equation,

$$\frac{1}{2}\frac{\mathrm{d}Z}{\mathrm{d}\zeta} + \zeta Z + 1 = 0. \tag{9.3.11}$$

This differential equation has several equivalent alternative solutions, one of which is

$$Z(\zeta) = \mathrm{i}\frac{k}{|k|}\sqrt{\pi}\mathrm{e}^{-\zeta^2} + \mathrm{i}2\mathrm{e}^{-\zeta^2}\int_0^{\mathrm{i}\zeta}\mathrm{e}^{-z^2}\,\mathrm{d}z. \tag{9.3.12}$$

By expanding e^{-z^2} and $\mathrm{e}^{-\zeta^2}$ in a power series, integrating term by term, and collecting the results together in a single power series, it can be shown that, for $|\zeta| \ll 1$, the plasma dispersion function can be represented by the expansion

$$Z(\zeta) = \mathrm{i}\frac{k}{|k|}\sqrt{\pi}\mathrm{e}^{-\zeta^2} - 2\zeta + \frac{4}{3}\zeta^3 - \frac{8}{15}\zeta^5 + \cdots \tag{9.3.13}$$

By using Eq. (9.3.11) to eliminate the term $\zeta Z(\zeta)$ from Eq. (9.3.6), the dispersion relation can be written in the following convenient form:

$$D(k,p) = 1 - \frac{1}{(k\lambda_{\mathrm{D}})^2}\frac{1}{2}Z'(\zeta) = 0, \tag{9.3.14}$$

where $Z'(\zeta)$ is the derivative of the plasma dispersion function. The derivative of the plasma dispersion function is obtained by simply differentiating the large- and small-argument expansions term by term, which for $|\zeta| \gg 1$ gives

$$Z'(\zeta) = -\mathrm{i}\frac{k}{|k|}\sqrt{\pi}2\zeta\mathrm{e}^{-\zeta^2} + \frac{1}{\zeta^2} + \frac{3}{2}\frac{1}{\zeta^4} + \frac{15}{4}\frac{1}{\zeta^6} + \cdots \tag{9.3.15}$$

and for $|\zeta| \ll 1$ gives

$$Z'(\zeta) = -\mathrm{i}\frac{k}{|k|}\sqrt{\pi}2\zeta\mathrm{e}^{-\zeta^2} - 2 + 4\zeta^2 - \frac{8}{3}\zeta^4 + \cdots \tag{9.3.16}$$

To illustrate the use of these expansions, we next evaluate the dispersion relation in the high phase velocity limit. Using the first three terms of the large-argument expansion equation (9.3.15), substituting $\zeta = x + \mathrm{i}y$, and expanding in powers of the small quantity y/x, the dispersion relation becomes

$$Ð(k,p) = 1 - \frac{1}{(k\lambda_{\mathrm{D}})^2}\left[\frac{1}{2x^2} + \frac{3}{4x^4} - \mathrm{i}\left(\frac{y}{x^3} + \frac{k}{|k|}\sqrt{\pi}x\,\mathrm{e}^{-x^2}\right)\right] = 0. \tag{9.3.17}$$

Setting the real and imaginary parts of the above equation equal to zero and using the definitions (Eq. (9.3.7)) for x and y, it can be shown that the frequency and growth rate are given by

$$\omega^2 = \omega_{\mathrm{p}}^2 + 3\left(\frac{\kappa T}{m}\right)k^2 \tag{9.3.18}$$

and

$$\gamma = -\sqrt{\frac{\pi}{8}}\frac{\omega_{\mathrm{p}}}{|k\lambda_{\mathrm{D}}|^3}\exp\left[-\frac{1}{2(k\lambda_{\mathrm{D}})^2} - \frac{3}{2}\right], \tag{9.3.19}$$

where, as in Section 9.1.1, the successive approximation method has been used to solve for ω^2. These results are identical to those obtained from the weak growth rate approximation; see Eq. (9.2.47).

9.4 The Dispersion Relation for a Multi-component Plasma

From inspection of the relevant equations, it is easy to see that the analysis of electrostatic waves in a hot multi-component plasma proceeds in a manner that is exactly the same as for a plasma of hot electrons and immobile ions, the only difference being that the dispersion relation now involves a sum over all species

$$Ð(k,p) = 1 - \sum_s \frac{\omega_{\mathrm{p}s}^2}{k^2}\int_C \frac{\partial F_{s0}/\partial v_z}{v_z - \mathrm{i}p/k}\,\mathrm{d}v_z = 0, \tag{9.4.1}$$

where the integration contour C must pass below the pole at $v_z = \mathrm{i}p/k$ when k is positive, and above the pole when k is negative.

To illustrate the effects introduced by ions, it is convenient to limit our initial discussion to the case of electrons and a single species of positively charged ions. In this case, the sum in the dispersion can be simplified by taking advantage of the fact that $n_{\mathrm{e}} = n_{\mathrm{i}}$, which allows us to write

$$\sum_s \omega_{\mathrm{p}s}^2 F_{s0}(v_z) = \omega_{\mathrm{pe}}^2\left[F_{\mathrm{e}0}(v_z) + \frac{m_{\mathrm{e}}}{m_{\mathrm{i}}}F_{\mathrm{i}0}(v_z)\right]. \tag{9.4.2}$$

One can then define an equivalent reduced distribution function:

$$F_0(v_z) = F_{\mathrm{e}0}(v_z) + \frac{m_{\mathrm{e}}}{m_{\mathrm{i}}}F_{\mathrm{i}0}(v_z). \tag{9.4.3}$$

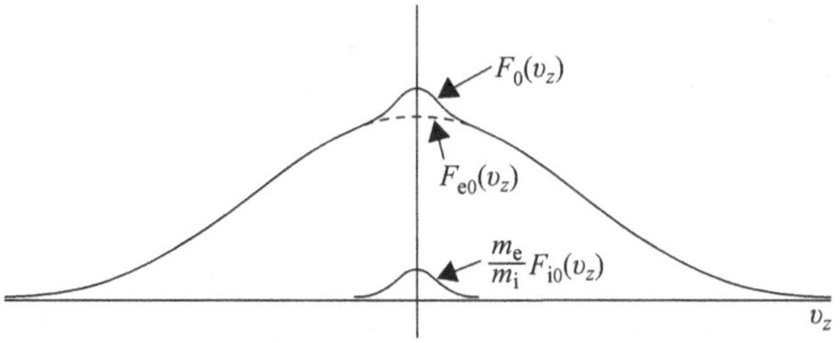

Figure 9.16 The equivalent reduced distribution function, $F_0(v_z)$, for a two-component electron–ion plasma.

Using the equivalent reduced distribution function, the dispersion relation can be written in a form that is identical to the dispersion relation for electrons and immobile ions:

$$D(k,p) = 1 - \frac{\omega_{pe}^2}{k^2}\int_C \frac{\partial F_0/\partial v_z}{v_z - ip/k}\,dv_z = 0, \tag{9.4.4}$$

It is interesting to note that because of the m_e/m_i factor, the contribution of the ions to the equivalent distribution function is strongly suppressed. However, this does not mean that the ions are unimportant. A typical plot of the equivalent distribution function (for $T_e \simeq T_i$) is shown in Figure 9.16. Note that the ions only produce a small bump in $F_0(v_z)$ near zero velocity. This bump becomes more sharply peaked as the ion temperature is decreased. As we shall soon see, the ratio of the electron temperature to the ion temperature is an important parameter in a hot electron–ion plasma.

9.4.1 Ion Acoustic Waves

Using the moment equations and the adiabatic equation of state, we demonstrated in Section 5.5.2 that ion motions lead to a new type of electrostatic wave called an ion acoustic wave. We next proceed to look for the roots of the multi-component dispersion relation that correspond to the ion acoustic wave.

Cauchy Electron and Ion Velocity Distributions

To simplify the analysis, we will first use Cauchy distributions for the zero-order electron and ion distribution functions

$$F_{e0}(v_z) = \frac{C_e}{\pi}\frac{1}{C_e^2 + v_z^2} \quad \text{and} \quad F_{i0}(v_z) = \frac{C_i}{\pi}\frac{1}{C_i^2 + v_z^2}. \tag{9.4.5}$$

Using the same procedure as discussed in Section 9.2.3, the integrals in the dispersion relation can be evaluated using the residue theorem. It is then easy to show that the dispersion relation is given by

$$Đ(k,p) = 1 + \frac{\omega_{\mathrm{pe}}^2}{(p + C_{\mathrm{e}}|k|)^2} + \frac{\omega_{\mathrm{pi}}^2}{(p + C_{\mathrm{i}}|k|)^2} = 0. \tag{9.4.6}$$

To find the roots that correspond to the Langmuir mode, we assume, as we did in Section 5.5.1, that the electron term is much larger than the ion term:

$$\frac{\omega_{\mathrm{pe}}^2}{(p + C_{\mathrm{e}}|k|)^2} \gg \frac{\omega_{\mathrm{pi}}^2}{(p + C_{\mathrm{i}}|k|)^2}. \tag{9.4.7}$$

The dispersion relation then becomes

$$1 + \frac{\omega_{\mathrm{pe}}^2}{(p + C_{\mathrm{e}}|k|)^2} = 0, \tag{9.4.8}$$

which after substituting $p = \gamma - \mathrm{i}\omega$ has solutions given by $\omega = \pm\omega_{\mathrm{pe}}$ and $\gamma = -C_{\mathrm{e}}|k|$, in agreement with our earlier results for Langmuir waves. One can easily demonstrate that the approximations used to obtain this solution are valid for any value of k. Thus, to a very good approximation, Langmuir waves are not affected by the presence of ions.

To find the roots that correspond to the ion acoustic mode, we assume that the electron and ion terms in the dispersion relation are both much greater than one, so we can omit the first term (i.e., the 1) in Eq. (9.4.6), which gives

$$\frac{\omega_{\mathrm{pe}}^2}{(p + C_{\mathrm{e}}|k|)^2} + \frac{\omega_{\mathrm{pi}}^2}{(p + C_{\mathrm{i}}|k|)^2} = 0. \tag{9.4.9}$$

After noting that $n_{\mathrm{e}} = n_{\mathrm{i}}$, the above equation simplifies to

$$(p + C_{\mathrm{i}}|k|)^2 = -\frac{m_{\mathrm{e}}}{m_{\mathrm{i}}}(p + C_{\mathrm{e}}|k|)^2. \tag{9.4.10}$$

Since we are looking for roots that have phase velocities much less than the electron thermal velocity, we assume that $p/|k| \ll C_{\mathrm{e}}$, so that

$$p + C_{\mathrm{i}}|k| = \pm\mathrm{i}\left(\frac{m_{\mathrm{e}}}{m_{\mathrm{i}}}\right)^{1/2} C_{\mathrm{e}}|k|, \tag{9.4.11}$$

or

$$\gamma - \mathrm{i}\omega = \pm\mathrm{i}\left(\frac{m_{\mathrm{e}}}{m_{\mathrm{i}}}\right)^{1/2} C_{\mathrm{e}}|k| - C_{\mathrm{i}}|k|. \tag{9.4.12}$$

If we use the minus sign, which gives the proper direction for the phase velocity, the phase velocity is given by

$$\frac{\omega}{k} = \pm\left(\frac{m_{\rm e}}{m_{\rm i}}\right)^{1/2} C_{\rm e}, \tag{9.4.13}$$

where we have inserted $\pm$ for $k/|k|$, and the growth rate is given by

$$\gamma = -C_{\rm i}|k|. \tag{9.4.14}$$

It is readily verified that the assumptions used to obtain the above solutions are satisfied whenever $k\lambda_{\rm D} \ll 1$. If we define the "temperature" of the Cauchy distributions by the relation $C_s^2 = \kappa T_s/m_s$, the phase velocity and growth rate for the ion acoustic wave simplify to

$$\frac{\omega}{k} = \pm\sqrt{\frac{\kappa T_{\rm e}}{m_{\rm i}}} \tag{9.4.15}$$

and

$$\gamma = -\sqrt{\frac{\kappa T_{\rm i}}{m_{\rm i}}}|k|. \tag{9.4.16}$$

As can be seen from Eq. (9.4.15), the phase velocity of the ion acoustic wave is proportional to the square root of the ratio of the electron temperature divided by the ion mass. This basic dependence is in agreement with our earlier results using the moment equations; see Eq. (5.5.25). However, Eq. (9.4.16) shows that the growth rate γ is negative, which means the waves are damped. The ratio of the damping rate $(-\gamma)$ to the wave frequency is given by

$$\frac{(-\gamma)}{\omega} = -\sqrt{\frac{T_{\rm i}}{T_{\rm e}}}. \tag{9.4.17}$$

As the above equation shows, if the electron and ion temperatures are comparable, the ion acoustic wave is strongly damped. This result is a marked departure from the results obtained using the moment equations, which indicated no damping at all. The damping of the ion acoustic wave is only small when $T_{\rm e} \gg T_{\rm i}$.

The strong dependence of the damping rate $(-\gamma)$ on $T_{\rm e}/T_{\rm i}$ has a simple interpretation. As shown by Eq. (9.4.15), the phase velocity of the ion acoustic wave is primarily dependent on the electron temperature. If $T_{\rm e} = T_{\rm i}$, the phase velocity is the same as the ion thermal velocity $C_{\rm i}$, which means that the slope of the ion distribution function at the phase velocity of the wave is very large, thereby causing very strong Landau damping. The location of the phase velocity relative to the ion and electron distribution is shown in Figure 9.17. As the electron

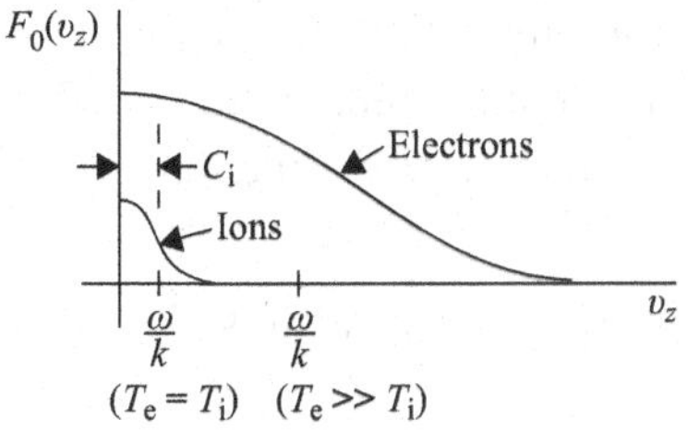

Figure 9.17 When $T_e \gg T_i$, the ion acoustic wave is weakly damped because the phase velocity, ω/k, is much greater than the ion thermal speed, C_i. When $T_e = T_i$, the damping is very large because $\omega/k \simeq C_i$.

temperature is increased, the phase velocity increases and the damping decreases, since the slope of the ion distribution decreases. Except for extremely high phase velocities ($T_e \gg T_i$), electrons contribute very little to the damping. This is because the slope of the electron velocity distribution is very small at low velocities. In Figure 9.17 the ion thermal speed has been greatly exaggerated in relation to the electron thermal speed. For realistic electron-to-ion mass ratios and temperatures, the electron thermal speed is usually a factor of fifty to one hundred times larger than the ion thermal speed.

Maxwellian Electron and Ion Distributions

To gain experience using the plasma dispersion function, it is instructive to repeat the above analysis using Maxwellian distributions for the electrons and ions. By a simple extension of Eq. (9.3.14), it is easy to show that the dispersion relation for a multi-component Maxwellian plasma is given by

$$D(k, p) = 1 - \sum_s \frac{1}{(k\lambda_{Ds})^2} \frac{1}{2} Z'(\zeta_s) = 0, \tag{9.4.18}$$

where $Z'(\zeta_s)$ is the derivative of the plasma dispersion function, and ζ_s and λ_{Ds} are defined by

$$\zeta_s = \sqrt{\frac{m_s}{2\kappa T_s}} \left(\frac{ip}{k}\right) \quad \text{and} \quad \lambda_{Ds}^2 = \frac{\epsilon_0 \kappa T_s}{n_s e^2}. \tag{9.4.19}$$

For a plasma consisting of electrons and one species of positive ion, the dispersion relation can be rewritten in the following form:

$$D(k, p) = 1 - \frac{1}{(k\lambda_{De})^2} \frac{1}{2} \left[Z'(\zeta_e) + \frac{T_e}{T_i} Z'(\zeta_i) \right] = 0, \tag{9.4.20}$$

where we have made use of the fact that $n_e = n_i$ and $\lambda_{De}^2 = (T_e/T_i)\lambda_{Di}^2$.

To obtain a completely general solution, the dispersion relation must be evaluated numerically. However, an approximate solution can be obtained for the ion acoustic mode if one assumes that the phase velocity is large compared to the ion thermal velocity and small compared to the electron thermal velocity. The large-argument expansion is then used for the ions

$$Z'(\zeta_i) = -i\frac{k}{|k|}\sqrt{\pi}2\zeta_i e^{-\zeta_i^2} + \frac{1}{\zeta_i^2}, \tag{9.4.21}$$

where we have only included the $1/\zeta_i^2$ term, and the small-argument expansion is used for the electrons

$$Z'(\zeta_e) = -i\frac{k}{|k|}\sqrt{\pi}2\zeta_e e^{-\zeta_e^2} - 2, \tag{9.4.22}$$

where we have only included the second term on the right of Eq. (9.3.16). Substituting these expansions into the dispersion relation, expanding $\zeta_i = x_i + iy_i$ in powers of y_i/x_i, and assuming $y_i \ll x_i$, the dispersion relation becomes

$$\begin{aligned} D(k,p) = 1 + \frac{1}{(k\lambda_{De})^2}\Bigg\{1 - \frac{1}{2}\left(\frac{T_e}{T_i}\right)\frac{1}{x_i^2}\left(1 - i\frac{2y_i}{x_i}\right) \\ + i\frac{k}{|k|}\sqrt{\pi}x_i\left[\sqrt{\frac{m_e}{m_i}}\sqrt{\frac{T_i}{T_e}} + \left(\frac{T_e}{T_i}\right)e^{-x_i^2}\right]\Bigg\} = 0, \end{aligned} \tag{9.4.23}$$

where we have made use of the relation $x_e = x_i\sqrt{m_e/m_i}\sqrt{T_i/T_e}$. Separating the real and imaginary parts, and writing x_i and y_i in terms of ω and γ then gives the following equations for the phase velocity and the growth rate:

$$\frac{\omega}{k} = \pm\sqrt{\frac{\kappa T_e}{m_i}\frac{1}{(1 + k^2\lambda_{De}^2)^{1/2}}} \tag{9.4.24}$$

and

$$\frac{\gamma}{\omega} = -\sqrt{\frac{\pi}{8}}\left[\sqrt{\frac{m_e}{m_i}} + \left(\frac{T_e}{T_i}\right)^{3/2}\exp\left(-\frac{T_e}{2T_i}\frac{1}{(1 + k^2\lambda_{De}^2)}\right)\right]\frac{1}{(1 + k^2\lambda_{De}^2)^{3/2}}. \tag{9.4.25}$$

Note that in the short-wavelength limit, $k\lambda_{De} \gg 1$, the frequency has an upper limit given by the ion plasma frequency, $\omega_{pi} = (\sqrt{\kappa T_e/m_i})/\lambda_{De}$, in agreement with the results obtained in Section 5.5.2 from the moment equations (see Figure 5.6). In the long-wavelength limit, $k\lambda_{De} \ll 1$, the equations for the phase velocity and growth rate simplify considerably and are given by

$$\frac{\omega}{k} = \pm\sqrt{\frac{\kappa T_e}{m_i}} \tag{9.4.26}$$

and

$$\frac{\gamma}{\omega} = -\sqrt{\frac{\pi}{8}}\left[\sqrt{\frac{m_e}{m_i}} + \left(\frac{T_e}{T_i}\right)^{3/2} \exp\left(-\frac{T_e}{2T_i}\right)\right]. \tag{9.4.27}$$

Just as for the Cauchy distribution, the phase velocity is proportional to the square root of the electron temperature divided by the ion mass. Also, the damping rate $(-\gamma)$ is very large when T_e is comparable to T_i. However, because of the exponential term, the damping decreases much more rapidly with increasing T_e/T_i than for a Cauchy distribution. The strong dependence on T_e/T_i is caused by the rapid decrease in the slope of the Maxwellian velocity distribution as the phase velocity increases. However, note that once the exponential term drops below $\sqrt{m_e/m_i}$, the damping no longer decreases as T_e/T_i increases. The $\sqrt{m_e/m_i}$ term is traceable to electron Landau damping via the second term on the right of Eq. (9.4.22) and remains constant, because the phase velocity maintains a fixed relationship with the electron thermal speed.

Although the analytic results described above give a good approximation at large T_e/T_i, they do not give good results when the electron temperature is comparable to the ion temperature. Figure 9.18 shows a comparison of the damping rate $(-\gamma)/\omega$ given by Eq. (9.4.27), labeled "to $1/\zeta_i^2$," with a full numerical solution of the plasma dispersion. As can be seen, the agreement with the numerical solution is not very good for T_e/T_i values below about 20. Improved results can be obtained by keeping terms through $1/\zeta_i^4$ in the expansion for the ion plasma dispersion

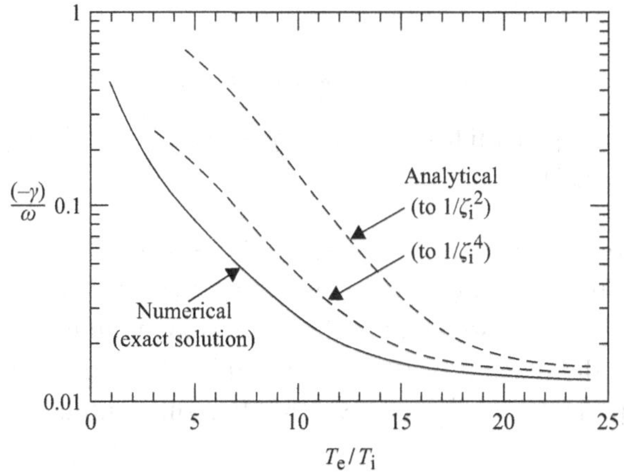

Figure 9.18 A comparison of the approximate analytical expression for the damping, $(-\gamma)/\omega$, of ion acoustic waves with the results of an exact numerical calculation.

function. The phase velocity and growth rate are then given by

$$\frac{\omega}{k} = \pm\sqrt{\frac{\kappa T_{\rm e} + 3\kappa T_{\rm i}}{m_{\rm i}}} \tag{9.4.28}$$

and

$$\frac{\gamma}{\omega} = -\sqrt{\frac{\pi}{8}}\left[\sqrt{\frac{m_{\rm e}}{m_{\rm i}}} + \frac{(3 + T_{\rm e}/T_{\rm i})^{5/2}}{(9 + T_{\rm e}/T_{\rm i})}\exp\left(-\frac{T_{\rm e}}{2T_{\rm i}} - \frac{3}{2}\right)\right], \tag{9.4.29}$$

where again we have assumed that $k\lambda_{\rm De} \ll 1$. Note that the phase velocity now has an additional term that is proportional to the ion temperature. This term accounts for the $\exp[-3/2]$ factor in the equation for the growth rate (9.4.29) and gives much better agreement with the exact numerical solution, as can be seen from the curve labeled "to $1/\zeta_{\rm i}^4$" in Figure 9.18.

9.5 Stability

In the previous sections we have encountered a variety of both stable and unstable distribution functions. It is now of interest to consider the general conditions for stability. Since the growth or decay of a disturbance in a plasma is determined by the roots of the dispersion relation, it follows that the plasma is stable if all the roots have negative values for $\text{Re}\{p\}$, and unstable if one or more roots have positive values for $\text{Re}\{p\}$. The condition $\gamma = 0$ (i.e., $\text{Re}\{p\} = 0$) marks the boundary between stability and instability and is called the marginal stability condition.

9.5.1 Gardner's Theorem

Before developing a general test for stability we first prove the following theorem due to Gardner (1963).

Theorem. *A single-humped velocity distribution is stable.*

To prove this theorem, we start by assuming that the theorem is not true. It follows then that there are solutions to the dispersion relation for which $\text{Re}\{p\} > 0$. Since $\text{Re}\{p\} > 0$, the integral in $D(k, p)$ (see Eq. (9.2.21)) can be expressed as an integral along the real velocity axis which, after substituting $p = \gamma - {\rm i}\omega$, can be written

$$D(k, p) = 1 - \frac{\omega_{\rm p}^2}{k^2}\int_{-\infty}^{\infty} \frac{\partial F_0/\partial v_z}{v_z - \omega/k - {\rm i}\gamma/k}\, {\rm d}v_z = 0. \tag{9.5.1}$$

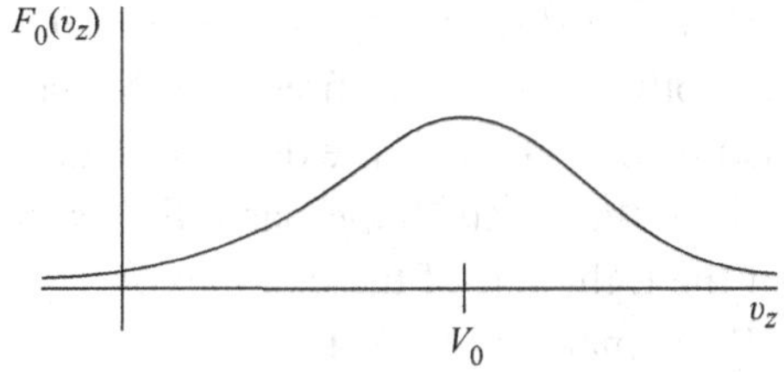

Figure 9.19 A velocity distribution with a single hump at velocity V_0.

Separating the integral into real and imaginary parts gives two equations:

$$Ð_{\mathrm{r}}(k,p) = 1 - \frac{\omega_{\mathrm{p}}^2}{k^2}\int_{-\infty}^{\infty}\frac{(\partial F_0/\partial v_z)(v_z-\omega/k)}{(v_z-\omega/k)^2+(\gamma/k)^2}\,\mathrm{d}v_z = 0 \tag{9.5.2}$$

and

$$Ð_{\mathrm{i}}(k,p) = -\frac{\omega_{\mathrm{p}}^2}{k^2}\frac{\gamma}{k}\int_{-\infty}^{\infty}\frac{\partial F_0/\partial v_z}{(v_z-\omega/k)^2+(\gamma/k)^2}\,\mathrm{d}v_z = 0. \tag{9.5.3}$$

Next, suppose that $F_0(v_z)$ has a single hump with a maximum at velocity V_0, as shown in Figure 9.19. Since $Ð_{\mathrm{r}}$ and $Ð_{\mathrm{i}}$ are both zero, any linear combination of these two terms is also zero. In particular, let us consider the combination

$$Ð_{\mathrm{r}} - \left(\frac{kV_0-\omega}{\gamma}\right)Ð_{\mathrm{i}} = 0. \tag{9.5.4}$$

It is easy to show, after substituting for $Ð_{\mathrm{r}}$ and $Ð_{\mathrm{i}}$, that the above equation simplifies to

$$1 + \frac{\omega_{\mathrm{p}}^2}{k^2}\int_{-\infty}^{\infty}\frac{(\partial F_0/\partial v_z)(V_0-v_z)}{(v_z-\omega/k)^2+(\gamma/k)^2}\,\mathrm{d}v_z = 0. \tag{9.5.5}$$

Since F_0 has only a single hump, $\partial F_0/\partial v_z$ is positive for $v_z < V_0$ and negative for $v_z > V_0$. Therefore, the integral is positive definite, which means that the above equation cannot be satisfied for any combination of ω and γ. Thus, we have arrived at a contradiction. The contradiction shows that the original assumption, $\mathrm{Re}\{p\} > 0$, is false. Therefore, a single-humped distribution is always stable.

9.5.2 The Nyquist Criterion

Having established that a single-humped velocity distribution is always stable, we next develop a general method for testing the stability of a distribution function with an arbitrary number of humps. This test is called the Nyquist criterion, after Nyquist (1932) who developed the method for analyzing the stability of electrical circuits. As shown earlier, the dispersion relation is a function of the complex frequency $p = \gamma - i\omega$. Both γ and ω are functions of the wave number k. The distribution function is unstable if some k exists for which $\gamma > 0$ and is stable if $\gamma < 0$ for all values of k. In the Nyquist approach, the procedure is to carry out a one-to-one mapping of the right half of the complex p-plane, $\mathrm{Re}\{p\} = \gamma > 0$, onto the complex $Đ$-plane. If the mapping includes the point $Đ = 0$, then the plasma is unstable, since a solution of $Đ = 0$ exists for which $\gamma > 0$. If the mapping does not include the point $Đ = 0$ for any value of k, then the distribution function is stable.

To carry out the required mapping, it is useful to note that the boundary of the right half of the complex p-plane is the marginal stability line, $\gamma = 0$. The corresponding marginal stability line in the complex $Đ$-plane can then be determined by computing $Đ(k, p)$ along the $\gamma = 0$ line in the complex p-plane for $-\infty < \omega < +\infty$. The $\gamma = 0$ curve in the complex $Đ$-plane is always a closed curve. By mapping one additional point that is not on the $\gamma = 0$ line, it is a simple matter to determine whether the right half of the complex p-plane maps into the interior or the exterior of the $\gamma = 0$ curve.

To illustrate the Nyquist mapping procedure, consider a Cauchy velocity distribution function. From the dispersion relation for the Cauchy distribution, Eq. (9.2.31), it is easy to show that the real and imaginary parts of the dispersion relation, evaluated along the line $\gamma = 0$, are given by

$$Đ_r = 1 + \frac{\omega_p^2(C^2k^2 - \omega^2)}{(C^2k^2 + \omega^2)^2} \tag{9.5.6}$$

and

$$Đ_i = 2\frac{\omega_p^2 C|k|\omega}{(C^2k^2 + \omega^2)^2}. \tag{9.5.7}$$

As ω varies from minus infinity to plus infinity, these equations generate a closed cardioid curve in the complex $Đ$-plane, as shown in the right panel of Figure 9.20.

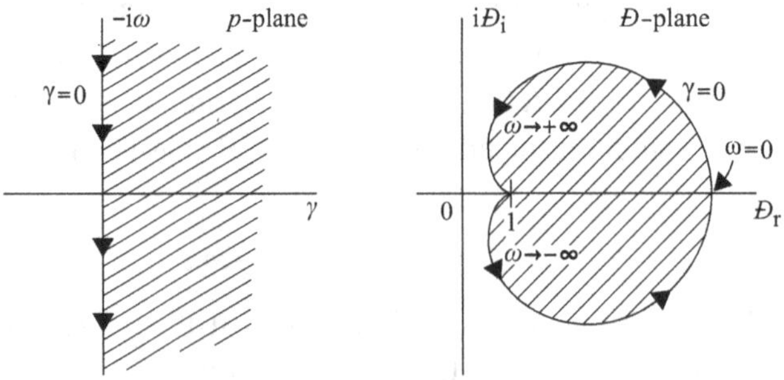

Figure 9.20 A mapping of the right half of the complex p-plane onto the complex $Đ$-plane for a Cauchy distribution function.

It is easy to show that the right half of the complex p-plane (indicated by the shaded region in the left panel) maps into the interior of the cardioid curve. Note that the interior of the cardioid curve does not include the point $Đ = 0$. Since there is no solution of $Đ = 0$ for which γ is positive, it follows that the distribution function is stable. This is the expected result. Since a Cauchy velocity distribution has only a single hump, by Gardner's theorem the distribution function is stable.

9.5.3 The Penrose Condition

Next we apply the Nyquist criterion to a distribution function with an arbitrary number of humps. To carry out the required mapping of the $\gamma = 0$ line onto the complex $Đ$-plane, we need an expression for $Đ(k, p)$ that is valid for a distribution function of arbitrary shape. Since γ is always small in the vicinity of the $\gamma = 0$ boundary, the Plemelj relation can be used to evaluate the real and imaginary parts of the dispersion relation, just as was done in the weak growth rate approximation (see Section 9.2.4). Using the Plemelj relation, it can be shown that the real and imaginary parts of $Đ(k, p)$, evaluated along the $\gamma = 0$ line, are given by

$$Đ_{\mathrm{r}} = 1 - \frac{\omega_{\mathrm{p}}^2}{k^2} P \int_{-\infty}^{\infty} \frac{F_0'(v_z)}{v_z - \omega/k} \, \mathrm{d}v_z \tag{9.5.8}$$

and

$$Đ_{\mathrm{i}} = -\pi \frac{k}{|k|} \frac{\omega_{\mathrm{p}}^2}{k^2} F_0'\left(\frac{\omega}{k}\right). \tag{9.5.9}$$

To understand the geometry of the $\gamma = 0$ curve in the complex $Đ$-plane, we start by investigating the shape near $\omega = \pm\infty$. Throughout this analysis, we assume that the wave number k is positive. A similar analysis is easily repeated for negative k.

Since the slope of the distribution function, $F_0'(\omega/k)$, must go to zero as the velocity goes to infinity, it follows that $Đ = 1$ at $\omega = \pm\infty$. Furthermore, since the distribution function is always positive, $F_0'(\omega/k)$ must be negative as $\omega/k \to +\infty$, and positive as $\omega/k \to -\infty$. Therefore, the $\gamma = 0$ curve must approach $Đ = 1$ from above the $Đ_{\mathrm{r}}$ axis as $\omega \to +\infty$, and from below the $Đ_{\mathrm{r}}$ axis as $\omega \to -\infty$, as shown in Figure 9.21. Having established the shape of the $\gamma = 0$ curve near $Đ = 1$, we next explore the shape at intermediate frequencies. Since $Đ_{\mathrm{i}}$ can be zero only when $F_0'(\omega/k)$ is zero, it follows that the $\gamma = 0$ curve can only cross the $Đ_{\mathrm{r}}$ axis at those frequencies ω_j for which the derivative of the distribution function is zero (i.e., $\omega_j = V_j k$ where $F'(V_j) = 0$). For a smooth continuous distribution function, there is always an odd number of such points. For a single-humped distribution function only one such point exists (for example, at $\omega = 0$ in Figure 9.20).

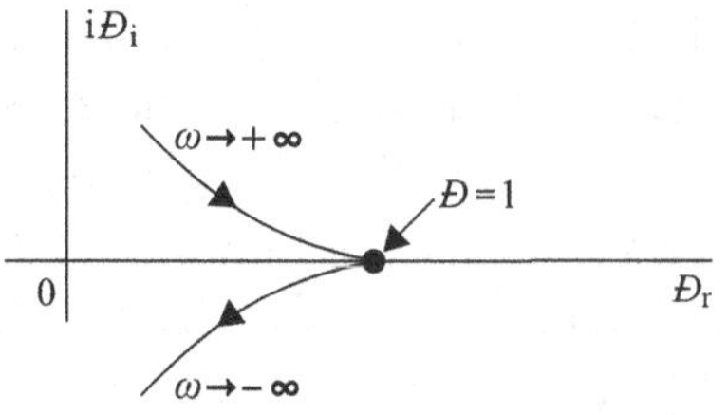

Figure 9.21 An expanded scale representation of the mapping of the line $\gamma = 0$ onto the complex $Đ$-plane in the vicinity of $Đ = 1$.

To proceed further, we use a result from conformal mapping theory called the winding theorem (Flanigan, 1983). The winding theorem states that if a closed contour C_p in the complex p-plane encloses n simple zeros of some mapping function $Đ(p)$, then the corresponding contour $C_Đ$ in the complex $Đ$-plane must make n turns around the origin. To prove the winding theorem, we start by noting from the residue theorem that the number of turns a contour $C_Đ$ makes around the origin ($Đ = 0$) is given by the integral

$$N_{\mathrm{w}} = \frac{1}{2\pi \mathrm{i}} \int_{C_Đ} \frac{\mathrm{d}Đ}{Đ}. \tag{9.5.10}$$

The number N_{w} is called the winding number. According to the usual convention for complex integration, the winding number is positive for a contour that loops around the origin in the counter-clockwise sense. By a simple change of variables, the above integral can be converted to a corresponding integral in the complex p-plane:

$$N_{\mathrm{w}} = \frac{1}{2\pi \mathrm{i}} \int_{C_p} \frac{1}{Đ} \frac{\partial Đ}{\partial p} \mathrm{d}p, \tag{9.5.11}$$

where the contour C_p is the corresponding mapping of the contour $C_Đ$ onto the complex p-plane. The relationship between the two contours for a representative mapping function is shown in Figure 9.22. To understand the relationship between the number of zeros in the complex p-plane and the winding number, let us assume that the mapping function $Đ(p)$ has n zeros, $p_1, p_2, \ldots, p_n$, within the contour C_p. Since the integral is independent of the shape of the contour, the contour can be deformed so that it shrinks closely around the n zeros. The integral can then be expressed as the sum of the integrals around each respective zero:

$$N_{\mathrm{w}} = \sum_{j=1}^{n} \frac{1}{2\pi \mathrm{i}} \int_{C_{pj}} \frac{1}{Đ} \frac{\partial Đ}{\partial p} \mathrm{d}p. \tag{9.5.12}$$

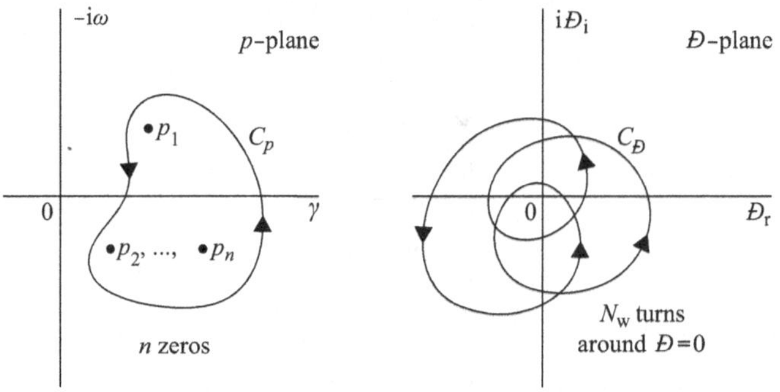

Figure 9.22 If a contour encloses n roots, $p_1, p_2, \ldots, p_n$ of $Đ(p) = 0$ in the complex p-plane, then the mapping of this contour onto the complex $Đ$-plane must encircle the origin $N_\mathrm{w} = n$ times.

Around each zero, the mapping function can be expanded in a Taylor series:

$$Đ = 0 + \left.\frac{\partial Đ}{\partial p}\right|_{p_j} (p - p_j) + \cdots, \tag{9.5.13}$$

where the first term is zero, since $Đ(p_j) = 0$. Similarly, the derivative of the mapping function can be expanded in a Taylor series, which to leading order is

$$\frac{\partial Đ}{\partial p} = \left.\frac{\partial Đ}{\partial p}\right|_{p_j} + \cdots \tag{9.5.14}$$

Substituting Eqs. (9.5.13) and (9.5.14) into Eq. (9.5.12), it follows that

$$N_\mathrm{w} = \sum_{j=1}^{n} \frac{1}{2\pi \mathrm{i}} \int_{C_{pj}} \frac{\mathrm{d}p}{p - p_j}. \tag{9.5.15}$$

From the residue theorem, each integral contributes a factor of $2\pi\mathrm{i}$, so the sum yields $N_\mathrm{w} = n$. Thus, we have shown that the winding number is equal to the number of zeros within the contour, which proves the theorem.

To see how the winding theorem is used to investigate the geometry of the $\gamma = 0$ curve in the complex $Đ$-plane, we start by selecting a contour C_p that passes downward along the $\gamma = 0$ line in the complex p-plane. The contour is closed at infinity along the outer boundary of the right half-plane, as shown in Figure 9.23. The reason for this choice of the contour is that the integral $\int \mathrm{d}Đ/Đ$ around the contour gives the total number of unstable roots. Note that any root of $Đ(k, p) = 0$ within this contour has $\mathrm{Re}\{p\} > 0$, and is therefore unstable. Since the portion of the contour at infinity maps to the point $Đ = 1$, the rest of the contour maps onto the $\gamma = 0$ curve. From the winding theorem, we can now see that the number of times that the $\gamma = 0$ curve winds around the origin ($Đ = 0$) must be equal to the number

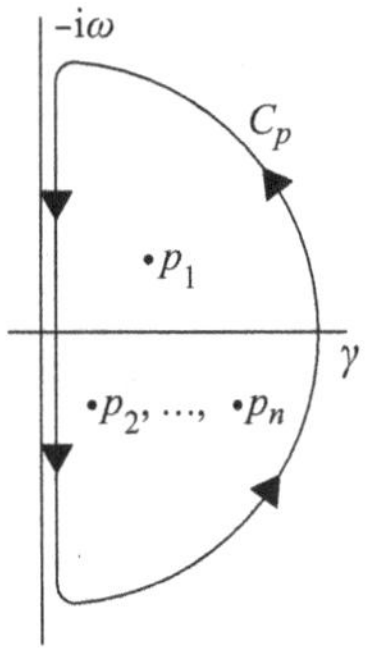

Figure 9.23 The contour selected to count the total number of zeros of $Đ(p)$ in the right half of the complex p-plane.

of unstable roots of $Đ(k, p) = 0$. Since the number of unstable roots can never be negative, if the $\gamma = 0$ curve encircles the origin ($Đ = 0$), then it must wind around the origin at least once in the counter-clockwise sense.

We now have enough tools to determine the geometry of the $\gamma = 0$ curve in the complex $Đ$-plane. If the velocity distribution function has a single hump, then it can cross the $Đ_{\rm r}$ axis only once. Since the $\gamma = 0$ curve starts from below the $Đ_{\rm r}$ axis at $\omega = -\infty$, this crossing must be in the upward direction, toward increasing $Đ_{\rm i}$. Since the winding number must be positive, the curve must cross the $Đ_{\rm r}$ axis to the right of $Đ = 0$, otherwise the curve would encircle the origin in the clockwise sense, which is prohibited. Since the crossing of the $Đ_{\rm r}$ axis must be to the right of $Đ = 0$, the curve does not encircle the origin, so the distribution function is stable. Thus, we have again proven Gardner's theorem, which states that a single-humped velocity distribution is always stable.

If the velocity distribution function has two humps, the $\gamma = 0$ curve crosses the $Đ_{\rm r}$ axis three times, corresponding to the three velocities V_1, V_2, and V_3, for which $F_0'(V_j) = 0$. Two of these crossings, V_1 and V_3, are in the upward direction, toward positive $Đ_{\rm i}$, and correspond to local maxima in $F_0(v_z)$. The remaining crossing, V_2, is in the downward direction and corresponds to a local minimum in $F_0(v_z)$. A representative $\gamma = 0$ curve and its relationship to $F_0(v_z)$ is shown in Figure 9.24.

From Eq. (9.5.8), it can be seen that the $Đ_{\rm r}$ values at which the $\gamma = 0$ line crosses the $Đ_{\rm r}$ axis are given by

$$Đ_{\rm r} = 1 - \frac{\omega_{\rm p}^2}{k^2} P \int_{-\infty}^{\infty} \frac{F_0'(v_z)}{v_z - V_j} \, {\rm d}v_z, \qquad (9.5.16)$$

where the V_j are the velocities for which $F_0'(V_j) = 0$. Since the right half of the complex p-plane maps into the interior of the $\gamma = 0$ curve, if any one of these crossings occurs to the left of the origin (i.e., $Đ_{\rm r} < 0$), then by the Nyquist criterion the distribution function is unstable. Since the wave number k can take on any

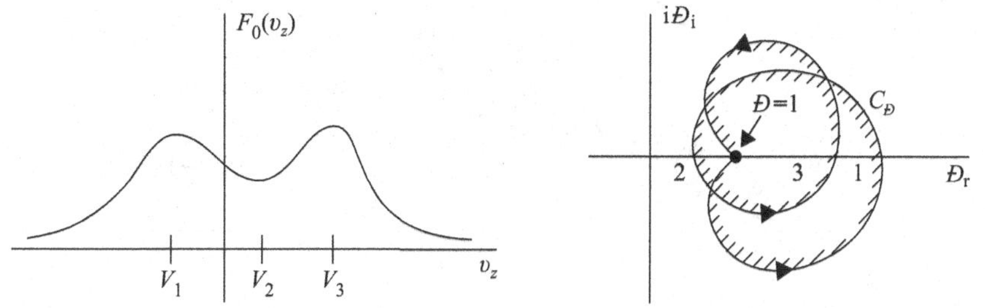

Figure 9.24 The mapping of the line $\gamma = 0$ onto the complex Đ-plane for a double-humped velocity distribution.

value whatever, including very small values, it is easy to see that the condition for a negative $Đ_\mathrm{r}$ value at a crossing of the $Đ_\mathrm{r}$ axis is that the integral be positive:

$$P\int_{-\infty}^{\infty} \frac{F_0'(v_z)}{v_z - V_j}\,\mathrm{d}v_z > 0, \tag{9.5.17}$$

where the V_j are the velocities at which $F_0'(V_j) = 0$. Therefore, the existence of a positive value for the above integral for at least one value of V_j is a necessary condition for instability. Since $F_0(V_j)$ is a constant, the above integral can be rewritten in the form

$$P\int_{-\infty}^{\infty} \frac{F_0'(v_z)}{v_z - V_j}\mathrm{d}v_z = P\int_{-\infty}^{\infty} \frac{\frac{\partial}{\partial v_z}[F_0(v_z) - F_0(V_j)]}{v_z - V_j}\mathrm{d}v_z > 0. \tag{9.5.18}$$

The integral on the right-hand side can be integrated by parts once to give the following condition for instability:

$$\int_{-\infty}^{\infty} \frac{F_0(v_z) - F_0(V_j)}{(v_z - V_j)^2}\,\mathrm{d}v_z > 0. \tag{9.5.19}$$

The above relation is called the Penrose condition, after Penrose (1960) who first showed that it is a necessary and sufficient condition for instability. Note that since the numerator in the integrand goes to zero at $v_z = V_j$, it is not necessary to specify that the integral be a principal value. It can be shown that the Penrose condition also applies for an arbitrary number of humps in the velocity distribution function.

The Penrose condition has a simple graphical interpretation. Consider the double-humped velocity distribution function shown in Figure 9.25. The integral in the Penrose condition gives the area above (+) and below (−) the line $F_0 = F_0(V_j)$, weighted by the function $1/(v_z - V_j)^2$. From Figure 9.25 one can see that for instability to occur it is not sufficient that the distribution be simply double humped. The two humps must be large enough that the positive contributions to the integral exceed the negative contributions.

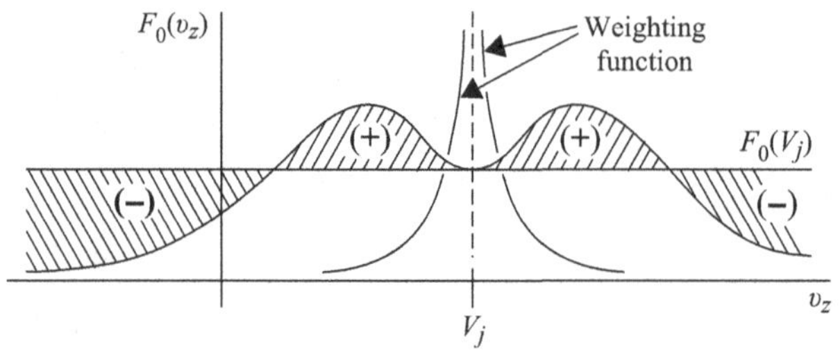

Figure 9.25 A schematic interpretation of the Penrose criterion.

9.5.4 Some Representative Instabilities

To illustrate the principles discussed above, we next consider various types of potentially unstable distribution functions. In all cases the instabilities involve distribution functions that have a finite temperature. In this respect they differ fundamentally from the cold, delta function, beam instabilities discussed earlier in this chapter. Whereas the cold beam instabilities could equally well have been analyzed using the fluid equation, the introduction of a finite temperature leads to an entirely new class of instabilities, called kinetic instabilities, that depend on the detailed shape of the distribution function.

The Counter-Streaming Beam Instability

A simple example that illustrates many of the above principles is provided by the counter-streaming Cauchy distribution

$$F_0(v_z) = \frac{C}{2\pi}\left[\frac{1}{C^2 + (v_z - V)^2} + \frac{1}{C^2 + (v_z + V)^2}\right]. \tag{9.5.20}$$

In the limit of zero "temperature" (i.e., $C = 0$), this distribution reduces to two oppositely directed delta function beams, which from our earlier analysis is known to be unstable. As the temperature increases, the beams spread in velocity, eventually turning into a single-humped distribution, which from Gardner's theorem is known to be stable. It is easy to show that the transition from a double-humped to a single-humped distribution takes place when $C = \sqrt{3}V$. Thus, the transition from unstable to stable must occur at some point between $C = 0$ and $C = \sqrt{3}V$.

The criterion for the onset of instability can be obtained by direct evaluation of the Penrose condition. There are three points where $F_0'(V_j) = 0$, near the two peaks associated with $v_z = \pm V$, and at $V_j = 0$. The points near the two peaks do not produce instability since the integral in the Penrose condition is in both cases clearly negative. For $V_j = 0$, the integral in the Penrose condition can be evaluated

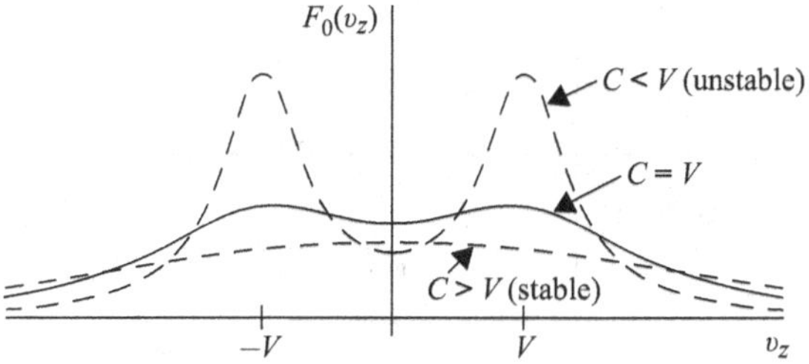

Figure 9.26 For a pair of counter-streaming Cauchy distributions with thermal velocity C and drift velocities V and $-V$, the plasma is stable if $C > V$, marginally stable if $C = V$, and unstable if $C < V$.

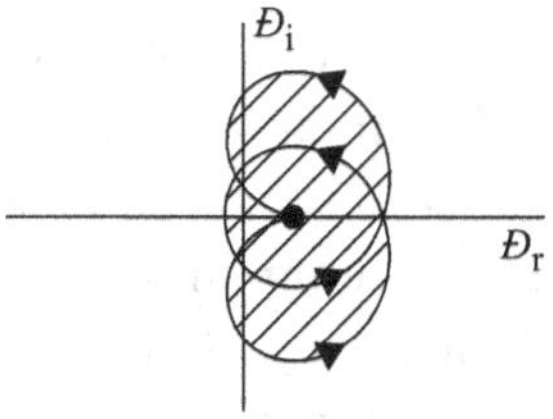

Figure 9.27 The mapping of the right half of the complex p-plane onto the complex $Đ$-plane for a counter-streaming Cauchy distribution with $C < V$. Since the mapping includes the origin ($Đ = 0$), the plasma is unstable.

directly and is given by

$$\int_{-\infty}^{\infty} \frac{F_0(v_z) - F_0(V_j)}{(v_z - V_j)^2} \mathrm{d}v_z = \frac{V^2 - C^2}{(V^2 + C^2)^2}. \tag{9.5.21}$$

The plasma is unstable whenever the above integral is positive, which occurs when $V > C$. The shape of the distribution function at the onset of instability ($V = C$) is shown by the solid curve in Figure 9.26.

By direct evaluation of $Đ_r$ and $Đ_i$, it is easy to show that the Nyquist plot for $C < V$ is as shown in Figure 9.27. The $\gamma = 0$ curve loops around the origin once which, by the Nyquist theorem, indicates the presence of an unstable root in the dispersion relation.

The "Bump-on-Tail" Instability

Using the weak growth rate approximation, we previously showed that the growth rate γ at high phase velocities is proportional to the slope of the distribution function evaluated at the phase velocity of the wave. Therefore, any distribution function that has a region of positive slope at high phase velocities leads to exponentially growing waves. Since the weak growth rate approximation is only valid at high phase velocities, well above the thermal velocity ($\omega/k \gg v_{\rm th}$), this analysis is applicable if the region of positive slope lies in the tail of the distribution

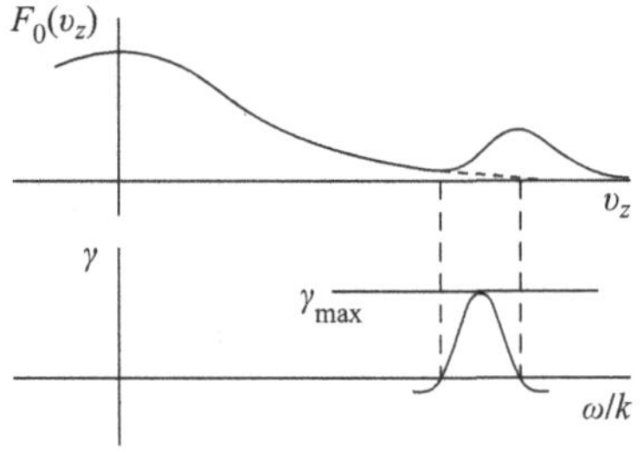

Figure 9.28 A bump-on-tail distribution produces growing waves over a narrow range of phase velocities, ω/k, where $F_0(v_z)$ has a positive slope.

function, well above the thermal velocity, such as shown in Figure 9.28. This type of instability is often called a "bump-on-tail" instability. Since the growth rate is proportional to $\partial F_0/\partial v_z$, it is easy to see that a range of phase velocities slightly below the peak is unstable. If the bump-on-tail overlaps with the main thermal distribution, the growth rate is made up of two terms, one negative (damping), caused by the tail of the main thermal distribution, and the other positive (growth), due to the positive slope of the bump. Thus, the overall growth rate is a competition between two effects. If the velocity at which the bump occurs gets too low, near the main part of the thermal distribution, or if the positive slope on the bump gets too small, the instability ceases. Notice that for a bump-on-tail distribution the growth rate is directly proportional to the density of the bump, $\gamma \propto n_b$. This dependence differs from the cold weak beam approximation discussed earlier, which has a growth rate proportional to $n_b^{1/3}$. The reason for this difference is that the physical mechanisms involved in the two instabilities are different. The bump-on-tail instability involves only those particles that are in resonance with the wave. Particles with velocities greater than the phase velocity contribute to the growth of the wave, and those with velocities less than the phase velocity contribute to damping. This type of behavior is typical of a kinetic instability. The cold weak beam instability, on the other hand, involves a fluid-like motion of the beam, with all of the particles oscillating in unison. This type of behavior is typical of a fluid instability.

Numerous examples of electrostatic instabilities produced by a bump-on-tail distribution can be found in both laboratory and space plasmas. A good example in a space plasma is given by a type III solar radio burst, which is produced by electrons emitted from the Sun by a solar flare. When the flare occurs, a very broad range of electron energies (up to several hundred keV) are impulsively released from the Sun, as shown in Figure 9.29. These electrons stream outward from the Sun along the solar wind magnetic field lines. At a given time after the flare, only those electrons with velocities greater than some cutoff velocity v_c can

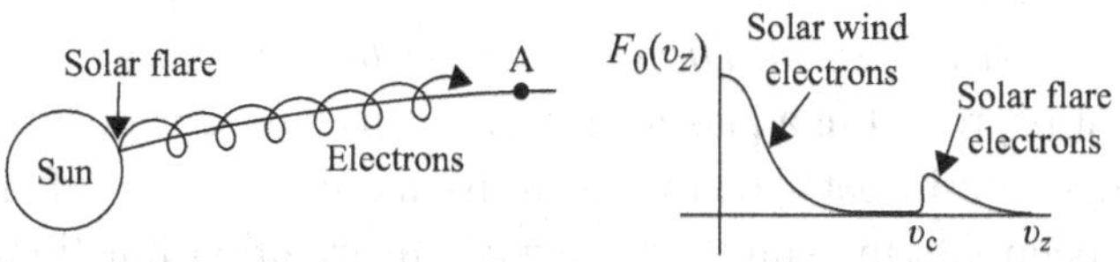

Figure 9.29 Electrons ejected by a solar flare produce a bump-on-tail distribution because of time of flight considerations, thereby causing the growth of Langmuir waves in the solar wind.

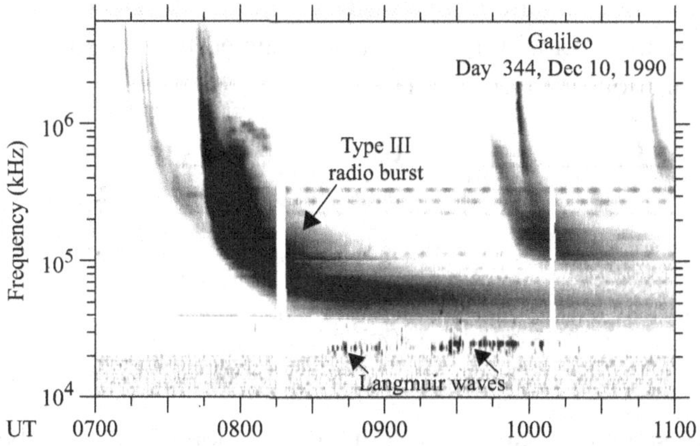

Figure 9.30 An example of a type III solar radio burst and its associated Langmuir waves (from Gurnett et al., 1993).

reach point A along the magnetic field line. Solar flare electrons of lower velocity cannot reach this point because the transit time is too long. This velocity cutoff automatically causes a bump-on-tail type of distribution function and leads to the growth of Langmuir waves (electron plasma oscillations) near the local electron plasma frequency. As the solar flare electron distribution evolves, a localized region of plasma oscillations is produced that moves outward from the Sun. Since the plasma oscillations are generated near the Sun, they cannot be detected directly from the Earth. However, the plasma oscillations produce electromagnetic radiation that can be detected on Earth. The radiation is generated by nonlinear interactions at the plasma frequency and at twice the plasma frequency (Ginzburg and Zhelezniakov, 1958). Since the electron plasma frequency decreases with increasing radial distance from the Sun, the frequency of the radio emission decreases with increasing time as the region of plasma oscillations moves outward from the Sun. This frequency–time variation is a characteristic feature of type III radio bursts. A spectrogram of an intense type III solar radio burst and its associated Langmuir waves is shown in Figure 9.30.

The Current-Driven Ion Acoustic Instability

When a current is present in a plasma, the electrons have a net drift relative to the ions. If the relative drift between the electrons and the ions is sufficiently large, a double hump occurs in the equivalent reduced distribution function, as shown in Figure 9.31. The ion acoustic mode can then be driven unstable. This instability is called the current-driven ion acoustic instability.

To provide a quantitative evaluation of the threshold current required to drive the ion acoustic instability, we first consider Cauchy distributions for the electrons and ions. To produce a current, the drift velocity of the electrons is assumed to be $U_{\rm e}$, and the drift velocity of the ions is assumed to be zero. The appropriate distribution functions are then given by

$$F_{\rm e0}(v_z) = \frac{C_{\rm e}}{\pi}\frac{1}{C_{\rm e}^2 + (v_z - U_{\rm e})^2} \tag{9.5.22}$$

and

$$F_{\rm i0}(v_z) = \frac{C_{\rm i}}{\pi}\frac{1}{C_{\rm i}^2 + v_z^2}. \tag{9.5.23}$$

Following the same procedure as described in Section 9.2.3, it can be shown that the dispersion relation is given by

$$Đ(k,p) = 1 + \frac{\omega_{\rm pe}^2}{(p + {\rm i}kU_{\rm e} + C_{\rm e}|k|)^2} + \frac{\omega_{\rm pi}^2}{(p + C_{\rm i}|k|)^2} = 0, \tag{9.5.24}$$

where the second term is due to the drifting electron distribution and the third term is due to the ion distribution. The $kU_{\rm e}$ term in the denominator of the electron term arises when the residue is evaluated at $v_z = U_{\rm e} - {\rm i}C_{\rm e}$, and is simply the Doppler shift produced by the electron drift. As before, for $k\lambda_{\rm De} \gg 1$, the ion acoustic root is obtained by assuming that the electron and ion terms are much larger than one,

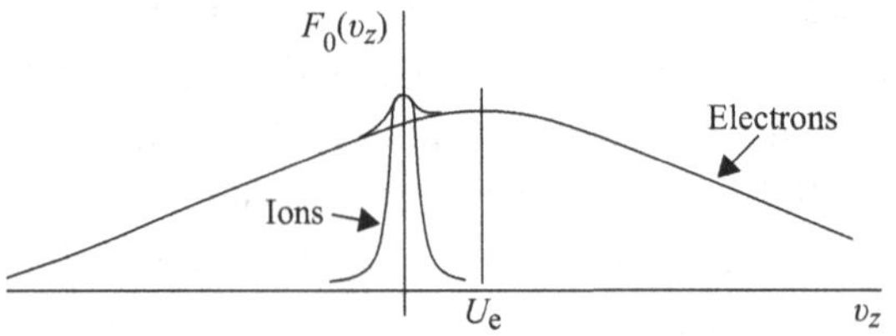

Figure 9.31 When a current is present, a double hump is produced in the equivalent reduced velocity distribution function. This double hump can lead to the growth of ion acoustic waves.

so that

$$\frac{\omega_{pe}^2}{(p + ikU_e + C_e|k|)^2} = \frac{-\omega_{pi}^2}{(p + C_i|k|)^2}. \quad (9.5.25)$$

Solving the above equation for p then gives

$$p + C_i|k| = \pm i\left(\frac{m_e}{m_i}\right)^{1/2}(p + ikU_e + C_e|k|), \quad (9.5.26)$$

which can be rewritten in the form

$$p\left(1 \mp i\left(\frac{m_e}{m_i}\right)^{1/2}\right) = -\left(C_i \pm \left(\frac{m_e}{m_i}\right)^{1/2} U_e \frac{k}{|k|}\right)|k| \pm i\left(\frac{m_e}{m_i}\right)^{1/2} C_e|k|. \quad (9.5.27)$$

Using the minus sign, which ensures that the phase velocity is in the proper direction, and neglecting the i $\sqrt{m_e/m_i}$ term on the left, which is small, it is easy to show that the phase velocity is given by

$$\frac{\omega}{k} = \pm\sqrt{\frac{\kappa T_e}{m_i}}, \quad (9.5.28)$$

where we used $C_e^2 = \kappa T_e/m_e$ and inserted $\pm$ for $k/|k|$. The corresponding growth rate is given by

$$\gamma = -\left(C_i - U_e\frac{k}{|k|}\left(\frac{m_e}{m_i}\right)^{1/2}\right)|k|. \quad (9.5.29)$$

As can be seen, if the electron drift velocity, U_e, is sufficiently large, the growth rate can become positive, indicating instability. Note that the unstable waves occur for k in the same direction as the electron drift (k and U_e must have the same sign). The threshold drift velocity at which the instability starts is given by

$$|U_e^*| = \sqrt{\frac{m_i}{m_e}}C_i = \left(\frac{T_i}{T_e}\right)^{1/2} C_e. \quad (9.5.30)$$

This equation shows that if the electron and ion temperatures are the same ($T_e = T_i$), the threshold drift velocity is the electron thermal speed, which is usually very large. Therefore, a very large current is required to drive the ion acoustic mode unstable. The instability threshold can be reduced considerably if the electron temperature is increased relative to the ion temperature, i.e., $T_e \gg T_i$. This reduction is caused by the decreased ion Landau damping that occurs when $T_e \gg T_i$ (see Section 9.4.1).

The ion acoustic instability can also be analyzed for a Maxwellian distribution of electron and ion velocities. Using a procedure similar to that used in Section 9.4.1, it is straightforward, but somewhat tedious, to show that the growth rate in the long-wavelength limit ($k\lambda_{\mathrm{De}} \ll 1$) is given by

$$\frac{\gamma}{\omega} = -\sqrt{\frac{\pi}{8}}\left[\sqrt{\frac{m_{\mathrm{e}}}{m_{\mathrm{i}}}}\left(1 - \frac{k}{|k|}\frac{U_{\mathrm{e}}}{C_{\mathrm{A}}}\right) + \left(\frac{T_{\mathrm{e}}}{T_{\mathrm{i}}}\right)^{3/2} \exp\left(-\frac{T_{\mathrm{e}}}{2T_{\mathrm{i}}}\right)\right], \tag{9.5.31}$$

where $C_{\mathrm{A}} = \sqrt{\kappa T_{\mathrm{e}}/m_{\mathrm{i}}}$ is the ion acoustic speed (see Problem 9.19). In the derivation of this equation, the plasma dispersion function for the ion term has been expanded to order $1/\zeta_{\mathrm{i}}^2$, as in Eq. (9.4.21). The threshold electron drift velocity, U_{e}^*, at which the instability starts, is obtained by setting the term in the large brackets to zero and solving for U_{e}, which gives

$$|U_{\mathrm{e}}^*| = \sqrt{\frac{\kappa T_{\mathrm{e}}}{m_{\mathrm{e}}}}\left[\sqrt{\frac{m_{\mathrm{e}}}{m_{\mathrm{i}}}} + \left(\frac{T_{\mathrm{e}}}{T_{\mathrm{i}}}\right)^{3/2} \exp\left(-\frac{T_{\mathrm{e}}}{2T_{\mathrm{i}}}\right)\right]. \tag{9.5.32}$$

If $T_{\mathrm{e}} = T_{\mathrm{i}}$, the threshold is again very large, on the order of the electron thermal speed, but decreases rapidly as $T_{\mathrm{e}}/T_{\mathrm{i}}$ increases. A comparison of the threshold drift velocity given by Eq. (9.5.32) with the results from a full numerical solution is shown as a function of $T_{\mathrm{e}}/T_{\mathrm{i}}$ in Figure 9.32. As can be seen, Eq. (9.5.32) provides a good approximation for large $T_{\mathrm{e}}/T_{\mathrm{i}}$, but greatly over-estimates the threshold at low $T_{\mathrm{e}}/T_{\mathrm{i}}$ values. The discrepancy between the analytic and numerical results at low $T_{\mathrm{e}}/T_{\mathrm{i}}$ values can be reduced considerably by keeping terms to order $1/\zeta_{\mathrm{i}}^4$ in the expansion for the ion plasma dispersion function, similar to the derivation of Eqs. (9.4.28) and (9.4.29). This exercise is left to the reader.

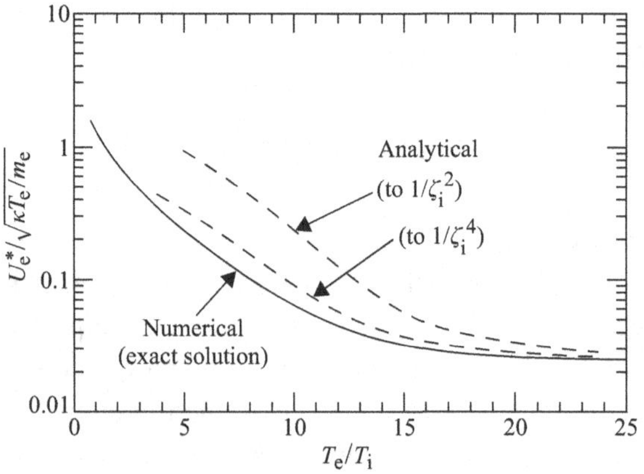

Figure 9.32 A comparison of the approximate analytical solution for the threshold electron drift velocity required to produce unstable ion acoustic waves with the results of an exact numerical calculation.

Problems

9.1. A plasma has a "spherical shell" distribution function given by

$$f_0(\mathbf{v}) = \frac{n_0}{4\pi c^2}\,\delta(|\mathbf{v}| - c),$$

where c is a constant.

(a) Using the Fourier analysis approach, show that the dispersion relation for electrostatic waves in this plasma is $\omega^2 = \omega_{\mathrm{p}}^2 + c^2k^2$.

(b) What is the region of validity of this dispersion relation?

9.2. A plasma consists of two homogeneous streams of electrons with equal densities and equal but opposite velocities ($V_1 = -V_2$).

(a) Show that the critical wave number for electrostatic instability is

$$k_{\mathrm{c}} = \sqrt{2}\frac{\omega_{\mathrm{p1}}}{V_1}.$$

(b) Show that the maximum growth rate is

$$\gamma = \frac{1}{2}\omega_{\mathrm{p1}}.$$

9.3. Buneman (1959) analyzed the maximum growth rate of an electron beam streaming at velocity $\mathbf{V}$ through an equal density of ions initially at rest. The dispersion relation for this system is

$$D(k,\omega) = 1 - \frac{\omega_{\mathrm{pi}}^2}{\omega^2} - \frac{\omega_{\mathrm{pe}}^2}{(kV - \omega)^2} = 0.$$

(a) If we assume $\omega_{\mathrm{pi}}^2 \ll \omega^2$, show that the dispersion relation can be written

$$kV = \omega + \omega_{\mathrm{pe}} + \frac{1}{2}\frac{\omega_{\mathrm{pe}}\omega_{\mathrm{pi}}^2}{\omega^2}.$$

(b) Show that the above dispersion relation has a solution for real k of the form

$$\omega = (\omega_{\mathrm{pe}}\omega_{\mathrm{pi}}^2\cos\theta)^{1/3}\,\mathrm{e}^{\mathrm{i}\theta},$$

where θ is any arbitrary angle.

(c) Show that the imaginary part of ω reaches a maximum when $\theta = \pi/3$.

(d) Show that, when $\theta = \pi/3$,

$$\gamma = \omega_{\mathrm{pe}}\left(\frac{m_{\mathrm{e}}}{2m_{\mathrm{i}}}\right)^{1/3}\frac{\sqrt{3}}{2} \quad \text{and} \quad \omega_{\mathrm{r}} = \omega_{\mathrm{pe}}\left(\frac{m_{\mathrm{e}}}{2m_{\mathrm{i}}}\right)^{1/3}\frac{1}{2}.$$

(e) Is assumption (a) satisfied?

9.4. The Laplace transform of $f(t)$ is defined by

$$\tilde{f}(p) = \int_0^\infty f(t)\,\mathrm{e}^{-pt}\mathrm{d}t.$$

Show that the Laplace transform of $f(t) = \cosh(at)$ is given by

$$\tilde{f}(p) = \frac{p}{p^2 - a^2}.$$

9.5. The inverse Laplace transform of $\tilde{f}(p)$ is given by

$$f(t) = \frac{1}{2\pi \mathrm{i}} \int_{\sigma - \mathrm{i}\infty}^{\sigma + \mathrm{i}\infty} \tilde{f}(p)\,\mathrm{e}^{pt}\,\mathrm{d}p,$$

where the integration along the line $p = \sigma$ must pass to the right of any poles of $\tilde{f}(p)$. Use the residue theorem to evaluate the inverse Laplace transform of

$$\tilde{f}(p) = \frac{1}{p^2 - a^2}.$$

9.6. A plasma has a reduced one-dimensional distribution given by the Cauchy distribution

$$F_0(v_z) = \frac{C}{\pi} \frac{1}{C^2 + v_z^2}.$$

(a) Using the Laplace transform approach, show that the dispersion relation is

$$D(k, p) = 1 + \frac{\omega_{\mathrm{p}}^2}{(p + C|k|)^2} = 0.$$

(b) Show that the roots of this dispersion relation (with $p = \gamma - \mathrm{i}\omega$) give $\gamma = -C|k|$ and $\omega = \pm\omega_{\mathrm{p}}$.

9.7. In the previous problem we showed that the Cauchy distribution

$$F_0(v_z) = \frac{C}{\pi} \frac{1}{v_z^2 + C^2}$$

gives a dispersion relation

$$D(k, p) = 1 + \frac{\omega_{\mathrm{p}}^2}{(p + C|k|)^2},$$

which has a solution $\omega = \pm\omega_{\mathrm{p}}$ and $\gamma = -C|k|$. Show that in the high phase velocity limit (i.e., $|k| \to 0$), the weak growth rate approximation,

$$Đ_{\mathrm{r}}(k,\omega) = 0 \quad \text{and} \quad \gamma = \frac{-Đ_{\mathrm{i}}}{\partial Đ_{\mathrm{r}}/\partial\omega},$$

gives the same result.

9.8. To model a hot beam, one can use a shifted Cauchy distribution of the form

$$F_0(v_z) = \frac{C}{\pi}\frac{1}{C^2 + (v_z - U)^2},$$

where U is the beam velocity. Show that the dispersion relation for this plasma is

$$Đ(k,p) = 1 + \frac{\omega_{\mathrm{p}}^2}{(p + C|k| + \mathrm{i}kU)^2} = 0.$$

9.9. The reduced one-dimensional kappa distribution is given by

$$F_0(v_z) = \frac{\Gamma(\kappa+1)}{\Gamma\left(\frac{1}{2}\right)\Gamma\left(\kappa+\frac{1}{2}\right)\kappa^{1/2}C}\left[1 + \frac{v_z^2}{\kappa C^2}\right]^{-(\kappa+1)},$$

where Γ is the gamma function and C is the thermal velocity.

(a) Show that in the limit $\kappa \to \infty$, the kappa distribution becomes a Gaussian distribution.

(b) Show that for small κ, the kappa distribution approaches a Cauchy distribution.

9.10. The plasma dispersion function is defined by the equation

$$Z(\zeta) = \frac{1}{\sqrt{\pi}}\int_C \frac{\mathrm{e}^{-z^2}}{z-\zeta}\,\mathrm{d}z,$$

where the contour C runs along the real ζ axis, and passes around the pole at $\zeta = z$ when $\mathrm{Re}\{\zeta\} < 0$. Explain why this integral cannot be evaluated by extending the contour around the upper half-plane in the same way as was done for the Cauchy distribution function.

9.11. If the pole at $\zeta = x + \mathrm{i}y$ is very close to the z axis (i.e., $|y| \ll |x|$), the plasma dispersion function is given by

$$Z(\zeta) = \mathrm{i}\frac{k}{|k|}\sqrt{\pi}\mathrm{e}^{-\zeta^2} + \frac{1}{\sqrt{\pi}}P\int_{-\infty}^{\infty}\frac{\mathrm{e}^{-z^2}}{z-\zeta}\,\mathrm{d}z.$$

By writing

$$\frac{1}{z-\zeta} = \frac{-1}{\zeta(1-z/\zeta)} = -\frac{1}{\zeta}\left[1+\left(\frac{z}{\zeta}\right)+\left(\frac{z}{\zeta}\right)^2+\cdots\right]$$

and integrating term by term, show that in the limit of large ζ the plasma dispersion function is given by the following power series:

$$Z(\zeta) = \mathrm{i}\frac{k}{|k|}\sqrt{\pi}\mathrm{e}^{-\zeta^2} - \left[\frac{1}{\zeta}+\frac{1}{2\zeta^3}+\frac{3}{4\zeta^5}+\cdots\right].$$

9.12. It is easy to show that the plasma dispersion function, $Z(\zeta)$, is the solution of the following first-order differential equation:

$$\frac{\mathrm{d}Z}{\mathrm{d}\zeta} + 2\zeta Z = -2.$$

Show that the following functions satisfy this differential equation:

(a) $Z(\zeta) = \frac{1}{\sqrt{\pi}}\int_{-\infty}^{\infty}\frac{\mathrm{e}^{-z^2}}{z-\zeta}\mathrm{d}z,$

(b) $Z(\zeta) = 2\mathrm{i}\,\mathrm{e}^{-\zeta^2}\int_{-\infty}^{\mathrm{i}\zeta}\mathrm{e}^{-z^2}\mathrm{d}z,$

(c) $Z(\zeta) = \mathrm{i}\sqrt{\pi}\mathrm{e}^{-\zeta^2}[1+\mathrm{erf}(\mathrm{i}\zeta)],$

where

$$\mathrm{erf}(\mathrm{i}\zeta) = \frac{2}{\sqrt{\pi}}\int_0^{\mathrm{i}\zeta}\mathrm{e}^{-z^2}\mathrm{d}z.$$

9.13. Using the erf representation for $Z(\zeta)$ in (b) above, show that for small ζ,

$$Z(\zeta) = \mathrm{i}\sqrt{\pi}\mathrm{e}^{-\zeta^2} - 2\zeta + \frac{4}{3}\zeta^3 - \frac{8}{15}\zeta^5 + \cdots$$

Hint: Use a Taylor series expansion for $\exp[-\zeta^2]$ and $\exp[-z^2]$, integrate term by term, and then collect like powers of ζ.

9.14. For a plasma consisting of electrons and one species of positive ion, both with Maxwellian velocity distributions, the dispersion relation can be written

$$Đ(k,p) = 1 - \frac{1}{(k\lambda_{\mathrm{De}})^2}\frac{1}{2}\left[Z'(\zeta_{\mathrm{e}}) + \frac{T_{\mathrm{e}}}{T_{\mathrm{i}}}Z'(\zeta_{\mathrm{i}})\right] = 0,$$

where T_{e} is the electron temperature and T_{i} is the ion temperature. Use the large-argument expansion of the plasma dispersion function for the ions,

$$Z'(\zeta_{\mathrm{i}}) = -\mathrm{i}\frac{k}{|k|}\sqrt{\pi}2\zeta_{\mathrm{i}}\mathrm{e}^{-\zeta_{\mathrm{i}}^2} + \frac{1}{\zeta_{\mathrm{i}}^2},$$

and the small-argument expansion for the electrons,

$$Z'(\zeta_e) = -i\frac{k}{|k|}\sqrt{\pi}2\zeta_e e^{-\zeta_e^2} - 2.$$

Substitute these expansions into the dispersion relation, expand in powers of y_i/x_i assuming $y_i \ll x_i$, and show that

$$\frac{\omega}{k} = \pm\sqrt{\frac{\kappa T_e}{m_i}\frac{1}{(1+k^2\lambda_{De}^2)^{1/2}}}$$

and

$$\frac{\gamma}{\omega} = -\sqrt{\frac{\pi}{8}}\left[\sqrt{\frac{m_e}{m_i}} + \left(\frac{T_e}{T_i}\right)^{3/2}\exp\left(-\frac{T_e}{2T_i}\frac{1}{(1+k^2\lambda_{De}^2)}\right)\right]\frac{1}{(1+k^2\lambda_{De}^2)^{3/2}}.$$

9.15. A Maxwellian plasma consists of equal densities of electrons and ions with temperatures T_e and T_i. The dispersion relation for this plasma can be written

$$D(k,p) = 1 - \frac{1}{(k\lambda_{De})^2}\frac{1}{2}Z'(\zeta_e) - \frac{1}{(k\lambda_{Di})^2}\frac{1}{2}Z'(\zeta_i) = 0.$$

By using the low phase velocity expansion for the electrons ($\zeta_e \ll 1$) and the high phase velocity expansion for the ions ($\zeta_i \gg 1$), and keeping all terms through $1/\zeta_i^4$ in the expansion for $Z(\zeta_i)$ and only the -2 term for $Z(\zeta_e)$, show that if $k\lambda_{De} \ll 1$ and $T_e \gg T_i$, then to a good approximation

$$\frac{\omega}{k} = \pm\sqrt{\frac{\kappa T_e + 3\kappa T_i}{m_i}}$$

and

$$\frac{\gamma}{\omega} = -\sqrt{\frac{\pi}{8}}\left[\sqrt{\frac{m_e}{m_i}} + \frac{(3+T_e/T_i)^{5/2}}{(9+T_e/T_i)}\exp\left(-\frac{T_e}{2T_i} - \frac{3}{2}\right)\right].$$

9.16. A plasma consists of two electron beams of equal "temperature" and density moving with velocities U and $-U$. The reduced one-dimensional distribution function is given by a counter-streaming Cauchy distribution:

$$F_0(v_z) = \frac{C}{2\pi}\left[\frac{1}{C^2+(v_z-U)^2} + \frac{1}{C^2+(v_z+U)^2}\right].$$

(a) Find the dispersion relation.

(b) Solve the dispersion relation for $p + C|k|$, where $p = \gamma - i\omega$. Your solution should reduce to the case of counter-streaming delta function beams in the limit $C \to 0$.

(c) Show that the solution for $p + C|k|$ in (b) is either purely real or purely imaginary. Which of these cases could possibly lead to instability? What is the frequency ω in this case?

(d) Show that the condition for instability is $U > C$. Hint: It is useful to know that the fastest growth rate occurs in the vicinity of $k = 0$. Prove this fact by expanding your solution to (b) in powers of k and examining the solution for the growth rate γ in the vicinity of $k = 0$.

(e) Sketch the shape of $F_0(v_z)$ when $U = C$.

9.17. A plasma consists of two electron beams with a reduced one-dimensional distribution function given by a counter-streaming Cauchy distribution:

$$F_0(v_z) = \frac{C}{2\pi}\left[\frac{1}{C^2 + (v_z - U)^2} + \frac{1}{C^2 + (v_z + U)^2}\right].$$

(a) Show that the distribution function is double humped only if $C < \sqrt{3}U$.

(b) Using the Penrose criterion, show that instability occurs only if $C < U$.

9.18. Following the procedure developed by Nyquist, make a plot of the mapping of the imaginary p axis onto the complex $Đ$-plane for three cases, $C < U$, $C = U$, and $C > U$ in Problem 9.17.

9.19. A plasma consists of a Maxwellian distribution of ions at temperature T_i and electrons at temperature T_e. The average velocity of the ions is zero and the electrons are drifting with an average velocity U_e.

(a) Make an appropriate modification of the derivation involving the plasma dispersion function to take into account the drifting electrons.

(b) Using the low phase velocity approximation for the electrons and the high phase velocity approximation for the ions, obtain an approximate expression for the normalized frequencies x_i and y_i of the ion acoustic mode. Assume that $k\lambda_{De} \ll 1$.

(c) Show that the growth rate is approximately

$$\frac{\gamma}{\omega} = -\sqrt{\frac{\pi}{8}}\left[\sqrt{\frac{m_e}{m_i}}\left(1 - \frac{U_e}{C_A}\right) + \left(\frac{T_e}{T_i}\right)^{3/2} \exp\left(-\frac{T_e}{2T_i}\right)\right],$$

where

$$C_A = \sqrt{\frac{\kappa T_e}{m_i}}$$

is the ion acoustic speed.

(d) Show that the instability condition for the current-driven ion acoustic mode is

$$|U_\mathrm{e}| > \sqrt{\frac{\kappa T_\mathrm{e}}{m_\mathrm{i}}}\left[1 + \sqrt{\frac{m_\mathrm{i}}{m_\mathrm{e}}}\left(\frac{T_\mathrm{e}}{T_\mathrm{i}}\right)^{3/2}\exp\left(-\frac{T_\mathrm{e}}{2T_\mathrm{i}}\right)\right].$$

(e) Plot $|U_\mathrm{e}|$ as a function of $T_\mathrm{e}/T_\mathrm{i}$. Where does the approximation fail?

References

Arfken, G. 1970. *Mathematical Methods for Physicists*. New York: Academic Press, pp. 311–315.

Armstrong, T. P. 1967. Numerical studies of the nonlinear Vlasov equation. *Phys. Fluids* **10**, 1269–1280.

Buneman, O. 1959. Dissipation of currents in ionized media. *Phys. Rev.* **115**, 503–517.

Flanigan, F. J. 1983. *Complex Variables: Harmonic and Analytic Functions*. Mineola, NY: Dover Publications, pp. 272–275. Originally published in 1972.

Fried, B. D., and Conte, S. D. 1961. *The Plasma Dispersion Function*. New York: Academic Press, pp. 2–3.

Gardner, C. S. 1963. Bound on the energy available from a plasma. *Phys. Fluids* **6**, 839–840.

Ginzburg, V. L., and Zhelezniakov, V. V. 1958. On the possible mechanism of sporadic solar radio emission (radiation in an isotropic plasma). *Sov. Astron. AJ* **2**, 653–666.

Gould, R. W., O'Neil, T. M., and Malmberg, J. H. 1967. Plasma wave echo. *Phys. Rev. Lett.* **19**, 219–222.

Gurnett, D. A., Hospodarsky, G. B., Kurth, W. S., Williams, D. J., and Bolton, S. J. 1993. Fine structure of Langmuir waves produced by a solar electron event. *J. Geophys. Res.* **98**, 5631–5637.

Landau, L. 1946. On the vibration of the electron plasma. *J. Phys. (USSR)* **10** (1), 85–94.

Malmberg, J. H., and Wharton, C. B. 1966. Dispersion of electron plasma waves. *Phys. Rev. Lett.* **17**, 175–178.

Nicholson, D. R. 1983. *Introduction to Plasma Theory*. Malabar, FL: Krieger Publishing, pp. 87–96.

Nyquist, H. 1932. Regeneration theory. *Bell System Tech. J.* **11**, 126–147.

O'Neil, T. M. 1965. Collisionless damping of nonlinear plasma oscillations. *Phys. Fluids* **8**, 2255–2262.

Penrose, O. 1960. Electrostatic instabilities of a uniform non-Maxwellian plasma. *Phys. Fluids* **3**, 258–265.

Vlasov, A. A. 1945. On the kinetic theory of an assembly of particles with collective interaction. *J. Phys. (USSR)* **9**, 25–44.

Further Reading

Chen, F. F. 1990. *Introduction to Plasma Physics and Controlled Fusion*. New York: Plenum Press, Chapter 7.

Nicholson, D. R. 1983. *Introduction to Plasma Theory*. Malabar, FL: Krieger Publishing, Chapter 6.

Stix, T. H. 1992. *Waves in Plasmas*. New York: American Institute of Physics, Chapter 8.

Swanson, D. G. 1989. *Plasma Waves*. New York: Academic Press, Section 4.2.

10

Waves in a Hot Magnetized Plasma

In this chapter we discuss the propagation of small-amplitude waves in a hot magnetized plasma. Just as for a cold plasma, the presence of a static zero-order magnetic field in a hot plasma leads to a wide variety of new phenomena. Because the zero-order motions of the particles in a magnetized plasma consist of circular orbits around the magnetic field, some type of resonance can be expected when the wave frequency is equal to the cyclotron frequency. In a cold plasma, this resonance is the same for all particles of a given charge-to-mass ratio, and gives rise to the well-defined cyclotron resonances described in Chapter 4. In a hot plasma, the frequency "felt" by a particle is Doppler-shifted by the thermal motion of the particle along the static magnetic field. For a given parallel velocity, resonance occurs when the frequency in the guiding-center frame of reference of the particle is at the cyclotron frequency, i.e., $\omega' = \omega - k_{\|}v_{\|} = \omega_{\mathrm{c}}$. Because of the thermal spread in the particle velocities, the resonance is no longer sharp, as it was in a cold plasma, but is now broadened by the thermal motion. The resonant interaction also produces damping, called cyclotron damping, in a manner somewhat analogous to Landau damping. If the cyclotron radius of the particle is a significant fraction of the wavelength, the phase shift introduced by the periodic cyclotron motion of the particles back and forth along the perpendicular component of the wave vector produces a phase modulation at the cyclotron frequency. As we will show, this phase modulation leads to a series of resonances at harmonics of the cyclotron frequency. The general cyclotron resonance condition then becomes $\omega' = \omega - k_{\|}v_{\|} = n\,\omega_{\mathrm{c}}$, where n is an integer.

Given the complexity of the cyclotron resonance interactions described above, it should come as no surprise that the mathematical analysis is more difficult than for a cold plasma. Nevertheless, it is possible to derive a general dispersion relation for the propagation of small-amplitude waves in a hot magnetized plasma. Because of the complexity of the analysis, we will restrict our discussion to certain special cases that are of general interest.

10.1 Linearization of the Vlasov Equation

Before proceeding with the derivation of the dispersion relation, it is useful to discuss some general features of the linearized Vlasov equation when a zero-order magnetic field is present. Since the wave amplitudes are assumed to be small, the Vlasov equation is linearized in the usual way by assuming that the velocity distribution function consists of a constant homogeneous zero-order term, $f_{s0}(\mathbf{v})$, plus a small perturbation, $f_{s1}(\mathbf{v})$, so that

$$f_s(\mathbf{v}) = f_{s0}(\mathbf{v}) + f_{s1}(\mathbf{v}). \tag{10.1.1}$$

Since the plasma is magnetized, we must now assume that the magnetic field consists of a constant homogeneous zero-order term plus a small first-order perturbation, i.e., $\mathbf{B} = \mathbf{B}_0 + \mathbf{B}_1$. For the electric field, it is sufficient to assume that the zero-order term is zero, so that $\mathbf{E} = \mathbf{E}_1$. This assumption is justified on the grounds that, except for exceedingly strong electric fields ($E_0 \geq cB_0$), it is always possible to transform to a frame of reference in which $\mathbf{E}_0 = 0$. Using these definitions it can be shown that the zero- and first-order linearized expansions of the Vlasov equation (5.2.15) are given by

$$\mathbf{v} \times \mathbf{B}_0 \cdot \nabla_{\mathbf{v}} f_{s0} = 0 \tag{10.1.2}$$

and

$$\frac{\partial f_s}{\partial t} + \mathbf{v} \cdot \nabla f_s + \frac{e_s}{m_s}(\mathbf{v} \times \mathbf{B}_0) \cdot \nabla_{\mathbf{v}} f_s + \frac{e_s}{m_s}[\mathbf{E} + \mathbf{v} \times \mathbf{B}] \cdot \nabla_{\mathbf{v}} f_{s0} = 0, \tag{10.1.3}$$

where as usual we have dropped the 1 subscript on the first-order terms. Because of the anticipated azimuthal symmetry with respect to the zero-order magnetic field, it is useful to introduce cylindrical velocity space coordinates $(v_\perp, \phi, v_\parallel)$, with $\mathbf{B}_0$ along the z axis as shown in Figure 10.1. To understand the meaning of the condition imposed on the zero-order distribution function by Eq. (10.1.2), we start by writing this equation in rectangular (x, y, z) coordinates:

$$(\mathbf{v} \times \mathbf{B}_0) \cdot \nabla_{\mathbf{v}} f_{s0} = \left(v_y \frac{\partial f_{s0}}{\partial v_x} - v_x \frac{\partial f_{s0}}{\partial v_y} \right) B_0 = 0. \tag{10.1.4}$$

Next, we transform this equation to cylindrical velocity space coordinates by using the equations $v_x = v_\perp \cos\phi, v_y = v_\perp \sin\phi$, and $v_z = v_\parallel$, which when substituted into the above equation gives

$$(\mathbf{v} \times \mathbf{B}_0) \cdot \nabla_{\mathbf{v}} f_{s0} = \left(v_\perp \sin\phi \frac{\partial f_{s0}}{\partial v_x} - v_\perp \cos\phi \frac{\partial f_{s0}}{\partial v_y} \right) B_0 = 0. \tag{10.1.5}$$

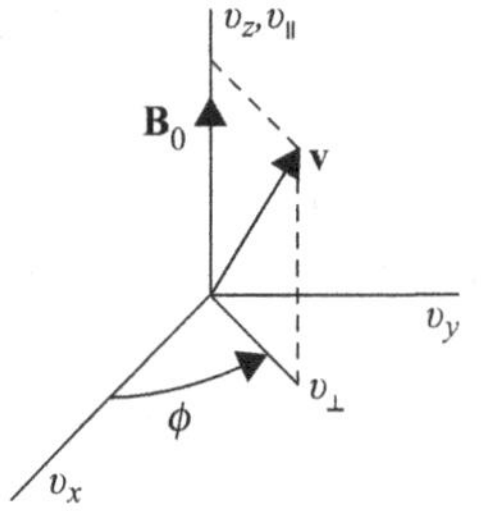

Figure 10.1 Cylindrical velocity space coordinates $v_\perp, v_\parallel$, and ϕ.

Note from the transformation equations that

$$\frac{\partial v_x}{\partial \phi} = -v_\perp \sin\phi \quad \text{and} \quad \frac{\partial v_y}{\partial \phi} = v_\perp \cos\phi. \tag{10.1.6}$$

Solving these equations for $\sin\phi$ and $\cos\phi$, and substituting them into Eq. (10.1.5) then gives

$$(\mathbf{v} \times \mathbf{B}_0) \cdot \nabla_{\mathbf{v}} f_{s0} = -\left[\frac{\partial v_x}{\partial \phi} \frac{\partial f_{s0}}{\partial v_x} + \frac{\partial v_y}{\partial \phi} \frac{\partial f_{s0}}{\partial v_y} \right] B_0 = 0. \tag{10.1.7}$$

Finally, using the chain rule to differentiate $f_{s0}(v_x, v_y)$ with respect to ϕ, one can see that the quantity in the large brackets on the right-hand side is simply $\partial f_{s0}/\partial \phi$. Equation (10.1.2) then becomes

$$(\mathbf{v} \times \mathbf{B}_0) \cdot \nabla_{\mathbf{v}} f_{s0} = -B_0 \frac{\partial f_{s0}}{\partial \phi} = 0. \tag{10.1.8}$$

Since $\mathbf{B}_0$ is assumed to be non-zero, it follows that $\partial f_{s0}/\partial \phi = 0$. This condition is called the gyrotropic condition, and shows that the zero-order distribution function must be azimuthally symmetric with respect to the magnetic field. This symmetry arises because the particles move in uniform circular motions around the magnetic field. It should be noted that the gyrotropic condition is satisfied only if there is no $\mathbf{E}_0 \times \mathbf{B}_0$ drift (i.e., only if $\mathbf{E}_0 = 0$).

Returning to the first-order linearized Vlasov equation (10.1.3), we note that the third term in this equation has exactly the same form as Eq. (10.1.8), except f_{s0} is replaced by f_s. Making this substitution and changing $e_s B_0/m_s$ to ω_{cs}, Eq. (10.1.3) becomes

$$\frac{\partial f_s}{\partial t} + \mathbf{v} \cdot \nabla f_s - \omega_{cs} \frac{\partial f_s}{\partial \phi} + \frac{e_s}{m_s} [\mathbf{E} + \mathbf{v} \times \mathbf{B}] \cdot \nabla_{\mathbf{v}} f_{s0} = 0. \tag{10.1.9}$$

Following the same procedure as in the previous chapter, the next step is to Fourier-transform this equation in space ($\nabla \rightarrow \mathrm{i}\mathbf{k}$) and in time ($\partial/\partial t \rightarrow -\mathrm{i}\omega$), which

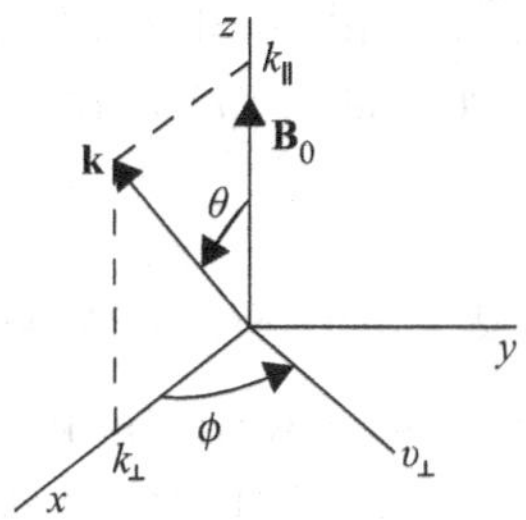

Figure 10.2 The wave vector, **k**, is assumed to lie in the x,z plane at an angle θ with respect to the magnetic field, $\mathbf{B}_0$. The parallel and perpendicular components of **k** are $k_\parallel$ and $k_\perp$.

gives

$$(-\mathrm{i}\omega + \mathrm{i}\mathbf{k}\cdot\mathbf{v})\tilde{f}_s - \omega_{cs}\frac{\partial \tilde{f}_s}{\partial \phi} + \frac{e_s}{m_s}[\tilde{\mathbf{E}} + \mathbf{v}\times\tilde{\mathbf{B}}]\cdot\nabla_{\mathbf{v}} f_{s0} = 0. \qquad (10.1.10)$$

At a later stage in the analysis, we will convert to Laplace transform notation by making the substitution $\omega \to \mathrm{i}p$. In order to reduce the number of field variables, the first-order magnetic field can be eliminated by using Faraday's law, which when Fourier-transformed can be written $\tilde{\mathbf{B}} = \mathbf{k}\times\tilde{\mathbf{E}}/\omega$. Substituting $\tilde{\mathbf{B}} = \mathbf{k}\times\tilde{\mathbf{E}}/\omega$ into the above equation then yields

$$(-\mathrm{i}\omega + \mathrm{i}\mathbf{k}\cdot\mathbf{v})\tilde{f}_s - \omega_{cs}\frac{\partial \tilde{f}_s}{\partial \phi} = -\frac{e_s}{m_s}\left[\tilde{\mathbf{E}} + \mathbf{v}\times\left(\frac{\mathbf{k}\times\tilde{\mathbf{E}}}{\omega}\right)\right]\cdot\nabla_{\mathbf{v}} f_{s0}, \qquad (10.1.11)$$

where we have organized the terms involving $\tilde{f}_s$ on the left-hand side and the terms involving the electric field on the right-hand side. Since the plasma is gyrotropic, without loss of generality we can assume that **k** lies in the x,z plane with components $k_\parallel$ and $k_\perp$, parallel and perpendicular to the static magnetic field, as shown in Figure 10.2. From the geometry it is easy to see that

$$\mathbf{k}\cdot\mathbf{v} = k_\parallel v_\parallel + k_\perp v_\perp \cos\phi. \qquad (10.1.12)$$

After substituting this result into Eq. (10.1.11) and dividing by $-\omega_{cs}$, the first-order linearized Vlasov equation can be written

$$\frac{\partial \tilde{f}_s}{\partial \phi} - \mathrm{i}(\alpha_s + \beta_s\cos\phi)\tilde{f}_s = \frac{e}{m_s\omega_{cs}}\left[\tilde{\mathbf{E}} + \mathbf{v}\times\left(\frac{\mathbf{k}\times\tilde{\mathbf{E}}}{\omega}\right)\right]\cdot\nabla_{\mathbf{v}} f_{s0}, \qquad (10.1.13)$$

where to simplify the notation we have introduced the definitions

$$\alpha_s = \frac{k_\parallel v_\parallel - \omega}{\omega_{cs}} \quad \text{and} \quad \beta_s = \frac{k_\perp v_\perp}{\omega_{cs}}. \qquad (10.1.14)$$

For any given first-order electric field, $\tilde{\mathbf{E}}$, this linear differential equation can be solved to give the first-order perturbation, $\tilde{f}_s$, in the velocity distribution function.

10.2 Electrostatic Waves

Since the analysis of electromagnetic waves is more difficult, we first turn our attention to electrostatic waves. The resulting dispersion relation is called the Harris dispersion relation.

10.2.1 The Harris Dispersion Relation

The advantage of restricting our attention to electrostatic waves is that the electric field can be written as the gradient of a potential, $\mathbf{E} = -\nabla\Phi$, which when Fourier-transformed becomes $\tilde{\mathbf{E}} = -\mathrm{i}\mathbf{k}\tilde{\Phi}$. For electrostatic waves one also has $\nabla \times \mathbf{E} = 0$ which, when Fourier-transformed, becomes $\mathbf{k} \times \tilde{\mathbf{E}} = 0$. The term $\mathbf{v}\times(\mathbf{k}\times\tilde{\mathbf{E}})$ in Eq. (10.1.13) is then zero, which provides considerable simplification. After substituting $\tilde{\mathbf{E}} = -\mathrm{i}\mathbf{k}\tilde{\Phi}$ into Eq. (10.1.13), the first-order Vlasov equation becomes

$$\frac{\partial \tilde{f}_s}{\partial \phi} - \mathrm{i}(\alpha_s + \beta_s \cos\phi)\tilde{f}_s = \frac{-\mathrm{i}e_s}{m_s\omega_{cs}}\tilde{\Phi}\mathbf{k}\cdot\nabla_{\mathbf{v}} f_{s0}. \tag{10.2.1}$$

The above equation is a first-order linear differential equation of the form

$$\frac{\mathrm{d}f}{\mathrm{d}x} + P(x)f = Q(x), \tag{10.2.2}$$

which has a general solution

$$f = \mathrm{e}^{-\int^x P(x')\mathrm{d}x'}\left[\int^x Q(x')\mathrm{e}^{\int^{x'} P(x'')\mathrm{d}x''}\mathrm{d}x'\right], \tag{10.2.3}$$

where the term $\int P(x)\,\mathrm{d}x$ is called the integrating factor (Boyce and DiPrima, 1992). The integrating factor in this case is

$$\begin{aligned}\int^\phi P(\phi')\mathrm{d}\phi' &= -\mathrm{i}\int^\phi (\alpha_s + \beta_s\cos\phi')\mathrm{d}\phi' \\ &= -\mathrm{i}(\alpha_s\phi + \beta_s\sin\phi).\end{aligned} \tag{10.2.4}$$

The general solution of Eq. (10.2.1) is then given by

$$\tilde{f}_s = \frac{-\mathrm{i}e_s n_s\tilde{\Phi}}{m_s\omega_{cs}}\mathrm{e}^{\mathrm{i}(\alpha_s\phi + \beta_s\sin\phi)}\int^\phi \mathbf{k}\cdot\nabla_{\mathbf{v}}F_{s0}\,\mathrm{e}^{-\mathrm{i}(\alpha_s\phi' + \beta_s\sin\phi')}\mathrm{d}\phi', \tag{10.2.5}$$

where we have introduced the normalized zero-order distribution function, F_{s0}, via the definition $f_{s0} = n_{s0}F_{s0}$. Following the same procedure used in a hot unmagnetized plasma, the next step is to compute the charge density

$$\tilde{\rho}_q = \sum_s e_s \int_{-\infty}^{\infty} \tilde{f}_s \mathrm{d}^3 v \tag{10.2.6}$$

and substitute the charge density into Poisson's equation, $\nabla^2 \Phi = -\rho_q/\epsilon_0$, which after Fourier transforming becomes

$$k^2 \tilde{\Phi} = \sum_s \frac{e_s}{\epsilon_0} \int_{-\infty}^{\infty} \tilde{f}_s \mathrm{d}^3 v. \tag{10.2.7}$$

The dispersion relation can then be obtained by substituting $\tilde{f}_s$ from Eq. (10.2.5) into the integral, canceling $\tilde{\Phi}$ on both sides of the equation, and dividing by k^2, which gives

$$\begin{aligned} D(k,\omega) = 1 + \sum_s \frac{\omega_{ps}^2}{k^2 \omega_{cs}} \int \mathrm{e}^{\mathrm{i}(\alpha_s \phi + \beta_s \sin\phi)} \\ \times \int^{\phi} \mathrm{i}\mathbf{k}\cdot\nabla_{\mathbf{v}} F_{s0}\, \mathrm{e}^{-\mathrm{i}(\alpha_s \phi' + \beta_s \sin\phi')} \mathrm{d}\phi'\, \mathrm{d}\phi v_\perp\, \mathrm{d}v_\perp\, \mathrm{d}v_\parallel = 0, \end{aligned} \tag{10.2.8}$$

where the velocity space volume element is written as $\mathrm{d}^3 v = \mathrm{d}\phi v_\perp\, \mathrm{d}v_\perp \mathrm{d}v_\parallel$. Unfortunately, the above equation is not yet in a useful form. To simplify the equation, we proceed by writing $\nabla_{\mathbf{v}} F_{s0}$ in cylindrical coordinates, noting from Eq. (10.1.8) that $\partial F_{s0}/\partial\phi = 0$, and taking the dot product with $\mathrm{i}\mathbf{k}$. The first term in the integrand is then given by

$$\mathrm{i}\mathbf{k}\cdot\nabla_{\mathbf{v}} F_{s0} = \mathrm{i}k_\parallel \frac{\partial F_{s0}}{\partial v_\parallel} + \mathrm{i}k_\perp \frac{\partial F_{s0}}{\partial v_\perp} \cos\phi'. \tag{10.2.9}$$

By writing $\cos\phi'$ in terms of complex exponentials, the $\mathrm{d}\phi'$ integral in Eq. (10.2.8) can be written

$$\begin{aligned} \int^{\phi} \cdots \mathrm{d}\phi' = \mathrm{i}k_\parallel \frac{\partial F_{s0}}{\partial v_\parallel} \int^{\phi} \mathrm{e}^{-\mathrm{i}(\alpha_s \phi' + \beta_s \sin\phi')} \mathrm{d}\phi' + \mathrm{i}k_\perp \frac{\partial F_{s0}}{\partial v_\perp} \\ \times \int^{\phi} \frac{1}{2}(\mathrm{e}^{\mathrm{i}\phi'} + \mathrm{e}^{-\mathrm{i}\phi'}) \mathrm{e}^{-\mathrm{i}(\alpha_s \phi' + \beta_s \sin\phi')}\, \mathrm{d}\phi'. \end{aligned} \tag{10.2.10}$$

The integrals in the above equation can be evaluated by using the following expansion:

$$\mathrm{e}^{-\mathrm{i}\beta_s \sin\phi'} = \sum_{n=-\infty}^{\infty} J_n(\beta_s) \mathrm{e}^{-\mathrm{i}n\phi'}, \tag{10.2.11}$$

where $J_n(\beta_s)$ is the nth order Bessel function (from Arfken, 1970). Using this expansion, the first integral on the right-hand side of Eq. (10.2.10) becomes

$$\int^{\phi} e^{-i(\alpha_s\phi' + \beta_s \sin\phi')} d\phi' = \sum_n J_n(\beta_s) \int^{\phi} e^{-i(\alpha_s + n)\phi'} d\phi'$$
$$= i \sum_n \frac{J_n(\beta_s)}{\alpha_s + n} e^{-i(\alpha_s + n)\phi}. \tag{10.2.12}$$

Similarly, the second integral on the right-hand side of Eq. (10.2.10) becomes

$$\int^{\phi} (e^{i\phi'} + e^{-i\phi'}) e^{-i(\alpha_s\phi' + \beta_s \sin\phi')} d\phi'$$
$$= \sum_n J_n(\beta_s) \int^{\phi} [e^{-i(\alpha_s - 1 + n)\phi'} + e^{-i(\alpha_s + 1 + n)\phi'}] d\phi'$$
$$= i \sum_n J_n(\beta_s) \left[\frac{e^{-i(\alpha_s - 1 + n)\phi}}{\alpha_s - 1 + n} + \frac{e^{-i(\alpha_s + 1 + n)\phi}}{\alpha_s + 1 + n} \right]. \tag{10.2.13}$$

Next, multiplying Eq. (10.2.10) by the term

$$e^{i(\alpha_s\phi + \beta_s \sin\phi)} = \sum_m J_m(\beta_s) e^{i(\alpha_s + m)\phi} \tag{10.2.14}$$

gives

$$e^{i(\alpha_s\phi + \beta_s \sin\phi)} \int^{\phi} \cdots d\phi' = -k_{\|} \frac{\partial F_{s0}}{\partial v_{\|}} \sum_{n,m} J_m J_n \left[\frac{e^{i(m-n)\phi}}{\alpha_s + n} \right]$$
$$- \frac{1}{2} k_{\perp} \frac{\partial F_{s0}}{\partial v_{\perp}} \sum_{n,m} J_m J_n \left[\frac{e^{i(m-n+1)\phi}}{\alpha_s + n - 1} + \frac{e^{i(m-n-1)\phi}}{\alpha_s + n + 1} \right]. \tag{10.2.15}$$

In the second summation, the index can be relabeled to give

$$\sum_{n,m} \frac{J_m J_{n+1}}{\alpha_s + n} e^{i(m-n)\phi} + \frac{J_m J_{n-1}}{\alpha_s + n} e^{i(m-n)\phi} = \sum_{n,m} J_m [J_{n+1} + J_{n-1}] \frac{e^{i(m-n)\phi}}{\alpha_s + n}. \tag{10.2.16}$$

Using the Bessel function recursion formula, $J_{n+1} + J_{n-1} = (2n/\beta_s)J_n$, this sum can be written in the more compact form

$$\sum_{m,n} \frac{2n\, J_m J_n}{\beta_s(\alpha_s + n)} e^{i(m-n)\phi}. \tag{10.2.17}$$

Combining all the terms then gives

$$e^{i(\alpha_s\phi+\beta_s\sin\phi)}\int^{\phi}\cdots d\phi' = -\sum_{n,m}\frac{J_m J_n}{\alpha_s+n}\left[k_{\|}\frac{\partial F_{s0}}{\partial v_{\|}}+\frac{n\omega_{cs}}{v_{\perp}}\frac{\partial F_{s0}}{\partial v_{\perp}}\right]e^{i(m-n)\phi}, \tag{10.2.18}$$

where in the second term on the right we have eliminated β_s by using $\beta_s = k_{\perp}v_{\perp}/\omega_{cs}$ from Eq. (10.1.14). Inserting this result into Eq. (10.2.8), we obtain

$$D(k,\omega) = 1-\sum_s\frac{\omega_{ps}^2}{k^2\omega_{cs}}\sum_{n,m}\int_{-\infty}^{\infty}\int_0^{\infty}v_{\perp}dv_{\perp}dv_{\|}\frac{J_m J_n}{\alpha_s+n}$$
$$\times\left[k_{\|}\frac{\partial F_{s0}}{\partial v_{\|}}+\frac{n\omega_{cs}}{v_{\perp}}\frac{\partial F_{s0}}{\partial v_{\perp}}\right]\int_0^{2\pi}e^{i(m-n)\phi}\,d\phi = 0. \tag{10.2.19}$$

The ϕ integral in the above equation is zero unless $m = n$, in which case it is 2π. Finally, reintroducing the definitions for α_s and β_s from Eq. (10.1.14), the dispersion relation can be written in the form

$$D(k,\omega) = 1-\sum_s\frac{\omega_{ps}^2}{k^2}\sum_n\int_{-\infty}^{\infty}\int_0^{\infty}\frac{J_n^2(k_{\perp}v_{\perp}/\omega_{cs})}{k_{\|}v_{\|}-\omega+n\omega_{cs}}\left[k_{\|}\frac{\partial F_{s0}}{\partial v_{\|}}+\frac{n\omega_{cs}}{v_{\perp}}\frac{\partial F_{s0}}{\partial v_{\perp}}\right]$$
$$\times 2\pi v_{\perp}dv_{\perp}dv_{\|} = 0. \tag{10.2.20}$$

This equation is called the Harris dispersion relation after Harris (1959), who first derived this result.

The Harris dispersion relation gives a very general result for small-amplitude electrostatic waves propagating in a hot magnetized plasma and is valid for any direction of propagation and any choice of plasma parameters. The origin of the cyclotron harmonics, $n\omega_{cs}$, in the dispersion relation can be traced to the term $\exp[i\beta_s\sin\phi]$ and has the following interpretation. First, note that the parameter $\beta_s = k_{\perp}v_{\perp}/\omega_{cs}$ is the product of the cyclotron radius, $\rho_c = v_{\perp}/\omega_c$, and the perpendicular wave number, $k_{\perp}$. Next, if the wavelength is comparable to the cyclotron radius, so that $k_{\perp}\rho_c \sim 1$, then a substantial phase shift is introduced in the applied electric field by the zero-order cyclotron motion of the particle. The origin of the phase shift is illustrated in Figure 10.3. To understand the significance of the phase shift, consider an applied electric field of the form $E_0\exp[i(k_{\perp}x-\omega t)]$. If the zero-order motion along the x axis is $x=\rho_c\sin\omega_c t$, the electric field at the position of the particle is given by

$$E = E_0 e^{i\beta_s\sin\omega_c t}e^{-i\omega t} = E_0\sum_{n=-\infty}^{\infty}J_n(\beta_s)e^{i(n\omega_c-\omega)t}, \tag{10.2.21}$$

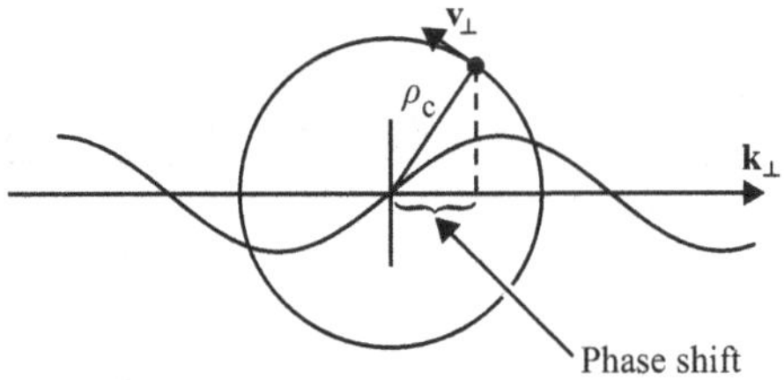

Figure 10.3 For a finite cyclotron radius, ρ_c, a phase shift is introduced by the cyclotron motion of the particle around the magnetic field. This phase shift is responsible for the resonances at $n\,\omega_c$.

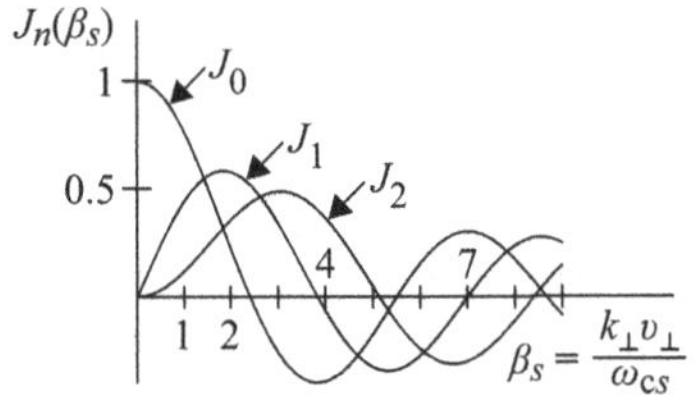

Figure 10.4 Plots of the zero-order, first-order, and second-order Bessel functions, $J_0(\beta_s), J_1(\beta_s)$, and $J_2(\beta_s)$.

where we have used the harmonic expansion given by Eq. (10.2.11) for the $\exp(i\beta_s \sin\omega_c t)$ term. Since the phase shift term, $\beta_s \sin\omega_c t$, is periodic with angular frequency ω_c, the above equation shows that the electric field force can be expressed as an expansion in harmonics of the cyclotron frequency, ω_c. The coefficients of this expansion are the Bessel functions $J_n(\beta_s)$. When $\omega \simeq n\,\omega_c$, the electric field force "felt" by the particle has a component in phase with the motion of the particle and a strong resonant interaction occurs. The Bessel function term gives the "strength" of this interaction, and is controlled by β_s. The functional forms of the first three Bessel functions are shown in Figure 10.4; see Watson (1995).

10.2.2 The Low-temperature, Long-Wavelength Limit

Before discussing the general solution of the Harris dispersion relation, it is useful to consider the low-temperature, long-wavelength limit. If the perpendicular temperature is sufficiently low ($v_\perp \to 0$) and/or the perpendicular wavelength is sufficiently long ($k_\perp \to 0$), the parameter $\beta_s = k_\perp v_\perp/\omega_{cs}$ can be regarded as a small quantity (i.e., $\beta_s \ll 1$). In this limit the Bessel functions in Eq. (10.2.11) can be approximated to first order in β_s by three terms, $J_0 \simeq 1, J_1 \simeq \beta_s/2$, and $J_{-1} \simeq -\beta_s/2$. All the remaining Bessel function terms are of higher order in β_s and can be omitted. Substituting these approximations into the dispersion relation, keeping only the $n = 0$ term associated with $\partial F_{s0}/\partial v_{\parallel}$ and the $n = \pm 1$ term associated with $\partial F_{s0}/\partial v_{\perp}$, and expanding the $v_{\parallel}$ integral in the small parameters $k_{\parallel}v_{\parallel}/\omega$ and $k_{\parallel}v_{\parallel}/(\omega \pm \omega_{\mathrm{c}})$, the integral can be written as a series of integrals involving moments in $v_{\parallel}$, very similar to the high phase velocity expansion in Eq. (9.1.18). Keeping only the zero-order moments (i.e., the zero temperature limit), it is easy to show that the Harris dispersion relation simplifies to

$$Đ_0(k,\omega) = \left[1 - \sum_s \frac{\omega_{\mathrm{p}s}^2}{\omega^2}\right]\cos^2\theta + \left[1 - \sum_s \frac{\omega_{\mathrm{p}s}^2}{\omega^2 - \omega_{\mathrm{c}s}^2}\right]\sin^2\theta = 0, \tag{10.2.22}$$

where $\cos\theta = k_{\parallel}/k$ and $\sin\theta = k_{\perp}/k$. The terms in the brackets can be immediately recognized as the functions P and S in cold plasma theory; see Eqs. (4.4.8) and (4.4.9). The dispersion relation can then be written

$$Đ_0(k,\omega) = P\cos^2\theta + S\sin^2\theta = 0. \tag{10.2.23}$$

For propagation parallel to the magnetic field, $\theta = 0$, the above equation reduces to $P = 0$, which gives the electron plasma oscillations encountered in cold plasma theory. For propagation perpendicular to the magnetic field, $\theta = \pi/2$, the dispersion relation reduces to $S = 0$, which gives the hybrid resonances. At intermediate angles the dispersion relation corresponds to resonance cones at $\tan^2\theta = -P/S$, i.e., Eq. (4.4.53). Thus, the low-temperature, long-wavelength limit of the Harris dispersion relation reduces to the electrostatic waves encountered in cold plasma theory.

10.2.3 The Bernstein Modes

To explore further the solutions of the Harris dispersion relation, we next consider the special case of an isotropic Maxwellian velocity distribution with the wave vector perpendicular to the magnetic field (i.e., $k_{\parallel} = 0$). To simplify the analysis, the ions are assumed to be immobile, so that only the electron motions need be considered. Keeping only the electron terms, the Harris dispersion relation equation (10.2.20), with $k_{\parallel} = 0, \omega_{\mathrm{c}} = |\omega_{\mathrm{ce}}| = -\omega_{\mathrm{ce}}$, and $\omega_{\mathrm{pe}} = \omega_{\mathrm{p}}$, can be written

$$Đ(k_{\perp},\omega) = 1 - \frac{\omega_{\mathrm{p}}^2}{k_{\perp}^2}\sum_{n=-\infty}^{\infty}\frac{n\omega_{\mathrm{c}}}{-\omega + n\omega_{\mathrm{c}}} \times \int_{-\infty}^{\infty}\int_0^{\infty} J_n^2\left(\frac{k_{\perp}v_{\perp}}{\omega_{\mathrm{c}}}\right)\frac{\partial F_0}{\partial v_{\perp}} 2\pi \mathrm{d}v_{\perp}\mathrm{d}v_{\parallel} = 0, \tag{10.2.24}$$

where we have changed n to $-n$ and used the fact that $J^2_{-n}(-\beta_s) = J^2_n(\beta_s)$. A further simplification is made by combining the positive $(+n)$ and negative $(-n)$ terms

$$\frac{n\omega_c}{-\omega + n\omega_c} + \frac{(-n)\omega_c}{-\omega + (-n)\omega_c} = \frac{-2n^2\omega_c^2}{\omega^2 - n^2\omega_c^2}, \tag{10.2.25}$$

which gives

$$Đ(k_\perp, \omega) = 1 + \frac{2\omega_p^2}{k_\perp^2} \sum_{n=1}^{\infty} \frac{n^2\omega_c^2}{\omega^2 - n^2\omega_c^2}$$

$$\times \int_{-\infty}^{\infty} \int_0^{\infty} J_n^2\left(\frac{k_\perp v_\perp}{\omega_c}\right) \frac{\partial F_0}{\partial v_\perp} 2\pi dv_\perp \, dv_\| = 0, \tag{10.2.26}$$

where the summation now extends only over the positive integers. In contrast to the situation for a Maxwellian velocity distribution function in an unmagnetized plasma, the integral in the dispersion relation can be performed in closed form. This is made possible by the following identity (from Watson, 1995)

$$\int_0^{\infty} J_n^2(au)e^{-u^2} u \, du = \frac{1}{2} \exp\left(-\frac{a^2}{2}\right) I_n\left(\frac{a^2}{2}\right), \tag{10.2.27}$$

where $I_n(x)$ is the modified Bessel function of order n. These functions can be computed using the power series expansion

$$I_n(x) = \sum_{m=0}^{\infty} \frac{1}{m!(m+|n|)!}\left(\frac{x}{2}\right)^{2m+|n|}. \tag{10.2.28}$$

The modified Bessel functions increase monotonically with increasing x as shown in Figure 10.5. After completing the required integration in Eq. (10.2.26), the dispersion relation can be written in the following form:

$$Đ(k_\perp, \omega) = 1 - \sum_{n=1}^{\infty} \frac{2\omega_p^2}{\beta_c^2\omega_c^2} \frac{\Gamma_n(\beta_c)}{(\omega/n\omega_c)^2 - 1} = 0, \tag{10.2.29}$$

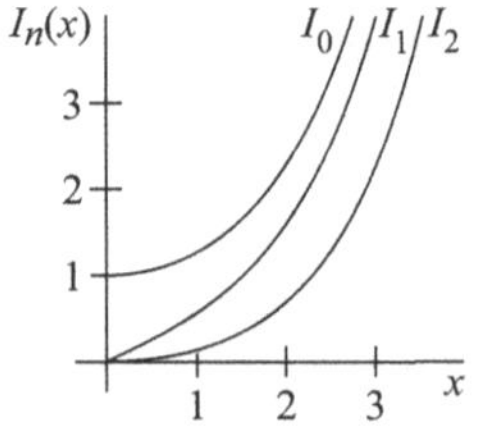

Figure 10.5 Plots of the zero-order, first-order, and second-order modified Bessel functions, $I_0(x)$, $I_1(x)$, and $I_2(x)$.

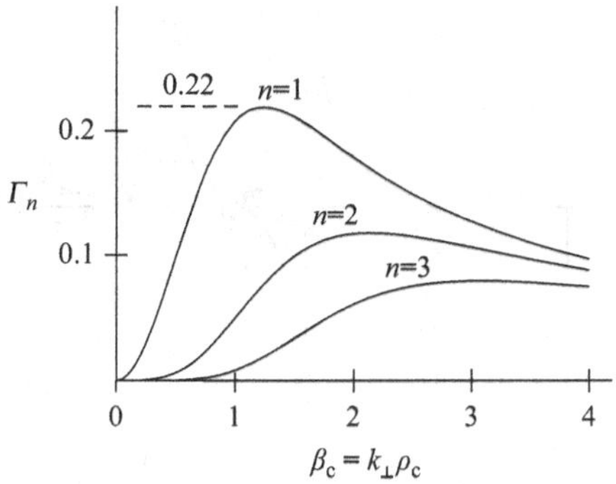

Figure 10.6 Plots of $\Gamma_n(\beta_c)$ for $n = 1$, 2, and 3.

where the function $\Gamma_n(\beta_c)$ in the above equation is defined by

$$\Gamma_n(\beta_c) = e^{-\beta_c^2} I_n(\beta_c^2), \tag{10.2.30}$$

and β_c is defined by

$$\beta_c = k_\perp \frac{\sqrt{\kappa T/m}}{\omega_c}. \tag{10.2.31}$$

Note that the quantity multiplying $k_\perp$ in the above equation is simply the thermal cyclotron radius, $\rho_c = \sqrt{\kappa T/m}/\omega_c$, i.e., the cyclotron radius of an electron moving at the thermal velocity, so that $\beta_c = k_\perp \rho_c$. Plots of $\Gamma_n(\beta_c)$ as a function of β_c are shown in Figure 10.6 for $n = 1$, 2, and 3. As can be seen, the Γ_n functions all have a single maximum that decreases and shifts to larger β_c values as n increases. For $n = 1$, the peak value is 0.22 at $\beta_c = 1.24$.

The dispersion relation given by Eq. (10.2.29) was first derived by Bernstein (1958), and the corresponding modes of propagation are called the Bernstein modes. The general nature of the roots of the dispersion relation can be determined by plotting $Đ(k_\perp, \omega)$ as a function of ω for a specific value of $\beta_c = k_\perp \rho_c$ and then looking for the points where $Đ = 0$. Notice that the thermal cyclotron radius, ρ_c, is a basic scale factor in the perpendicular wavelength. Note also that as $k_\perp \rho_c \to 0$, the dispersion relation, $Đ(k_\perp, \omega)$, approaches the cold plasma equation, S, given by Eq. (4.4.8). A representative plot of $Đ(k_\perp, \omega)$ is shown in Figure 10.7. Since $Đ(k_\perp, \omega)$ has a single-order pole at each of the cyclotron harmonics, the roots of the dispersion relation consist of an infinite number of frequencies, $\omega_1, \omega_2, \ldots$, with one root between each of the adjacent cyclotron harmonics. By solving the dispersion relation, $Đ(k_\perp, \omega) = 0$, as a function of $\beta_c = k_\perp \rho_c$, the frequency of each mode, ω_n, can be determined as a function of $k_\perp \rho_c$. This procedure must be done numerically. A typical plot of ω_n as a function of $\beta_c = k_\perp \rho_c$ is shown in Figure 10.8. For the parameters used, $\omega_p = 2.5\omega_c$, the ratio of the plasma

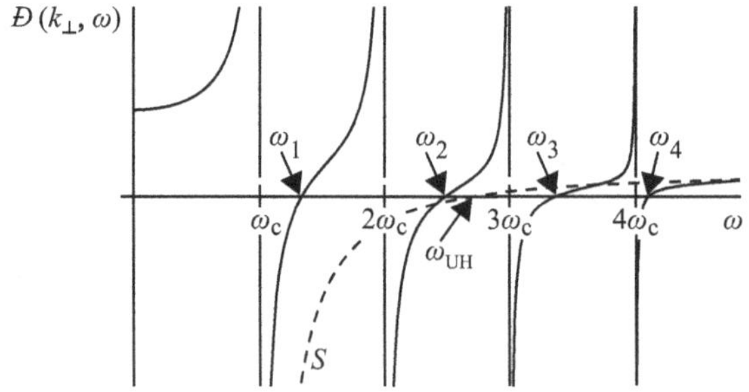

Figure 10.7 A plot of $\text{Đ}(k_\perp, \omega)$ as a function of frequency. The dashed line is the cold plasma limit, S.

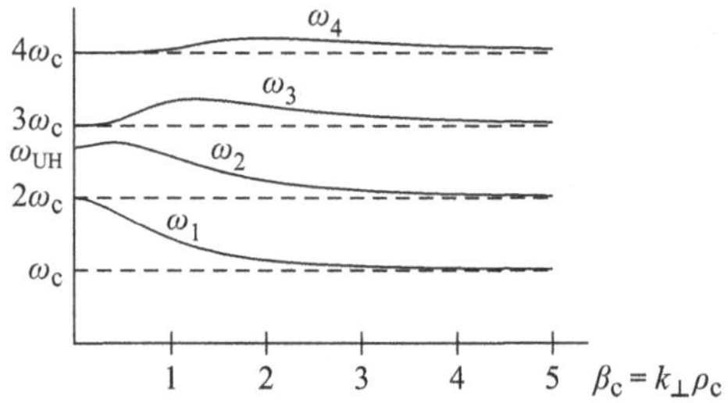

Figure 10.8 A plot of the solutions of $\text{Đ}(k_\perp, \omega) = 0$ as a function of the perpendicular wave number, $k_\perp$. These solutions are called the Bernstein modes.

frequency to the cyclotron frequency is such that the upper hybrid resonance frequency, ω_{UH}, is located between the second and third harmonics of the electron cyclotron frequency. Note that the branch of the dispersion relation between these two harmonics goes to ω_{UH} as $k_\perp \rho_c \to 0$. This is to be expected from our previous discussion of the low-temperature, long-wavelength limit. Also, note that the qualitative form of the dispersion curves changes at ω_{UH}. For frequencies below ω_{UH}, the allowed frequency range extends across the entire band between adjacent harmonics, whereas above ω_{UH} the allowed frequency range is restricted to a narrow frequency range just above the harmonic.

The existence of electrostatic waves near the harmonics of the cyclotron frequency is a totally new, hot plasma effect not predicted by either cold plasma theory or fluid theory. Although one might expect some type of cyclotron damping because the plasma is hot, for propagation perpendicular to the magnetic field ($k_\parallel = 0$) there is no damping. This occurs because for $k_\parallel = 0$ the resonance velocity, $v_\parallel = (\omega - n\omega_{cs})/k_\parallel$, is infinite, so no particles can resonate with the wave. The absence of damping means that there is no short-wavelength limit to the electrostatic waves, contrary to the results found in Chapter 9 for an unmagnetized plasma. However, this situation is a special case applicable only for propagation

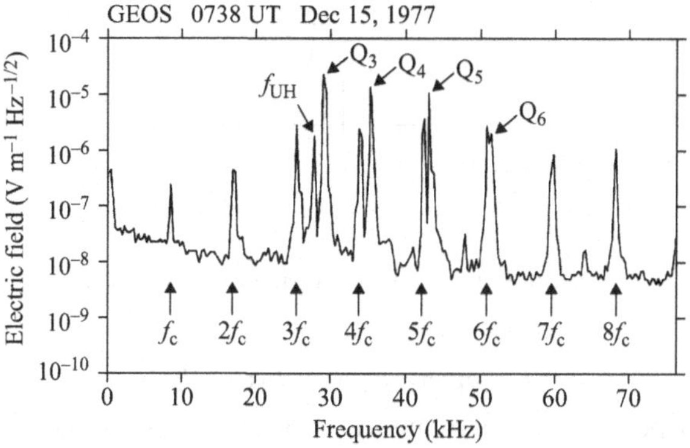

Figure 10.9 An ionospheric sounder response showing the various resonances associated with the Bernstein modes (data provided by P. Canu).

exactly perpendicular to the magnetic field. For any angle slightly away from perpendicular, the damping reappears. Wavelengths less than the Debye length are then strongly damped.

Both laboratory and space plasma measurements have clearly verified the existence of the Bernstein modes. In space experiments, the clearest examples occur in the records from satellite-borne ionospheric sounders, such as were discussed in Chapter 4. Since the Bernstein modes have no damping, once excited the waves decay very slowly and are easy to detect. The received signal typically consists of a series of "spikes" or "resonances" near harmonics of the electron cyclotron frequency, as shown in Figure 10.9. Although the sounding pulse excites waves over a broad range of frequencies, the waves detected after the initial transient has decayed are those that have zero group velocity, since only these waves stay in the vicinity of the spacecraft. As can be seen from Figure 10.8, the group velocity, $\partial\omega/\partial k_\perp$, goes to zero at (1) the harmonics of the electron cyclotron frequency, $n\omega_c$; (2) the upper hybrid resonance frequency, f_{UH}; and (3) a series of frequencies, f_Q, called the Q resonances. The Q resonances are the frequencies where the dispersion curves in Figure 10.8 reach a maximum (i.e., where $\partial\omega/\partial k_\perp = 0$). All three types of resonances can be seen in Figure 10.9.

10.2.4 Instabilities

Next we consider various types of electrostatic instabilities that can occur in a hot magnetized plasma. Two broad classes of instabilities can be identified: (1) anisotropy-driven instabilities, and (2) current-driven instabilities. We will discuss examples of each.

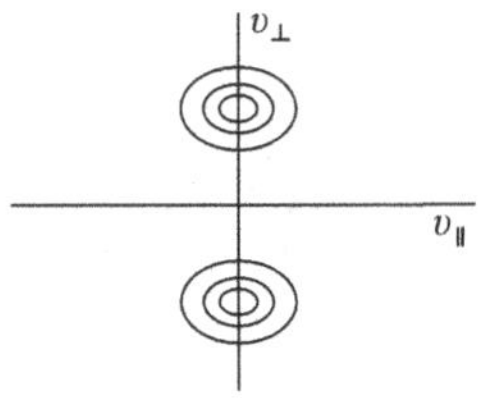

Figure 10.10 A representative ring-type velocity distribution.

Anisotropy-Driven Instabilities

For propagation perpendicular or nearly perpendicular to the magnetic field, an interesting class of instabilities occurs between the harmonics of the electron cyclotron frequency that are driven by anisotropic electron velocity distributions. To illustrate the existence of such instabilities, we start by considering a ring-type electron velocity distribution, such as that shown in Figure 10.10. Distribution functions of this type can be produced by injecting a beam of electrons into the plasma at pitch angles nearly perpendicular to the magnetic field. It is easy to show that such distributions are unstable. As a simple example, consider the following delta function distribution:

$$F_0(v_\perp, v_\parallel) = \frac{1}{2\pi V_0} \delta(v_\parallel) \delta(v_\perp - V_0), \tag{10.2.32}$$

which represents a narrow ring of particles in velocity space centered at $v_\parallel = 0$ and $v_\perp = V_0$. If we restrict our attention to waves propagating exactly perpendicular to the magnetic field ($k_\parallel = 0$), the dispersion relation can be evaluated in a straightforward way by substituting the above equation into Eq. (10.2.26) and integrating by parts once, which gives

$$Đ(k_\perp, \omega) = 1 - \frac{4\omega_p^2}{k_\perp V_0 \omega_c} \sum_{n=1}^{\infty} \frac{J_n(\beta_c) J_n'(\beta_c)}{(\omega/n\omega_c)^2 - 1} = 0, \tag{10.2.33}$$

where $\beta_c = k_\perp V_0/\omega_c$ is 2π times the ratio of the cyclotron radius to the wavelength. The signs of the various terms in the summation are determined by the sign of the product $J_n(\beta_c) J_n'(\beta_c)$. Representative plots of $J_n(\beta_c)$ and $J_n'(\beta_c)$ are shown in Figure 10.11 for $n = 2$. For small velocities ($\beta_c \ll 1$), the signs of the $J_n(\beta_c) J_n'(\beta_c)$ terms are all positive, and the solutions to the dispersion relation have the same general form as for the Bernstein modes, with one root between each cyclotron harmonic. However, as the ring velocity V_0 increases, β_c increases until eventually the sign of one of the $J_n(\beta_c) J_n'(\beta_c)$ terms becomes negative, as in the shaded regions of Figure 10.11. The plot of $Đ(k_\perp, \omega)$ versus ω then changes form, with the branches adjacent to this cyclotron harmonic curving alternately concave downward and then concave upward. As V_0 is increased further, a point is

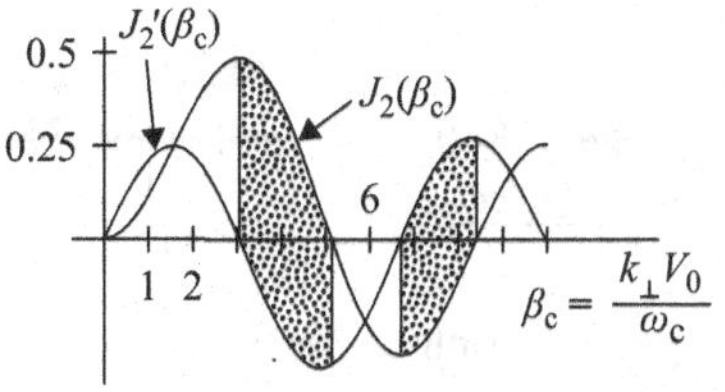

Figure 10.11 Plots of $J_2'(\beta_c)$ and $J_2(\beta_c)$. Instabilities can occur in the shaded regions where the product $J_n'(\beta_c)J_n(\beta_c)$ is negative. The exact instability condition depends on the ratio ω_p/ω_c.

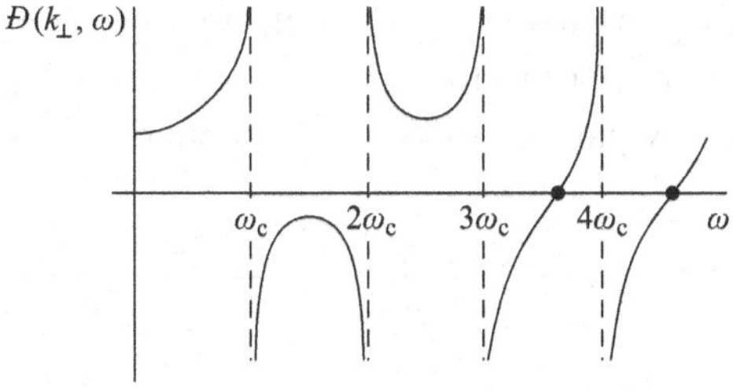

Figure 10.12 A plot of $D(k_\perp,\omega)$ for an unstable distribution. In this case the instability is associated with the $n = 2$ term, for which the product $J_2'(\beta_c)J_2(\beta_c)$ is negative.

reached where these branches no longer cross the $D = 0$ axis, as illustrated in Figure 10.12 for the $n = 2$ term. At this point the two real roots associated with these branches change to a complex conjugate pair, one of which is unstable. Because of the complexity of $D(k_\perp,\omega)$, the exact point at which the instability starts must be evaluated numerically. Detailed numerical studies by Tataronis and Crawford (1970) show that a minimum plasma density exists for an instability to occur, at $\omega_p/\omega_c \simeq 4.13$ for the 0 to ω_c band, $\omega_p/\omega_c \simeq 2.61$ for the ω_c to $2\omega_c$ band, etc. Tataronis and Crawford also explored a number of other similar velocity distributions, such as a spherical shell distribution and a combination of a Maxwellian and a ring distribution, all of which are unstable for certain choices of the parameters. In many respects these instabilities are like the cold beam instabilities discussed in Chapter 9 for unmagnetized plasmas. The ring distribution essentially produces two oppositely directed beams. The results are modified, of course, by the presence of magnetic field forces.

In addition to the ring-type instabilities described above, which are non-resonant fluid-like instabilities, a variety of other instabilities, very similar to the Bernstein modes, can be generated by resonant interactions. To investigate these instabilities, we must convert the Harris dispersion relation equation (10.2.20) to the Landau formalism. This conversion is accomplished by replacing ω by ip and changing the integration along the $v_\parallel$ axis such that it always passes below the pole at

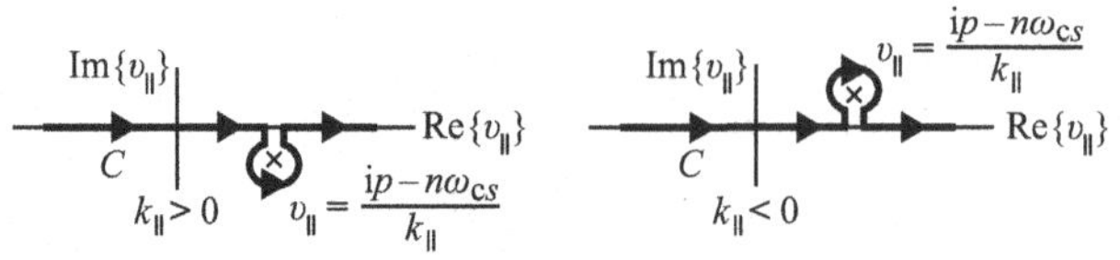

Figure 10.13 The $v_\|$ integration contour for the two cases $k_\| > 0$ and $k_\| < 0$.

$v_\| = (ip - n\omega_{cs})/k_\|$ for $k_\| > 0$, and above the pole for $k_\| < 0$, as shown in Figure 10.13. Following the approach developed in Chapter 9, we use the weak growth rate approximation, which involves separating the dispersion relation into its real and imaginary parts using the Plemelj relation (see Eq. (9.2.39)). Again keeping only the electron terms (with $\omega_c = -\omega_{ce}$ and $\omega_p = \omega_{pe}$, and changing $-n$ to n), the real and imaginary parts of the Harris dispersion relation are

$$Ð_r = 1 - \frac{\omega_p^2}{k^2} \sum_n P \int_{-\infty}^{\infty} \int_0^{\infty} \frac{J_n^2(k_\perp v_\perp/\omega_c)}{k_\| v_\| - \omega + n\omega_c} \times \left[k_\| \frac{\partial F_0}{\partial v_\|} + \frac{n\omega_c}{v_\perp} \frac{\partial F_0}{\partial v_\perp} \right] 2\pi v_\perp \, dv_\perp \, dv_\| \tag{10.2.34}$$

and

$$Ð_i = -\pi \frac{k_\|}{|k_\||} \frac{\omega_p^2}{k^2} \sum_n \int_0^{\infty} J_n^2\left(\frac{v_\perp k_\perp}{\omega_c}\right) \times \left[k_\| \frac{\partial F_0}{\partial v_\|} + \frac{n\omega_c}{v_\perp} \frac{\partial F_0}{\partial v_\perp} \right] 2\pi v_\perp \, dv_\perp \Bigg|_{v_{\|\mathrm{Res}} = \frac{\omega - n\omega_c}{k_\|}}, \tag{10.2.35}$$

where, as indicated, the integral in $Ð_i$ must be evaluated at the resonance velocity $v_{\|\mathrm{Res}} = (\omega - n\omega_c)/k_\|$. To achieve a finite resonance velocity, the wave vector cannot be exactly perpendicular to the magnetic field, even though in most cases it is nearly perpendicular. The growth rate can then be calculated from the weak growth rate approximation equation (9.2.37)

$$\gamma = \frac{-Ð_i}{\partial Ð_r/\partial\omega}, \tag{10.2.36}$$

where ω and k are determined by $Ð_r(\omega, k) = 0$.

Since $\partial Ð_r/\partial\omega$ is always positive for the Bernstein-like modes (see Figure 10.7), the sign of the growth rate is controlled by the sign of the integral in Eq. (10.2.35). Consider for the moment only waves with $k_\| > 0$. The growth rate is then positive if the sign of the integral is positive. The sign of the integral is controlled by the terms $\partial F_0/\partial v_\|$ and $\partial F_0/\partial v_\perp$. Any region of the distribution function where either

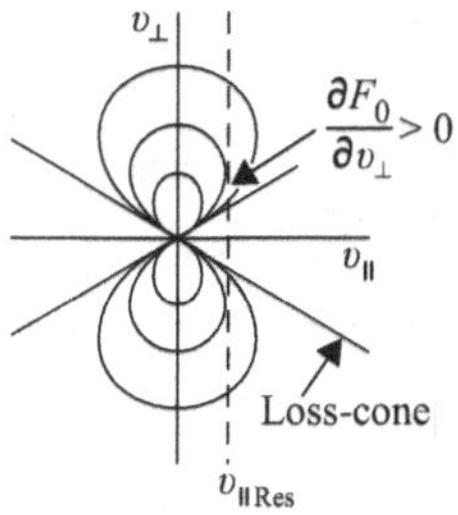

Figure 10.14 A loss-cone type of distribution function.

of these terms is positive is a potential source of instability. These regions are called free energy sources. If the integral over perpendicular velocities turns out to be positive for a particular **k** (assuming $k_\parallel > 0$), then the plasma is unstable. Note that both $k_\parallel$ and $k_\perp$ are involved in the integral. Also note that $k_\parallel$ plays a controlling role in determining the resonance velocity, $v_{\parallel\text{Res}} = (\omega - n\omega_c)/k_\parallel$. Our problem is to understand the nature of this control.

In the previous chapter we have already encountered instabilities associated with regions where $\partial F_0/\partial v_\parallel$ is positive. This free energy source is characteristic of a beam. The presence of a magnetic field reveals a new type of free energy source that is associated with regions where $\partial F_0/\partial v_\perp$ is positive. This free energy source is characteristic of an anisotropy. Although many different velocity distributions have an anisotropy, the prototypical distribution function that illustrates this type of free energy source is a loss-cone distribution, such as shown in Figure 10.14. For an instability to occur, the parallel wave number must be such that the parallel resonance velocity, $v_{\parallel\text{Res}}$, indicated by the vertical dashed line, passes through the region where $\partial F_0/\partial v_\perp$ is positive. Both positive and negative values of $\partial F_0/\partial v_\perp$ occur along this resonance line. As can be seen from Eq. (10.2.35), the integral over perpendicular velocities is of the form

$$\int_0^\infty J_n^2\left(\frac{k_\perp v_\perp}{\omega_c}\right)\frac{\partial F_0}{\partial v_\perp}\mathrm{d}v_\perp\Bigg|_{v_{\parallel\text{Res}}=\frac{\omega-n\omega_c}{k_\parallel}}, \tag{10.2.37}$$

which involves a convolution of the terms $J_n^2(k_\perp v_\perp/\omega_c)$ and $\partial F_0/\partial v_\perp$. For an instability to occur, the integral must be sufficiently positive (assuming $k_\parallel > 0$) to overcome any negative contribution from the $k_\parallel \partial F_0/\partial v_\parallel$ term in Eq. (10.2.35). A representative plot of the various terms in Eq. (10.2.37) is shown in Figure 10.15 for a loss-cone distribution. As can be seen from this illustration, the perpendicular wave number, $k_\perp$, controls the peaks in the Bessel function. The largest positive value for the integral usually occurs when $k_\perp$ is such that the first maximum in $J_n^2(k_\perp v_\perp/\omega_c)$ coincides with the peak in $\partial F_0/\partial v_\perp$, which is near the edge of the

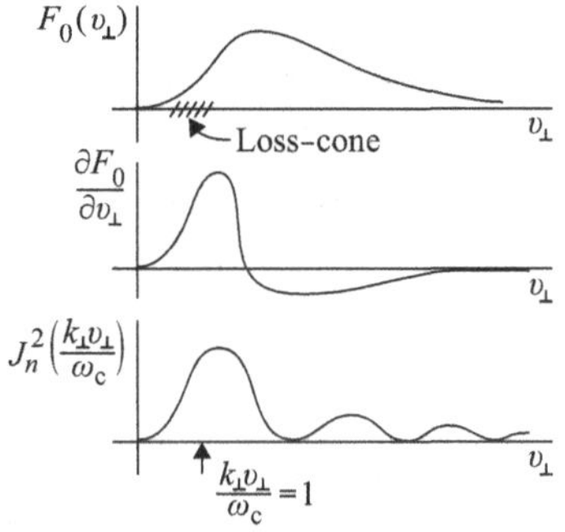

Figure 10.15 Plots of $F_0(v_\perp)$, $\partial F_0/\partial v_\perp$, and $J_n^2(k_\perp v_\perp/\omega_c)$ for a representative loss-cone distribution.

loss cone. Thus, the Bessel function plays a key role in determining the range of unstable perpendicular wave numbers for each harmonic. We can make a rough estimate of the wave normal angle by noting that, in the unstable region (i.e., where $\partial F_0/\partial v_\perp > 0$), one has $k_\perp v_\perp/\omega_c \sim 1$ and $v_\parallel = (\omega - n\omega_c)/k_\parallel$. Since $v_\perp/v_\parallel$ is typically of order one in this region, it follows that $\tan\theta = k_\perp/k_\parallel \approx \omega_c/(\omega - n\omega_c)$. As with the Bernstein modes, the frequency is usually close to the cyclotron harmonic (see Figure 10.9), especially at the higher harmonics. The wave vector direction is then at a large angle to the magnetic field, typically 70° to 80°.

Electrostatic electron cyclotron instabilities driven by cyclotron resonance interactions near the loss cone are frequently observed in both laboratory and space plasmas. The instabilities usually occur as a series of emission lines between harmonics of the electron cyclotron frequency. These waves are often called $(n + 1/2)f_c$ emissions, even though the frequencies are only somewhere between the n and $(n+1)$ cyclotron harmonics. An example showing the spectrum of a series of electrostatic $(n + 1/2)f_c$ emissions detected in Earth's magnetosphere is shown in Figure 10.16. Similar emissions also sometimes occur near the upper hybrid resonance, at frequencies such that $(n + 1/2)f_c \simeq f_{UH}$. For a further discussion of these waves, see Ashour-Abdalla and Kennel (1978), Rönnmark (1983) and Horne (1989).

In addition to the electron cyclotron instabilities, another important class of instabilities occurs near the harmonics of the ion cyclotron frequency. These waves are called electrostatic ion cyclotron waves and are driven by anisotropies in the ion distribution function. Ion effects are, of course, already included in the general form of the Harris dispersion relation equation (10.2.20). Unfortunately, in the analysis of electrostatic ion cyclotron waves, it is not possible to omit the electron contribution in the same way that the ion contribution can be omitted in the analysis of electron cyclotron waves. However, the electron contribution to the dispersion relation can be greatly simplified by making use of suitable approximations. Since

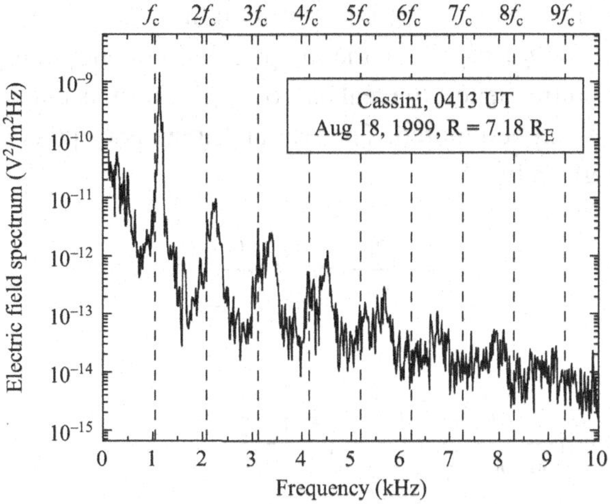

Figure 10.16 A frequency–time spectrogram showing a series of electrostatic electron cyclotron emissions observed in the outer regions of Earth's magnetosphere.

Landau damping is expected to strongly damp waves with parallel phase velocities near the electron thermal speed, we can assume that any unstable waves that might occur must have parallel phase velocities much greater than the electron thermal velocity. Therefore, we can use the high phase velocity approximation, $k_{\|}v_{\|}/\omega \ll 1$ for the electron terms. Also, for comparable electron and ion temperatures, the electron cyclotron radius is much less than the ion cyclotron radius. Since the ion cyclotron waves are expected to have wavelengths comparable to the ion cyclotron radius, we can also use the long-wavelength approximation ($k_{\perp}\rho_{ce} \ll 1$), for the electron Bessel function terms, which eliminates all of the electron terms except for $n = 0$ and $n = \pm 1$. It can then be shown that the electron contribution to the dispersion relation is identical to the low-temperature, long-wavelength limit equation (10.2.22) derived earlier, i.e.,

$$Ð_e \simeq Ð_{e0} = \left[1 - \frac{\omega_{pe}^2}{\omega^2}\right]\cos^2\theta + \left[1 - \frac{\omega_{pe}^2}{\omega^2 - \omega_{ce}^2}\right]\sin^2\theta. \tag{10.2.38}$$

If we assume that the ion cyclotron waves have frequencies near the low-order harmonics of the ion cyclotron frequency, we can also make use of the approximations $\omega \ll \omega_{ce}$ and $\omega \ll \omega_{pe}$. Assuming that the plasma is also moderately dense ($\omega_{pe} \gg \omega_{ce}$), Eq. (10.2.38) simplifies to

$$Ð_e = -\frac{\omega_{pe}^2}{\omega^2}\cos^2\theta + \frac{\omega_{pe}^2}{\omega_{ce}^2}\sin^2\theta. \tag{10.2.39}$$

To explore the conditions for instability, we again use the weak growth rate approximation. Following the usual procedure, the Plemelj relation is used to give the real and imaginary parts of the ion contribution to the dispersion relation, with the result that $Đ_{ion} = Đ_r + iĐ_i$. Since the ion plasma frequency is much smaller than the electron plasma frequency, the real part of the ion term is much smaller than the electron term and can be ignored. The real and imaginary parts of the dispersion relation are then given by

$$Đ_r = -\frac{\omega_{pe}^2}{\omega^2}\cos^2\theta + \frac{\omega_{pe}^2}{\omega_{ce}^2}\sin^2\theta \tag{10.2.40}$$

and

$$Đ_i = -\pi \frac{k_\parallel}{|k_\parallel|}\frac{\omega_{pi}^2}{k^2}\sum_n \int_0^\infty J_n^2\left(\frac{k_\perp v_\perp}{\omega_{ci}}\right) \times \left[k_\parallel \frac{\partial F_{i0}}{\partial v_\parallel} + \frac{n\omega_{ci}}{v_\perp}\frac{\partial F_{i0}}{\partial v_\perp}\right] 2\pi v_\perp \, dv_\perp \Bigg|_{v_{\parallel Res} = \frac{\omega - n\omega_{ci}}{k_\parallel}} . \tag{10.2.41}$$

As usual, the growth rate is given by

$$\gamma = \frac{-Đ_i}{\partial Đ_r/\partial\omega}, \tag{10.2.42}$$

where ω and k are determined by $Đ_r(k,\omega) = 0$.

Since $\partial Đ_r/\partial\omega$ is easily shown to be positive from Eq. (10.2.40), the growth rate is positive whenever the integral in $Đ_i$ is positive (again considering only $k_\parallel > 0$). As with the electron cyclotron instabilities, the possible free energy sources can be classified according to whether $\partial F_{i0}/\partial v_\parallel$ or $\partial F_{i0}/\partial v_\perp$ is positive. The prototypical free energy sources are then a "beam" and a "loss cone," just as in the case of the electron cyclotron instabilities. Of these, the loss-cone free energy source ($\partial F_{i0}/\partial v_\perp > 0$) is of special importance for both laboratory magnetic mirror machines and planetary radiation belts, since they always have a loss cone. The region of the ion distribution function that causes the instability is essentially the same as in Figure 10.14, except that the parallel resonance velocity is now given by $v_{\parallel Res} = (\omega - n\omega_{ci})/k_\parallel$, and the resonance interaction is with the ions.

Since ω is expected to be much less than ω_{ce}, the condition $Đ_r = 0$, where $Đ_r$ is given by Eq. (10.2.40), shows that the wave normal angle, θ, given by $\tan\theta = \omega_{ce}/\omega$, must be very close to $\pi/2$. The parallel wave number $k_\parallel$ must then be quite small, which is consistent with the high phase velocity approximation used for the electrons. In order for $v_{\parallel Res} = (\omega - n\omega_{ci})/k_\parallel$ to be in the range of the ion velocities, which are much smaller than the electron thermal velocities, it also follows that $\omega - n\omega_{ci}$ must be very small. The frequencies of the unstable modes are then very close to the harmonics of the ion cyclotron frequency, $\omega \simeq n\omega_{ci}$. The wave normal

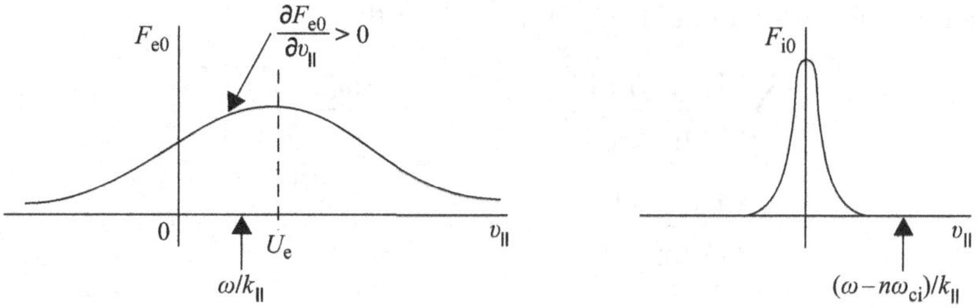

Figure 10.17 Representative reduced one-dimensional electron and ion velocity distribution functions, F_{e0} and F_{i0}, used to illustrate the current-driven electrostatic ion cyclotron instability.

angle is given approximately by

$$\tan\theta = \frac{\omega_{ce}}{\omega} \simeq \frac{\omega_{ce}}{n\omega_{ci}} = \frac{m_i}{nm_e}. \qquad (10.2.43)$$

For an electron–proton plasma, the wave normal angle for the $n = 1$ mode is approximately 89.97°, which is very nearly perpendicular to the magnetic field. The range of perpendicular wave numbers that are unstable is again determined by the first maximum in $J_n^2(k_\perp v_\perp/\omega_{ci})$, which must coincide with the peak in $\partial F_{i0}/\partial v_\perp$ in order to give a net positive contribution to the $v_\perp$ integral, very similar to the situation for the electrostatic electron cyclotron instabilities (see Figure 10.15).

Current-Driven Instabilities

Electrostatic ion cyclotron waves can also be driven unstable by currents flowing along the static magnetic field. Such magnetic field-aligned currents are a common feature of magnetized plasmas and require that the electrons drift relative to the ions, as shown in Figure 10.17. The free energy source for the current-driven electrostatic ion cyclotron instability is the region of positive $\partial F_{e0}/\partial v_\parallel$ in the electron distribution, very similar to the current-driven ion acoustic instability (see Section 9.5.4). The existence of a current-driven ion cyclotron instability was first demonstrated by Drummond and Rosenbluth (1962), using a Maxwellian electron and ion distribution functions of the form

$$F_{s0} = \left(\frac{m_s}{2\pi\kappa T_s}\right)^{3/2} \exp\left\{-\frac{m_s[(v_\parallel - U_s)^2 + v_\perp^2]}{2\kappa T_s}\right\}, \qquad (10.2.44)$$

where U_s is the drift velocity along the magnetic field. For a plasma consisting of electrons with a drift velocity U_e and one species of positive ion with a drift velocity $U_i = 0$, it is easy to show, using Eqs. (9.3.5) and (10.2.27), that the Harris dispersion relation equation (10.2.20) can be written

$$Đ(k,p) = 1 - \sum_n \frac{\Gamma_n(\beta_{ce})}{(k\lambda_{De})^2}\left[\frac{1}{2}Z'(\zeta_e - \zeta_{U_e}) - \frac{n\omega_{ce}}{k_\parallel}\sqrt{\frac{m_e}{2\kappa T_e}}Z(\zeta_e - \zeta_{U_e})\right]$$
$$- \sum_n \frac{\Gamma_n(\beta_{ci})}{(k\lambda_{Di})^2}\left[\frac{1}{2}Z'(\zeta_i) - \frac{n\omega_{ci}}{k_\parallel}\sqrt{\frac{m_i}{2\kappa T_i}}Z(\zeta_i)\right] = 0, \qquad (10.2.45)$$

where $Z(\zeta)$ is the plasma dispersion function described in Section 9.3, $\Gamma_n(\beta_{cs})$ is defined by Eq. (10.2.30), and for notational convenience, we introduce the following definitions:

$$\zeta_s = \sqrt{\frac{m_s}{2\kappa T_s}}\frac{\omega - n\omega_{cs} + i\gamma}{k_\parallel} \quad \text{and} \quad \zeta_{U_e} = \sqrt{\frac{m_e}{2\kappa T_e}}U_e. \qquad (10.2.46)$$

Since the wavelength of the electrostatic ion cyclotron waves is expected to be on the order of the ion cyclotron radius, which is usually much larger than the electron cyclotron radius, we can assume that $\beta_{ce} \ll 1$, so $\Gamma_0(\beta_{ce}) = 1$ and $\Gamma_n(\beta_{ce}) = 0$ for $n \neq 0$. Also, since the ion cyclotron radius is usually much greater than the Debye length, we also assume that $k\lambda_{De} \ll 1$. The dispersion relation equation (10.2.45) then simplifies to

$$\frac{1}{2}Z'(\zeta_e - \zeta_{U_e}) + \frac{T_e}{T_i}\sum_n \Gamma_n(\beta_{ci})\left[\frac{1}{2}Z'(\zeta_i) - \frac{n\omega_{ci}}{k_\parallel}\sqrt{\frac{m_i}{2\kappa T_i}}Z(\zeta_i)\right] = 0, \qquad (10.2.47)$$

where we have made use of the relation

$$\frac{1}{(k\lambda_{Di})^2} = \frac{T_e}{T_i}\frac{1}{(k\lambda_{De})^2}. \qquad (10.2.48)$$

To proceed further, we assume that the unstable $k_\parallel$ values are such that the Landau resonance velocity for the electrons, $v_{\parallel Res} = \omega/k_\parallel$, is in the region of positive slope slightly below U_e (thereby providing a free energy source), and that the ion cyclotron resonance velocity, $v_{\parallel Res} = (\omega - \omega_{ci})/k_\parallel$, is well above the ion thermal speed (thereby avoiding strong ion cyclotron damping). These assumptions correspond to assuming that $|\zeta_e - \zeta_{U_e}| \ll 1$ and $|\zeta_i| \gg 1$. We can then use the small argument expansion for the plasma dispersion function in the electron term,

$$Z'(\zeta_e - \zeta_{U_e}) \simeq -i\frac{k_\parallel}{|k_\parallel|}\sqrt{\pi}2(\zeta_e - \zeta_{U_e}) - 2, \qquad (10.2.49)$$

and the large-argument expansion in the ion term,

$$Z(\zeta_i) = i\frac{k_\parallel}{|k_\parallel|}\sqrt{\pi}e^{-\zeta_i^2} - \frac{1}{\zeta_i} \quad \text{and} \quad Z'(\zeta_i) = -i\frac{k_\parallel}{|k_\parallel|}\sqrt{\pi}\,2\zeta_i\,e^{-\zeta_i^2} + \frac{1}{\zeta_i^2}. \qquad (10.2.50)$$

Using these expansions, Eq. (10.2.47) becomes

$$\frac{1}{2}\left[-i\frac{k_{\|}}{|k_{\|}|}\sqrt{\pi}\,2(\zeta_e-\zeta_{U_e})-2\right]+\frac{T_e}{T_i}\sum_n\Gamma_n(\beta_{ci})\frac{1}{2}\left(-i\frac{k_{\|}}{|k_{\|}|}\sqrt{\pi}2\zeta_i e^{-\zeta_i^2}\right)$$
$$-\frac{T_e}{T_i}\sum_n\Gamma_n(\beta_{ci})\frac{n\omega_{ci}}{k_{\|}}\sqrt{\frac{m_i}{2\kappa T_i}}\left(i\frac{k_{\|}}{|k_{\|}|}\sqrt{\pi}e^{-\zeta_i^2}-\frac{1}{\zeta_i}\right)=0. \quad (10.2.51)$$

Writing $\zeta_i = x_i + iy_i$, where

$$x_i = \sqrt{\frac{m_i}{2\kappa T_i}}\frac{\omega-n\omega_{ci}}{k_{\|}} \quad \text{and} \quad y_i = \sqrt{\frac{m_i}{2\kappa T_i}}\frac{\gamma}{k_{\|}}, \quad (10.2.52)$$

and assuming that the growth rate is small (i.e., $|y_i| \ll |x_i|$), it is easy to show that the real and imaginary parts of the dispersion relation equation (10.2.51) yield two equations. The real part gives an equation for the wave frequencies,

$$\frac{\omega-n\omega_{ci}}{n\omega_{ci}}=\frac{T_e}{T_i}\Gamma_n(\beta_{ci}), \quad (10.2.53)$$

and the imaginary part (after eliminating the x_i term using the real part of the dispersion relation) gives the growth rate

$$\frac{\gamma}{n\omega_{ci}}=\frac{k_{\|}}{|k_{\|}|}\Gamma_n(\beta_{ci})\frac{T_e}{T_i}\sqrt{\frac{\pi}{2}}$$
$$\times\left[\frac{U_e}{C_e}-\left\{\frac{1}{C_e}+\frac{T_e}{T_i}\frac{\Gamma_n(\beta_{ci})}{C_i}\exp\left[-\frac{1}{2}\left(\frac{\omega-n\omega_{ci}}{k_{\|}C_i}\right)^2\right]\right\}\frac{\omega}{k_{\|}}\right], \quad (10.2.54)$$

where C_e and C_i are the electron and ion thermal speeds, as defined by Eq. (2.1.3). Note that the growth rate is proportional to $\Gamma_n(\beta_{ci})$, which has the form shown in Figure 10.6.

Since Γ_1 has the largest peak value (0.22 at $\beta_{ci} = 1.24$), the $n = 1$ mode has the highest growth rate. Also note that since the peak values of Γ_n decrease rapidly with increasing n and are always less than 0.22, the frequencies of all the modes are restricted to a narrow range of frequencies slightly above the corresponding harmonic, $n\omega_{ci}$, of the ion cyclotron frequency.

The critical drift velocity, U_e^*, for the onset ($\gamma = 0$) of the current-driven electrostatic ion cyclotron instability can be obtained by setting the term in the outermost bracket in Eq. (10.2.54) to zero, which gives

$$\frac{U_e^*}{C_e}=\frac{\omega}{k_{\|}C_e}+\frac{T_e}{T_i}\frac{0.22\omega}{k_{\|}C_i}\exp\left[-\frac{1}{2}\left(\frac{\omega-n\omega_{ci}}{k_{\|}C_i}\right)^2\right], \quad (10.2.55)$$

where we have used $\Gamma_1(1.24) = 0.22$ for $\Gamma_n(\beta_{ci})$, since the $n = 1$ mode has the highest growth rate and is therefore the most unstable. The critical drift velocity varies with $k_\parallel$. For small $k_\parallel$ (i.e., $\omega/k_\parallel$ near but less than U_e), the threshold is very high, corresponding to the fact that the slope of the electron velocity distribution, which provides the free energy source for the instability, is small in this region (see Figure 10.17). As $k_\parallel$ increases and the slope of the electron velocity distribution function at $v_{\parallel\text{Res}} = \omega/k_\parallel$ increases, the threshold drift velocity decreases. However, as $k_\parallel$ continues to increase, the ion cyclotron resonance velocity, $v_{\parallel\text{Res}} = (\omega - n\,\omega_{ci})/k_\parallel$, begins to move downward into the main part of the ion distribution function (see Figure 10.17), causing an increase in ion cyclotron damping, which eventually raises the threshold drift velocity. These variations lead to a broad minimum in the threshold drift velocity. Although this minimum is difficult to analyze analytically, it is easily evaluated using numerical techniques. Drummond and Rosenbluth (1962) have shown that the threshold drift velocity is given approximately by

$$U_e^* = [14(T_i/T_e) + 3]C_i. \tag{10.2.56}$$

For moderate electron-to-ion temperature ratios ($T_e/T_i \sim 1$), the threshold electron drift velocity for the current-driven electrostatic ion cyclotron instability is very low, much lower than the corresponding threshold for the current-driven ion acoustic instability given by Eq. (9.5.32) for an unmagnetized plasma. Because of the low threshold for this instability, electrostatic ion cyclotron waves are often observed in both laboratory and space plasmas in regions where field-aligned

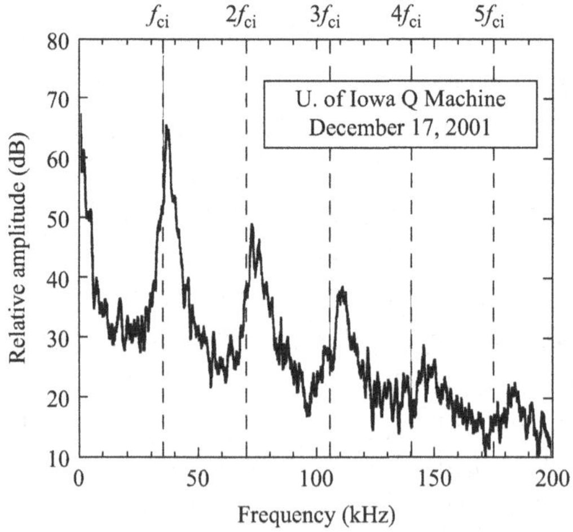

Figure 10.18 A spectrum of current-driven electrostatic ion cyclotron waves observed in a laboratory Q machine, using a cesium plasma (plot provided by R. Merlino).

currents are present. An example of a series of current-driven electrostatic ion cyclotron waves detected in a laboratory Q machine is shown in Figure 10.18.

10.3 Electromagnetic Waves

Since an electromagnetic wave has both an electric field and a magnetic field, it is not possible to describe the electric field by an electrostatic potential, as was done in the previous section. Instead, the full set of Maxwell's equations must be used. The basic procedure used to analyze electromagnetic wave propagation in a dielectric medium has been described in Chapter 4, and is based on finding non-trivial solutions to the homogeneous equation

$$\mathbf{k}\times(\mathbf{k}\times\tilde{\mathbf{E}})+\frac{\omega^2}{c^2}\overleftrightarrow{\mathbf{K}}\cdot\tilde{\mathbf{E}}=0, \quad (10.3.1)$$

where $\tilde{\mathbf{E}}$ is the electric field of the wave and $\overleftrightarrow{\mathbf{K}}$ is the dielectric tensor. Non-trivial solutions of this system of linear equations are possible only when the determinant of the matrix in the homogeneous equation (10.3.1) is zero, which gives the dispersion relation. Before we can proceed further, we must compute the dielectric tensor. The procedure used to compute the dielectric tensor consists of first solving the linearized Vlasov equation for the first-order distribution function, $\tilde{f}_s$, for some given electric field, $\tilde{\mathbf{E}}$, and then using $\tilde{f}_s$ to compute the current density

$$\tilde{\mathbf{J}}=\sum_s e_s\int_{-\infty}^{\infty}\mathbf{v}\tilde{f}_s\,\mathrm{d}^3v. \quad (10.3.2)$$

By expressing the current density in the form $\tilde{\mathbf{J}}=\overleftrightarrow{\sigma}\cdot\tilde{\mathbf{E}}$, the conductivity tensor $\overleftrightarrow{\sigma}$ can be identified. Once the conductivity tensor is known, the dielectric tensor can be computed using $\overleftrightarrow{\mathbf{K}}=\overleftrightarrow{\mathbf{1}}-\overleftrightarrow{\sigma}/(\mathrm{i}\omega\epsilon_0)$; see Eq. (4.2.6).

10.3.1 The Dispersion Relation

Following the above procedure, the first step in deriving the dispersion relation is to solve the linearized Vlasov equation (10.1.13) for the first-order distribution, $\tilde{f}_s$, for an arbitrary wave electric field, $\tilde{\mathbf{E}}$. The procedure is basically the same as for the electrostatic waves described in the previous section. However, the solution is more complicated because of the $\mathbf{v}\times(\mathbf{k}\times\tilde{\mathbf{E}})$ term introduced by the magnetic wave field. This term can be expanded using the BAC minus CAB rule, which gives

$$\mathbf{v}\times(\mathbf{k}\times\tilde{\mathbf{E}})=\mathbf{k}(\mathbf{v}\cdot\tilde{\mathbf{E}})-(\mathbf{v}\cdot\mathbf{k})\tilde{\mathbf{E}}. \quad (10.3.3)$$

Substituting the above term into Eq. (10.1.13) gives the following linear differential equation:

$$\frac{\partial \tilde{f}_s}{\partial \phi} - \mathrm{i}(\alpha_s + \beta_s \cos\phi)\tilde{f}_s$$
$$= \frac{e_s}{m_s \omega_{cs}} \left[\left(1 - \frac{k_{\|} v_{\|}}{\omega} - \frac{k_{\perp} v_{\perp}}{\omega} \cos\phi \right) \tilde{\mathbf{E}} + \frac{\mathbf{k}}{\omega} (\mathbf{v} \cdot \tilde{\mathbf{E}}) \right] \cdot \nabla_{\mathrm{v}} f_{s0}. \tag{10.3.4}$$

As before, to solve this equation for $\tilde{f}_s$ we must express all of the terms on the right-hand side in cylindrical velocity space coordinates ($v_\perp$, ϕ, and $v_\|$). Starting with rectangular (v_x, v_y, v_z) coordinates, the dot product between $\tilde{\mathbf{E}}$ and $\nabla_{\mathrm{v}} f_{s0}$ in the first term in the brackets can be written

$$\begin{aligned} \tilde{\mathbf{E}} \cdot \nabla_{\mathrm{v}} f_{s0} &= \widetilde{E}_x \frac{\partial f_{s0}}{\partial v_x} + \widetilde{E}_y \frac{\partial f_{s0}}{\partial v_y} + \widetilde{E}_z \frac{\partial f_{s0}}{\partial v_z} \\ &= \widetilde{E}_x \frac{\partial v_\perp}{\partial v_x} \frac{\partial f_{s0}}{\partial v_\perp} + \widetilde{E}_y \frac{\partial v_\perp}{\partial v_y} \frac{\partial f_{s0}}{\partial v_\perp} + \widetilde{E}_z \frac{\partial f_{s0}}{\partial v_\|}. \end{aligned} \tag{10.3.5}$$

From the transformation equation $v_\perp = v_x \cos\phi + v_y \sin\phi$, it is easy to show that

$$\frac{\partial v_\perp}{\partial v_x} = \frac{v_x}{v_\perp} = \cos\phi \quad \text{and} \quad \frac{\partial v_\perp}{\partial v_y} = \frac{v_y}{v_\perp} = \sin\phi. \tag{10.3.6}$$

Substituting these relations into Eq. (10.3.5) then gives

$$\tilde{\mathbf{E}} \cdot \nabla_{\mathrm{v}} f_{s0} = \frac{\partial f_{s0}}{\partial v_\perp} (\widetilde{E}_x \cos\phi + \widetilde{E}_y \sin\phi) + \frac{\partial f_{s0}}{\partial v_\|} \widetilde{E}_z. \tag{10.3.7}$$

Similarly, the dot product between $\mathbf{v}$ and $\tilde{\mathbf{E}}$ in the second term in the brackets of Eq. (10.3.4) can be written

$$\begin{aligned} \mathbf{v} \cdot \tilde{\mathbf{E}} &= v_x \widetilde{E}_x + v_y \widetilde{E}_y + v_z \widetilde{E}_z \\ &= v_\perp (\widetilde{E}_x \cos\phi + \widetilde{E}_y \sin\phi) + v_\| \widetilde{E}_z. \end{aligned} \tag{10.3.8}$$

Substituting Eqs. (10.3.7) and (10.3.8) into Eq. (10.3.4), and changing the zero-order distribution function to the normalized zero-order distribution function using $f_{s0} = n_{s0} F_{s0}$ gives the equation

$$\frac{\partial \tilde{f}_s}{\partial \phi} - \mathrm{i}(\alpha_s + \beta_s \cos\phi)\tilde{f}_s$$
$$= \frac{e_s n_s}{m_s \omega_{cs}} \left[(A_s \widetilde{E}_x + D_s \widetilde{E}_z) \cos\phi + A_s \widetilde{E}_y \sin\phi + \frac{\partial F_{s0}}{\partial v_\|} \widetilde{E}_z \right], \tag{10.3.9}$$

where to simplify the notation we have introduced the definitions

$$A_s = \frac{\partial F_{s0}}{\partial v_\perp} + \frac{k_\parallel}{\omega}\left(v_\perp \frac{\partial F_{s0}}{\partial v_\parallel} - v_\parallel \frac{\partial F_{s0}}{\partial v_\perp}\right) \tag{10.3.10}$$

and

$$D_s = -\frac{k_\perp}{\omega}\left(v_\perp \frac{\partial F_{s0}}{\partial v_\parallel} - v_\parallel \frac{\partial F_{s0}}{\partial v_\perp}\right). \tag{10.3.11}$$

This linear differential equation can be solved using the same integrating factor, Eq. (10.2.4), used to solve Eq. (10.2.1). This result is

$$\tilde{f}_s = \mathrm{i}\frac{e_s n_s}{m_s \omega_{cs}} \sum_{m,n} \frac{\mathrm{e}^{\mathrm{i}(m-n)\phi}}{\alpha_s + n} J_m(\beta_s) \times \left[\left(\frac{n}{\beta_s} J_n(\beta_s)\widetilde{E}_x + \mathrm{i}J_n'(\beta_s)\widetilde{E}_y\right) A_s + J_n(\beta_s) B_s \widetilde{E}_z\right], \tag{10.3.12}$$

where the Bessel function identities

$$J_{n+1}(\beta_s) + J_{n-1}(\beta_s) = \frac{2n}{\beta_s} J_n(\beta_s) \quad \text{and} \quad J_{n+1}(\beta_s) - J_{n-1}(\beta_s) = -2J_n'(\beta_s) \tag{10.3.13}$$

have been used, and the term B_s is defined by

$$B_s = \frac{\partial F_{s0}}{\partial v_\parallel} + \frac{n}{\beta_s} D_s = \frac{\partial F_{s0}}{\partial v_\parallel} - \frac{n\omega_{cs}}{\omega v_\perp}\left(v_\perp \frac{\partial F_{s0}}{\partial v_\parallel} - v_\parallel \frac{\partial F_{s0}}{\partial v_\perp}\right). \tag{10.3.14}$$

Next, we compute the current by evaluating the integral

$$\tilde{\mathbf{J}} = \sum_s e_s \int_0^{2\pi} \int_{-\infty}^{\infty} \int_0^{\infty} \mathbf{v} \tilde{f}_s v_\perp \mathrm{d}v_\perp \, \mathrm{d}v_\parallel \, \mathrm{d}\phi, \tag{10.3.15}$$

where $\mathbf{v} = \hat{\mathbf{x}} v_\perp \cos\phi + \hat{\mathbf{y}} v_\perp \sin\phi + \hat{\mathbf{z}} v_\parallel$. The result after integrating over $\mathrm{d}\phi$ can be organized in matrix form as

$$\begin{bmatrix} \widetilde{J}_x \\ \widetilde{J}_y \\ \widetilde{J}_z \end{bmatrix} = \begin{bmatrix} \sigma_{xx} & \sigma_{xy} & \sigma_{xz} \\ \sigma_{yx} & \sigma_{yy} & \sigma_{yz} \\ \sigma_{zx} & \sigma_{zy} & \sigma_{zz} \end{bmatrix} \begin{bmatrix} \widetilde{E}_x \\ \widetilde{E}_y \\ \widetilde{E}_z \end{bmatrix}, \tag{10.3.16}$$

where the 3×3 matrix is the conductivity tensor, $\overleftrightarrow{\sigma}$. The integral in Eq. (10.3.15) is most easily performed by expressing $\cos\phi$ and $\sin\phi$ in complex exponential form and noting that

$$\int_0^{2\pi} \mathrm{e}^{\mathrm{i}(m-n)\phi}\,\mathrm{d}\phi = 2\pi\delta_{m,n} \quad \text{and} \quad \int_0^{2\pi} \mathrm{e}^{\mathrm{i}(m-n\pm 1)\phi}\,\mathrm{d}\phi = 2\pi\delta_{m,n\pm 1}, \quad (10.3.17)$$

where $\delta_{m,n}$ is the Kronecker delta.

The Kronecker deltas have the effect of collapsing the double sum over m and n into a single sum over n, very similar to the procedure used to derive the Harris dispersion relation equation (10.2.20). The conductivity tensor is then found to be

$$\overleftrightarrow{\sigma} = \mathrm{i}\sum_s \frac{e_s^2 n_s}{m_s\omega_{cs}} \sum_n \int_{-\infty}^{\infty}\int_0^{\infty} \frac{2\pi v_\perp \mathrm{d}v_\perp \mathrm{d}v_\parallel}{\alpha_s + n}$$

$$\times \begin{bmatrix} A_s \dfrac{n^2 v_\perp}{\beta_s^2} J_n^2 & \mathrm{i}A_s \dfrac{n v_\perp}{\beta_s} J_n J_n' & B_s \dfrac{n v_\perp}{\beta_s} J_n^2 \\ -\mathrm{i}A_s \dfrac{n v_\perp}{\beta_s} J_n J_n' & A_s v_\perp J_n' J_n' & -\mathrm{i}B_s v_\perp J_n J_n' \\ A_s \dfrac{n v_\parallel}{\beta_s} J_n^2 & \mathrm{i}A_s v_\parallel J_n J_n' & B_s v_\parallel J_n^2 \end{bmatrix}, \quad (10.3.18)$$

where the Bessel function identities given in Eq. (10.3.13) have been used.

Having computed the conductivity tensor, it is then easy to compute the dielectric tensor, which is given by $\overleftrightarrow{\mathbf{K}} = \overleftrightarrow{\mathbf{1}} - \overleftrightarrow{\sigma}/(\mathrm{i}\omega\epsilon_0)$. After substituting the appropriate expressions for α_s, β_s, A_s, and B_s from Eqs. (10.1.14), (10.3.10), and (10.3.14) into Eq. (10.3.18), the dielectric tensor elements are found to be

$$K_{xx} = 1 - \sum_s \frac{\omega_{ps}^2}{\omega} \sum_n \int_{-\infty}^{\infty}\int_0^{\infty} \frac{n^2 J_n^2(\beta_s)}{\beta_s^2(k_\parallel v_\parallel - \omega + n\omega_{cs})}$$

$$\times \left[\left(1 - \frac{k_\parallel v_\parallel}{\omega}\right)\frac{\partial F_{s0}}{\partial v_\perp} + \frac{k_\parallel v_\perp}{\omega}\frac{\partial F_{s0}}{\partial v_\parallel}\right] 2\pi v_\perp^2\,\mathrm{d}v_\perp\,\mathrm{d}v_\parallel, \quad (10.3.19)$$

$$K_{xy} = -\mathrm{i}\sum_s \frac{\omega_{ps}^2}{\omega} \sum_n \int_{-\infty}^{\infty}\int_0^{\infty} \frac{n J_n(\beta_s) J_n'(\beta_s)}{\beta_s(k_\parallel v_\parallel - \omega + n\omega_{cs})}$$

$$\times \left[\left(1 - \frac{k_\parallel v_\parallel}{\omega}\right)\frac{\partial F_{s0}}{\partial v_\perp} + \frac{k_\parallel v_\perp}{\omega}\frac{\partial F_{s0}}{\partial v_\parallel}\right] 2\pi v_\perp^2\,\mathrm{d}v_\perp\,\mathrm{d}v_\parallel, \quad (10.3.20)$$

$$K_{xz} = -\sum_s \frac{\omega_{ps}^2}{\omega} \sum_n \int_{-\infty}^{\infty}\int_0^{\infty} \frac{n J_n^2(\beta_s)}{\beta_s(k_\parallel v_\parallel - \omega + n\omega_{cs})}$$

$$\times\left[\frac{\partial F_{s0}}{\partial v_{\|}} - \frac{n\omega_{cs}}{\omega v_{\perp}}\left(v_{\perp}\frac{\partial F_{s0}}{\partial v_{\|}} - v_{\|}\frac{\partial F_{s0}}{\partial v_{\perp}}\right)\right] 2\pi v_{\perp}^2 \, dv_{\perp} \, dv_{\|}, \tag{10.3.21}$$

$$K_{yx} = i\sum_s \frac{\omega_{ps}^2}{\omega}\sum_n \int_{-\infty}^{\infty}\int_0^{\infty} \frac{nJ_n(\beta_s)J_n'(\beta_s)}{\beta_s(k_{\|}v_{\|} - \omega + n\omega_{cs})}$$

$$\times\left[\left(1 - \frac{k_{\|}v_{\|}}{\omega}\right)\frac{\partial F_{s0}}{\partial v_{\perp}} + \frac{k_{\|}v_{\perp}}{\omega}\frac{\partial F_{s0}}{\partial v_{\|}}\right] 2\pi v_{\perp}^2 \, dv_{\perp} \, dv_{\|}, \tag{10.3.22}$$

$$K_{yy} = 1 - \sum_s \frac{\omega_{ps}^2}{\omega}\sum_n \int_{-\infty}^{\infty}\int_0^{\infty} \frac{J_n'(\beta_s)J_n'(\beta_s)}{(k_{\|}v_{\|} - \omega + n\omega_{cs})}$$

$$\times\left[\left(1 - \frac{k_{\|}v_{\|}}{\omega}\right)\frac{\partial F_{s0}}{\partial v_{\perp}} + \frac{k_{\|}v_{\perp}}{\omega}\frac{\partial F_{s0}}{\partial v_{\|}}\right] 2\pi v_{\perp}^2 \, dv_{\perp} \, dv_{\|}, \tag{10.3.23}$$

$$K_{yz} = i\sum_s \frac{\omega_{ps}^2}{\omega}\sum_n \int_{-\infty}^{\infty}\int_0^{\infty} \frac{J_n(\beta_s)J_n'(\beta_s)}{(k_{\|}v_{\|} - \omega + n\omega_{cs})}$$

$$\times\left[\frac{\partial F_{s0}}{\partial v_{\|}} - \frac{n\omega_{cs}}{\omega v_{\perp}}\left(v_{\perp}\frac{\partial F_{s0}}{\partial v_{\|}} - v_{\|}\frac{\partial F_{s0}}{\partial v_{\perp}}\right)\right] 2\pi v_{\perp}^2 \, dv_{\perp} \, dv_{\|}, \tag{10.3.24}$$

$$K_{zx} = -\sum_s \frac{\omega_{ps}^2}{\omega}\sum_n \int_{-\infty}^{\infty}\int_0^{\infty} \frac{nJ_n^2(\beta_s)}{\beta_s(k_{\|}v_{\|} - \omega + n\omega_{cs})}$$

$$\times\left[\left(1 - \frac{k_{\|}v_{\|}}{\omega}\right)\frac{\partial F_{s0}}{\partial v_{\perp}} + \frac{k_{\|}v_{\perp}}{\omega}\frac{\partial F_{s0}}{\partial v_{\|}}\right] 2\pi v_{\perp}v_{\|} \, dv_{\perp} \, dv_{\|}, \tag{10.3.25}$$

$$K_{zy} = -i\sum_s \frac{\omega_{ps}^2}{\omega}\sum_n \int_{-\infty}^{\infty}\int_0^{\infty} \frac{J_n(\beta_s)J_n'(\beta_s)}{(k_{\|}v_{\|} - \omega + n\omega_{cs})}$$

$$\times\left[\left(1 - \frac{k_{\|}v_{\|}}{\omega}\right)\frac{\partial F_{s0}}{\partial v_{\perp}} + \frac{k_{\|}v_{\perp}}{\omega}\frac{\partial F_{s0}}{\partial v_{\|}}\right] 2\pi v_{\perp}v_{\|} \, dv_{\perp} \, dv_{\|}, \tag{10.3.26}$$

and

$$K_{zz} = 1 - \sum_s \frac{\omega_{ps}^2}{\omega}\sum_n \int_{-\infty}^{\infty}\int_0^{\infty} \frac{J_n^2(\beta_s)}{(k_{\|}v_{\|} - \omega + n\omega_{cs})}$$

$$\times\left[\frac{\partial F_{s0}}{\partial v_{\|}} - \frac{n\omega_{cs}}{\omega v_{\perp}}\left(v_{\perp}\frac{\partial F_{s0}}{\partial v_{\|}} - v_{\|}\frac{\partial F_{s0}}{\partial v_{\perp}}\right)\right] 2\pi v_{\perp}v_{\|} \, dv_{\perp} \, dv_{\|}. \tag{10.3.27}$$

The homogeneous equation for the electric field is obtained by substituting the dielectric tensor elements into Eq. (10.3.1), which becomes

$$\begin{bmatrix} K_{xx}-\dfrac{c^2k^2}{\omega^2}\cos^2\theta & K_{xy} & K_{xz}+\dfrac{c^2k^2}{\omega^2}\sin\theta\cos\theta \\ K_{yx} & K_{yy}-\dfrac{c^2k^2}{\omega^2} & K_{yz} \\ K_{zx}+\dfrac{c^2k^2}{\omega^2}\sin\theta\cos\theta & K_{zy} & K_{zz}-\dfrac{c^2k^2}{\omega^2}\sin\theta \end{bmatrix}\begin{bmatrix} \widetilde{E}_x \\ \widetilde{E}_y \\ \widetilde{E}_z \end{bmatrix}=0. \quad (10.3.28)$$

The dispersion relation is then given by the determinant of the matrix. Although the determinant can, in principle, be evaluated, the resulting equation is too complicated to be usefully displayed, since each term consists of an infinite series of integrals involving Bessel functions. To evaluate the dispersion relation it is usually necessary to use either numerical methods or highly specialized assumptions. However, one special case that can be analyzed analytically is when the wave vector is parallel to the magnetic field ($\theta = 0$). Since parallel propagation illustrates many of the important features that occur for arbitrary wave normal angles, we will analyze this case in some detail.

10.3.2 Parallel Propagation

From inspection of the homogeneous equation (10.3.28), it is evident that the dispersion relation simplifies considerably if the wave vector is parallel to the magnetic field (i.e., $\theta = 0$). There are three such simplifications. First, it is easy to see that the terms involving $\sin\theta$ are all zero. Second, since $\beta_s = k_\perp v_\perp/\omega_{cs} = 0$ for parallel propagation, it is also easy to see that the K_{xz} and K_{zx} terms are zero, since all of the Bessel function terms, $nJ_n^2(\beta_s)$, go to zero as $\beta_s \to 0$. Third, it turns out that the K_{yz} and K_{zy} terms are also zero. That these terms are zero can be seen by noting that, as β_s goes to zero, all of the $J_n(\beta_s)J_n'(\beta_s)$ terms go to zero (note that $J_0(\beta_s) \simeq 1$ and $J_0'(\beta_s) = -J_1(\beta_s) \simeq -\beta_s/2$, as $\beta_s \to 0$). The homogeneous equation (10.3.28) then simplifies to

$$\begin{bmatrix} K_{xx}-\dfrac{c^2k^2}{\omega^2} & K_{xy} & 0 \\ K_{yx} & K_{yy}-\dfrac{c^2k^2}{\omega^2} & 0 \\ 0 & 0 & K_{zz} \end{bmatrix}\begin{bmatrix} \widetilde{E}_x \\ \widetilde{E}_y \\ \widetilde{E}_z \end{bmatrix}=0. \quad (10.3.29)$$

By taking the determinant of the matrix in the above equation, the dispersion relation is easily shown to be

$$Ð(k,\omega) = \left[\left(K_{xx} - \frac{c^2k^2}{\omega^2}\right)\left(K_{yy} - \frac{c^2k^2}{\omega^2}\right) - K_{xy}K_{yx}\right]K_{zz} = 0. \tag{10.3.30}$$

From the homogeneous equation (10.3.29), it can be seen that the roots of the dispersion relation separate into three types: a series of electrostatic modes with $\tilde{\mathbf{E}} = (0,0,\tilde{E}_z)$; a series of electromagnetic modes with eigenvectors $\tilde{\mathbf{E}} = (\tilde{E}_0, \mathrm{i}\tilde{E}_0, 0)$; and a series of electromagnetic modes with eigenvectors $\tilde{\mathbf{E}} = (\tilde{E}_0, -\mathrm{i}\tilde{E}_0, 0)$. The electrostatic modes are the same as those described in Chapter 9 for an unmagnetized plasma. For these modes, the zero-order magnetic field plays no role, since the particle motions are along the magnetic field and there is no $\mathbf{v} \times \hat{\mathbf{B}}_0$ force. The analysis of the remaining electromagnetic modes can be simplified considerably by noting that $K_{xy} = -K_{yx}$ and $K_{xx} = K_{yy}$. The latter occurs because, in the limit $\beta_s \to 0$, the $n^2 J_n^2(\beta_s)/\beta_s^2$ terms in K_{xx} and the $J_n'(\beta_s)J_n'(\beta_s)$ terms in K_{yy} both reduce to $1/4$ for $n = \pm 1$, and zero for $n = 0$. With these simplifications, the electromagnetic part of the dispersion reduces to

$$Ð(k,\omega) = K_{xx} - \frac{c^2k_\parallel^2}{\omega^2} \pm \mathrm{i}K_{xy} = 0, \tag{10.3.31}$$

which, after substituting the appropriately simplified expressions for K_{xx} and K_{xy}, can be written in the form

$$Ð(k,\omega) = 1 - \frac{c^2k_\parallel^2}{\omega^2} - \sum_s \frac{\omega_{ps}^2}{\omega} \times \int_{-\infty}^{\infty}\int_0^{\infty} \frac{\dfrac{\partial F_{s0}}{\partial v_\perp} + \dfrac{k_\parallel}{\omega}\left(v_\perp \dfrac{\partial F_{s0}}{\partial v_\parallel} - v_\parallel \dfrac{\partial F_{s0}}{\partial v_\perp}\right)}{k_\parallel v_\parallel - \omega \pm \omega_{cs}} \pi v_\perp^2 \,\mathrm{d}v_\perp \mathrm{d}v_\parallel = 0. \tag{10.3.32}$$

After factoring out $k_\parallel/\omega$ from both the numerator and the denominator, the above equation can be written in the following, somewhat simpler, form:

$$Ð(k,\omega) = 1 - \frac{c^2k_\parallel^2}{\omega^2} - \sum_s \frac{\omega_{ps}^2}{\omega^2} \int_{-\infty}^{\infty} \frac{G_{s0}(v_\parallel)}{v_\parallel - \dfrac{\omega^2 \pm \omega_{cs}}{k_\parallel}} \mathrm{d}v_\parallel = 0, \tag{10.3.33}$$

where, for convenience, we have defined a reduced one-dimensional distribution function

$$G_{s0}(v_\parallel) = \int_0^{\infty} \left[\left(\frac{\omega}{k_\parallel} - v_\parallel\right)\frac{\partial F_{s0}}{\partial v_\perp} + v_\perp \frac{\partial F_{s0}}{\partial v_\parallel}\right] \pi v_\perp^2 \,\mathrm{d}v_\perp. \tag{10.3.34}$$

Finally, the above equations can be converted to Laplace transform notation by changing ω to $\mathrm{i}p$ and changing the $v_\parallel$ integration contour such that it passes below the pole at $v_\parallel = (\mathrm{i}p \pm \omega_{cs})/k_\parallel$ for $k_\parallel > 0$, and above the pole for $k_\parallel < 0$, as in Figure 10.13.

The Low-temperature, Long-Wavelength Limit

As with the electrostatic dispersion relation, it is useful to consider the low-temperature, long-wavelength limit, so that we can relate the dispersion relation to the results from cold plasma theory in Chapter 4. This can be done by expanding the integrand in Eq. (10.3.33) in powers of $k_\parallel v_\parallel/(\omega \pm \omega_{cs})$ and considering the low-temperature, long-wavelength limit of the resulting series of integrals. However, a simpler approach is to assume that the plasma is cold, with a distribution function given by

$$F_{s0} = \frac{1}{2\pi v_\perp}\delta(v_\perp)\,\delta(v_\parallel). \tag{10.3.35}$$

After substituting this distribution function into Eq. (10.3.34) and integrating by parts once, the reduced one-dimensional distribution function simplifies to

$$G_{s0}(v_\parallel) = -\left(\frac{\omega}{k_\parallel} - v_\parallel\right)\delta(v_\parallel). \tag{10.3.36}$$

Completing the $v_\parallel$ integration in Eq. (10.3.33) then gives the dispersion relation

$$Đ(k,\omega) = 1 - \frac{c^2k_\parallel^2}{\omega^2} - \sum_s \frac{\omega_{ps}^2}{\omega(\omega \pm \omega_{cs})} = 0. \tag{10.3.37}$$

Noting that $n^2 = c^2k_\parallel^2/\omega^2$, the above equation is seen to be identical to the cold plasma dispersion relation obtained in Chapter 4 for electromagnetic waves propagating parallel to the magnetic field ($\theta = 0$). The plus sign corresponds to right-hand polarized waves ($n^2 = R$), and the minus sign corresponds to left-hand polarized waves ($n^2 = L$).

Cyclotron Damping

Next we show that the imaginary part of the dispersion relation introduced by the resonance at $v_\parallel = (\mathrm{i}p \pm \omega_{cs})/k_\parallel$ causes electromagnetic waves to be damped, similar in some respects to Landau damping. This damping is called cyclotron damping. In the weak damping limit, the frequency and growth rate can be determined from Eq. (10.3.33) in the usual way using the Plemelj formula, which gives

$$Đ_{\mathrm{r}} = 1 - \frac{c^2k_\parallel^2}{\omega^2} - \sum_s \frac{\omega_{ps}^2}{\omega^2} P\int_{-\infty}^{\infty} \frac{G_{s0}(v_\parallel)}{v_\parallel - \dfrac{\omega \pm \omega_{cs}}{k_\parallel}}\,\mathrm{d}v_\parallel = 0 \tag{10.3.38}$$

and

$$\gamma = \frac{-Ð_{\rm i}}{\partial Ð_{\rm r}/\partial\omega} = \pi\frac{k_\parallel/|k_\parallel|}{\partial Ð_{\rm r}/\partial\omega}\sum_s \frac{\omega_{{\rm p}s}^2}{\omega^2}G_{s0}(v_\parallel)\bigg|_{v_\parallel = v_{\parallel\rm Res}}, \quad (10.3.39)$$

where $G_{s0}(v_\parallel)$ is given by Eq. (10.3.34), and $v_{\parallel\rm Res} = (\omega \pm \omega_{cs})/k_\parallel$ is the cyclotron resonance velocity. The term $k_\parallel/|k_\parallel|$ takes care of the usual sign change that occurs in the residue as $k_\parallel$ changes sign.

Equation (10.3.39) shows that the growth or damping of the wave is determined by the sign of $\sum_s \omega_{{\rm p}s}^2 G_{s0}(v_{\parallel\rm Res})$. For the purpose of evaluating $G_{s0}(v_{\parallel\rm Res})$, it is convenient to separate this function into two integrals as follows:

$$G_{s0}(v_\parallel) = \frac{\omega}{k_\parallel}\int_0^\infty \frac{\partial F_{s0}}{\partial v_\perp}\pi v_\perp^2\,{\rm d}v_\perp - \int_0^\infty \left(v_\parallel\frac{\partial F_{s0}}{\partial v_\perp} - v_\perp\frac{\partial F_{s0}}{\partial v_\parallel}\right)\pi v_\perp^2\,{\rm d}v_\perp. \quad (10.3.40)$$

The first integral in the above equation can be integrated by parts once to give

$$G_{s0}(v_\parallel) = -\frac{\omega}{k_\parallel}\int_0^\infty F_{s0}2\pi v_\perp\,{\rm d}v_\perp - \int_0^\infty \left(v_\parallel\frac{\partial F_{s0}}{\partial v_\perp} - v_\perp\frac{\partial F_{s0}}{\partial v_\parallel}\right)\pi v_\perp^2\,{\rm d}v_\perp, \quad (10.3.41)$$

which, when substituted into Eq. (10.3.39), gives the following equation for the growth rate:

$$\gamma = \pi\frac{1}{\partial Ð_{\rm r}/\partial\omega}\sum_s\frac{\omega_{{\rm p}s}^2}{\omega^2}\left[-\frac{\omega}{|k_\parallel|}\int_0^\infty F_{s0}2\pi v_\perp\,{\rm d}v_\perp - \frac{k_\parallel}{|k_\parallel|}\int_0^\infty\left(v_\parallel\frac{\partial F_{s0}}{\partial v_\perp} - v_\perp\frac{\partial F_{s0}}{\partial v_\parallel}\right)\pi v_\perp^2\,{\rm d}v_\perp\right]\Bigg|_{v_\parallel = v_{\parallel\rm Res}}. \quad (10.3.42)$$

To help understand the various terms in the above equation, we next consider the special case of an isotropic velocity distribution function. For an isotropic distribution, the distribution function is a function only of the magnitude of the velocity, $F_{s0} = F_{s0}(v)$, where $v = (v_\parallel^2 + v_\perp^2)^{1/2}$. For such a distribution function it is easy to show that

$$\frac{\partial F_{s0}}{\partial v_\perp} = \frac{\partial v}{\partial v_\perp}\frac{\partial F_{s0}}{\partial v} = \frac{v_\perp}{v}\frac{\partial F_{s0}}{\partial v} \quad \text{and} \quad \frac{\partial F_{s0}}{\partial v_\parallel} = \frac{\partial v}{\partial v_\parallel}\frac{\partial F_{s0}}{\partial v} = \frac{v_\parallel}{v}\frac{\partial F_{s0}}{\partial v}. \quad (10.3.43)$$

After substituting these expressions into Eq. (10.3.42), it is evident that the two terms in the second integral cancel. The growth rate is then given by

$$\gamma = -\pi\frac{\omega/|k_\parallel|}{\partial Ð_{\rm r}/\partial\omega}\sum_s\frac{\omega_{{\rm p}s}^2}{\omega^2}\int_0^\infty F_{s0}2\pi v_\perp\,{\rm d}v_\perp\bigg|_{v_\parallel = v_{\parallel\rm Res}}, \quad (10.3.44)$$

which is negative, indicating damping, since, as we will show, $\partial D_r/\partial\omega$ is always positive. This damping is called cyclotron damping. Although cyclotron damping has many similarities to Landau damping, it differs from Landau damping in one important respect. As can be seen from the above equation, cyclotron damping is directly proportional to the distribution function evaluated at the parallel resonance velocity, whereas Landau damping is proportional to the slope of the distribution function at the resonance velocity. This difference indicates that cyclotron damping is fundamentally different from Landau damping in that all particles in the vicinity of the resonance contribute to the damping, not just those below the resonance velocity as in the case of Landau damping.

If we limit our consideration to resonance velocities well above the thermal velocity, the real part of the dispersion relation, Eq. (10.3.38), reduces to the same equations ($n^2 = R$ and $n^2 = L$) obtained in a cold plasma theory, i.e., Eq. (10.3.37). Thus, if the cyclotron resonance velocity is not too close to the main part of the distribution function, the waves propagate exactly as in cold plasma theory, except for the damping, which is directly proportional to the number of particles in resonance with the wave. For example, consider the whistler mode, for which the ion motions can be ignored. If the electrons have a Maxwellian velocity distribution with temperature T, the damping $(-\gamma)$ is given by

$$(-\gamma) = \pi \frac{\omega/|k_{\|}|}{\partial D_r/\partial\omega} \frac{\omega_p^2}{\omega^2} \left(\frac{m}{2\pi\kappa T}\right)^{1/2} \exp\left(-\frac{m v_{\|\mathrm{Res}}^2}{2\kappa T}\right)\Bigg|_{v_{\|\mathrm{Res}}=\frac{\omega-\omega_c}{k_{\|}}}. \tag{10.3.45}$$

A plot of the index of refraction squared, $n^2 = c^2k^2/\omega^2$, and the damping rate $(-\gamma)$ for the right-hand polarized whistler mode, $n^2 = R$, is shown in Figure 10.19. Because the damping rate $(-\gamma)$ has an exponential dependence on $-(\omega - \omega_c)^2$, cyclotron damping is strong only near the electron cyclotron frequency. Note

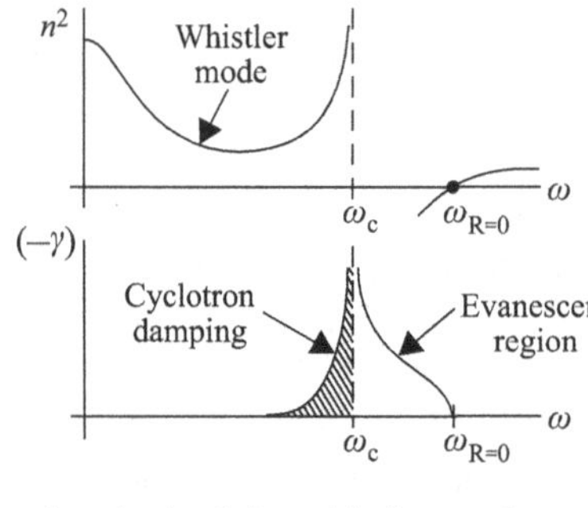

Figure 10.19 The damping $(-\gamma)$ of the whistler mode as a function of frequency. Cyclotron damping becomes very strong just below the electron cyclotron frequency (the hatched region). The whistler mode cannot propagate in the evanescent region from ω_c to $\omega_R = 0$.

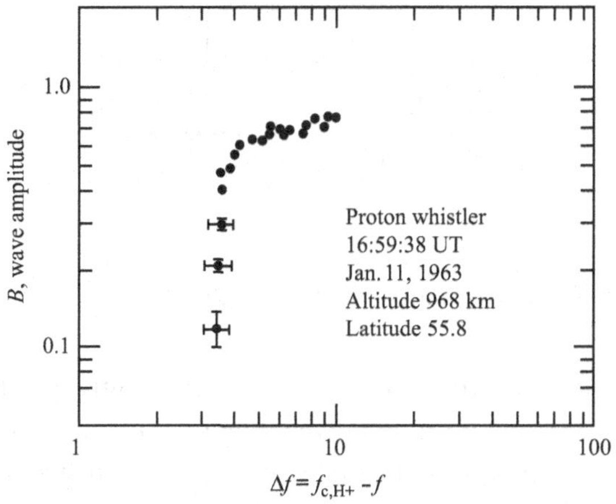

Figure 10.20 An illustration showing the abrupt onset of ion cyclotron damping as the frequency of a proton whistler approaches the proton cyclotron frequency (Gurnett and Brice, 1966).

that the frequency range over which significant damping occurs decreases with decreasing temperature, and goes to zero for a cold plasma ($T = 0$). Also note that in the frequency range between ω_c and $\omega_{R=0}$, the imaginary part of the frequency $(-\gamma)$ is non-zero even when the plasma is cold. This is the evanescent region discussed in Chapter 4. Therefore, the reasons for the non-zero values of $(-\gamma)$ are qualitatively different above and below the cyclotron frequency. Below the cyclotron frequency, the damping $(-\gamma)$ is large because the plasma absorbs energy from the wave via cyclotron damping, whereas in the region above the cyclotron frequency the wave cannot propagate because of dynamical considerations (the displacement current in Maxwell's equation has the wrong sign).

Similar considerations also apply to the left-hand polarized ion cyclotron mode. The damping is then called ion cyclotron damping. Ion cyclotron whistlers provide a good example of ion cyclotron damping. They typically exhibit an abrupt cutoff in the wave amplitude as the wave frequency asymptotically approaches the ion cyclotron frequency (see Figure 4.20). The abrupt decrease in the wave amplitude for a proton whistler is shown in Figure 10.20 as the difference between the wave frequency and the proton cyclotron frequency, Δf, goes to zero.

Instabilities

Since the first term in Eq. (10.3.42) always leads to damping, it is clear that any electromagnetic instabilities that occur for parallel propagation must arise from the second term, which is of the form

$$-\frac{k_\parallel}{|k_\parallel|}\int_0^\infty \left[v_\parallel \frac{\partial F_{s0}}{\partial v_\perp} - v_\perp \frac{\partial F_{s0}}{\partial v_\parallel}\right] \pi v_\perp^2 \, dv_\perp . \tag{10.3.46}$$

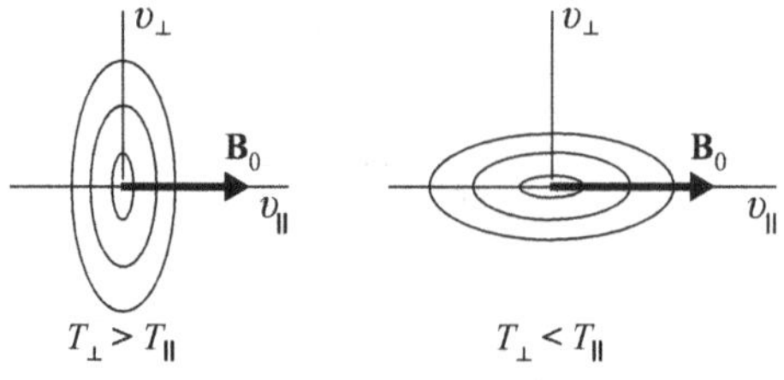

Figure 10.21 A bi-Maxwellian distribution for two cases, $T_\perp > T_\parallel$ and $T_\perp < T_\parallel$.

To produce an instability, this term must be sufficiently positive to offset the damping caused by the first term. Since the integrand in the above equation is zero for an isotropic distribution, it is clear that an anisotropic distribution is required to drive an instability. The type of anisotropy required can be illustrated by evaluating Eq. (10.3.46) for a bi-Maxwellian velocity distribution function (Eq. (5.1.16)). The result for a bi-Maxwellian is easily shown to be

$$-\frac{k_\parallel}{|k_\parallel|}\int_0^\infty \left[v_\parallel \frac{\partial F_{s0}}{\partial v_\perp} - v_\perp \frac{\partial F_{s0}}{\partial v_\parallel}\right] \pi v_\perp^2 \, \mathrm{d}v_\perp = -\frac{k_\parallel}{|k_\parallel|} v_\parallel A_s \int_0^\infty F_{s0} 2\pi v_\perp \, \mathrm{d}v_\perp, \quad (10.3.47)$$

where $A_s = (T_{s\perp}/T_{s\parallel} - 1)$ is a parameter called the anisotropy. A positive anisotropy corresponds to a disk-shaped velocity distribution, and a negative anisotropy corresponds to a cigar-shaped velocity distribution (see Figure 10.21). In principle, either type of anisotropy can lead to an instability. To determine the controlling factors, we substitute Eq. (10.3.47) into Eq. (10.3.42), which gives the following equation for the growth rate:

$$\gamma = \pi \frac{1}{\partial D_\mathrm{r}/\partial \omega} \sum_s \frac{\omega_{\mathrm{p}s}^2}{\omega^2} \left[-\frac{\omega}{|k_\parallel|} - \frac{k_\parallel}{|k_\parallel|} v_{\parallel\mathrm{Res}} A_s\right] \int_0^\infty F_{s0} 2\pi v_\perp \, \mathrm{d}v_\perp \Bigg|_{v_\parallel = v_{\parallel\mathrm{Res}}}. \quad (10.3.48)$$

This equation shows that the possible existence of an instability depends on the sign of $k_\parallel v_{\parallel\mathrm{Res}} A_s$. For an instability to occur, the product $k_\parallel v_{\parallel\mathrm{Res}} A_s$ must be negative and sufficiently large to overcome the damping effect of the $-\omega/|k_\parallel|$ term. If $k_\parallel v_{\parallel\mathrm{Res}}$ is positive, then instability can occur only if A_s is negative and, if $k_\parallel v_{\parallel\mathrm{Res}}$ is negative, then instability can occur only if A_s is positive. To proceed further, it is necessary to examine some specific modes of propagation.

Whistler-Mode Instabilities

As an example of an electromagnetic instability driven by an anisotropy, we next consider the case of whistler-mode waves, for which ion motions can be ignored. Using the (+) sign in Eq. (10.3.37), and the high-density approximation introduced in Chapter 4, the real part of the dispersion relation for the whistler mode is given

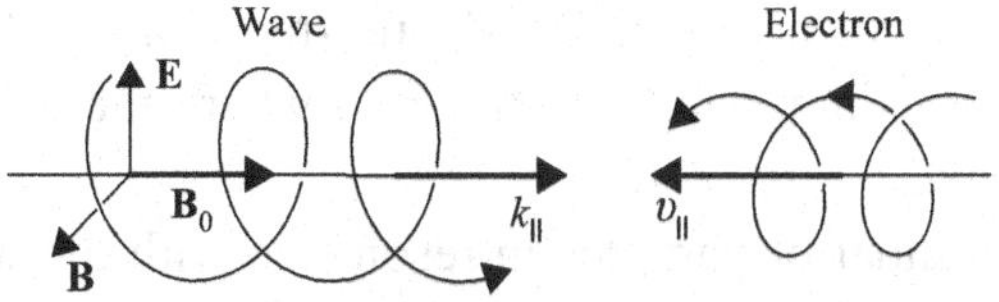

Figure 10.22 For the whistler mode, the guiding center of resonant electrons moves in the opposite direction of the wave, i.e., $k_{\|}v_{\|} < 0$. Note that both the wave and the electrons are rotating in the same (right-hand) sense with respect to the zero-order magnetic field.

approximately by

$$Ð_{\rm r} = -\frac{c^2 k_{\|}^2}{\omega^2} + \frac{\omega_{\rm p}^2}{\omega(\omega_{\rm c} - \omega)} = 0, \qquad (10.3.49)$$

where we have neglected the 1 in Eq. (10.3.37) and used $\omega_{\rm p} = \omega_{\rm pe}$ and $\omega_{\rm c} = \omega_{\rm ce}$. Solving for the parallel wave number, the resonance velocity is given by

$$v_{\|{\rm Res}} = -\frac{(\omega_{\rm c} - \omega)}{k_{\|}} = -c\frac{k_{\|}}{|k_{\|}|}\frac{(\omega_{\rm c} - \omega)^{3/2}}{\omega^{1/2}\omega_{\rm p}}. \qquad (10.3.50)$$

Since the frequency of the whistler mode is always less than the electron cyclotron frequency, the above equation shows that $k_{\|}$ and $v_{\|}$ must have opposite signs. The reason they must have opposite signs is that the wave frequency must be Doppler-shifted up to the electron cyclotron frequency, i.e., $\omega' = \omega - k_{\|}v_{\|} = \omega_{\rm c}$. To do this, the guiding center of the resonant electrons must move in the opposite direction of the wave vector, as shown in Figure 10.22. Since $k_{\|}v_{\|{\rm Res}}$ is negative, whistler-mode instabilities are always driven by a positive anisotropy in the electron velocity distribution, i.e., $T_{\perp}/T_{\|} > 1$ in the case of a bi-Maxwellian velocity distribution.

For resonance to occur, the wave field in the guiding center frame of the particle must be polarized in the same sense as the rotation of the resonant particle. For parallel propagation ($\theta = 0$), the wave field of the whistler mode rotates in the same right-hand sense as the electrons. The polarization condition is then clearly satisfied. Whistler-mode waves can also resonate with ions, which rotate in the opposite (left-hand) sense, assuming positively charged ions. However, the guiding center of the ion must then be moving in the same direction and faster than the wave, so that the polarization of the wave field is reversed (i.e., right-hand to left-hand) in the guiding-center frame of the ion. Using the (+) sign in the general formula for the cyclotron resonance velocity, it can be shown that the parallel resonance velocity for an ion is given by $v_{\|{\rm Res}} = (\omega + \omega_{\rm ci})/k_{\|}$, which confirms that the resonance velocity is greater than the phase velocity of the wave. This equation also shows that $k_{\|}v_{\|{\rm Res}}$ is positive, so a negative ion anisotropy is required to drive the instability. Such ion-driven whistler-mode instabilities are rare, since the resonance velocities are usually very large, where there are few particles to resonate with the wave.

To estimate the number of particles in resonance with the wave, it is useful to compute the parallel resonant energy, which for electrons is given by

$$W_{\|\text{Res}} = \frac{1}{2} m_e v_{\|\text{Res}}^2. \tag{10.3.51}$$

Using Eq. (10.3.50), it can be shown that the parallel resonant energy for the whistler mode is given to a good approximation by the equation

$$W_{\|\text{Res}} = W_c \left(1 - \frac{\omega}{\omega_c}\right)^3 \frac{\omega_c}{\omega}, \tag{10.3.52}$$

where

$$W_c = \frac{B^2}{2\mu_0 n_0} \tag{10.3.53}$$

is a characteristic energy for whistler-mode resonance with electrons. Note that the characteristic resonant energy, $B^2/(2\mu_0 n_0)$, can be simply interpreted as the magnetic energy density per electron. For the characteristic energy to be non-relativistic, $W_c \ll (1/2)mc^2$, the electron plasma frequency must be substantially greater than the electron cyclotron frequency ($\omega_p \gg \omega_c$). Cyclotron resonance can also occur for relativistic energies, but the equations must be modified to account for relativistic effects. The frequency dependence of the parallel resonance energy is plotted in Figure 10.23 for the non-relativistic case. As can be seen, the resonance energy becomes very low near the cyclotron frequency, and very high at low frequencies. Since there are usually more electrons at low energies than at high energies, cyclotron resonant interactions are usually more important as the wave frequency approaches the cyclotron frequency.

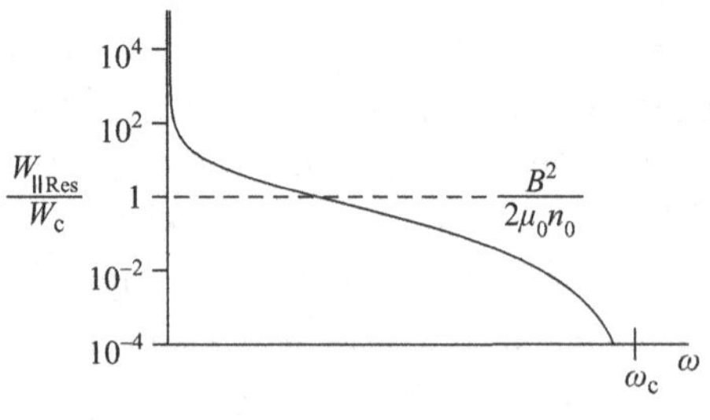

Figure 10.23 The parallel resonance energy, $W_{\|}$, of the whistler mode as a function of frequency.

Having determined the resonance velocity, the growth rate of the whistler mode can be computed from

$$\gamma = \frac{-\mathcal{D}_{\rm i}}{\partial \mathcal{D}_{\rm r}/\partial \omega} = \pi \frac{k_{\parallel}/|k_{\parallel}|}{\partial \mathcal{D}_{\rm r}/\partial \omega} \frac{\omega_{\rm p}^2}{\omega^2} G_0(v_{\parallel \rm Res}), \tag{10.3.54}$$

where $G_0(v_{\parallel \rm Res})$ is given by Eq. (10.3.41). Computing $\partial \mathcal{D}_{\rm r}/\partial \omega$ from Eq. (10.3.49), one obtains

$$\partial \mathcal{D}_{\rm r}/\partial \omega = \frac{\omega_{\rm p}^2 \omega_{\rm c}}{\omega^2(\omega_{\rm c} - \omega)^2}, \tag{10.3.55}$$

which, as previously noted, is always positive.

If the velocity distribution has a loss cone, as is often the case in a magnetic mirror machine or a planetary radiation belt, it is usually more convenient to express $F_0(\mathbf{v})$ as a function of the magnitude of the velocity, v, and the pitch angle, α, rather than as a function of the perpendicular velocity, $v_{\perp}$, and the parallel velocity, $v_{\parallel}$. By transforming to spherical velocity space coordinates, it can be shown that the integrand in the second term of Eq. (10.3.41) can be written in the following simple form:

$$\left[v_{\parallel} \frac{\partial F_0}{\partial v_{\perp}} - v_{\perp} \frac{\partial F_0}{\partial v_{\parallel}} \right] = (\mathbf{v} \times \nabla_{\mathbf{v}} F_0)_{\phi} = \frac{\partial F_0}{\partial \alpha}, \tag{10.3.56}$$

where the pitch angle α is measured from the $+v_{\parallel}$ axis, as shown in Figure 10.24. Note that for an isotropic distribution, $\partial F_0/\partial \alpha$ is zero, which is consistent with our earlier results concerning an isotropic distribution.

From Eqs. (10.3.55) and (10.3.42) it can then be shown that the growth rate of the whistler mode is given by

$$\frac{\gamma}{\omega_{\rm c}} = \pi \left(1 - \frac{\omega}{\omega_{\rm c}}\right)^2 \times \left[-\frac{\omega}{|k_{\parallel}|} \int_0^{\infty} F_0 2\pi v_{\perp} {\rm d}v_{\perp} + \frac{k_{\parallel}}{|k_{\parallel}|} \int_0^{\infty} (-\partial F_0/\partial \alpha) \pi v_{\perp}^2 {\rm d}v_{\perp} \right] \Bigg|_{v_{\parallel} = v_{\parallel \rm Res}}, \tag{10.3.57}$$

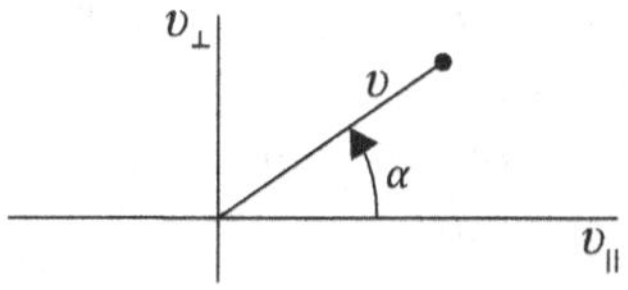

Figure 10.24 The pitch angle, α, is the angle between the velocity vector and the $+v_{\parallel}$ axis.

where $F_0(v,\alpha)$ is the electron distribution function. This equation can be put in a more convenient form by dividing by the first integral and substituting $v_\perp = |\tan\alpha||v_{\|\mathrm{Res}}|$ for one of the $v_\perp$ terms in the second integral. With these changes, Eq. (10.3.57) can be written in the form

$$\frac{\gamma}{\omega_c} = \pi\left(1-\frac{\omega}{\omega_c}\right)^2 |v_{\|\mathrm{Res}}|\left(A - \frac{\omega}{|k_\| v_{\|\mathrm{Res}}|}\right)\int_0^\infty F_0 2\pi v_\perp \mathrm{d}v_\perp\bigg|_{v_\| = v_{\|\mathrm{Res}}}, \quad (10.3.58)$$

where the anisotropy A is now defined by

$$A = \frac{\int_0^\infty (k_\|/|k_\||)(-\partial F_0/\partial\alpha)|\tan\alpha|\pi v_\perp \mathrm{d}v_\perp}{\int_0^\infty F_0 2\pi v_\perp \mathrm{d}v_\perp}\bigg|_{v_\| = v_{\|\mathrm{Res}}}, \quad (10.3.59)$$

and $|k_\||\,|v_{\|\mathrm{Res}}|$ has been combined into one term, $|k_\| v_{\|\mathrm{Res}}|$.

The quantity $\omega/|k_\| v_{\|\mathrm{Res}}|$ in Eq. (10.3.58) can be evaluated from the cyclotron resonance condition and is given by

$$\frac{\omega}{|k_\| v_{\|\mathrm{Res}}|} = \frac{\omega}{\omega_c - \omega}. \quad (10.3.60)$$

The growth rate then becomes

$$\gamma = \pi\omega_c\left(1-\frac{\omega}{\omega_c}\right)^2 |v_{\|\mathrm{Res}}|\left(A - \frac{\omega}{\omega_c - \omega}\right)\int_0^\infty F_0 2\pi v_\perp \mathrm{d}v_\perp\bigg|_{v_\| = v_{\|\mathrm{Res}}}. \quad (10.3.61)$$

It is evident that the whistler mode is unstable whenever the anisotropy, A, is positive. The marginal stability condition ($\gamma = 0$) gives an upper frequency limit for the range of unstable frequencies and is given by

$$\omega^* = \frac{A}{1+A}\omega_c. \quad (10.3.62)$$

At frequencies greater than ω^*, cyclotron damping, given by the first term in Eq. (10.3.42), is too strong for instability to occur. The growth rate in the unstable region depends strongly on the resonance velocity, which controls both the anisotropy and the number of particles in resonance with the wave.

It is easily demonstrated that a loss-cone type of distribution function has a positive anisotropy and is therefore unstable. As shown in Figure 10.25, for $k_\| > 0$ the pitch angles involved in the resonant interaction occur in a region of velocity space where $\partial F/\partial\alpha < 0$, which from Eq. (10.3.59) implies a positive anisotropy, which is unstable. Electrons trapped in planetary radiation belts are often modeled by a distribution function of the form

$$F_0 \sim \frac{\sin^m\alpha}{v^n}. \quad (10.3.63)$$

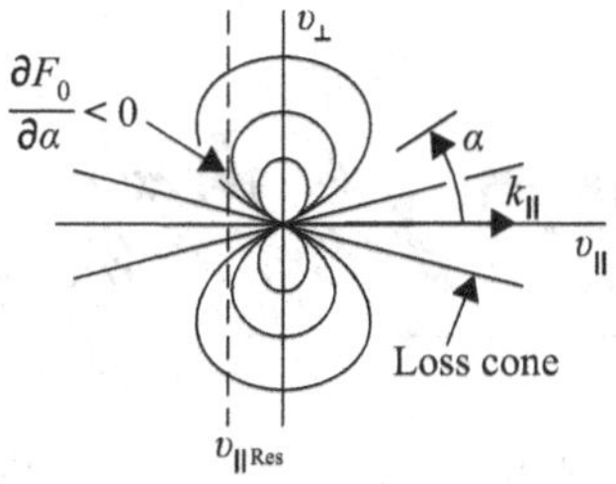

Figure 10.25 A loss-cone electron velocity distribution showing the cyclotron resonance velocity, $v_{\parallel Res}$, for a whistler-mode wave propagating in the $k_\parallel$ direction.

For this type of distribution function, it is easy to show that the anisotropy factor is given by $A = m/2$. Note that a distribution function with a small m is nearly isotropic, whereas a distribution function with a large m is highly anisotropic.

Since electrons trapped in a planetary magnetic field always have a loss cone, whistler-mode instabilities of the type described above are a common feature of planetary magnetospheres, and play an important role in the pitch-angle scattering and loss of radiation belt electrons (Kennel and Petschek, 1966). Simple angular momentum considerations show that the generation of right-hand polarized electromagnetic angular momentum by growing whistler-mode waves must be associated with a reduction in the right-hand angular momentum of the electrons. The reduction in the angular momentum decreases the pitch angle and drives the resonant particles toward the loss cone, where they are lost from the system.

Two types of whistler-mode emissions occur in planetary radiation belts, hiss and chorus. Hiss has an almost featureless spectrum, and is believed to be generated by a marginally stable electron velocity distribution that arises from a balance between wave growth, which is driven by the loss-cone anisotropy, and pitch-angle scattering, which acts to scatter particles toward the loss cone, hence reducing the anisotropy. In contrast to hiss, chorus is highly structured and typically consists of many discrete tones, often rising in frequency, as shown in Figure 10.26. When played through a speaker, chorus is often described as sounding like the whistling songs that birds make when they wake up in the morning, called the "dawn chorus" in England. The complex frequency–time structure of chorus is believed to be caused by the trapping of resonant electrons in the rotating wave field. This trapping is similar in some respects to the electrostatic particle trapping discussed in Section 9.2.5, except the particles are trapped in the rotating electromagnetic wave field rather than the longitudinal electrostatic field. For a numerical simulation of the nonlinear trapping process believed to be responsible for the frequency–time structure of chorus, see Nunn et al. (1997), and

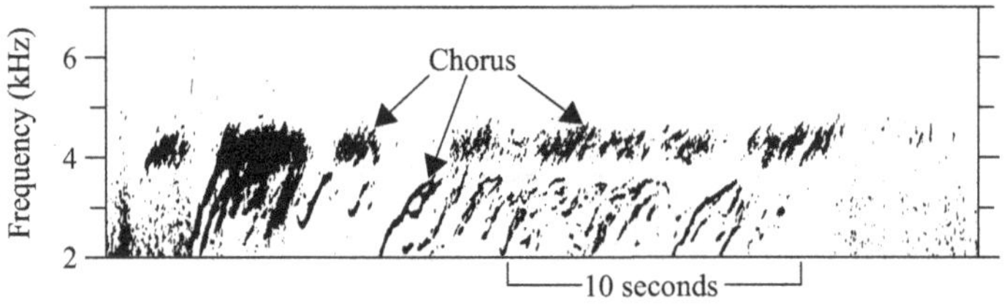

Figure 10.26 A frequency–time spectrogram of whistler-mode chorus observed in Earth's magnetosphere, sometimes also called *dawn chorus* (Allcock, 1957).

for a discussion of the loss of radiation belt electrons due to interaction with these waves, see Horne and Thorne (2003). In addition to causing pitch angle scattering, numerical simulations have also shown that cyclotron resonant interactions with chorus can cause significant acceleration of radiation belt electrons (Omura et al., 2007).

Electromagnetic Ion Cyclotron Mode Instabilities

An instability very similar to the whistler-mode instability also occurs for electromagnetic ion cyclotron waves. In this case, the primary resonant interaction occurs between the left-hand polarized ion cyclotron wave and positively charged ions, which rotate in the left-hand sense relative to the magnetic field. For a two-component plasma consisting of electrons and one species of positive ions, it can be shown from Eq. (10.3.37) that the real part of the dispersion relation is given to a good approximation by

$$Đ_{\rm r} = -\frac{c^2 k_{\|}^2}{\omega^2} + \frac{\omega_{\rm pi}^2}{\omega_{\rm ci}(\omega_{\rm ci} - \omega)} = 0, \tag{10.3.64}$$

where we have neglected the 1 in Eq. (10.3.37) and made use of the fact that $\omega \ll |\omega_{\rm ce}|$. This equation provides a good approximation for the index of refraction of the ion cyclotron mode over the frequency range $\omega \lesssim \omega_{\rm ci}$, including the Alfvén wave regime near $\omega \simeq 0$. Note that the functional form of the frequency dependence is somewhat different than for the whistler mode. This occurs because the electron term in Eq. (10.3.37) cannot be neglected for the ion cyclotron mode, whereas the ion term can be neglected for the whistler mode.

From Eq. (10.3.64), it is straightforward to show that the resonance velocity for the ion cyclotron mode is given by

$$v_{\|\rm Res} = -\frac{(\omega_{\rm ci} - \omega)}{k_{\|}} = -c\frac{k_{\|}}{|k_{\|}|}\frac{\omega_{\rm ci}^{1/2}(\omega_{\rm ci} - \omega)^{3/2}}{\omega\,\omega_{\rm pi}}. \tag{10.3.65}$$

As with the whistler mode, $k_{\|} v_{\|\rm Res}$ is negative, which implies that a positive (loss-cone type) anisotropy is required to drive the instability. Using the above equation it is easy to show that the parallel resonant energy is given by

$$W_{\|\text{Res}} = W_{\text{c}}\left(1 - \frac{\omega}{\omega_{\text{ci}}}\right)^3\left(\frac{\omega_{\text{ci}}}{\omega}\right)^2, \tag{10.3.66}$$

where the characteristic energy is again defined by $W_{\text{c}} = B^2/(2\mu_0 n_0)$. Note that the only essential difference compared to the whistler mode is that the resonant energy increases more rapidly at low frequencies because of the ω^{-2} term.

After evaluating $\partial D_{\text{r}}/\partial\omega$ from the real part of the dispersion relation, which turns out to be

$$\partial D_{\text{r}}/\partial\omega = \frac{\omega_{\text{pi}}^2(2\omega_{\text{ci}} - \omega)}{\omega\,\omega_{\text{ci}}(\omega_{\text{ci}} - \omega)^2}, \tag{10.3.67}$$

the growth rate can be written in the following form:

$$\gamma = \pi\frac{(\omega_{\text{ci}} - \omega)^2}{(2\omega_{\text{ci}} - \omega)}\frac{\omega_{\text{ci}}}{\omega}|v_{\|\text{Res}}|\left(A - \frac{\omega}{(\omega_{\text{ci}} - \omega)}\right)\int_0^\infty F_{\text{i}0}2\pi v_\perp \text{d}v_\perp\bigg|_{v_\| = v_{\|\text{Res}}}, \tag{10.3.68}$$

where the ion anisotropy factor is defined in the same way as in Eq. (10.3.59), except F_0 is now the ion distribution function, $F_{\text{i}0}(v,\alpha)$.

The electromagnetic ion cyclotron instability occurs whenever the ion anisotropy is sufficiently positive to offset the ion cyclotron damping term. Thus, the instability is driven by a loss-cone type of distribution, very similar to the situation with the whistler mode. Electromagnetic ion cyclotron waves have been observed in Earth's magnetosphere, and resonant interactions with these waves are believed to be one of the main mechanisms by which energetic ions are lost from planetary radiation belts. It is also worth noting that even small concentrations of minor ions (a few percent or less) can have important effects on the propagation and growth of ion cyclotron waves; see Summers and Thorne (2003).

10.3.3 Oblique Propagation

It is evident from an inspection of the homogeneous equation (10.3.28) that the dispersion relation for oblique propagation is quite complicated. Each term in the matrix consists of an infinite sum of integrals involving Bessel functions. Also, there are no terms that are zero, such as occurred for parallel propagation. Since the dispersion relation, which is given by the determinant of the matrix in Eq. (10.3.28), has so many terms, it is not possible to obtain a general analytical solution for the roots of the dispersion relation, as was possible for a cold magnetized plasma (see Chapter 4). Therefore, to obtain useful results, it is necessary either (1) to make very specific assumptions, or (2) to resort to numerical solutions. Because of the very restrictive nature of the assumptions needed to obtain usable analytical solutions, oblique propagation will not be discussed in detail. However, we can make some very general comments on the main features involved.

As one can see from the dielectric tensor elements in Eqs. (10.3.19) through (10.3.27), all of the integrals involve a resonance denominator of the form $k_{\parallel}v_{\parallel} - \omega + n\omega_{cs}$. Therefore, the integrals can be evaluated using contour integration just as in the case of electrostatic waves. If one uses the weak growth rate approximation, as in Section 9.2.4, then the integrals split up into real and imaginary parts, similar to Eqs. (10.2.34) and (10.2.35). The real part controls the propagation of the wave, i.e., $\omega(\mathbf{k})$, and the imaginary part controls the growth (or damping) of the wave, i.e., γ. For parallel propagation the Bessel function integrals are all zero, except the $n = 0$ and $n = \pm 1$ terms, and simply yield the cold plasma electromagnetic cyclotron modes. For oblique propagation, the higher order ($|n| > 1$) Bessel function integrals are, in general, non-zero, which introduces a series of new electromagnetic modes, one for each n, comparable in some respects to the electrostatic Bernstein modes. These modes have no counterpart in cold plasma theory. The numerical values of the integrals associated with these Bernstein-like electromagnetic modes are controlled by the $\beta_s = k_\perp v_\perp/\omega_{cs}$ term in the Bessel functions. If the thermal cyclotron radius is small compared to the perpendicular wavelength, such that $\beta_s \ll 1$, then these integrals are small and the corresponding modes are usually of negligible importance. On the other hand, if β_s is of order unity, these terms and their corresponding modes cannot be neglected. Sometimes, to isolate a specific mode, all the terms in the Bessel function series are omitted except for the one of interest. This approximation is usually valid only very close to the corresponding cyclotron harmonic, i.e., $\omega \sim n\omega_{cs}$.

Since the imaginary parts of the dielectric tensor elements control the growth rate, one can see that the $\partial F_0/\partial v_{\parallel}$ and $\partial F_0/\partial v_\perp$ terms provide the free energy source for any resonant instability that might occur. Therefore, the types of distribution functions that can give rise to wave growth for obliquely propagating electromagnetic waves are exactly the same as those that have already been discussed for electrostatic waves, namely beam and loss-cone types of velocity distributions. However, there is one difference. For electrostatic waves it is only the electric field that is involved in the resonant interaction, whereas for electromagnetic waves both the electric and the magnetic fields are involved. In fact, for some cases, such as the whistler mode in the high-density limit, the wave magnetic field plays the dominant role. For a further discussion of oblique electromagnetic wave propagation in a hot magnetized plasma, see Kennel and Wong (1967) and Stix (1992).

Problems

10.1. The Harris dispersion relation is given by

$$Ð(k,\omega) = 1 - \sum_s \frac{\omega_{ps}^2}{k^2} \sum_{n=-\infty}^{\infty} \int_{-\infty}^{\infty} \int_0^{\infty} \frac{J_n^2(\beta_s)}{k_\| v_\| - \omega + n\omega_{cs}} \times \left[k_\| \frac{\partial F_{0s}}{\partial v_\|} + \frac{n\omega_{cs}}{v_\perp} \frac{\partial F_{0s}}{\partial v_\perp} \right] 2\pi v_\perp \mathrm{d}_\perp \mathrm{d}v_\| = 0,$$

where $\beta_s = k_\perp v_\perp / \omega_{cs}$.
Show that the low-temperature, high phase velocity limit of this equation is

$$Ð_0(k,\omega) = P\cos^2\theta + S\sin^2\theta = 0,$$

where $P = 1 - \sum_s(\omega_{ps}^2/\omega^2)$ and $S = 1 - \sum_s \omega_{ps}^2/(\omega^2 - \omega_{cs}^2)$ are the functions defined in cold plasma theory.

10.2. For the case of propagation perpendicular to the magnetic field ($k_\| = 0$) and a Maxwellian electron distribution, show that the Harris dispersion relation reduces to the following equation:

$$Ð(k,\omega) = 1 - \sum_{n=1}^{\infty} \frac{2\omega_p^2}{\beta_c^2\omega_c^2} \frac{e^{-\beta_c^2} I_n(\beta_c^2)}{(\omega/n\omega_c)^2 - 1} = 0,$$

where $\beta_c = k_\perp \rho_c$ and $\rho_c = \sqrt{\kappa T/m}/\omega_c$. The solutions to this dispersion relation are called the Bernstein modes.

10.3. Using a computer:
(a) develop a computer program to solve for the roots of the Bernstein mode dispersion relation in Problem 10.2 above;
(b) plot the first four roots of $Ð(k_\perp,\omega) = 0$ as a function $\beta_c = k_\perp\rho_c$ for $\omega_p = (5/2)\omega_c$ (this value for ω_p gives $\omega_{UH} = 2.7\omega_c$);
(c) compute the upper hybrid resonance frequency using the cold plasma equation $\omega_{UH} = \sqrt{\omega_c^2 + \omega_p^2}$; does your program give the proper value for ω_{UH} in the limit $k_\perp\rho_c \to 0$?
(d) find the ratio of the frequencies of the first three Q resonances to the electron cyclotron frequency (the Q resonances are the frequencies where $\partial\omega/\partial k_\perp = 0$).

10.4. Starting with the Harris dispersion relation and a Maxwellian distribution function

$$F_{s0} = \left(\frac{m_s}{2\pi\kappa T_s}\right)^{3/2} e^{-m_s(v_\perp^2 + v_\|^2)/2\kappa T_s},$$

show that the dispersion relation is given by

$$Ɖ(k,p) = 1 - \sum_{n,s} \frac{\Gamma_n(k_\perp \rho_{cs})^2}{(k\lambda_{Ds})^2}\left[\frac{1}{2}Z'(\zeta_s) - \frac{m\omega_{cs}}{k_\parallel}\sqrt{\frac{m_s}{2\kappa T_s}}Z(\zeta_s)\right] = 0,$$

where

$$\Gamma_n(x) = \exp[-x^2]I_n(x^2), \quad \rho_{cs} = \sqrt{\frac{\kappa T_s}{m_s}}\frac{1}{\omega_{cs}}, \quad \text{and}$$

$$\zeta_s = \sqrt{\frac{m_s}{2\kappa T_s}}\frac{\omega - n\omega_{cs} + i\gamma}{k_\parallel},$$

and $\zeta_s = x_s + iy_s$ is given by

$$x_s = \sqrt{\frac{m_s}{2\kappa T_s}}\frac{\omega - n\omega_{cs}}{k_\parallel} \quad \text{and} \quad y_s = \sqrt{\frac{m_s}{2\kappa T_s}}\frac{\gamma}{k_\parallel}.$$

10.5. If we add a drift U_s to the parallel component of the Maxwellian, such that

$$F_{s0} = \left(\frac{m_s}{2\pi\kappa T_s}\right)^{3/2} e^{-m_s[(v_\parallel - U_s)^2 + v_\perp^2]/2\kappa T_s},$$

show that the resulting dispersion relation can be obtained from Problem 10.4 above by making the substitution $\zeta_s \to (\zeta_s - \zeta_{U_s})$, where

$$\zeta_{U_s} = \sqrt{\frac{m_s}{2\kappa T_s}}U_s.$$

10.6. Show that the growth rate for electrostatic waves propagating in a two-component electron ion plasma with drifting Maxwellian distributions,

$$F_{s0} = \left(\frac{m_s}{2\pi\kappa T_s}\right)^{3/2} e^{-m_s[(v_\parallel - U_s)^2 + v_\perp^2]/2\kappa T_s},$$

is given by

$$\frac{\gamma}{\omega_{ci}} = -\frac{k_\parallel}{|k_\parallel|}\sqrt{\frac{\pi}{2}}\sqrt{\frac{m_e}{\kappa T_e}}\left(\frac{T_e}{T_i}\right)\Gamma_1(k_\perp \rho_{ci})\left(\frac{\omega}{k_\parallel} - U_e\right),$$

where

$$\Gamma_n(x) = \exp[-x^2]I_n(x^2), \quad \rho_{cs} = \sqrt{\frac{\kappa T_s}{m_s}}\frac{1}{\omega_{cs}}, \quad \text{and}$$

$$\zeta_s = \sqrt{\frac{m_s}{2\kappa T_s}}\frac{\omega - n\omega_{cs} + i\gamma}{k_\parallel},$$

and $\zeta_{U_s} = \sqrt{m_s/2\kappa T_s}U_s$.

Determine the frequency, growth rate, and electron drift velocity threshold for the $n = 1$ current-driven ion cyclotron mode.

10.7. In the analysis of whistler-mode instabilities, it is convenient to approximate the distribution function in the resonance interaction region by a function of the form

$$F_0 = k\frac{\sin^m \alpha}{v^n},$$

where k is a constant, α is the pitch angle, and v is the magnitude of the velocity.

(a) Make a sketch of contours of constant F as a function of α for $m = 0, m = 1$, and $m = 4$.

(b) Show that the anisotropy factor

$$A = \left.\frac{\int_0^\infty (k_\parallel/|k_\parallel|)(-\partial F_0/\partial\alpha)|\tan\alpha|\pi v_\perp \mathrm{d}v_\perp}{\int_0^\infty F_0 2\pi v_\perp \mathrm{d}v_\perp}\right|_{v_\parallel = v_{\parallel\mathrm{Res}}}$$

is given by $A = m/2$ for this distribution function. Note that since $v_{\parallel\mathrm{Res}}$ is negative, α is in the range from $\pi/2$ to π, so $|\tan\alpha| = -\tan\alpha$.

(c) Using the marginal stability condition $\omega^* = A\omega_c/(1 + A)$, make a plot of the unstable frequency range as a function of m.

10.8. A distribution function consists of a bi-Maxwellian given by

$$F_0 = \left(\frac{m}{2\pi\kappa T_\parallel}\right)^{1/2}\left(\frac{m}{2\pi\kappa T_\perp}\right)\exp\left[-\frac{mv_\parallel^2}{2\kappa T_\parallel} - \frac{mv_\perp^2}{2\kappa T_\perp}\right],$$

where $T_\parallel$ and $T_\perp$ are the parallel and perpendicular temperatures. Show that the anisotropy factor is

$$A = \frac{T_\perp}{T_\parallel} - 1.$$

Again, to obtain the above result you must make use of the fact that $v_{\parallel\mathrm{Res}}$ is negative, so that $|\tan\alpha| = -\tan\alpha$.

10.9. Show that the index of refraction of the ion cyclotron mode is given to a good approximation by the equation

$$n^2 = \frac{\omega_{\mathrm{pi}}^2}{\omega_{\mathrm{ci}}(\omega_{\mathrm{ci}} - \omega)},$$

and that the ion resonance energy is given by

$$W_{\mathrm{Res}} = \frac{B^2}{2\mu_0 n_{\mathrm{i}}}\left(1 - \frac{\omega}{\omega_{\mathrm{ci}}}\right)^3 \left(\frac{\omega_{\mathrm{ci}}}{\omega}\right)^2.$$

References

Allcock, G. McK. 1957. A study of the audio-frequency radio phenomena known as "dawn chorus", *Australian J. Phys.* **10**, 286–298.

Arfken, G. 1970. *Mathematical Methods for Physicists*. New York: Academic Press, pp. 488.

Ashour-Abdalla, M., and Kennel, C. F. 1978. Nonconvective and convective electron cyclotron harmonic instabilities. *J. Geophys. Res.* **83**, 1531–1543.

Bernstein, I. B. 1958. Waves in a plasma in a magnetic field. *Phys. Rev.* **109**, 10–21.

Boyce, W. E., and DiPrima, R. C. 1992. *Elementary Differential Equations and Boundary Value Problems*, Fifth Edition. New York: Wiley.

Drummond, W. E., and Rosenbluth, M. N. 1962. Anomalous diffusion arising from microinstabilities in a plasma. *Phys. Fluids* **5**, 1507–1513.

Gurnett, D. A., and Brice, N. M. 1966. Ion temperature in the ionosphere obtained from cyclotron damping of proton whistlers. *J. Geophys. Res.* **71**, 3639–3652.

Harris, E. G. 1959. Unstable plasma oscillations in a magnetic field. *Phys. Rev. Lett.* **2**, 34–36.

Horne, R. B. 1989. Path integrated growth of electrostatic waves: the generation of terrestrial myriametric radiation, *J. Geophys. Res.* **94**, 8895.

Horne, R. B., and Thorne, R. M. 2003. Relativistic electron acceleration and precipitation during resonant interaction with whistler-mode chorus. *Geophys. Res. Lett.* **30** (10), 1527.

Kennel, C. F., and Petschek, H. E. 1966. Limit on stably trapped particle fluxes. *J. Geophys. Res.* **71**, 1–28.

Kennel, C. F., and Wong, H. V. 1967. Resonant particle instabilities in a uniform magnetic field. *J. Plasma Phys.* **1** (1), 75–80.

Nunn, D., Omura, Y., Matsumoto, H., Nagano, I., and Yagitani, S. 1997. The numerical simulation of VLF chorus and discrete emissions observed on the Geotail spacecraft using a Vlasov code. *J. Geophys. Res.* **102** (A12), 27083–27097.

Omura, Y., Furuya, N., and Summers, D. 2007. Relativistic turning acceleration of resonant electrons by coherent whistler waves in a dipole magnetic field. *J. Geophys. Res.* **112**, A06236.

Rönnmark, K. 1983. Computation of the dielectric tensor of a Maxwellian plasma. *Plasma Phys.* **25**, 699–701.

Stix, T. H. 1992. *Waves in Plasmas*. New York: American Institute of Physics, pp. 237–293.

Summers, D., and Thorne, R. M. 2003. Relativistic electron pitch-angle scattering by electromagnetic ion cyclotron waves during geomagnetic storms. *J. Geophys. Res.* **108** (A4), 1143.

Tataronis, J. A., and Crawford, F. W. 1970. Cyclotron harmonic wave propagation and instabilities, I. Perpendicular propagation. *J. Plasma Phys.* **4**, 231–248.

Watson, G. N. 1995. *A Treatise on the Theory of Bessel Functions*, Cambridge Mathematical Library Edition. Cambridge: Cambridge University Press, p. 395. Originally published in 1922.

Further Reading

Chen, F. F. 1990. *Introduction to Plasma Physics and Controlled Fusion.* New York: Plenum Press, Chapter 7.

Nicholson, D. R. 1983. *Introduction to Plasma Theory*. Malabar, FL: Krieger Publishing, Chapter 6.

Stix, T. H. 1992. *Waves in Plasmas*. New York: American Institute of Physics, Chapters 10 and 11.

Swanson, D. G. 1989. *Plasma Waves*. San Diego, CA: Academic Press, Section 4.3.

11

Nonlinear Effects

In this chapter we give an introduction to various types of nonlinear effects that can occur in a plasma. Almost all of the basic equations in plasma physics have nonlinear terms. For example, these include the $(\mathbf{E} + \mathbf{v} \times \mathbf{B}) \cdot \nabla_{\mathbf{v}} f$ term in the Vlasov equation (5.2.15), the $(\mathbf{U} \cdot \nabla)\mathbf{U}$ term in the convective derivative equation (5.4.26), the $\rho_{\mathrm{m}}\mathbf{U}$ term in the MHD mass continuity equation (6.1.31), and the $\mathbf{J} \times \mathbf{B}$ term in the MHD momentum equation (6.1.32). All of these terms represent potential sources of nonlinear effects. There are many more. In our analysis of waves in the previous chapters, we always assumed that the wave amplitude was small, so that the governing equations could be linearized. This assumption provides a remarkably accurate description of many types of small-amplitude waves. However, if the wave amplitude becomes large, as always occurs for an instability, the linearization assumption breaks down. Nonlinear effects must then be taken into account. There are many such nonlinear effects, more than we can possibly discuss in this introductory textbook. In order to limit the scope of the discussion, we will concentrate on four quite different types of nonlinear analyses that have a wide range of applications. These are quasi-linear theory, wave–wave interactions, Langmuir wave solitons, and time-stationary electrostatic potentials. For a more comprehensive discussion of nonlinear effects in plasmas, the reader is referred to one of the classic books on nonlinear effects, such as Kadomtsev (1965), Sagdeev and Galeev (1969), and Davidson (1972).

11.1 Quasi-linear Theory

In quasi-linear theory, it is assumed that the plasma is only weakly unstable, and that the instability leads to a broad spectrum of waves that modifies the background plasma in a self-consistent way via nonlinear interactions. To illustrate the basic features of quasi-linear theory, we limit our discussion to the temporal evolution of electrostatic waves in a one-dimensional unmagnetized collisionless plasma, in which all of the spatial variations are in the z direction. The governing equations for such a system are the Vlasov equation,

$$\frac{\partial f_s}{\partial t} + v_z \frac{\partial f_s}{\partial z} - \frac{e_s}{m_s}\frac{\partial \Phi}{\partial z}\frac{\partial f_s}{\partial v_z} = 0, \tag{11.1.1}$$

and Poisson's equation,

$$\frac{\partial^2 \Phi}{\partial z^2} = -\sum_s \frac{e_s}{\epsilon_0}\int_{-\infty}^{\infty} f_s \, \mathrm{d}v_z, \tag{11.1.2}$$

where Φ is the electrostatic potential and f_s is the velocity distribution function of the sth species. We start by assuming that the plasma is initially in a homogeneous equilibrium state, $f_s = f_{s0}(v_z)$, with no wave electric field, i.e., $\partial \Phi_0/\partial z = 0$. It is easy to see that such an equilibrium solution identically satisfies Eqs. (11.1.1) and (11.1.2). The first term on the left of Eq. (11.1.1) vanishes because f_{s0} is time-independent, the second term vanishes because f_{s0} is spatially uniform, and the last term vanishes because the equilibrium is field free. Equation (11.1.2) is satisfied because the plasma is quasi-neutral, i.e.,

$$\sum_s e_s \int_{-\infty}^{\infty} f_{s0} \, \mathrm{d}v_z = 0. \tag{11.1.3}$$

At $t = 0$, we perturb the equilibrium by introducing a small perturbation $f_{s1}(z, v_z, t = 0)$ so that the total distribution function can be written as the sum of two terms:

$$f_s(z, v_z, t = 0) = f_{s0}(v_z) + f_{s1}(z, v_z, t = 0). \tag{11.1.4}$$

In Chapters 9 and 10, we approached the problem of wave propagation in hot plasmas by linearizing Eqs. (11.1.1) and (11.1.2) about the equilibrium velocity distribution, $f_{s0}(v_z)$, and determining the time evolution of the first-order perturbation, $f_{s1}(z, v_z, t)$. In doing so, we assumed that the equilibrium distribution function $f_{s0}(v_z)$ at $t = 0$ remains unchanged as the perturbation evolves from its initial value $f_{s1}(z, v_z, t = 0)$ at $t = 0$ to its value $f_{s1}(z, v_z, t)$ at an arbitrary time t. Clearly, this assumption cannot be valid for all times. For instance, if the plasma is subject to a linear, exponentially growing instability, f_{s1} will grow exponentially, invalidating the assumption that f_{s1} remains small. Then the question is how the growing disturbance $f_{s1}(z, v_z, t)$ changes the initial equilibrium $f_{s0}(v_z)$, and how that, in turn, acts back on $f_{s1}(z, v_z, t)$. As discussed below, the quasi-linear approximation provides one way to approach this problem.

We start the discussion of quasi-linear theory by defining the spatial average of the time-dependent distribution function $f_{s1}(z, v_z, t)$ as

$$\langle f_s \rangle (v_z, t) = \frac{1}{2L} \int_{-L}^{L} f_s(z, v_z, t) \mathrm{d}z, \tag{11.1.5}$$

where the spatial integration is carried out over the entire length $2L$ of a one-dimensional plasma. We further require that the spatially averaged distribution function be identical to the zero-order distribution function at $t = 0$,

$$\langle f_s \rangle (v_z, t = 0) = f_{s0}(v_z), \tag{11.1.6}$$

but allow $\langle f_s \rangle$ to deviate from $f_{s0}(v_z)$ for $t > 0$. The total distribution function is then written as the sum of the averaged distribution function $\langle f_s \rangle$ plus a fluctuation, f_{s1}, i.e.,

$$f_s(z, v_z, t) = \langle f_s \rangle (v_z, t) + f_{s1}(z, v_z, t). \tag{11.1.7}$$

Note that at $t = 0$, Eq. (11.1.7) reduces identically to Eq. (11.1.4). The main difference between linear theory and quasi-linear theory lies in what one linearizes about: in linear theory, one linearizes about a time-independent equilibrium distribution function $f_{s0}(v_z)$, whereas in quasi-linear theory one linearizes about a spatially averaged distribution function $\langle f_{s0} \rangle (v_z, t)$, which is allowed to vary slowly in time.

Taking the spatial average of both sides of Eq. (11.1.7) in the manner of Eq. (11.1.5), it follows that the spatial average of the first-order perturbation of the velocity distribution function is zero:

$$\langle f_{s1}(z, v_z, t) \rangle = 0. \tag{11.1.8}$$

For electrostatic perturbations, we can write the perturbed electric field as

$$E_1(z, t) = -\frac{\partial}{\partial z} \Phi_1(z, t). \tag{11.1.9}$$

Taking the spatial average of Eq. (11.1.9) and requiring that $\Phi_1(z, t)$ decays to zero at the system boundaries, we can show that the spatial average of the first-order electric field is also zero:

$$\langle E_1(z, t) \rangle = -\frac{1}{2L} \int_{-L}^{L} \frac{\partial \Phi_1(z, t)}{\partial z} \mathrm{d}z = 0. \tag{11.1.10}$$

11.1.1 The Quasi-linear Diffusion Equation

Next, we develop an equation called the quasi-linear diffusion equation, which describes the time evolution of the average distribution function. This equation is obtained by taking the spatial average of the Vlasov equation (11.1.1), which can be written

$$\frac{\partial}{\partial t}\langle f_s\rangle + \left\langle v_z \frac{\partial f_s}{\partial z}\right\rangle = \frac{e_s}{m_s}\left\langle \frac{\partial \Phi}{\partial z}\frac{\partial f_s}{\partial v_z}\right\rangle. \tag{11.1.11}$$

The second term on the left-hand side of the above equation can be rewritten as

$$\left\langle v_z \frac{\partial f_s}{\partial z}\right\rangle = \left\langle \frac{\partial}{\partial z}(v_z f_s)\right\rangle = 0, \tag{11.1.12}$$

because the integrand is a perfect differential and we require that the perturbation $f_{s1}(z, v_z, t)$ vanish at the boundaries of the system. One way to ensure that this condition is satisfied is to take L to be very large and require that f_s decay to zero at large values of z. The term on the right-hand side of Eq. (11.1.11) can then be written

$$\frac{e_s}{m_s}\left\langle \frac{\partial \Phi}{\partial z}\frac{\partial f_s}{\partial v_z}\right\rangle = \frac{e_s}{m_s}\left\langle \left(\frac{\partial \Phi_0}{\partial z} + \frac{\partial \Phi_1}{\partial z}\right)\left(\frac{\partial \langle f_s\rangle}{\partial v_z} + \frac{\partial f_{s1}}{\partial v_z}\right)\right\rangle. \tag{11.1.13}$$

Since $\partial\Phi_0/\partial z = 0$ in equilibrium and $\langle \partial\Phi_1/\partial z\rangle = 0$ by Eq. (11.1.10), we obtain

$$\frac{e_s}{m_s}\left\langle \frac{\partial \Phi}{\partial z}\frac{\partial f_s}{\partial v_z}\right\rangle = \frac{e_s}{m_s}\left\langle \frac{\partial \Phi_1}{\partial z}\frac{\partial f_{s1}}{\partial v_z}\right\rangle = \frac{e_s}{m_s}\frac{\partial}{\partial v_z}\left\langle f_{s1}\frac{\partial \Phi_1}{\partial z}\right\rangle. \tag{11.1.14}$$

Using Eqs. (11.1.12) and (11.1.14), we can then reduce Eq. (11.1.11) to

$$\frac{\partial}{\partial t}\langle f_s\rangle = \frac{e_s}{m_s}\frac{\partial}{\partial v_z}\left\langle f_{s1}\frac{\partial \Phi_1}{\partial z}\right\rangle. \tag{11.1.15}$$

To derive an equation for $f_{s1}(z, v_z, t)$, we subtract the averaged equation (11.1.15) from the full Vlasov equation (11.1.1), to give

$$\begin{aligned}\left(\frac{\partial}{\partial t} + v_z\frac{\partial}{\partial z}\right) f_{s1}(z, v_z, t) = {} & \frac{e_s}{m_s}\frac{\partial \Phi_1}{\partial z}\frac{\partial \langle f_s\rangle}{\partial v_z} \\ & + \frac{e_s}{m_s}\frac{\partial}{\partial v_z}\left[\frac{\partial \Phi_1}{\partial z} f_{s1} - \left\langle \frac{\partial \Phi_1}{\partial z} f_{s1}\right\rangle\right].\end{aligned} \tag{11.1.16}$$

Both Eqs. (11.1.15) and (11.1.16) are exact. However, they are not a closed set, because a time evolution equation for the second-order fluctuations $(\partial\Phi_1/\partial z) f_{s1}$ (and its average) is needed. Any effort to determine such a time evolution equation will inevitably bring in third-order fluctuations. In turn, a time-dependent equation for third-order fluctuations will involve fluctuations of the fourth order. This process, when continued, produces a hierarchy of indefinitely higher order. Therefore, there is a closure problem, similar to the closure problem encountered with the moment equations in Section 5.4.4. In order to truncate this hierarchy, it is necessary to introduce an approximation. The approximation used in quasi-linear theory is simply to neglect the last term on the right-hand side of Eq. (11.1.16), which is assumed to be small. This approximation yields the equation

$$\left(\frac{\partial}{\partial t}+v_z\frac{\partial}{\partial z}\right)f_{s1}(z,v_z,t)=\frac{e_s}{m_s}\frac{\partial\Phi_1}{\partial z}\frac{\partial\langle f_s\rangle}{\partial v_z}. \tag{11.1.17}$$

We note that the above equation involves only fluctuating quantities of first order. The first-order electrostatic potential is obtained from Poisson's equation

$$\frac{\partial^2\Phi_1}{\partial z^2}=-\sum_s\frac{e_s}{\epsilon_0}\int_{-\infty}^{\infty}f_{s1}\,\mathrm{d}v_z, \tag{11.1.18}$$

which is exact. Equations (11.1.15), (11.1.17), and (11.1.18) now constitute a closed set of equations.

Equation (11.1.17) is formally similar to the time evolution equation for f_{s1} in linear theory, except that the average distribution function $\langle f_s\rangle$ takes the place of the equilibrium distribution function f_{s0}. To solve these equations, we expand the spatial fluctuations, $\Phi_1(z,t)$ and $f_{s1}(z,v_z,t)$, in terms of Fourier transforms:

$$\Phi_1(z,t)=\int_{-\infty}^{\infty}\tilde{\Phi}_1(k,t)\,\mathrm{e}^{\mathrm{i}kz}\,\mathrm{d}k \tag{11.1.19}$$

and

$$f_{s1}(z,v_z,t)=\int_{-\infty}^{\infty}\tilde{f}_{s1}(k,v_z,t)\,\mathrm{e}^{\mathrm{i}kz}\,\mathrm{d}k. \tag{11.1.20}$$

Substituting the above two equations into Eqs. (11.1.17) and (11.1.18), and using the usual operator substitution ($\partial/\partial_z\rightarrow \mathrm{i}k$), we obtain

$$\left(\frac{\partial}{\partial t}+\mathrm{i}kv_z\right)\tilde{f}_{s1}(k,v_z,t)=\frac{e_s}{m_s}\,\mathrm{i}k\tilde{\Phi}_1\,(k,t)\frac{\partial\langle f_s\rangle}{\partial v_z} \tag{11.1.21}$$

and

$$k^2\tilde{\Phi}_1(k,t)=\sum_s\frac{e_s}{\epsilon_0}\int_{-\infty}^{\infty}\tilde{f}_{s1}(k,v_z,t)\,\mathrm{d}v_z. \tag{11.1.22}$$

To determine the time dependence of $\tilde{\Phi}_1(k,t)$ and $\tilde{f}_{s1}(k,v_z,t)$, we assume that these functions can be written in the form

$$\tilde{\Phi}_1(k,t)=\hat{\Phi}_1(k)\mathrm{e}^{-\mathrm{i}\omega(k,t)t} \tag{11.1.23}$$

and

$$\tilde{f}_{s1}(k, v_z, t) = \hat{f}_{s1}(k, v_z)e^{-i\omega(k,t)t}, \tag{11.1.24}$$

where $\omega(k,t)$ is a complex frequency. If $\langle f_s \rangle$ in Eq. (11.1.21) were f_{s0}, then $\omega(k,t)$ would be independent of time. However, unlike f_{s0}, the average distribution function $\langle f_s \rangle$ changes as a function of time, in accordance with Eq. (11.1.15). Each realization of $\langle f_s \rangle$ in time produces its own frequency, ω, making ω a function of time as well. However, the time evolution of $\langle f_s \rangle$, and hence of ω, is slower than that of f_{s1} because, according to Eq. (11.1.15), $\partial\langle f_s \rangle/\partial t$ is second order in the fluctuations while $\partial f_{s1}/\partial t$ is first order. Hence, it is reasonable to assume that ω remains approximately constant on the faster time-scale of fluctuations that evolve according to Eqs. (11.1.21) and (11.1.22). Substituting Eqs. (11.1.23) and (11.1.24) into Eqs. (11.1.21) and (11.1.22), we obtain

$$\hat{f}_{s1}(k, v_z) = -\frac{e_s}{m_s} \frac{k\hat{\Phi}_1(k)}{\omega - kv_z} \frac{\partial\langle f_s \rangle}{\partial v_z} \tag{11.1.25}$$

and

$$k^2\hat{\Phi}_1(k) = \sum_s \frac{e_s}{\epsilon_0} \int_{-\infty}^{\infty} \hat{f}_{s1}(k, v_z)\,dv_z. \tag{11.1.26}$$

We will assume that the wave is unstable: that is, the complex frequency ($\omega = \omega_r + i\gamma$) has a positive imaginary part ($\gamma > 0$). This makes a more detailed Landau analysis for damped modes, as discussed in Section (9.2), unnecessary. The generalization to include a Landau analysis, which covers stable as well as unstable modes, will not be carried out here.

Before we proceed to derive the quasi-linear diffusion equation, it is useful to introduce some identities involving Fourier representations. These follow from the requirement that the electrostatic potential must be real for all t, i.e.,

$$\Phi_1(z,t) = \Phi_1^*(z,t), \tag{11.1.27}$$

where * denotes a complex conjugate. Written in terms of Fourier transforms, the above equation becomes

$$\int_{-\infty}^{\infty} \hat{\Phi}_1(k)\,e^{ikz}\,e^{-i\omega^*(k,t)t}dk = \int_{-\infty}^{\infty} \hat{\Phi}_1^*(k)\,e^{-ikz}\,e^{i\omega^*(k,t)t}\,dk. \tag{11.1.28}$$

We note that k is a dummy variable of integration. So changing k to $-k$ will not change the value of the integral on the right-hand side. We can then rewrite the integral on the right-hand side as

$$\int_{-\infty}^{\infty} \hat{\Phi}_1^*(k)\,\mathrm{e}^{-\mathrm{i}kz}\,\mathrm{e}^{\mathrm{i}\omega^*(k,t)t}\mathrm{d}k = \int_{\infty}^{-\infty} \hat{\Phi}_1^*(-k)\,\mathrm{e}^{\mathrm{i}kz}\,\mathrm{e}^{\mathrm{i}\omega^*(-k,t)t}\mathrm{d}(-k)$$
$$= \int_{-\infty}^{\infty} \hat{\Phi}_1^*(-k)\,\mathrm{e}^{\mathrm{i}kz}\,\mathrm{e}^{\mathrm{i}\omega^*(-k,t)}\mathrm{d}k. \qquad (11.1.29)$$

Comparing the right-hand side of the above equation with the left-hand side of the previous equation, we obtain the identities

$$\hat{\Phi}_1(k) = \hat{\Phi}_1^*(-k) \qquad (11.1.30)$$

and

$$\omega(k,t) = -\omega^*(-k,t). \qquad (11.1.31)$$

The above equation implies that

$$\omega_{\mathrm{r}}(k,t) = -\omega_{\mathrm{r}}(-k,t), \quad \gamma(k,t) = \gamma(-k,t). \qquad (11.1.32)$$

To obtain the quasi-linear diffusion equation from Eq. (11.1.15), consider the expression

$$\frac{\partial}{\partial v_z}\left\langle f_{s1}\frac{\partial \Phi_1}{\partial z}\right\rangle = \frac{\partial}{\partial v_z}\frac{1}{2L}\int_{-L}^{L}\frac{\partial \Phi_1}{\partial z}f_{s1}\mathrm{d}z$$
$$= \frac{\partial}{\partial v_z}\frac{1}{2L}\int_{-L}^{L}\left[\left\{\frac{\partial}{\partial z}\int_{-\infty}^{\infty}\tilde{\Phi}_1(k,t)\,\mathrm{e}^{\mathrm{i}kz}\mathrm{d}k\right\}\right.$$
$$\left.\times\int_{-\infty}^{\infty}\tilde{f}_{s1}(k',v_z,t)\,\mathrm{e}^{\mathrm{i}k'z}\,\mathrm{d}k'\right]\mathrm{d}z, \qquad (11.1.33)$$

where we have used the Fourier transformation equations (11.1.19) and (11.1.20). If we now change the order of differentiation with respect to z and the integration with respect to k in the first integrand (in parentheses), take the limit of a large box ($L \to \infty$), and use the identity

$$\lim_{L\to\infty}\int_{-L}^{L}\mathrm{e}^{\mathrm{i}(k+k')z}\mathrm{d}z = 2\pi\delta(k+k'), \qquad (11.1.34)$$

where $\delta(k+k')$ is the Dirac delta function; we obtain

$$\frac{\partial}{\partial v_z}\left\langle f_{s1}\frac{\partial \Phi_1}{\partial_z}\right\rangle = -\frac{\pi}{L}\frac{\partial}{\partial v_z}\int_{-\infty}^{\infty}\mathrm{i}k\tilde{\Phi}_1(-k,t)\tilde{f}_{s1}(k,v_z,t)\,\mathrm{d}k. \qquad (11.1.35)$$

Using Eqs. (11.1.23) and (11.1.24), we can write the above equation as

$$\frac{\partial}{\partial v_z}\left\langle f_{s1}\frac{\partial \Phi_1}{\partial z}\right\rangle = -\frac{\pi}{L}\frac{\partial}{\partial v_z}\int_{-\infty}^{\infty} \mathrm{i}k\hat{\Phi}_1(-k)\hat{f}_{s1}(k,v_z)\mathrm{e}^{-\mathrm{i}[\omega(k,t)+\omega(-k,t)]t}\mathrm{d}k$$
$$= -\frac{\pi}{L}\frac{\partial}{\partial v_z}\int_{-\infty}^{\infty} \mathrm{i}k\hat{\Phi}_1(-k)\hat{f}_{s1}(k,v_z)\mathrm{e}^{2\gamma(k,t)t}\mathrm{d}k, \tag{11.1.36}$$

where we have used identities (11.1.32) to obtain

$$\omega(k,t)+\omega(-k,t) = \omega_{\mathrm{r}}(k,t)+\mathrm{i}\gamma(k,t)+\omega_{\mathrm{r}}(-k,t)+\mathrm{i}\gamma(-k,t)$$
$$= 2\mathrm{i}\gamma(k,t). \tag{11.1.37}$$

Substituting expression (11.1.25) for $\hat{f}_{s1}(k,v_z)$ into Eq. (11.1.36), we obtain

$$\frac{\partial}{\partial v_z}\left\langle f_{s1}\frac{\partial \Phi_1}{\partial z}\right\rangle = \frac{\pi}{L}\frac{e_s}{m_s}\frac{\partial}{\partial v_z}\int_{-\infty}^{\infty} \mathrm{i}k^2\frac{\hat{\Phi}_1(k)\hat{\Phi}_1(-k)}{\omega-kv_z}\frac{\partial}{\partial v_z}\langle f_s\rangle\mathrm{e}^{2\gamma(k,t)t}\mathrm{d}k, \tag{11.1.38}$$

which, when substituted into the right-hand side of Eq. (11.1.15), gives

$$\frac{\partial}{\partial t}\langle f_s\rangle = \frac{\pi}{L}\left(\frac{e_s}{m_s}\right)^2\frac{\partial}{\partial v_z}\left[\int_{-\infty}^{\infty}\hat{E}_1(k)\hat{E}_1(-k)\mathrm{e}^{2\gamma(k,t)t}\frac{\mathrm{i}}{\omega-kv_z}\frac{\partial}{\partial v_z}\langle f_s\rangle\right]\mathrm{d}k, \tag{11.1.39}$$

where $\hat{E}_1(k) = -\mathrm{i}k\hat{\Phi}_1(k)$. The above equation can be written more compactly in terms of the averaged electrostatic energy, which is defined as

$$\langle W_{\mathrm{E}}\rangle = \frac{\epsilon_0}{4L}\int_{-L}^{L}|E_1(z,t)|^2\mathrm{d}z, \tag{11.1.40}$$

where $\epsilon_0|E_1(z,t)|^2/2$ is the electrostatic energy density. Using the identity (11.1.34), it can be shown that

$$\langle W_{\mathrm{E}}\rangle = \frac{\pi\epsilon_0}{2L}\int_{-\infty}^{\infty}\tilde{E}_1(k,t)\tilde{E}_1(-k,t)\,\mathrm{d}k = \frac{\pi\epsilon_0}{2L}\int_{-\infty}^{\infty}\tilde{E}_1(k,t)\tilde{E}_1^*(k,t)\,\mathrm{d}k$$
$$= \frac{\pi\epsilon_0}{2L}\int_{-\infty}^{\infty}|\tilde{E}_1(k,t)|^2\,\mathrm{d}k$$
$$= \int_{-\infty}^{\infty}\mathscr{E}(k,t)\mathrm{d}k, \tag{11.1.41}$$

where $\mathscr{E}(k,t) = (\pi\epsilon_0/2L)|\tilde{E}_1(k,t)|^2$ is called the *spectral density* of the electric field. We can also write $\mathscr{E}(k,t)$ in the equivalent form:

$$\mathscr{E}(k,t) = \frac{\pi\epsilon_0}{2L}|\hat{E}_1(k)|^2\,\mathrm{e}^{2\gamma(k,t)t}. \tag{11.1.42}$$

The above equation implies that the spectral density $\mathscr{E}(k,t)$ obeys the differential equation

$$\frac{\partial \mathscr{E}(k,t)}{\partial t} = 2\gamma(k,t)\mathscr{E}(k,t), \tag{11.1.43}$$

where the time dependence of $\gamma(k,t)$ is suppressed in integrating $\mathscr{E}(k,t)$, because $\gamma(k,t)$ varies much more slowly with time than the fluctuating energy $\mathscr{E}(k,t)$. Using the identity $\hat{E}_1(-k) = \hat{E}_1^*(k)$ and Eq. (11.1.42), Eq. (11.1.39) can be written in the form of a diffusion equation,

$$\frac{\partial}{\partial t}\langle f_s\rangle(v_z,t) = \frac{\partial}{\partial v_z}\left[D_{\rm q}(v_z,t)\frac{\partial}{\partial v_z}\langle f_z\rangle(v_z,t)\right], \tag{11.1.44}$$

which is called the quasi-linear diffusion equation, where

$$D_{\rm q}(v_z,t) = \frac{2}{\epsilon_0}\left(\frac{e_s}{m_s}\right)^2 \int_{-\infty}^{\infty} \frac{{\rm i}\mathscr{E}(k,t)}{\omega - kv_z}\,{\rm d}k \tag{11.1.45}$$

is the diffusion coefficient. Since Eq. (11.1.44) is the time evolution equation for a real distribution function $\langle f_s\rangle(v_z,t)$, the coefficient $D_{\rm q}(v_z,t)$, given by the above equation, must also be real. This can be seen by writing

$$\frac{{\rm i}\mathscr{E}(k,t)}{\omega - kv_z} = \frac{{\rm i}\mathscr{E}(k,t)}{\omega_{\rm r} + {\rm i}\gamma - kv_z} = \frac{{\rm i}\mathscr{E}(k,t)[\omega_{\rm r} - kv_z - {\rm i}\gamma]}{(\omega_{\rm r} - kv_z)^2 + \gamma^2}. \tag{11.1.46}$$

Using the fact that $\mathscr{E}(k,t)$ is even in k (i.e., $\mathscr{E}(k,t) = \mathscr{E}(-k,t)$) and $\omega_{\rm r}(k,t)$ is odd in k (by identity (11.1.32)), it follows that the imaginary part of the integral in Eq. (11.1.45) vanishes identically because the integrand is odd in k. We are thus left with the diffusion coefficient

$$D_{\rm q}(v_z,t) = \frac{2}{\epsilon_0}\left(\frac{e_s}{m_s}\right)^2 \int_{-\infty}^{\infty} \frac{\mathscr{E}(k,t)\gamma(k,t)}{[\omega_{\rm r}(k,t) - kv_z]^2 + \gamma^2(k,t)}\,{\rm d}k, \tag{11.1.47}$$

which is manifestly real and positive for growing modes.

Equations (11.1.42) and (11.1.44), with the diffusion coefficient specified by Eq. (11.1.47), are the basic equations of quasi-linear theory.

11.1.2 Application to the Bump-on-Tail Instability

To illustrate an application of quasi-linear theory, we next discuss the time evolution of the electron bump-on-tail instability. Such a distribution function was first discussed in Section 9.5.4, and is shown in Figure 11.1. The phase velocities, ω/k, of the unstable Langmuir waves that arise from this distribution function lie in the region where $f_{\rm e0} = \langle f_{\rm e}\rangle(v_z, t=0)$ is an increasing function of v_z, i.e., where

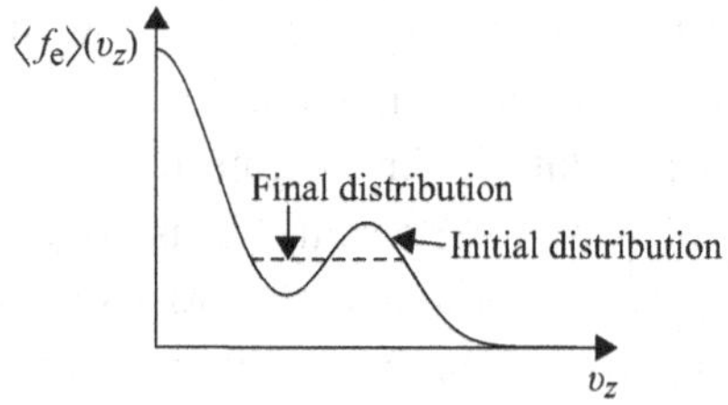

Figure 11.1 The electrostatic instability caused by a double hump in the electron distribution function (solid line) leads to wave–particle interactions that act to flatten the distribution function (dashed line).

$\partial f_{e0}/\partial v_z > 0$. Assuming that the growth rate of the unstable waves is small, i.e., $|\gamma/\omega_r| \ll 1$, it can be shown that

$$\frac{\gamma}{(\omega_r - kv_z)^2 + \gamma^2} \simeq \pi\delta(\omega_r - kv_z). \tag{11.1.48}$$

Substituting the above equation into the quasi-linear diffusion coefficient given by Eq. (11.1.47), we obtain

$$D_q(v_z, t) = \frac{2\pi}{\epsilon_0}\left(\frac{e}{m_e}\right)^2 \int_{-\infty}^{\infty} \mathscr{E}(k)\ \delta(\omega_r - kv_z)\,\mathrm{d}k. \tag{11.1.49}$$

The quasi-linear diffusion equation (11.1.44) then becomes

$$\frac{\partial}{\partial t}\langle f_e\rangle(v_z, t) = \frac{\partial}{\partial v_z}\left\{\frac{2\pi}{\epsilon_0 v_z}\left(\frac{e}{m_e}\right)^2 \mathscr{E}\left(k = \frac{\omega_r}{v_z}\right)\frac{\partial\langle f_e\rangle}{\partial v_z}(v_z, t)\right\}. \tag{11.1.50}$$

The diffusion coefficient in this case is non-zero in the region where the velocity v_z of the particles is resonant with the phase velocity ω_r/k of the waves, and is zero everywhere else in velocity space. Such a diffusion coefficient tends to smooth out velocity derivatives in the region of velocity space where $\partial f_{e0}/\partial v_z > 0$. However, this flattening process involves a larger region than the region where strictly $\partial f_{e0}/\partial v_z > 0$, because the particles removed from the unstable region must, by particle conservation, be taken up by the stable region nearby. The net outcome of this process is the flattening final distribution function shown in Figure 11.1. In the final state, the Langmuir waves reach marginal stability, i.e., $\gamma(k, t \to \infty) = 0$, and no longer grow in time.

11.1.3 Chaotic Velocity Space Diffusion

Quasi-linear theory predicts that an unstable particle distribution leads to the growth of a broad spectrum of waves, which in turn diffuses the particles in velocity space in such a way as to eliminate the region of positive slope that caused the instability. It is interesting to investigate exactly how this diffusion takes place. In Section 9.2.5 we discussed the trapping and phase mixing of particles in phase space when one wave is present. To investigate the nonlinear effects that occur when a broad spectrum of waves is present, let us consider what happens when we superpose an infinite number of sinusoidal waves of the same amplitude and wavelength. For such a superposition, the equations of motion are

$$\frac{\mathrm{d}z}{\mathrm{d}t} = v_z \tag{11.1.51}$$

and

$$m\frac{\mathrm{d}v_z}{\mathrm{d}t} = qE \sum_{n=-N}^{N} \cos(kz - n\omega t), \tag{11.1.52}$$

where $N \to \infty$. To simplify the equations, we define dimensionless variables $\xi = kz, \tau = \omega t$, and $v = v_z/(\omega/k)$. The above equations can then be rewritten in the form

$$\frac{\mathrm{d}\xi}{\mathrm{d}\tau} = v \tag{11.1.53}$$

and

$$\frac{\mathrm{d}v}{\mathrm{d}\tau} = \epsilon^2 \sum_{n=-N}^{N} \cos(\xi - n\tau), \tag{11.1.54}$$

where $\epsilon^2 = qkE/(m\omega^2)$ is a parameter that measures the depth of the potential well formed by a given wave. In the limit $N \to \infty$, it can be shown that the summation in the above equation can be written

$$\sum_{n=-N}^{N} \cos(\xi - n\tau) \overset{N\to\infty}{\Rightarrow} 2\pi \cos\xi \sum_{n=-\infty}^{\infty} \delta(\tau - 2\pi n), \tag{11.1.55}$$

where $\delta(\tau - 2\pi n)$ is the Dirac delta function. In other words, the charged particle is subject to a sequence of instantaneous kicks when $\tau = 2\pi n$, which produces a series of discontinuous changes in its velocity. In between each kick, the particle moves with a uniform velocity. The process can be described rigorously by a series of difference equations, obtained by integrating the differential equations (11.1.53) and (11.1.54) with respect to time until just after the $(m+1)$st kick. Defining a new variable $z_n = \xi_n/2\pi - 1/4$, it can be shown that Eqs. (11.1.54) and (11.1.55) reduce to the so-called Standard Map, given by the difference equations

$$z_{m+1} = z_m + v_m \tag{11.1.56}$$

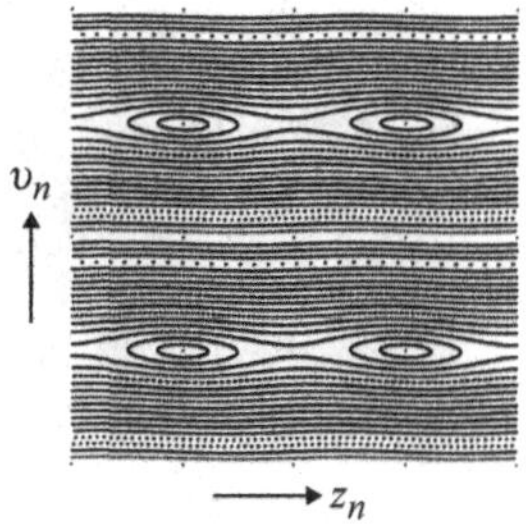

Figure 11.2 Phase space of the Standard Map for $\epsilon = 0.10$.

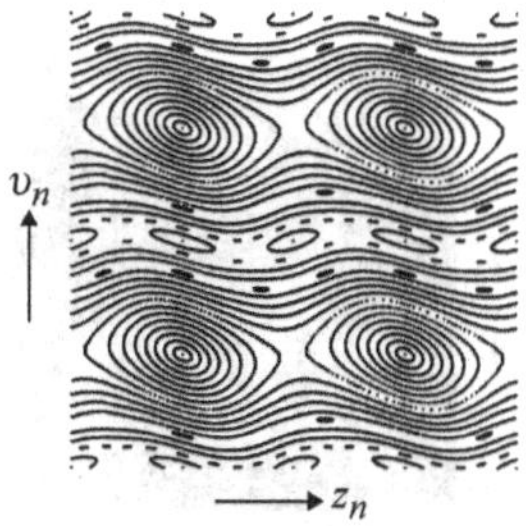

Figure 11.3 Phase space of the Standard Map for $\epsilon = 0.30$.

and

$$v_{m+1} = v_m - 2\pi\epsilon^2 \sin(2\pi z_{m+1}). \tag{11.1.57}$$

The Standard Map, which is originally due to Chirikov (1969) and Taylor (1969), is one of the most widely studied examples of chaos in Hamiltonian systems, and its application to quasi-linear theory is discussed by Stix (1992). Because this map is already in finite difference form, it can be integrated easily on a computer. The computing strategy is as follows. To start, choose a set of initial conditions (z_0, v_0). For each initial condition, a series of iterants $(z_1, v_1), (z_2, v_2), \ldots$, is then calculated and plotted as points in $z - v$ phase space. Figure 11.2 shows such a plot for $\epsilon = 0.10$. The v_n values cover a range suitable for showing two of the infinite number of waves. We note the strong similarity to Figure 9.13, except that there are now trapping regions associated with each of the waves, only two of which are shown. As the wave amplitude increases, ϵ becomes larger, and the trapping regions increase in size. This trend can be seen in Figure 11.3, which is for $\epsilon = 0.30$. Well before the trapping regions become large enough to touch each other, we note the appearance of smaller trapping regions between the separatrices of the main trapping regions associated with the two waves.

For an even larger wave amplitude, $\epsilon = 0.50$, shown in Figure 11.4, the phase space becomes a mixture of some regions that contain trapped particles and other regions that contain particles that appear to be randomly located. As ϵ becomes

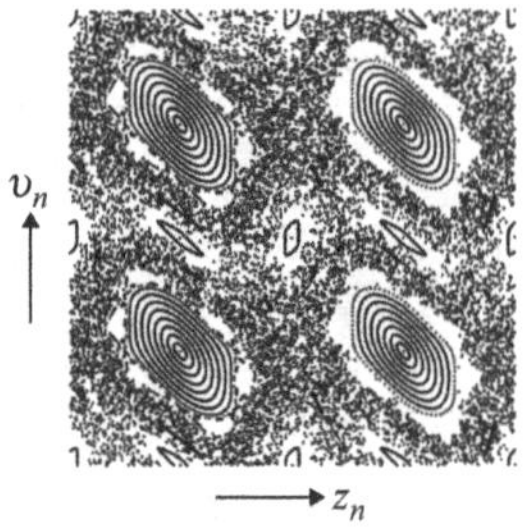

Figure 11.4 Phase space of the Standard Map for $\epsilon = 0.50$.

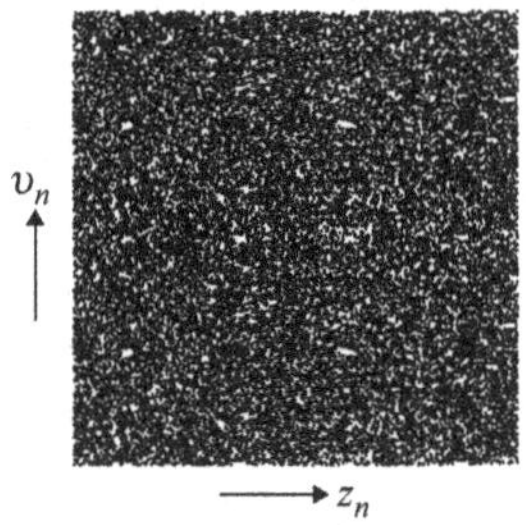

Figure 11.5 Phase space of the Standard Map for $\epsilon = 1.00$.

even larger, the trapping regions are destroyed and the particle trajectories are no longer constrained by nice, well-behaved functions that are constants of motion. Instead, the trajectories wander over and tend to fill most of the phase space, as shown in Figure 11.5 for $\epsilon = 1.00$. Such behavior is characteristic of chaos, similar to the example discussed in Section 3.10. It is worth emphasizing again that the onset of chaos does *not* mean that the particle motions are random. Since the mapping is strictly deterministic, the final location of the particle in phase space is unique, no matter how large the value of ϵ may be. However, even small differences in the initial conditions or the wave amplitudes and phases can lead to very widely diverging final results when the particle motion is chaotic. In many situations of physical interest, such as a plasma comprising many particles, it is impossible to specify exactly the initial conditions of all the particles. The particle trajectories in phase space then become unpredictable, and hence random. This simple demonstration of chaotic particle motions provides a powerful justification for the basic assumption of quasi-linear theory, which is that the growth of waves over a range of phase velocities leads to the rapid diffusion of particles in velocity space and an eventual flattening of the distribution function in the unstable region of velocity space.

A good example of such velocity space diffusion can be seen in the spectra of solar flare electrons responsible for type III solar radio bursts. As discussed

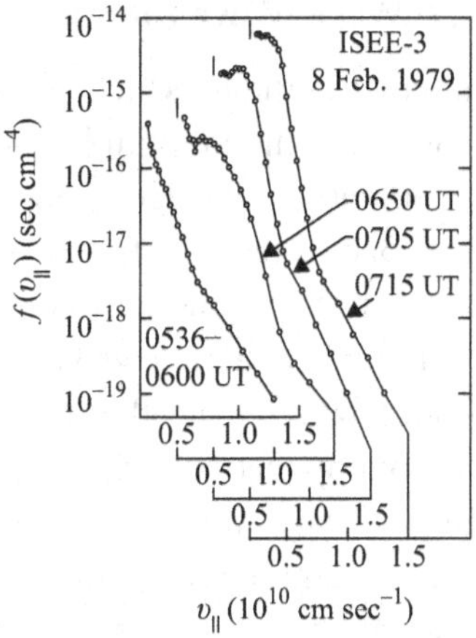

Figure 11.6 A series of reduced one-dimensional electron velocity distribution functions showing the region of positive slope, $\partial f_0/\partial v_{\|} > 0$, produced by electrons from a solar flare (from Lin et al., 1986). The distribution labeled 0536–0600 UT is the pre-event solar wind distribution.

in Section 9.5.4, a type III radio burst is produced by mode conversion from Langmuir waves generated by energetic electrons from a solar flare. Because of time of flight considerations, a sharp velocity cutoff is introduced in the solar flare electron velocity distribution. This cutoff automatically leads to a region of positive slope, $\partial f_0/\partial v_{\|} > 0$, and the growth of Langmuir waves; see Figure 9.29. An example of a series of electron velocity distribution functions observed in the solar wind near the orbit of Earth following a solar flare is shown in Figure 11.6. This event produced strong Langmuir waves and an intense type III radio burst comparable to the example shown in Figure 9.30. Although the region of positive slope caused by the bump-on-tail distribution can be seen in the spectra at 0650 and 0705 UT, the velocity space cutoff is not nearly as sharp as would be expected from the time of flight cutoff (see Figure 9.29), and has been almost completely eliminated by the end of the event, at 0715 UT. This flattening of the electron velocity distribution function is caused by quasi-linear diffusion. In fact, regions with a substantial positive slope are seldom observed, since as soon as such a region occurs, waves quickly grow and flatten the distribution function. Quasi-linear diffusion is also believed to play a crucial role in the propagation of solar flare electrons by limiting the amplitude of the Langmuir waves. In the absence of quasi-linear diffusion, Langmuir waves would grow to such large amplitudes that they would completely disrupt the bump-on-tail distribution, thereby preventing the electrons from propagating to great distances from the Sun, >1 AU, where they are often observed.

11.2 Wave–Wave Interactions

When two or more waves interact in the presence of a medium obeying nonlinear equations, the effect is to modulate the amplitudes of the interacting waves. As we shall show, amplitude modulation produces new frequency components that were not originally present. If one of the new frequency components is at or near a normal mode of the plasma, then that mode may gain or lose energy, leading to a coupled system of mutually interacting waves. Such coupled systems are important because they produce waves that otherwise would not be predicted by a linear theory. There are many applications where wave–wave interactions play an important role. For example, wave–wave interactions are directly relevant to the goal of achieving laser fusion, a process in which intense laser beams are used to heat small pellets of material to fusion temperatures. In such attempts, it is found that much of the laser energy is converted to Langmuir waves, which then heat electrons rather than ions via Landau damping. In a similar process, powerful ground-based radio transmitters have been used to perform various ionospheric heating experiments via the coupling of the radiated electromagnetic wave to Langmuir waves. In the following we illustrate some of the fundamental processes by which these and other nonlinear wave–wave interactions take place.

11.2.1 Amplitude Modulation

Before considering the details of such interactions, it is useful to first discuss the most basic type of nonlinear interaction that can occur in a plasma, namely amplitude modulation. Amplitude modulation is generated by the product of two first-order waveforms. That a product is involved arises from the fact that the first nonlinear term in a Taylor series expansion of the nonlinear characteristic of the medium consists of the square of the sum of the two waveforms, which necessarily involves the product of the two waveforms. As a commonly occurring example that will be encountered later, consider the product of a constant amplitude electric field oscillation $E\cos(\omega_1 t)$ and a small-amplitude density oscillation $n_{\mathrm{A}}\cos(\omega_2 t)$ superposed on a constant background density n_0. Such a density can be written $n_0[1+m\cos\omega_2 t]$, where the parameter $m = n_{\mathrm{A}}/n_0$ is called the modulation index. Using the double-angle trigonometric identity,

$$\cos A \cos B = (1/2)[\cos(A-B) + \cos(A+B)] \tag{11.2.1}$$

(see Appendix B), it follows that the product of the two waveforms is given by

$$n_0 E\cos(\omega_1 t)[1+m\cos(\omega_2 t)] = n_0 E[\cos(\omega_1 t) + \frac{m}{2}\cos(\omega_1 - \omega_2)t + \frac{m}{2}\cos(\omega_1 + \omega_2)t]. \tag{11.2.2}$$

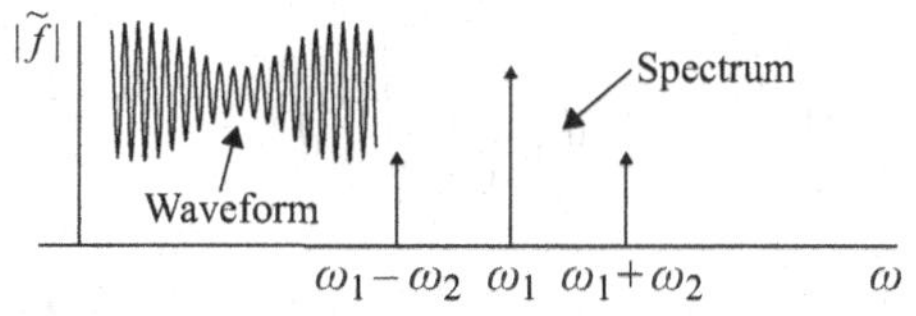

Figure 11.7 The frequency spectrum that arises from amplitude modulation of a waveform $A\cos\omega_1 t$ by a factor $n_0[1+m\cos\omega_2 t]$.

From the above result, it is apparent that the product of the two waveforms has introduced two new frequencies, $(\omega_1 - \omega_2)$ and $(\omega_1 + \omega_2)$, that were not present in the original waveforms. The new frequencies are called sidebands, and are symmetrically located on either side of ω_1, as shown in Figure 11.7. Note that amplitude modulation is linear in the sense that the amplitudes of the sidebands $(m/2)$ are directly proportional to the modulation index m. Such modulation effects are quite common. For example, because the response of the human ear is nonlinear, it is common to hear a "beat" at the frequency difference, $\Delta\omega = \omega_1 - \omega_2$, between two audio frequencies ω_1 and ω_2.

Because a plasma has many normal modes, the possible wave–wave interactions that can occur in a plasma are much more numerous than in a neutral gas, which has only one mode of propagation. In principle, any number of wave modes can participate in a wave–wave interaction. However, the coupling between multiple modes is proportional to the product of the coupling coefficients and decreases rapidly as the number of interacting modes increases. So, we will restrict our attention to the most basic of such interactions, which is the three-wave interaction. As discussed above, since the product of two cosine functions, $\cos(\omega_1 t)\cos(\omega_2 t)$, generates two frequencies, $\omega_1 - \omega_2$ and $\omega_1 + \omega_2$, it is apparent that a three-wave interaction involves two possible combinations of three frequencies:

$$\omega_0 = \omega_1 \pm \omega_2. \tag{11.2.3}$$

The above equation is called a frequency matching condition. For the product of two propagating waves, $\cos(\mathbf{k}_1 \cdot \mathbf{r} - \omega_1 t)\cos(\mathbf{k}_2 \cdot \mathbf{r} - \omega_2 t)$, a similar matching condition also exists for the spatially dependent part of the waveforms. Therefore, to the above frequency matching condition we must also add a corresponding matching condition for the wave vectors, specifically

$$\mathbf{k}_0 = \mathbf{k}_1 \pm \mathbf{k}_2. \tag{11.2.4}$$

Note that the assignment of indices in the above equations is completely arbitrary, so the matching conditions could very well be written simply as $\omega_0 = \omega_1 + \omega_2$ and

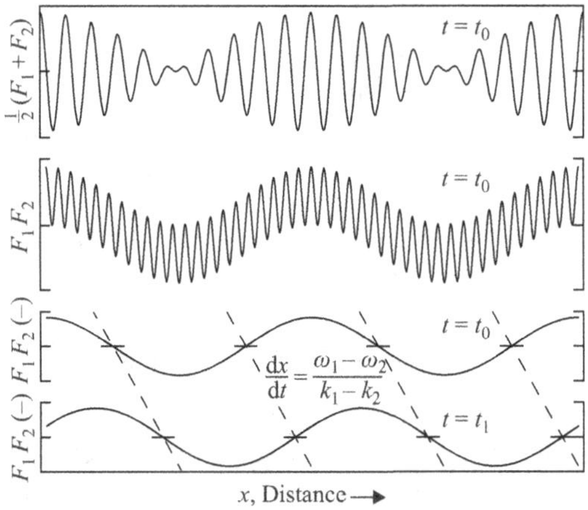

Figure 11.8 The top panel shows the spatial interference pattern that arises from the superposition (sum) of two propagating waves, $F_1 = A_1 \cos(k_{x1}x - \omega_1 t)$ and $F_2 = A_2 \cos(k_{x2}x - \omega_2 t)$, at a fixed time, $t = t_0$. The second panel shows the waveform that arises from taking the product of the two waves, $F_1 F_2$. The third panel shows the waveform of the long-wavelength component obtained after "filtering" out the high-frequency/short-wavelength component by omitting the $\cos(A + B)$ term in Eq. (11.2.1). The bottom panel shows that at a later time $t = t_1$ the long-wavelength component has propagated to the right at a phase speed given by $dx/dt = (\omega_1 - \omega_2)/(k_{x1} - k_{x2})$.

$\mathbf{k}_0 = \mathbf{k}_1 + \mathbf{k}_2$, without regard for the ± signs. As we will show, the exact choice of signs depends on the physics of the interaction. If the above two equations are multiplied by Planck's constant, $\hbar$, one can recognize that these matching conditions correspond respectively, to conservation of energy and momentum in quantum mechanics. Thus, just as an electromagnetic wave can be regarded as consisting of particles called photons, plasma waves are sometimes regarded as consisting of particles called plasmons.

To further illustrate the significance of the above matching conditions, Figure 11.8 shows the various waveforms that can arise by combining two propagating waves, $F_1 = A_1 \cos(k_{x1}x - \omega_1 t)$ and $F_2 = A_2 \cos(k_{x2}x - \omega_2 t)$. The k_x and ω values have been selected to be slightly different in order to produce waveforms with two distinctly different spatial and temporal scales. The top panel shows the superposition (sum) of the two waveforms as a function of x at a fixed time $t = t_0$. A clearly defined "interference pattern" can be seen between the two waves. Although defined regions of constructive and destructive interference exist, no new frequencies are generated. Superposition is simply the linear sum of the two frequency components, ω_1 and ω_2. The second panel shows the product, $F_1 F_2$, of the two waveforms, again at the same time, $t = t_0$. Here the results are quite different. Two distinctly different frequency components, i.e., beats, can be seen, one at a long wavelength, $k_{x0} = k_{x1} - k_{x2}$, that corresponds to the low-frequency component, $\omega_0 = \omega_1 - \omega_2$, and the other at a short wavelength, $k_{x0} = k_{x1} + k_{x2}$, that corresponds to the high-frequency component, $\omega_0 = \omega_1 + \omega_2$.

Experimentally, the two components in the second panel of Figure 11.8 can be isolated by a process called "filtering." Filtering of experimental measurements is often accomplished by electronically averaging over a suitable long or short time-scale, or by specially designed electronic filters. Mathematically, filtering can be achieved by simply omitting either the $(A - B)$ term or the $(A + B)$ term in the trigonometric identity for $\cos(A)\cos(B)$. The third panel of this figure shows the waveform that results if we keep only the low-frequency $(\omega_1 - \omega_2)$ term, which we indicate by the symbol $F_1F_2(-)$. As can be seen, the result is a long-wavelength/low-frequency wave that perfectly matches the phase of the interference pattern in the top panel. The bottom panel, which is at a slightly later time $t = t_1$, shows that the phase of this low-frequency component has advanced to the right with increasing time. The speed of this advance is easily shown to be $\mathrm{d}x/\mathrm{d}t = (\omega_1 - \omega_2)/(k_{x1} - k_{x2})$. For the short-wavelength/high-frequency $(\omega_1 + \omega_2)$ component (not shown) the corresponding phase speed is $\mathrm{d}x/\mathrm{d}t = (\omega_1 + \omega_2)/(k_{x1} + k_{x2})$. Whether a coherent resonant interaction occurs depends on the presence of a third mode with $\omega_0 = \omega_1 \pm \omega_2$ and $k_{x0} = k_{x1} \pm k_{x2}$ that matches one of these two phase speeds. If the phase speeds match, then a coherent resonant response occurs for that mode. It is this type of nonlinear three-wave interaction that we are seeking.

11.2.2 Three Coupled Harmonic Oscillators

To get a broader understanding of the physics involved in such nonlinear interactions, we next discuss the case of three weakly coupled harmonic oscillators. For a specific example of such a system, consider three pendulums coupled by identical springs, as shown in Figure 11.9. If there is no coupling, we assume that the small-amplitude oscillation frequencies of the pendulums are ω_0, ω_1, and ω_2, and that the instantaneous displacements of the pendulum masses from their equilibrium position are x_0, x_1, and x_2. To examine the effects of the springs, let's first assume that the springs are linear and that the effective force acting on each mass is given by Hooke's law, $F = -k\Delta x$, where k is the spring constant, and Δx is the change in the length of the spring relative to its equilibrium position. If we focus on the 0th pendulum, one can see that there are two spring forces that must be considered. The effective force acting on mass m is then given by $F/m = -(k/m)[(x_0 - x_1) + (x_0 - x_2)]$. To obtain the equation of motion we simply add this force to the right-hand side of the harmonic oscillator equation for the

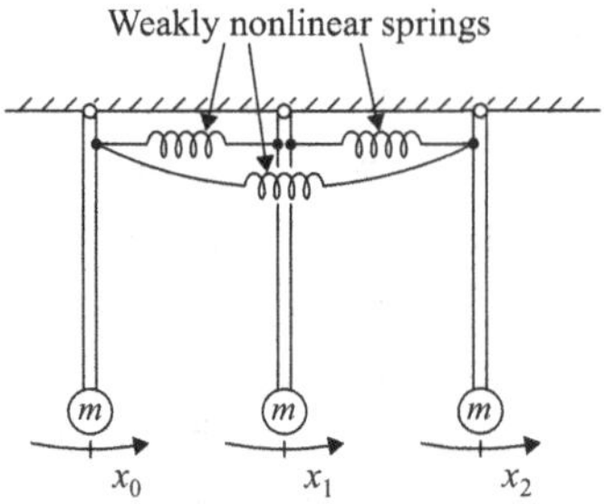

Figure 11.9 A coupled oscillator system consisting of three pendulums connected by three identical nonlinear springs that provide the coupling between the oscillators.

uncoupled motion of the pendulum, so that

$$\ddot{x}_0 + \omega_0^2 x_0 = -\frac{k}{m}(2x_0 - x_1 - x_2), \tag{11.2.5}$$

where $\cdot\cdot$ represents d^2/dt^2. Of the three terms on the right-hand side, the first term can be moved to the left-hand side and combined with the $\omega_0^2 x_0$ term. This term causes a shift in the oscillation frequency of the pendulum due to the presence of the springs. Assuming that the coupling is very weak we ignore this correction, although it could be easily included. The second and third terms, which are at frequencies ω_1 and ω_2, do not produce a coherent response because they are not at the natural oscillation frequency, ω_0, of the 0th pendulum. From this analysis, we see that linear coupling plays no role in the coherent three-wave interaction that we are seeking.

Next, consider the effects caused by the nonlinear characteristics of the spring. If the spring force is represented by a Taylor series expansion in powers of Δx, the next most important term is the squared term, Δx^2. Of course, it could be that the expansion coefficient for this term is zero, in which case the next most important term would be the cubic term. However, for our purposes we assume that the most important nonlinear term is the squared term. The force per unit mass acting on the pendulum is then given by the equation $F/m = -(\lambda/2)(2x_0 - x_1 - x_2)^2$, where λ is the nonlinear coupling coefficient. The sign of λ can be either positive or negative, depending on the characteristic of the spring. Adding this force to the right-hand side of the harmonic oscillator equation then gives the following equation of motion for the pendulum:

$$\ddot{x}_0 + \omega_0^2 x_0 = -\frac{\lambda}{2}(4x_0^2 - 4x_0x_1 - 4x_0x_2 + x_1^2 + 2x_1x_2 + x_2^2), \tag{11.2.6}$$

where we have expanded the squared term to show all of the cross-products. Simple inspection of these cross-products shows that there is only one term, $2x_1x_2$, that can produce a three-wave interaction. To produce a three-wave interaction the driving frequencies involved in the cross-product term must satisfy the frequency matching condition $\omega_0 = \omega_1 \pm \omega_2$. None of the other product terms satisfy this condition. For example, by considering $x_0 = A_0 \cos(\omega_0 t)$ in the above equation it can be seen that the x_0^2 term generates a driving frequency of $2\omega_0$, which is not at the resonant frequency ω_0 of the pendulum. The factor of two in the frequency follows from the trigonometric identity $\cos^2(A) = (1/2)[1 + \cos(2A)]$; see Appendix B. Similarly, x_1^2 and x_2^2 generate frequencies of $2\omega_1$ and $2\omega_2$, and the products $x_0 x_1$ and $x_0 x_2$ generate frequencies of $\omega_0 \pm \omega_1$ and $\omega_0 \pm \omega_2$, none of which are at ω_0. Extending the above analysis to the other two pendulums, it follows that the nonlinear interactions of the three pendulums can be described by the following system of coupled nonlinear equations:

$$\ddot{x}_0 + \omega_0^2 x_0 = -\lambda x_1 x_2, \tag{11.2.7}$$

$$\ddot{x}_1 + \omega_1^2 x_1 = -\lambda x_0 x_2, \tag{11.2.8}$$

and

$$\ddot{x}_2 + \omega_2^2 x_2 = -\lambda x_0 x_1. \tag{11.2.9}$$

It is interesting to note, using Hamilton's equation $\dot{p}_i = -\partial H/\partial \dot{x}_i$, that the above system of equations can be obtained from the Hamiltonian

$$H = \sum_{i=0}^{2} \left(\frac{p_i^2}{2} + \frac{1}{2}\omega_i^2 x_i^2 \right) + \lambda x_0 x_1 x_2, \tag{11.2.10}$$

where the interaction term is $\lambda x_0 x_1 x_2$. This Hamiltonian is often used to describe the nonlinear coupling of harmonic oscillators in a variety of applications, such as molecular vibrations. It also served as the starting point for a discussion of wave–wave interactions in the well-known book on nonlinear plasma theory by Sagdeev and Galeev (1969).

To explore the solutions of the above system of equations we note that if there is no coupling ($\lambda = 0$) then the solutions are simply $x_j = A_j \cos(\omega_j t + \phi_j)$, where ω_j is the natural resonant frequency of the jth harmonic oscillator and ϕ_j is the corresponding phase. To obtain a solution for non-zero coupling, we shall assume that the coupling constant λ is sufficiently small that the solutions can be represented by a two-time-scale approximation, in which the amplitudes $A_j(t)$ vary slowly compared to the rapid oscillations given by the $\cos(\omega_j t + \phi_j)$ terms. The various ω_j are assumed to be the same as in the uncoupled system. To proceed further, one must decide what kind of representation to use for x_j. From a linear analysis point of view it would seem natural to use a complex representation, $x_j = \text{Re}\{A_j \exp(\text{i}\omega_j t)\}$, as we have done in the past. However, in nonlinear analyses, where one must deal with the product of two functions, the real part of the product of two complex functions is not the product of the real parts, i.e., $\text{Re}\{A_j \exp(\text{i}\omega_j t) B_k \exp(\text{i}\omega_k t)\} \neq \text{Re}\{A_j \exp(\text{i}\omega_j t)\}\text{Re}\{B_k \exp(\text{i}\omega_k t)\}$. The same is true for Fourier transforms. So, one must be careful with complex representations when the product of two functions is involved. Correct results can be obtained by adding the complex conjugate in such a way that the resulting function is always real. However, such calculations quickly become very complicated when products are involved. Similarly, one could use the cosine/sine representation $x_j = A_j \cos\omega_j t + B_j \sin\omega_j t$, but again the calculations become similarly complicated. Here, we find it convenient to simply substitute various combinations of $\cos\omega_j t$ and $\sin\omega_j t$ into the above equations and construct a table of the allowed solutions and their associated parameters.

To begin the detailed analysis of Eqs. (11.2.7)–(11.2.9), we start with the combination

$$x_0 = A_0(t)\sin\omega_0 t, \tag{11.2.11}$$

$$x_1 = A_1(t)\cos\omega_1 t, \tag{11.2.12}$$

and

$$x_2 = A_2(t)\cos\omega_2 t. \tag{11.2.13}$$

With this combination we first need to compute $\dot{x}_0$ and $\ddot{x}_0$, which are respectively

$$\dot{x}_0 = A_0\omega_0\cos\omega_0 t + \dot{A}_0\sin\omega_0 t \tag{11.2.14}$$

and

$$\ddot{x}_0 = -A_0\omega_0^2\sin\omega_0 t + \dot{A}_0\omega_0\cos\omega_0 t + \dot{A}_0\omega_0\cos\omega_0 t + \ddot{A}_0\sin\omega_0 t, \tag{11.2.15}$$

where we have omitted the time-dependent (t) notation on the amplitude terms. Substituting the above equations into the left-hand side of Eq. (11.2.7), we note that terms involving ω_0^2 cancel, so that

$$\ddot{x}_0 + \omega_0^2 f_0 = 2\dot{A}_0\omega_0\cos\omega_0 t + \ddot{A}_0\sin\omega_0 t. \tag{11.2.16}$$

At this point we assume that the time-scale for the A_0 amplitude variation is sufficiently slow that we can ignore the second term on the right in the above equation, i.e., $\ddot{A}_0 \ll 2\dot{A}_0\omega_0$. Equation (11.2.7) then becomes

$$2\dot{A}_0\omega_0\cos\omega_0 t = -\lambda A_1 A_2\cos\omega_1 t\cos\omega_2 t. \tag{11.2.17}$$

Using the double angle trigonometric identity in Eq. (11.2.1), the product of the two cosine terms on the right can be rewritten to give the equation

$$2\dot{A}_0\omega_0\cos\omega_0 t = -\frac{\lambda}{2}A_1A_2[\cos(\omega_1+\omega_2)t+\cos(\omega_1-\omega_2)t]. \tag{11.2.18}$$

If we now think of averaging the above equation over a long time-scale, one can see that to get a consistent relationship between the amplitudes we must require that ω_0 be either $\omega_1+\omega_2$ or $\omega_1-\omega_2$, i.e., a resonant interaction. If we choose $\omega_0=\omega_1+\omega_2$ and multiply the above equation by $\cos(\omega_0+\omega_1)t$ and average, $\langle\rangle$, on both sides of the equation, the result is

$$2\dot{A}_0\omega_0\langle\cos(\omega_0 t)\cos(\omega_1+\omega_2)t\rangle = -(\lambda/2)A_1A_2\left[\langle\cos^2(\omega_1+\omega_2)t\rangle\right], \tag{11.2.19}$$

where the other resonant term, $\langle\cos(\omega_0+\omega_1)t\cos(\omega_0-\omega_1)t\rangle$, averages to zero because the two frequencies are incommensurate. Note from the choice $\omega_0=\omega_1+\omega_2$ that the term on the left-hand side of the above equation is given by

$$\langle\cos(\omega_0 t)\cos(\omega_1+\omega_2)t\rangle = \langle\cos^2(\omega_1+\omega_2)t\rangle, \tag{11.2.20}$$

which cancels on both sides of the equation. It follows then that for a long-term resonant interaction the rate of change of the amplitude of the 0th oscillator must be given by

$$\dot{A}_0 = -\frac{\lambda}{4\omega_0}A_1A_2. \tag{11.2.21}$$

By trying various combinations of $A_j\cos\omega_j t$ and $A_j\sin\omega_j t$ for x_0, x_1, and x_2, and using the double-angle formulas given in Appendix B, a tabulation of allowed solutions to Eqs. (11.2.7)–(11.2.9) can be constructed and is given in Table 11.1.

Note in constructing this table that only three combinations of sine and cosine functions give allowable solutions. The allowed solutions are listed in the columns of Table 11.1 as Case 1, Case 2, and Case 3. Upon comparing these, one can see that in each case one of the solutions for $\dot{A}$ is negative, whereas the other two are positive. For instance, in Case 1 the allowed solutions are

$$\dot{A}_0 = -\frac{\lambda}{4\omega_0}A_1A_2,\quad \dot{A}_1 = \frac{\lambda}{4\omega_1}A_0A_2,\quad \dot{A}_2 = \frac{\lambda}{4\omega_2}A_0A_1. \tag{11.2.22}$$

These solutions can be put in a somewhat more symmetric form by defining new frequency-normalized amplitudes as $a_j = A_j\omega_j^{1/2}$. Note the close similarity in this normalization to the first adiabatic invariant $A\omega^{1/2}$; see Section (3.8.1). With this newly defined amplitude, the above equation can be written in the following

Table 11.1 *Solutions to the Coupled Harmonic Oscillator Equations*

	Case 1	Case 2	Case 3
$x_0 =$	$A_0 \sin \omega_0 t$	$A_0 \cos \omega_0 t$	$A_0 \cos \omega_0 t$
$x_1 =$	$A_1 \cos \omega_1 t$	$A_1 \sin \omega_1 t$	$A_1 \cos \omega_1 t$
$x_2 =$	$A_2 \cos \omega_2 t$	$A_2 \cos \omega_2 t$	$A_2 \sin \omega_2 t$
$\dot{A}_0 =$	$-\frac{\lambda}{4}\frac{A_1 A_2}{\omega_0}$	$\frac{\lambda}{4}\frac{A_1 A_2}{\omega_0}$	$\frac{\lambda}{4}\frac{A_1 A_2}{\omega_0}$
$\dot{A}_1 =$	$\frac{\lambda}{4}\frac{A_0 A_2}{\omega_1}$	$-\frac{\lambda}{4}\frac{A_0 A_2}{\omega_1}$	$\frac{\lambda}{4}\frac{A_0 A_2}{\omega_1}$
$\dot{A}_2 =$	$\frac{\lambda}{4}\frac{A_0 A_1}{\omega_2}$	$\frac{\lambda}{4}\frac{A_0 A_1}{\omega_2}$	$-\frac{\lambda}{4}\frac{A_0 A_1}{\omega_2}$

symmetrical form:

$$\dot{a}_j = \pm \frac{\lambda}{4} \frac{a_k a_\ell}{(\omega_j \omega_k \omega_\ell)^{1/2}}, \tag{11.2.23}$$

where the ± sign must be determined from Table 11.1.

By multiplying the above equation by a_j with the appropriate signs from the solutions in Table 11.1 and integrating, it follows that

$$-\frac{\mathrm{d}a_j^2}{\mathrm{d}t} = \frac{\mathrm{d}a_k^2}{\mathrm{d}t} = \frac{\mathrm{d}a_\ell^2}{\mathrm{d}t}. \tag{11.2.24}$$

This relationship simply shows that if one of the oscillators is gaining (or losing) energy, which is proportional to a_j^2, then this gain (or loss) must be compensated by a change in energy of the other two oscillators.

From the basic relations derived above it is possible to prove that the total energy of the system is conserved. To do this we start by assuming that the total energy is conserved. Since the total energy of a harmonic oscillator is proportional to $\omega^2 A^2$ (see Section 3.8.1), conservation of energy requires that

$$\frac{\mathrm{d}}{\mathrm{d}t}(\omega_0^2 A_0^2 + \omega_1^2 A_1^2 + \omega_2^2 A_2^2) = 0. \tag{11.2.25}$$

Using $a_j = A_j \omega_j^{1/2}$, this equation can be converted to

$$\frac{\mathrm{d}}{\mathrm{d}t}(\omega_0 a_0^2 + \omega_1 a_1^2 + \omega_2 a_2^2) = 0, \tag{11.2.26}$$

which, after carrying out the differentiation, becomes

$$2[\omega_0 a_0 \dot{a}_0 + \omega_1 a_1 \dot{a}_1 + \omega_2 a_2 \dot{a}_2] = 0. \tag{11.2.27}$$

Then, by using Eq. (11.2.23) with the appropriate signs from the solutions in Table 11.1, it can be shown that the sum in the bracket simplifies to

$$2\frac{\lambda}{4}\frac{a_0 a_1 a_2}{(\omega_0\omega_1\omega_2)^{1/2}}[-\omega_0+\omega_1+\omega_2]=0, \tag{11.2.28}$$

which is identically zero if $\omega_0=\omega_1+\omega_2$. Thus, the original assumption is true and the total energy is indeed conserved.

Since the above derivations could equally well have been done by making the choice $\omega_0=\omega_1-\omega_2$, it is clear that there are two resonance conditions $\omega_0=\omega_1-\omega_2$ and $\omega_0=\omega_1+\omega_2$, both of which are consistent with conservation of energy. Although the above results are for three time-dependent harmonic oscillators, by changing ω to k we can think of the temporal equations (11.2.7)–(11.2.9) as also applying to the spatial dependence of the three normal wave modes, where x_0, x_1, and x_2 are the field components. After changing d/dt to d/dx, then to d/dy, and finally to d/dz, the above derivation can be repeated for each component, thereby showing that the three wave vectors must satisfy the matching condition $\mathbf{k}_0=\mathbf{k}_1\pm\mathbf{k}_2$. Thus, by analyzing a specific physical system we have re-derived the matching conditions discussed in the introduction.

11.2.3 Three-Wave Coupling in a Hot Unmagnetized Plasma

Next, we show how the analysis techniques described above can be applied to the nonlinear interaction of waves in a hot unmagnetized plasma. We will assume that the plasma consists of electrons and one species of positive ions. In such a plasma there are only three linear small-amplitude modes; see Chapter 5. These are the transverse electromagnetic mode (T), the Langmuir mode (L), and the ion acoustic mode (A). Following our previous discussion of coherent three-wave coupling, the three modes must satisfy the matching conditions $\omega_T=\omega_L\pm\omega_A$ and $\mathbf{k}_T=\mathbf{k}_L\pm\mathbf{k}_A$. Here, we choose to use the + sign and assume that all of the wave vectors are parallel. This assumption is not particularly restrictive. If there is a transverse component to the wave vector, all we need to do is move into a frame of reference where the two transverse components add to zero. The wave vector matching condition then reduces to the sum of the components, i.e., $k_T=k_L+k_A$. We also assume that the wavelengths are sufficiently long, $k\lambda_D\ll 1$, that there is no Landau damping of the Langmuir wave; and that the electron-to-ion temperature ratio is sufficiently large, $T_e/T_i\gg 1$, that there is also no Landau damping of the ion acoustic wave. For these conditions the dispersion relations of the three modes are given by

$$\omega_{\rm T}^2 = \omega_{\rm pe}^2 + c^2 k_{\rm T}^2 \qquad \text{Transverse EM mode (T),} \tag{11.2.29}$$

$$\omega_{\rm L}^2 = \omega_{\rm pe}^2 + \frac{3\kappa T_{\rm e}}{m_{\rm e}} k_{\rm L}^2 \qquad \text{Langmuir mode (L),} \tag{11.2.30}$$

and

$$\omega_{\rm A}^2 = C_{\rm A}^2 k_{\rm A}^2 \qquad \text{Ion acoustic mode (A),} \tag{11.2.31}$$

where $C_{\rm A} = (3\kappa T_{\rm e}/m_{\rm i})^{1/2}$ is the ion acoustic speed, and we have assumed that the electrons have only one degree of freedom, so that $\gamma_{\rm e} = 3$; see Eqs. (5.5.20) and (5.5.25). From the first two equations it is evident that the frequencies of both the transverse EM mode and the Langmuir mode are always greater than the electron plasma frequency. In contrast, the frequency of the ion acoustic mode is at a much lower frequency, well below the ion plasma frequency; see Figure 5.6. Because of the large separation between these frequencies, the frequency difference between the transverse EM mode and the Langmuir mode is very small, $(\omega_{\rm T} - \omega_{\rm L})/\omega_{\rm pe} < (m_{\rm e}/m_{\rm i})^{1/2}$, typically a few percent or less. Also, because of the long-wavelength assumption, $k\lambda_{\rm D} \ll 1$, the frequencies of both of these modes must be very close to the electron plasma frequency. Later, we will take advantage of the large frequency separation between the ion acoustic mode and the two high frequency modes. Specifically, we will use a two-time-scale approximation in which the ion acoustic wave is treated as a long time-scale/low-frequency oscillation, and the transverse EM and Langmuir waves are treated as short time-scale/high-frequency oscillations.

The frequency and wave number matching conditions for the above system of equations can be conveniently analyzed by making the $\omega - k$ plot shown in Figure 11.10. Note that in order to make a visually useful illustration the frequencies shown are not realistic. (For a realistic representation, the ion acoustic mode would have to be very near the bottom of the diagram). With a little study one can see that the frequency and wave vector matching conditions are satisfied by constructing a parallelogram outward from the origin with the remaining three vertices each located on one of the three modes. Also, note that because the speed of light is normally much greater than the electron thermal speed, for this three-wave interaction the ion acoustic wave must propagate in a direction opposite to that of the Langmuir wave.

Next we discuss the nonlinear terms that are involved in the coupling of the three modes. Three wave equations must be considered: one for the transverse EM mode, one for the Langmuir mode, and one for the ion acoustic mode. Because the transverse EM and Langmuir modes are at high frequencies near the

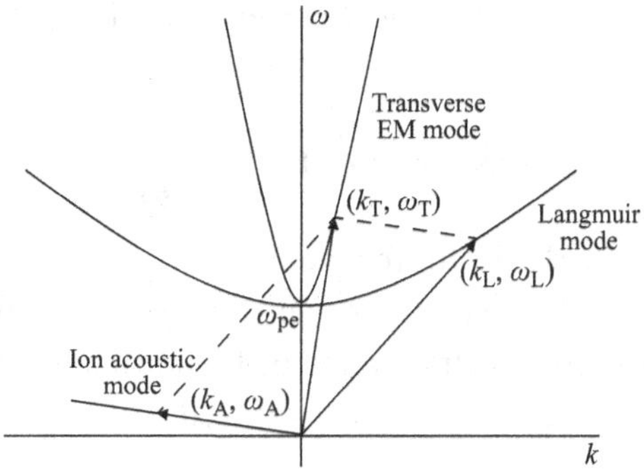

Figure 11.10 A $\omega - k$ diagram of the three plasma wave modes that can propagate in a hot unmagnetized plasma with one species. The parallelogram shows the geomagnetic construction needed to satisfy the matching conditions $\omega_T = \omega_L + \omega_A$ and $\mathbf{k}_T = \mathbf{k}_L + \mathbf{k}_A$.

electron plasma frequency, whereas the ion acoustic mode is at a much lower frequency, these two limiting cases will be discussed separately. For the two high-frequency modes only the electron motions need to be considered. To see where nonlinear terms arise, it is useful to refer back to Section 4.3, where we first introduced the process of linearization. In the example discussed there, the current, $\mathbf{J} = \sum e_s n_s \mathbf{U}_s$, was expanded into a first-order linear term $\sum e_s n_{s0} \mathbf{U}_{s1}$ and a second-order nonlinear term $\sum e_s n_{n1} \mathbf{U}_{s1}$. In the ensuing linearized analysis the linear term was incorporated into the homogeneous equation for the electric field, and the nonlinear term was discarded. However, now we must keep the nonlinear term, and this term now becomes a source for what was previously the homogeneous equation for the electric field.

In a more complicated system of equations there may be several nonlinear terms. Usually they consist of the product of two small first-order terms, as in the above example where the product $n_{s1}\mathbf{U}_{s1}$ becomes the nonlinear source term. If there is more than one nonlinear term, the usual procedure is to determine which term is the largest and to keep only that term. Sometimes this can be done by analysis. For example, in the long-wavelength limit, the term $(\mathbf{U}_s \cdot \nabla)\mathbf{U}_s$ is smaller than $(n_s\mathbf{U}_s)$. This is because ∇ varies inversely with the wavelength whereas $(n_s\mathbf{U}_s)$ does not. Sometimes one cannot easily determine which nonlinear term is dominant. In such cases, one must just guess and compare the result for each possibility. Once the dominant term is identified, the linear terms are usually organized on the left-hand side and the dominant nonlinear term is placed on the right-hand side of an otherwise linear system of equations.

To apply the above procedure we refer to Eq. (4.3.21), which was derived to give the electric field produced by an external current source. Simple inspection shows that this equation can be modified to accommodate a nonlinear current source by simply replacing $\tilde{\mathbf{J}}_{\text{ext}}$ with the nonlinear current $\tilde{\mathbf{J}}_{\text{NL}}$. The resulting wave equation then reads

$$c^2\mathbf{k}\times(\mathbf{k}\times\tilde{\mathbf{E}}) + \omega^2\overleftrightarrow{\mathbf{K}}\cdot\tilde{\mathbf{E}} = \frac{-\mathrm{i}\omega}{\epsilon_0}\tilde{\mathbf{J}}_{\text{NL}}. \tag{11.2.32}$$

This equation is particularly useful for the high-frequency modes because at high frequencies it is only necessary to consider the electron motions. To obtain the corresponding equations for the transverse EM mode and the Langmuir mode, one can then substitute the dielectric tensors derived in Chapters 4 and 5 for these modes,

$$\overleftrightarrow{\mathbf{K}} = \overleftrightarrow{\mathbf{1}}\left[1 - \frac{\omega_{\text{pe}}^2}{\omega^2}\right] \quad \text{and} \quad \overleftrightarrow{\mathbf{K}} = \overleftrightarrow{\mathbf{1}}\left[1 - \frac{\omega_{\text{pe}}^2}{\omega^2}(1 + 3\lambda_{\text{De}}^2 k^2)\right], \tag{11.2.33}$$

into Eq. (11.2.32). These substitutions then give two wave equations, one for the transverse EM mode

$$\left(\frac{\partial^2}{\partial t^2} - c^2\nabla^2 + \omega_{\text{pe}}^2\right)\mathbf{E}_{\text{T}} = -\frac{(-e)}{\epsilon_0}\frac{\partial}{\partial t}(n_{\text{e}}\mathbf{U}_{\text{e}}), \tag{11.2.34}$$

and the other for the Langmuir mode

$$\left(\frac{\partial^2}{\partial t^2} - \frac{3\kappa \mathrm{T}_{\text{e}}}{m_{\text{e}}}\nabla^2 + \omega_{\text{pe}}^2\right)\mathbf{E}_{\text{L}} = -\frac{(-e)}{\epsilon_0}\frac{\partial}{\partial t}(n_{\text{e}}\mathbf{U}_{\text{e}}), \tag{11.2.35}$$

where we have changed $\omega_{\text{pe}}^2\lambda_{\text{De}}^2$ to $\kappa T_{\text{e}}/m_{\text{e}}$ in the second equation. Note that the nonlinear source current $(-e)(n_{\text{e}}\mathbf{U}_{\text{e}})$ is now on the right-hand side of both equations. In a more detailed derivation of the above two equations, there are other nonlinear terms that must also be considered. However, in the long-wavelength limit, $k\lambda_{\text{D}} \ll 1$, it can be shown that the product $n_{\text{e}}\mathbf{U}_{\text{e}}$ is the dominant nonlinear term. To avoid using Fourier transforms we have converted $\mathrm{i}\mathbf{k}$ to ∇ and $-\mathrm{i}\omega$ to $\partial/\partial t$ using the usual operator substitutions.

To proceed further with the analysis of the above equations, we must next relate the nonlinear source terms $(\partial/\partial t)(n_{\text{e}}\mathbf{U}_{\text{e}})$ in Eqs. (11.2.34) and (11.2.35) to the wave electric fields. Since there are two terms that arise from the time derivative of $n_{\text{e}}\mathbf{U}_{\text{e}}$, it is worth examining each of these two terms, which are given by

$$\frac{\partial}{\partial t}(n_{\text{e}}\mathbf{U}_{\text{e}}) = \frac{\partial n_{\text{e}}}{\partial t}\mathbf{U}_{\text{e}} + n_{\text{e}}\frac{\partial \mathbf{U}_{\text{e}}}{\partial t}. \tag{11.2.36}$$

The first $(\partial n_{\text{e}}/\partial t)\mathbf{U}_{\text{e}}$ term on the right-hand side of the above equation can be ignored because the electron density contribution to the three-wave interaction must be that of the ion acoustic wave, i.e., $n_{\text{e}} = n_{\text{A}}$, which is very slowly varying. Therefore, this term is much smaller than the second term, $n_{\text{A}}(\partial \mathbf{U}_{\text{e}}/\partial t)$, which varies rapidly due to the high-frequency electron oscillations. Note that in this term we must use the density n_A of the ion acoustic wave. To evaluate the electron acceleration term, $\partial \mathbf{U}_e/\partial t$, we use the electron momentum equation (5.5.7), which gives

$$\frac{\partial \mathbf{U}_e}{\partial t} = \frac{(-e)}{m_e}\mathbf{E}, \tag{11.2.37}$$

where we have ignored the ∇P_e and $(\mathbf{U}_e \cdot \nabla)\mathbf{U}_e$ terms because $k\lambda_D \ll 1$. Substituting these results into Eqs. (11.2.34) and (11.2.35) then gives the equations

$$\left(\frac{\partial^2}{\partial t^2} - c^2\nabla^2 + \omega_{pe}^2\right)\mathbf{E}_T = -\omega_{pe}^2\frac{n_A}{n_0}\mathbf{E}_L \tag{11.2.38}$$

and

$$\left(\frac{\partial^2}{\partial t^2} - \frac{3\kappa T_e}{m_e}\nabla^2 + \omega_{pe}^2\right)\mathbf{E}_L = -\omega_{pe}^2\frac{n_A}{n_0}\mathbf{E}_T \tag{11.2.39}$$

for the transverse EM mode and the Langmuir mode, respectively, where ω_{pe}^2 has been substituted for $n_0e^2/\epsilon_0 m_e$, and n_0 is the zero-order background density. In the equation for the transverse EM mode, note that for the electric field $\mathbf{E}$ on the right-hand side we have inserted the electric field of the Langmuir wave, $\mathbf{E}_L$, with no contribution from the other two modes. That we did not include the contribution to the electric field from the ion acoustic mode, $\mathbf{E}_A$, is because the frequency of this mode is much too low. The frequency of this mode is much too low to drive either of the two high-frequency modes, i.e., $\omega_A \ll (\omega_T, \omega_L)$. Similarly, just as we did for the x_0x_1 and x_0x_2 terms in the pendulum analysis, we did not include the electric field of the transverse EM wave, $\mathbf{E}_T$, because in a three-wave interaction this electric field does not couple to itself, but rather to the other two modes. So, it is only the electric field of the Langmuir wave $\mathbf{E}_L$ that is relevant on the right-hand side of Eq. (11.2.38). The same reasoning applies to the right-hand side of Eq. (11.2.39), though for $\mathbf{E}_T$ instead of $\mathbf{E}_L$. As noted before, the density n_A of the ion acoustic wave has been substituted for n_e on the right-hand side. This is the only contribution to the electron density that is relevant in a three-wave interaction. The result is that the density of the ion acoustic wave acts to "modulate" the amplitude of the high-frequency electric field oscillations, which is the motivation for the waveform modulation example given by Eq. (11.2.2) at the beginning of this section.

Next, we turn our attention to the ion acoustic mode. Since the ion acoustic mode involves both electron and ion motions, we must use both the electron and ion momentum equations to analyze how the transverse EM and Langmuir wave interacts with the ion acoustic wave. We start with the electron momentum equation, which from Eq. (5.5.7) is given by

$$n_e m_e \left[\frac{\partial \mathbf{U}_e}{\partial t} + (\mathbf{U}_e \cdot \nabla)\mathbf{U}_e \right] = (-e)n_e \mathbf{E} - 3\kappa T_e \nabla n_e, \tag{11.2.40}$$

where we have substituted $3\kappa T_e \nabla n_e$ for the electron pressure. Note that the term $(\mathbf{U}_e \cdot \nabla)\mathbf{U}_e$ in the brackets is a nonlinear second-order term. Although we ignored this term in the previous analysis of the two high-frequency modes, for the ion acoustic mode this term provides the mechanism for coupling the high-frequency transverse EM mode and the Langmuir mode to the low-frequency oscillations of the ion acoustic mode, and cannot be ignored. To see how this coupling occurs we use the vector identity for $\nabla(\mathbf{F} \cdot \mathbf{G})$ given in Appendix D to show that

$$(\mathbf{U}_e \cdot \nabla)\mathbf{U}_e = (1/2)\nabla(\mathbf{U}_e^2) - \mathbf{U}_e \times (\nabla \times \mathbf{U}_e). \tag{11.2.41}$$

For the interaction between the Langmuir wave and the transverse EM wave, the second term on the right-hand side of the above equation can be omitted. The reasons are as follows. Since the Langmuir wave is electrostatic, the electron velocity is proportional to the gradient of the electrostatic wave potential, $\mathbf{U}_e \sim \nabla\Phi$, so that the identity $\nabla \times \nabla\Phi = 0$ gives $\nabla \times \mathbf{U}_e = 0$. For the transverse electromagnetic wave this second term is non-zero, but is perpendicular to the velocity of the Langmuir wave oscillations by virtue of the cross-product in $\mathbf{U}_e \times (\nabla \times \mathbf{U}_e)$. Thus, the $\mathbf{U}_e \times (\nabla \times \mathbf{U}_e)$ term in Eq. (11.2.41) makes no contribution to the $\mathbf{U}_{eT} \cdot \mathbf{U}_{eL}$ coupling discussed below.

Using $(\mathbf{U}_e \cdot \nabla)\mathbf{U}_e = (1/2)\nabla(\mathbf{U}_e^2)$, the electron momentum equation (11.2.40) can then be written

$$n_e m_e \frac{\partial \mathbf{U}_e}{\partial t} = (-e)n_e \mathbf{E} - 3\kappa T_e \nabla n_e - n_e m_e \frac{1}{2}\nabla(\mathbf{U}_e^2), \tag{11.2.42}$$

where we have moved the $(\mathbf{U}_e \cdot \nabla)\mathbf{U}_e$ term to the right-hand side of the equation, recognizing that this term now becomes the nonlinear source term. To show how this term drives the ion acoustic mode, we note that the relevant electron velocity is the sum of the velocities of the two high-frequency modes, $\mathbf{U}_e = \mathbf{U}_{eT} + \mathbf{U}_{eL}$. Substituting this expression for $\mathbf{U}_e$ into $(1/2)\nabla(\mathbf{U}_e^2)$ then gives

$$\frac{1}{2}\nabla(\mathrm{U}_e^2) = \frac{1}{2}\nabla(\mathbf{U}_{eT} + \mathbf{U}_{eL})^2, \tag{11.2.43}$$

which can then be expanded into three terms:

$$\frac{1}{2}\nabla(\mathbf{U}_e^2) = \frac{1}{2}\nabla(\mathbf{U}_{eT}^2 + 2\mathbf{U}_{eT} \cdot \mathbf{U}_{eL} + \mathbf{U}_{eL}^2). \tag{11.2.44}$$

Just as in the pendulum analysis, of these three terms only the cross-product term, $\mathbf{U}_{eT} \cdot \mathbf{U}_{eL}$, contributes to the nonlinear coupling. To evaluate this coupling we must relate the electron velocities in the cross-product term to the electric fields of the transverse EM mode and the Langmuir mode. By integrating Eq. (11.2.37) it is easy to see that the electron velocities produced by the transverse EM and Langmuir electric fields are given by

$$\mathbf{U}_{\mathrm{eT}} = \frac{(-e)}{m_{\mathrm{e}}\omega_{\mathrm{T}}}\mathbf{E}_{\mathrm{T}} \qquad \text{and} \qquad \mathbf{U}_{\mathrm{eL}} = \frac{(-e)}{m_{\mathrm{e}}\omega_{\mathrm{L}}}\mathbf{E}_{\mathrm{L}}, \tag{11.2.45}$$

where we have ignored the $\pi/2$ phase shift that is common to both terms. This phase shift is negligible on the long time-scale of the ion acoustic oscillation. Substituting these velocities into the electron momentum equation (11.2.42) then gives the following modified electron momentum equation:

$$n_0 m_{\mathrm{e}}\frac{\partial \mathbf{U}_{\mathrm{e}}}{\partial t} = (-e)n_0\mathbf{E} - 3\kappa T_{\mathrm{e}}\nabla n_{\mathrm{A}} - \frac{n_0 e^2}{m_{\mathrm{e}}\omega_{\mathrm{T}}\omega_{\mathrm{L}}}\nabla(\mathbf{E}_{\mathrm{T}}\cdot\mathbf{E}_{\mathrm{L}}), \tag{11.2.46}$$

where we have changed n_{e} to the average background density n_0. This substitution is made because the density of the ion acoustic wave, n_{A}, is regarded as a small first-order perturbation superposed on the background plasma density. Comparing the nonlinear force term on the right-hand side with Eq. (3.7.7), one can see that this term has the same basic form as the ponderomotive force, with the product $(\mathbf{E}_{\mathrm{T}}\cdot\mathbf{E}_{\mathrm{L}})$ playing a role similar to E_0^2, and with $\omega_{\mathrm{T}}\omega_{\mathrm{L}}$ playing a role similar to ω^2.

To show how the ponderomotive force is coupled to the ion acoustic mode, we next derive the wave equation for the ion acoustic mode. Since the frequency of the ion acoustic wave is very low and the electron mass is very small, the electron acceleration term on the left-hand side of the electron momentum equation (11.2.46) can be ignored. The low-frequency slowly varying electric field force that acts on the ion acoustic mode is then given by

$$(-e)\mathbf{E}_{\mathrm{A}} = \frac{3\kappa T_{\mathrm{e}}}{n_0}\nabla n_{\mathrm{A}} + \frac{e^2}{m_{\mathrm{e}}\omega_{\mathrm{T}}\omega_{\mathrm{L}}}\nabla(\mathbf{E}_{\mathrm{T}}\cdot\mathbf{E}_{\mathrm{L}}), \tag{11.2.47}$$

where the ponderomotive force explicitly appears as an additive term on the right-hand side of the equation.

We next turn our attention to the ion motions. For the ion momentum equation, several useful simplifications can be made. Because of the previously mentioned assumption that $T_{\mathrm{i}} \ll T_{\mathrm{e}}$, the ion pressure gradient term can be ignored. Also, since the ponderomotive force acting on the ions is inversely proportional to the ion mass, this force is very small and can be ignored. The ion momentum equation then simplifies to

$$\frac{\partial \mathbf{U}_{\mathrm{i}}}{\partial t} = \frac{e}{m_{\mathrm{i}}}\mathbf{E}_{\mathrm{A}}. \tag{11.2.48}$$

Using Eq. (11.2.47) to eliminate the $e\mathbf{E}_A$ term from the above equation, we then obtain an equation for the rate of change of the ion velocity:

$$\frac{\partial \mathbf{U}_i}{\partial t} = -\frac{3\kappa T_e}{m_i n_0}\nabla n_A - \frac{e^2}{m_e m_i \omega_T \omega_L}\nabla(\mathbf{E}_T \cdot \mathbf{E}_L). \tag{11.2.49}$$

To derive the wave equation for the ion acoustic mode, we next need to use the ion continuity Eq. (5.5.6). Although this equation introduces a nonlinear second-order term of the form $(\mathbf{U}_i \cdot \nabla)n_i$, if the ion density gradient is small, which it is when $k\lambda_D \ll 1$, this term can be ignored. The linearized version of the ion continuity equation then becomes

$$\frac{\partial n_A}{\partial t} + n_0 \nabla \cdot \mathbf{U}_i = 0. \tag{11.2.50}$$

Next, take the derivative of this equation with respect to time,

$$\frac{\partial^2 n_A}{\partial t^2} + n_0 \nabla \cdot \frac{\partial \mathbf{U}_i}{\partial t} = 0, \tag{11.2.51}$$

and use Eq. (11.2.49) to eliminate the $\partial \mathbf{U}_i/\partial t$ term, which gives the equation

$$\frac{\partial^2 n_A}{\partial t^2} - n_0 \nabla \cdot \left(\frac{3\kappa T_e}{m_i n_0}\nabla n_A + \frac{e^2}{m_e m_i \omega_T \omega_L}\nabla(\mathbf{E}_T \cdot \mathbf{E}_L) \right) = 0. \tag{11.2.52}$$

Noting that $3\kappa T_e/m_i = C_A^2$ is the ion acoustic speed, the above equation can be reorganized in the form of a wave equation with the ponderomotive force acting as the source term on the right:

$$\left(\frac{\partial^2}{\partial t^2} - C_A^2 \nabla^2 \right) n_A = \frac{n_0 e^2}{m_e m_i \omega_T \omega_L}\nabla^2(\mathbf{E}_T \cdot \mathbf{E}_L). \tag{11.2.53}$$

The above equation, together with Eqs. (11.2.38) and (11.2.39), constitutes a system of three coupled equations that can be solved for $\mathbf{E}_T$, $\mathbf{E}_L$, and n_A. The coupling between the high-frequency electric fields and the ion acoustic mode can be readily understood in terms of the waveforms plotted in Figure 11.8. The waveforms F_1 and F_2 correspond to the electric fields $\mathbf{E}_T$ and $\mathbf{E}_L$, and the ponderomotive force corresponds to the product $F_1F_2(-)$ in panel three. Resonant coupling between the two high-frequency modes occurs when the phase velocity of the ponderomotive force matches the phase velocity of the ion acoustic wave, i.e., $\mathrm{d}x/\mathrm{d}t = (\omega_T - \omega_L)/(k_T - k_L) = \omega_A/k_A$. This resonance condition only occurs when the frequency and wave number matching conditions are both satisfied, i.e., $\omega_A = \omega_T - \omega_L$ and $k_A = k_T - k_L$. Inspection of Eqs. (11.2.38), (11.2.39), and (11.2.53) shows that the three coupled equations for the waves in a hot non-magnetized plasma have the same basic form as the equations for three coupled harmonic oscillators given by Eqs. (11.2.7)–(11.2.9). In fact, if $\mathbf{E}_\mathrm{T}$ and $\mathbf{E}_\mathrm{L}$ are assumed to be parallel, and $-k_\mathrm{A}^2$ is substituted for ∇^2, scale factors can be found that convert these equations exactly to the equations for three weakly coupled pendulums; see Problem 11.5. Solutions to these equations can be readily obtained using numerical methods and typically consist of complex time-dependent transfers of energy between the three modes.

11.2.4 Driven Waves and Parametric Decay

The above discussion of the nonlinear coupling of the transverse EM mode, the Langmuir mode, and the ion acoustic mode assumes that the solutions are determined via some initially imposed set of boundary conditions. Although mathematically correct, this is not a very realistic model of an actual application. More often, one of the modes is driven by some external process, with the other two modes being produced via nonlinear coupling. This process is often called "parametric decay," because, as we will show, the driven wave produces a parametric instability that causes the driven wave to "decay" into the other two modes. In such cases the driving wave at frequency ω_0 is sometimes called the "pump" wave, and the newly generated waves at frequencies ω_1 and ω_2, where $\omega_0 = \omega_1 + \omega_2$, are called "daughter" waves. Usually the electric field of the pump wave is much stronger than the daughter waves. Although one could analyze such interactions on a case-by-case basis, the previous discussion shows that three-wave interaction in a hot unmagnetized plasma is formally identical to the interactions between three coupled harmonic oscillators. Therefore, we can use the coupled harmonic oscillator equations as a model for understanding what happens when one of the modes is strongly driven. We also include damping, since waves in plasmas always experience some damping, either due to collisional effects or Landau damping. As we will see, wave damping has a strong influence on the conditions under which daughter waves are produced.

For a model of such a driven system, we start with a modified version of coupled oscillator equations (11.2.7)–(11.2.9) in which one of the waves, the pump, is driven at a constant amplitude, $x_0 = A_0 \sin \omega_0 t$, while the other two waves are represented by $x_1 = A_1 e^{\gamma t} \cos \omega_1 t$, $x_2 = A_2 e^{\gamma t} \cos \omega_2 t$, where γ is an exponential growth rate that is included to allow for possible instabilities. Because the amplitude A_0 of the pump is assumed to be constant, the coupled oscillator equations simplify to two linear coupled equations:

$$\ddot{x}_1 + \nu_1 \dot{x}_1 + \omega_1^2 x_1 = -\lambda(A_0 \sin \omega_0 t) x_2, \tag{11.2.54}$$

$$\ddot{x}_1 + \nu_2 \dot{x}_2 + \omega_2^2 x_2 = -\lambda(A_0 \sin \omega_0 t) x_1, \tag{11.2.55}$$

where λ is the nonlinear coupling coefficient and damping coefficients ν_1 and ν_2 have been introduced to represent the damping. In an actual application, the damping coefficients would have to be suitably linked to the collisional or Landau damping characteristics of the plasma. To see what conditions these equations place on the growth rate, we proceed as we did in the analysis of the coupled harmonic oscillators by computing $\dot{x}_1$ and $\ddot{x}_2$, which are

$$\dot{x}_1 = A_1 e^{\gamma t}(-\omega_1)\sin\omega_1 t + \gamma A_1 e^{\gamma t}\cos\omega_1 t \tag{11.2.56}$$

and

$$\begin{aligned}\ddot{x}_1 = A_1 e^{\gamma t}(-\omega_1^2)\cos\omega_1 t + \gamma A_1(-\omega_1)e^{\gamma t}\sin\omega_1 t \\ + \gamma A_1(-\omega_1)e^{\gamma t}\sin\omega_1 + \gamma^2 A_1 e^{\gamma t}\cos\omega_1 t.\end{aligned} \tag{11.2.57}$$

Assuming that the growth rate is small, $\gamma \ll \omega_1$, we omit the last term in $\ddot{x}$, which after substituting into Eq. (11.2.54) and canceling the common factor of $e^{\gamma t}$ gives

$$\gamma\nu_1 A_1\cos\omega_1 t - (2\gamma + \nu_1)\omega_1 A_1\sin\omega_1 t = -\lambda A_0 A_2\sin\omega_0 t\cos\omega_2 t. \tag{11.2.58}$$

Similarly, computing $\dot{x}_2$ and $\ddot{x}_2$, and substituting these into Eq. (11.2.55) gives

$$\gamma\nu_2 A_2\cos\omega_2 t - (2\gamma + \nu_2)\omega_2 A_2\sin\omega_2 t = -\lambda A_0 A_2\sin\omega_0 t\cos\omega_1 t. \tag{11.2.59}$$

Using the double-angle formula $\sin A\cos B = (1/2)[\sin(A + B) + \sin(A - B)]$, multiplying the first equation (11.2.58) by $\sin\omega_1 t$ and the second equation (11.2.59) by $\sin\omega_2 t$, averaging both equations over time, and invoking the frequency matching condition $\omega_0 = \omega_1 + \omega_2$ yields two equations that must be satisfied simultaneously:

$$(2\gamma + \nu_1)\omega_1 A_1 = \frac{\lambda}{2}A_0 A_2 \quad\text{and}\quad (2\gamma + \nu_2)\omega_2 A_2 = \frac{\lambda}{2}A_0 A_1. \tag{11.2.60}$$

Eliminating A_1 and A_2 from the above pair of equations then gives a quadratic equation that can be solved for the growth rate:

$$\left[4\gamma^2 + 2\gamma(\nu_1 + \nu_2) + \nu_1\nu_2\right]\omega_1\omega_2 = \frac{\lambda^2}{4}A_0^2. \tag{11.2.61}$$

Simple inspection shows that for a sufficiently large pump wave amplitude, A_0, the system is unstable, $\gamma > 0$, and the daughter waves x_1 and x_2 will grow exponentially. The instability threshold can be obtained by setting $\gamma = 0$. The resulting equation shows that the pump wave must have an amplitude of at least

$$A_0^2 = \frac{4\nu_1\nu_2\omega_1\omega_2}{\lambda^2} \tag{11.2.62}$$

Table 11.2 *Representative wave–wave interactions*

Interaction	Matching Condition	Comment
$T \rightarrow L + A$	$\omega_T = \omega_L + \omega_A$	EM decay instability
$L + A \rightarrow T$	$\omega_T = \omega_L + \omega_A$	EM coalescense
$L \rightarrow T + A$	$\omega_L = \omega_T + \omega_A$	Langmuir decay instability
$T \rightarrow T' + L$	$\omega_T = \omega_{T'} + \omega_L$	Raman scattering
$T \rightarrow T' + A$	$\omega_T = \omega_{T'} + \omega_A$	Brillouin scattering
$L \rightarrow L' + A$	$\omega_L = \omega_{L'} + \omega_A$	Electrostatic Langmuir decay
$L' + A \rightarrow L$	$\omega_L = \omega_{L'} + \omega_A$	Electrostatic Langmuir coalescence
$L + L' \rightarrow T$	$\omega_L + \omega_{L'} = \omega_T \simeq 2\omega_{pe}$	Requires backscatter L′

for the daughter waves to be generated. Below this threshold the pump wave does not have enough power to offset energy loss caused by the damping terms ν_1 and ν_2. The existence of this threshold, below which the nonlinearly driven daughter waves are not generated, has been confirmed in laboratory experiments by Stenzel and Wong (1972).

A well-studied example of the above process consists of the decay of a high-frequency transverse EM pump wave into a Langmuir wave and an ion acoustic wave, commonly denoted as $T \rightarrow L + A$. In addition to this decay mode, there are many other ways in which parametric decay can occur, especially in a plasma that has a zero-order background magnetic field. For any combination of plasma waves of type W_n that can lead to a parametric decay of the form $W_0 \rightarrow W_1 + W_2$, there also exists an inverse interaction of the form $W_1 + W_2 \rightarrow W_0$. Such processes are called "coalescence."

Many other parametric decay mechanisms exist that have not been discussed here. For example, plasmas often have quasi-static density structures (i.e., zero frequency waves) that can reflect plasma waves. Beam-driven Langmuir waves are easily reflected from small density irregularities of this type, thereby producing two oppositely propagating Langmuir waves, L and L′. Such oppositely propagating Langmuir waves, which have $\mathbf{k}_1 + \mathbf{k}_2 \approx 0$, can interact coherently, $L + L' \rightarrow T$, to produce radio emissions at $\omega_L + \omega_{L'} \approx 2\omega_{pe}$. Type III solar radio emissions, discussed earlier in Section 9.5.4, are produced by this parametric decay mechanism, as well as by the $L + A \rightarrow T$ decay process. It is beyond the scope of this book to discuss all of these many types of wave–wave interactions. Some of the best known examples of wave–wave interactions are listed in Table 11.2, along with their common names. For additional information on these nonlinear wave–wave interactions the reader is referred to the "Further Reading" list at the end this chapter.

11.3 Langmuir Wave Solitons

In the previous section, we showed that when a transverse EM pump wave exceeds a specific threshold, the system becomes parametrically unstable and daughter waves begin to grow exponentially in time. Here we focus on the question of what limits the amplitude of these waves, which for an unmagnetized plasma are a Langmuir wave and an ion acoustic wave. This question has a variety of possible answers. The amplitudes could be limited by a cascade of further three-wave or multiple-wave interactions that ultimately lead to turbulent energy dissipation similar to that discussed in Section 11.1.3, by wave–particle interactions that convert the wave energy to thermal energy via nonlinear Landau damping, or by a variety of other nonlinear processes. In this section we discuss a highly nonlinear process in which the ponderomotive force of the Langmuir wave becomes so large that it repels the plasma from regions of strong electric fields, thereby forming intense isolated Langmuir wave packets. As we will show, these wave packets, which are called solitons, may or may not be stable depending on the amount of damping that is present in the plasma. In two or three dimensions, if the structures are intense enough they can "collapse" to small scales on the order of tens of Debye lengths, thereby forming structures called "collapsed wave packets" or "cavitons." As the collapse proceeds, the intense electric fields in the packet act to strongly heat the plasma electrons, leading to dissipation of the wave energy. Because the analysis becomes complicated in two or three dimensions, we restrict our detailed analysis of such collapse processes to one dimension.

11.3.1 The Zakharov Equations

We start the discussion of Langmuir wave solitons by deriving two well-known equations called the *Zakharov equations* (Zakharov, 1972). These equations are based on a two-time-scale approach in which the relevant variables are separated into short time-scale/high-frequency variations and long time-scale/low-frequency variations, much as we did in Section 11.2.3 for three-wave interactions. For the high-frequency variations we assume that the Langmuir wave is driven via the parametric decay of a strong transverse EM pump wave as discussed in the previous section, although it also could be driven by an electron beam-driven instability as discussed in Chapter 9, or by any other suitable process. If the Langmuir wave is driven by the parametric decay of a constant amplitude EM pump wave, the pump wave plays no role in the temporal evolution because it is assumed to be of constant amplitude. Therefore, the temporal evolution is now controlled by a two-wave process involving only the Langmuir wave and the ion acoustic wave. As we did before, we assume that the Langmuir mode and the ion acoustic mode have negligible damping, i.e., $k\lambda_{\mathrm{D}} \ll 1$ and $T_{\mathrm{e}}/T_{\mathrm{i}} \gg 1$.

Since only electron motions are involved in the high-frequency Langmuir oscillations, the relevant wave equation for the Langmuir mode is Eq. (11.2.39), which is repeated below:

$$\left(\frac{\partial^2}{\partial t^2} - \frac{3\kappa T_e}{m_e}\nabla^2 + \omega_{pe}^2\right)\mathbf{E}_L = -\omega_{pe}^2\frac{n_A}{n_0}\mathbf{E}_L, \tag{11.3.1}$$

with the important change that on the right-hand side we have inserted the electric field, $\mathbf{E}_L$, of the Langmuir wave, rather than the electric field, $\mathbf{E}_T$, of the transverse electromagnetic wave pump wave. The pump wave is of fixed amplitude and plays no role in the nonlinear interaction. It is now the electric field of the Langmuir wave that provides the high-frequency contribution to the nonlinear source current, with the current being modulated by the density, n_A, of the ion acoustic wave.

To make the two-time-scale approximation more explicit, we assume that the electric field of the Langmuir wave is given by

$$\mathbf{E}_L(\mathbf{r},t) = \mathbf{E}_0(r,t)e^{-i\omega_{pe}t}, \tag{11.3.2}$$

where the amplitude $\mathbf{E}_0(\mathbf{r},t)$ is assumed to be a slowly varying function of time, and the exponential term represents the rapidly varying Langmuir wave oscillation at the electron plasma frequency, ω_{pe}. For the moment we ignore the shift in ω_{pe} caused by the electron pressure term, which is small, since $k\lambda_D \ll 1$. Substituting the above equation into Eq. (11.3.1) and carrying out the required differentiations then gives the following equation:

$$2i\omega_{pe}\frac{\partial \mathbf{E}_0}{\partial t} + \frac{3\kappa T_e}{m_e}\nabla^2\mathbf{E}_0 = \omega_{pe}^2\frac{n_A}{n_0}\mathbf{E}_0, \tag{11.3.3}$$

where we have omitted the $\partial^2\mathbf{E}_0/\partial t^2$ term, which is assumed to be small because of the slowly varying electric field amplitude. The above equation is called the *first Zakharov equation*. It controls the long time-scale evolution of the Langmuir wave electric field amplitude, $\mathbf{E}_0$.

We now turn to the ion acoustic mode. Because the interaction of the Langmuir wave with the ion acoustic mode is fundamentally different than in the three-wave analysis discussed in the previous section, we must partially repeat the derivation of the wave equation for the ion acoustic mode. To do this, we start with the electron momentum equation (11.2.40). After using Eq. (11.2.41) and the subsequent reasoning leading to the substitution $(\mathbf{U}_e \cdot \nabla)\mathbf{U}_e = (1/2)(\mathbf{U}_e^2)$, the momentum equation can be written

$$n_e m_e\frac{\partial \mathbf{U}_e}{\partial t} = (-e)n_e\mathbf{E} - 3\kappa T_e\nabla n_e - n_e m_e\frac{1}{2}\nabla(\mathbf{U}_e^2). \tag{11.3.4}$$

Next, substitute Eq. (11.2.45) for the electron velocity produced by the high-frequency Langmuir wave oscillations,

$$\mathbf{U}_{\rm e} = \frac{(-e)\mathbf{E}}{m_{\rm e}\omega_{\rm pe}}, \tag{11.3.5}$$

into the $(1/2)(\mathbf{U}_{\rm e}^2)$ term of Eq. (11.3.4), which gives

$$n_{\rm e}m_{\rm e}\frac{\partial \mathbf{U}_{\rm e}}{\partial t} = (-e)n_{\rm e}\mathbf{E} - 3\kappa T_{\rm e}\nabla n_{\rm e} - \frac{n_{\rm e}e^2}{2m_{\rm e}\omega_{\rm pe}^2}\nabla(\mathbf{E}^2). \tag{11.3.6}$$

Comparing the above equation with Eq. (3.7.7), one can now identify the last term on the right-hand side as the ponderomotive force. As before, because of the slow time variation of the ion acoustic mode we note that the $n_{\rm e}m_{\rm e}\partial\mathbf{U}_{\rm e}/\partial t$ term is small compared to the other terms and can be omitted. The electron momentum equation can then be rearranged to give an equation for the electric field force acting on the plasma, which is

$$(-e)\mathbf{E} = \frac{3\kappa T_{\rm e}}{n_0}\nabla n_{\rm e} + \frac{e^2}{2m_{\rm e}\omega_{\rm pe}^2}\nabla(\mathbf{E}^2). \tag{11.3.7}$$

Note that the ponderomotive force effectively adds to the electron pressure gradient force. Since our objective here is to identify the slowly varying electric field force that drives the ion acoustic wave, we need to average the rapidly oscillating $\mathbf{E}^2$ term over a time-scale that is long compared to the Langmuir oscillations, but short compared to the ion acoustic wave variations. This can be done by simply noting that $\langle\mathbf{E}^2\rangle = (1/2)|\mathbf{E}_0|^2$, where $\mathbf{E}_0$ is the now complex electric field amplitude defined by Eq. (11.3.2). The average electric field force $(-e)\mathbf{E}_{\rm A}$ acting on the ion acoustic wave is then given by

$$(-e)\mathbf{E}_{\rm A} = \frac{3\kappa T_{\rm e}}{n_0}\nabla n_{\rm e} + \frac{e^2}{4m_{\rm e}\omega_{\rm pe}^2}\nabla|\mathbf{E}_0|^2. \tag{11.3.8}$$

Following the same steps as in the derivation of Eq. (11.2.52), we then obtain the following wave equation for the ion acoustic mode:

$$\left(\frac{\partial^2}{\partial t^2} - C_{\rm A}^2\nabla^2\right)n_{\rm A} = \frac{\epsilon_0}{4m_{\rm i}}\nabla^2|\mathbf{E}_0|^2, \tag{11.3.9}$$

where we have used $\omega_{\rm pe}^2 = n_0e^2/(\epsilon_0 m_{\rm e})$ to eliminate the plasma frequency from Eq. (11.3.7). This equation is called the *second Zakharov equation*. Together with the first Zakharov equation (11.3.3), these two equations describe the self-consistent dynamical evolution of the Langmuir wave electric field amplitude, $\mathbf{E}_0$, and the deviation, $n_{\rm A}$, of the electron density from its equilibrium density, n_0.

Although the Zakharov equations are fully three-dimensional, to simplify the analysis we next restrict our discussion to one-dimensional solutions that involve only a single spatial coordinate, x. The ∇^2 term in these equations then becomes $\partial^2/\partial x^2$. At this point we find it convenient to change t, x, E_0, and n_A to normalized dimensionless variables using the following transformations:

$$2\left(\frac{m_e}{m_i}\right)\omega_{pe}t \rightarrow t, \qquad \frac{2}{\sqrt{3}}\left(\frac{m_e}{m_i}\right)^{1/2}\frac{x}{\lambda_{De}} \rightarrow x, \tag{11.3.10}$$

$$\frac{1}{4}\left(\frac{m_i}{m_e}\right)^{1/2}\left(\frac{\epsilon_0}{3n_0\kappa T_e}\right)^{1/2}E_0 \rightarrow E, \quad \text{and} \quad \frac{1}{4}\left(\frac{m_i}{m_e}\right)\frac{n_A}{n_0} \rightarrow n. \tag{11.3.11}$$

These transformations, first introduced by Nicholson (1983), have the effect of converting the Zakharov equations to the following pair of dimensionless equations (see Problem 11.6):

$$\mathrm{i}\frac{\partial E}{\partial t} + \frac{\partial^2 E}{\partial x^2} = nE \quad \text{(the first Zakharov equation)} \tag{11.3.12}$$

and

$$\frac{\partial^2 n}{\partial t^2} - \frac{\partial^2 n}{\partial x^2} = \frac{\partial^2}{\partial x^2}|E^2| \quad \text{(the second Zakharov equation).} \tag{11.3.13}$$

Next, consider a slowly varying solution of the above two equations. Specifically, ignore the temporal term in the second of the above equations, i.e., assume $\partial^2 n/\partial t^2 = 0$, but keep the less rapidly varying $\mathrm{i}\partial E/\partial t$ term in the first equation. The second equation can then be integrated twice to give

$$n = -|E^2|, \tag{11.3.14}$$

where we have assumed that the integration constants are zero. The minus sign in this equation indicates that the electron density is depressed from its equilibrium value by an amount that is in direct proportion to the square of the electric field strength, i.e., the ponderomotive force. Substituting this result into Eq. (11.3.12) then gives the following nonlinear differential equation:

$$\mathrm{i}\frac{\partial E}{\partial t} + \frac{\partial^2 E}{\partial x^2} + |E|^2 E = 0. \tag{11.3.15}$$

Because of its close similarity to the well-known Schrödinger equation in quantum mechanics, the above equation is called the nonlinear Schrödinger equation. The nonlinear Schrödinger equation has many solutions, most of which have spatially localized oscillatory solutions that asymptotically go to zero at $x = \pm\infty$. The oscillatory solutions can be described as though they were a particle moving in a pseudo-potential well, the details of which are discussed by Dauxois and Peyrard (2006). However, there is one particular solution of special interest that has no spatial oscillations. This solution is given by

$$E(x,t) = (2\Omega)^{1/2} \frac{\exp(i\Omega t)}{\cosh(\Omega^{1/2} x)}, \tag{11.3.16}$$

where Ω is a constant. That this equation is a valid solution can be verified by direct substitution into Eq. (11.3.15); see Problem 11.7. From Eq. (11.3.14) it can be seen that the corresponding dimensionless electron density is given by

$$n = \frac{-(2\Omega)}{\cosh^2(\Omega^{1/2} x)}, \tag{11.3.17}$$

which is negative, indicating that the density is depressed from its equilibrium value, n_0. Such isolated nonlinear structures are often called solitons; see Dauxois and Peyrard (2006).

To help understand the nature of the above solution, Figure 11.11 shows the electric field amplitude and electron density as a function of x for various values of the parameter Ω. As one can clearly see, as Ω increases the peak electric field amplitude at $x = 0$ increases and the electron density at $x = 0$ decreases. The density minimum occurs because the ponderomotive force repels plasma away from the region of highest electric field strength. Of course, there is a limit to the amount that the electron density can be depressed. Specifically, the magnitude of the decrease cannot exceed the density of the surrounding plasma, otherwise the

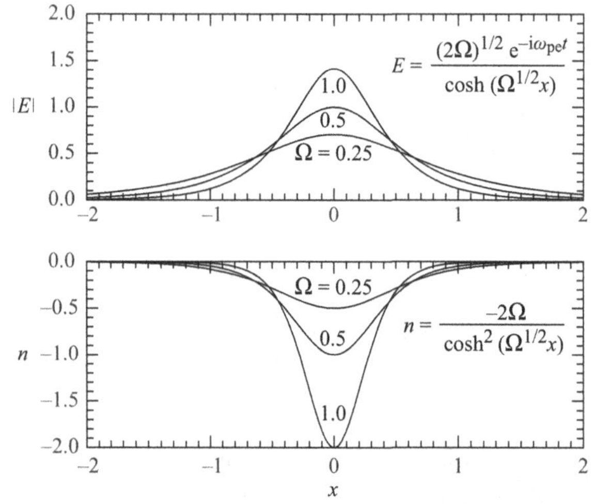

Figure 11.11 Plots of the normalized electric field (top) and the associated normalized electron density (bottom) for various values of the parameter Ω.

electron density would be negative. It is interesting to investigate the implications of this limiting condition. Setting $n_A/n_0 = -1$ in Eq. (11.3.11) and noting that $n = -E^2$, it is easy to show that in this limit $(1/2)\epsilon_0 E_0{}^2 = 6n_0\kappa T_e$. This simple calculation shows that there is an absolute upper limit to the electric field strength. In practice, such high field strengths never occur. Defining the scale size L of the wave packet by the condition $\Omega^{1/2}L = 1$, one can show using Eq. (11.3.10) that as $|n_A|$ approaches n_0 the normalized wave number $k\lambda_D$ approaches one. From Eq. (9.3.19) this means that the Langmuir wave experiences strong Landau damping well before this limiting condition is reached.

11.3.2 Soliton Collapse

The solution of the nonlinear Schrödinger equation described above was derived under the assumption that the solution is time-independent, i.e., $\partial^2 n/\partial t^2 = 0$. One can now ask the question of whether the solution that we obtained is a stable solution. To investigate this question, we start by assuming that E_0 is a steady-state solution given by Zakharov's first equation

$$\mathrm{i}\frac{\partial E_0}{\partial t} + \mathrm{i}\nu E_0 + \frac{\partial^2 E_0}{\partial x^2} + |E_0|^2 E_0 = 0, \tag{11.3.18}$$

where E_0 is assumed to be real and we have added a damping term, $\mathrm{i}\nu E_0$, anticipating that damping might play a role in the stability, as it did in the analysis of the parametric decay instability. To test the stability of this solution, we proceed by adding a small complex perturbation $\widetilde{\xi}$ to E_0 and then perform a linear stability analysis to see if the perturbed solution is stable. Following the usual Fourier transform approach to linear stability analysis, we substitute the perturbed solution $E_0 + \widetilde{\xi}$ into the above equation, which shows the effect of the small perturbation via the following equation:

$$\mathrm{i}\frac{\partial E_0}{\partial t} + \mathrm{i}\frac{\partial \widetilde{\xi}}{\partial t} + \mathrm{i}\nu E_0 + \mathrm{i}\nu\widetilde{\xi} + \frac{\partial^2 E_0}{\partial x^2} + \frac{\partial^2 \widetilde{\xi}}{\partial x^2} + |E_0 + \widetilde{\xi}|^2\left(E_0 + \widetilde{\xi}\right) = 0. \tag{11.3.19}$$

Expanding the terms involving the parameter $\widetilde{\xi}$ then gives the equation

$$\mathrm{i}\frac{\partial E_0}{\partial t} + \mathrm{i}\frac{\partial \widetilde{\xi}}{\partial t} + \mathrm{i}\nu E_0 + \mathrm{i}\nu\widetilde{\xi} + \frac{\partial^2 E_0}{\partial x^2} + \frac{\partial^2 \widetilde{\xi}}{\partial x^2} + \left(E_0^2 + E_0\widetilde{\xi} + E_0\widetilde{\xi}^* + |\widetilde{\xi}|^2\right)\left(E_0 + \widetilde{\xi}\right) = 0, \tag{11.3.20}$$

where for the squared term we used the rule $|F|^2 = FF^*$. Keeping only terms that are linear in $\widetilde{\xi}$ and $\widetilde{\xi}^*$, we then subtract the above equation from the steady-state version of Zakharov's first equation (11.3.18). The difference then gives the change to Zakharov's first equation caused by the perturbation, which is

$$\mathrm{i}\frac{\partial\tilde{\xi}}{\partial t}+\mathrm{i}\nu\tilde{\xi}+\frac{\partial^2\tilde{\xi}}{\partial x^2}+2E_0^2\tilde{\xi}+E_0^2\tilde{\xi}^*=0. \tag{11.3.21}$$

To test the stability of the zeroth order solution, E_0, we assume that the perturbation is a small amplitude propagating wave of the form $\tilde{\xi}=\tilde{\xi}_0\exp(\mathrm{i}kx+\gamma t)$, where γ is the growth rate. Substituting this perturbation into the above equation then yields two independent equations,

$$(\mathrm{i}\gamma+\mathrm{i}\nu-k^2+2E_0^2)\tilde{\xi}_0=-E_0^2\tilde{\xi}_0^* \tag{11.3.22}$$

and

$$(-\mathrm{i}\gamma-\mathrm{i}\nu-k^2+2E_0^2)\tilde{\xi}_0^*=-E_0^2\tilde{\xi}_0. \tag{11.3.23}$$

Eliminating the amplitude factors $\tilde{\xi}_0$ and $\tilde{\xi}_0^*$ between these two equations then gives the dispersion relation

$$(\gamma+\nu)^2=-k^4+4k^2|E_0|^2-3|E_0|^4. \tag{11.3.24}$$

From the dispersion relation one can show that the maximum growth rate occurs at $k_{\mathrm{max}}^2=2|E_0|^2$. Substituting this value for k_{max} back into the equation then gives the maximum growth rate, which is

$$\gamma_{\mathrm{max}}=|E_0|^2-\nu. \tag{11.3.25}$$

The above equation shows that the perturbation is unstable whenever $|E_0|^2>\nu$. Thus, just as for the parametric decay instability, there is a threshold amplitude for the Langmuir wave electric field above which the steady-state solution to the nonlinear Schrödinger equation is unstable. The threshold is determined by the damping coefficient, ν, which must ultimately be related to collisional or Landau damping via the appropriate equations.

If the system is weakly unstable the solution can be thought of as a continuous sequence of quasi-steady state solutions, each of which is given by Eq. (11.3.16), but with a steadily increasing Ω value and a correspondingly decreasing scale size given by $L\sim 1/\Omega^{1/2}$. This process is called collapse, and the resulting density structure is sometimes called a caviton. The collapse is caused by the outward-directed ponderomotive force of the Langmuir wave that digs a hole in the plasma density. As the density decreases, the electric field strength of the Langmuir wave increases even more, ultimately leading to the total collapse of the density structure. Since the Langmuir wave frequency remains approximately constant as the collapse progresses, the dispersion relation for the Langmuir wave,

Eq. (11.2.30), shows that as the density decreases the characteristic wave number, $k_c \sim 1/L$, of the Langmuir wave oscillations must increase. The increase in the wave number then causes the resonance velocity, $v_{\max} = \omega_{pe}/k_c$, to decrease, eventually leading to strong Landau damping and electron heating when the resonance velocity reaches the main part of the electron velocity distribution. A similar collapse process also occurs in three dimensions, but with the added feature that the electric field strength is further enhanced by the focusing of the Langmuir waves into the region of depressed plasma density. It is interesting to note that in the absence of the ponderomotive force a region of depressed density such as that described above would spread out due to the dispersion effects, which become increasingly important as the wave number increases; see Eq. (9.2.46). Thus, the nonlinear effects of the ponderomotive force, which are represented by the $|E|^2E$ term in the nonlinear Schrödinger equation, act to cancel the normal spreading effects caused by dispersion. Such cancellation effects are a common feature of all soliton-like phenomena that occur in nature (Dauxois and Peyrard, 2006).

The extent to which the soliton collapse actually occurs in real plasmas depends on many factors. In laboratory plasmas there is clear evidence that Langmuir wave collapse occurs when the wave is driven by a strong radio frequency EM pump; see Kim et al. (1974). Langmuir wave collapse is also believed to be

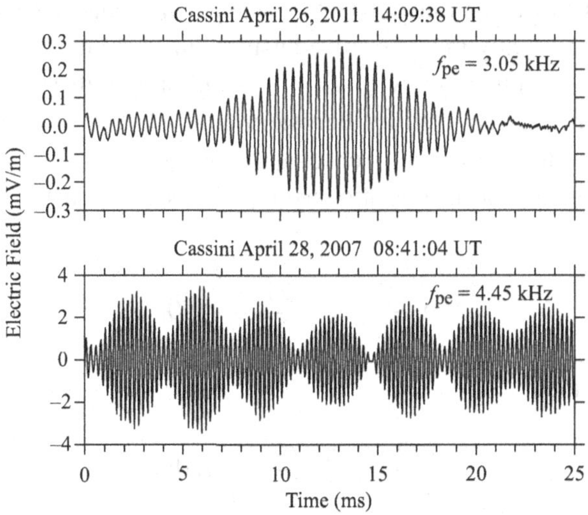

Figure 11.12 Langmuir wave packets generated by electron beams streaming outward ahead of Saturn's bow shock, as observed by the *Cassini* spacecraft. Sometimes isolated soliton-like wave packets are observed, as in the top panel. However, more commonly the Langmuir waves appear as an amplitude-modulated wave-train of the type shown in the bottom panel.

the mechanism by which electrons are strongly heated in laser-fusion plasmas (Sircombe et al., 2005; Atzeni and Meyer-ter-Vehn, 2009). In space plasmas there has been considerable interest in nonlinear effects associated with Langmuir waves observed in the solar wind. These waves are commonly observed in association with electron beams upstream of Earth's bow shock (Filbert and Kellogg, 1979), and with the electron beams from solar flares that are responsible for type III solar radio bursts (Lin et al., 1986). Although soliton-like wave packets are sometimes observed (see Figure 11.12), usually the Langmuir wave electric field energy densities do not reach the very high intensities typically needed for soliton collapse to occur (Cairns et al., 1998; Graham et al., 2012). Instead, the Langmuir waves observed in the solar wind are believed to be stabilized by a three-wave interaction (Cairns and Robinson, 1992) in which the beam-generated Langmuir wave decays into a daughter Langmuir wave and an ion acoustic wave, i.e., $L \to L' + A$. This three-wave decay process is believed to be responsible for amplitude-modulated electric field waveforms illustrated in the bottom panel of Figure 11.12.

11.4 Stationary Nonlinear Electrostatic Potentials

As the final example of nonlinear analysis techniques, we consider a case in which the assumption of linearization breaks down completely and the full measure of the nonlinearity inherent in the electrostatic Vlasov–Poisson system of Eqs. (11.1.1) and (11.1.2) must be taken into account. As the general time-dependent problem is very difficult and usually must be solved using numerical methods, we will focus our attention on time-stationary ($\partial/\partial t = 0$) solutions of these equations. These solutions were first discovered by Bernstein et al. (1957) and are known as the BGK solutions (after Bernstein, Greene, and Kruskal).

In one dimension, the time-stationary equations that must be solved are the Vlasov equation,

$$v_z \frac{\partial f_s}{\partial z} - \frac{e_s}{m_s} \frac{\partial \Phi}{\partial z} \frac{\partial f_s}{\partial v_z} = 0, \tag{11.4.1}$$

and Poisson's equation,

$$\frac{\partial^2 \Phi}{\partial z^2} = -\sum_s \frac{e_s}{\epsilon_0} \int_{-\infty}^{\infty} f_s \, \mathrm{d}v_z. \tag{11.4.2}$$

As shown in Section 5.3, a general solution of the Vlasov equation for a time-stationary electrostatic potential can be written in the form

$$f_s(v_z, z) = f_s(W_s), \tag{11.4.3}$$

where

$$W_s = \frac{1}{2}m_s v_z^2 + e_s\Phi(z) \tag{11.4.4}$$

is the total energy of a particle of type s. All that we now need to do is to solve Poisson's equation, which can be written

$$\frac{\partial^2\Phi}{\partial z^2} = -\sum_s \frac{e_s}{\epsilon_0}\int_{-\infty}^{\infty} f_s\left(\frac{1}{2}m_s v_z^2 + e_s\Phi\right) \mathrm{d}v_z. \tag{11.4.5}$$

If we assume that f_s is given, then the above equation becomes a differential equation for Φ. This is what we did in Section 2.2 when we specified that f_s be a Maxwellian, and with certain assumptions obtained the Debye–Hückel potential. Alternatively, if we assume that $\Phi(z)$ is given, then the above equation can be thought of as an integral equation for $f_s(W_s)$.

Next, we show the remarkable result that electron and ion distribution functions, f_e and f_i, can be found that satisfy Poisson's equation and the Vlasov equation for any arbitrary $\Phi(z)$. The only requirements are that $\Phi(z)$ be single valued, have a continuous second derivative, and be finite. Consider the potential $e\Phi(z)$ shown in Figure 11.13. The particles moving in this potential can be classified according to their total energy, $W_s = (1/2)m_s v_z^2 + e_s\Phi(z)$. Untrapped electrons and ions occur at energies below $e\Phi_{\rm min}$ and above $e\Phi_{\rm max}$, respectively. The trapped particles can be subdivided by energy into separate trapping regions ($\mathrm{A_e}, \mathrm{B_e}, \mathrm{C_e}$, and $\mathrm{D_e}$ for electrons, and $\mathrm{A_i}, \mathrm{B_i}$, and $\mathrm{C_i}$ for ions) bounded by different turning points. Since the particles trapped in the various subregions cannot reach adjacent subregions, the distribution functions $f(W_s)$ can be independently prescribed in each subregion. This freedom to specify the distribution function independently in the various subregions can be illustrated more clearly by considering constant energy contours in $z - v_z$ phase space, as shown for the ions in Figure 11.14. Since the total energy is

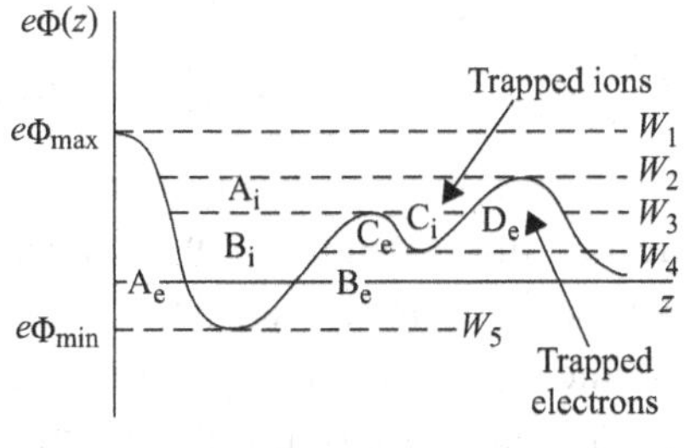

Figure 11.13 An assumed time-stationary electrostatic potential, $\Phi(z)$. In each region, bounded by total energies $W_1, W_2, \ldots$, a unique solution can be found for any arbitrary $\Phi(z)$, provided only that $\Phi(z)$ be single valued and have a continuous second derivative.

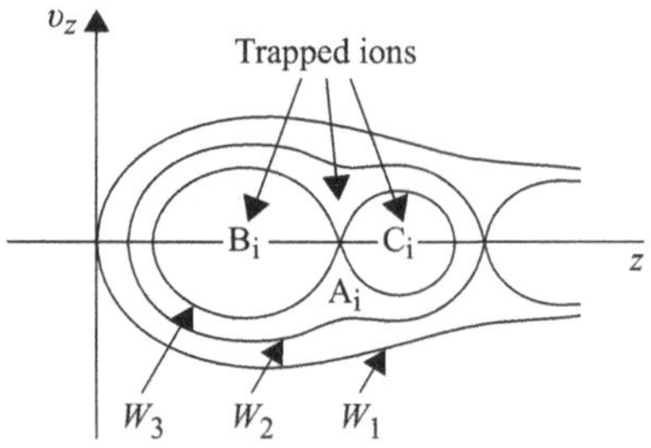

Figure 11.14 The ion phase-space (z, v_z) boundaries for the first three total energy levels, W_1, W_2, and W_3, in Figure 11.13.

a constant of the motion, the phase-space trajectories of the particles are contours of constant energy. Since $\mathrm{d}f_s/\mathrm{d}t = 0$ along a representative particle trajectory, it follows that the distribution function is constant along a contour of constant total energy. Hence, once the distribution function is specified on a particular energy contour, it is determined at all points along the contour. However, in the trapping regions, the distribution functions can be independently specified because the energy contours are closed in the individual trapping regions and do not link to other regions of phase space.

This flexibility in specifying the distribution function for the electrons and ions in their respective trapping regions can be used to produce the charge density required to satisfy Poisson's equation for any specified $\Phi(z)$, the only requirement being that f_e and f_i are always positive. From Eq. (11.4.4), we obtain $\mathrm{d}W_s = m_s v_z \mathrm{d}v_z$, which implies that for all particles with $v_z \geq 0$, we have

$$f_s(v_z)\mathrm{d}v_z = \frac{f_s(W_s)\mathrm{d}W_s}{\sqrt{2m_s(W_s - e_s\Phi)}}. \tag{11.4.6}$$

To find the distribution function $f_s(W_s)$ giving any specified charge density, we must solve an integral equation of the form

$$\rho_{qs}(z) = e_s \int_{e_s\Phi}^{e_s\Phi_{\max}} \frac{f_s(W_s)\mathrm{d}W_s}{\sqrt{2m_s(W_s - e_s\Phi)}} \tag{11.4.7}$$

over a region where $\Phi(z)$ is monotonic, such as that shown in Figure 11.15. In this equation $\Phi_{\max}$ is the maximum value of Φ in the region. The lower limit of integration is $e_s\Phi$ because no particle can have a total energy W_s less than $e_s\Phi$. Since $\Phi(z)$ is monotonic in any given subregion, $\rho_{qs}(z)$ can be specified as a function of $e_s\Phi$, i.e., $\rho_{qs}(z) = \rho_{s0}(e_s\Phi)$, so the above integral equation becomes

$$\rho_{s0}(e_s\Phi) = e_s \int_{e_s\Phi}^{e_s\Phi_{\max}} \frac{f_s(W_s)\mathrm{d}W_s}{\sqrt{2m_s(W_s - e_s\Phi)}}. \tag{11.4.8}$$

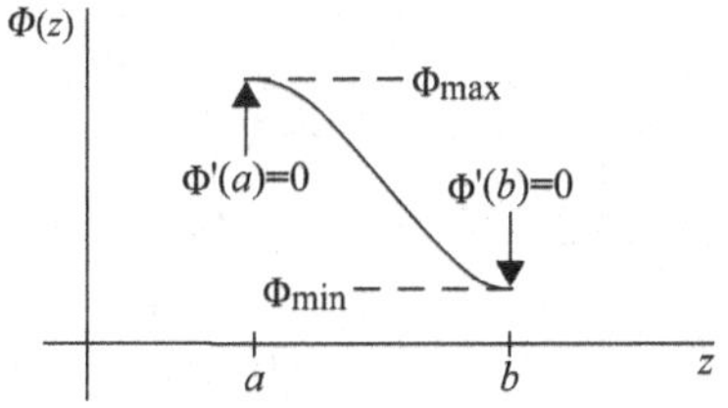

Figure 11.15 A typical electrostatic potential profile between successive maxima and minima. For each such segment, a unique solution can be found for $f_s(v_z, z)$.

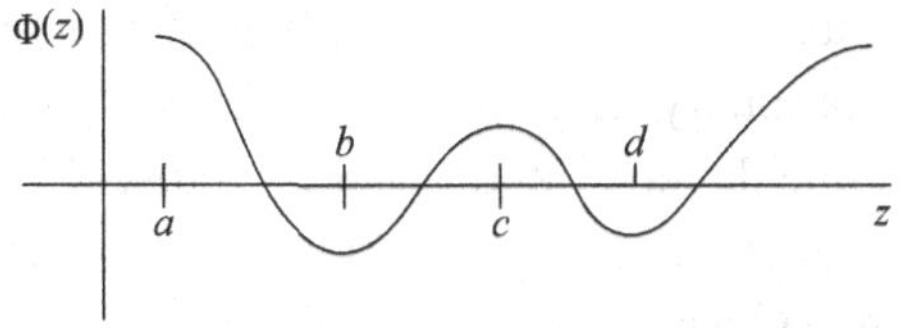

Figure 11.16 To find a complete solution for $f_s(v_z, z)$, the electrostatic potential $\Phi(z)$ is broken up into segments, $a-b, b-c, c-d$, etc., between successive maxima and minima.

We shall assume that $\rho_{qs}(a) = \rho_{s0}(e_s\Phi_{\rm max}) = 0$. It is easily verified that the following distribution function is an exact solution of the above integral equation:

$$f_s(W_s) = -(2m_s)^{1/2}\frac{1}{e_s\pi}\int_{W_s}^{e_s\Phi_{\rm max}} {\rm d}z\frac{{\rm d}\rho_{s0}(z)}{{\rm d}z}\frac{1}{\sqrt{z - W_s}}. \tag{11.4.9}$$

Using the above solution, the distribution function for the trapped particles can be determined in a piecewise fashion between successive maxima and minima of $\Phi(z)$, as illustrated in Figure 11.16. For example, between a and b, the ion distribution function, $f_{\rm i}(W_{\rm i})$, can be determined over the range $e\Phi(a) > W_{\rm i} > e\Phi(b)$; between b and c the electron distribution $f_{\rm e}(W_{\rm e})$ can be determined over the range $-e\Phi(b) > W_{\rm e} > -e\Phi(c)$, etc. Proceeding in this manner, a complete set of solutions for $f_{\rm i}(v_z)$ and $f_{\rm e}(v_z)$ can be obtained in all regions of the phase space for any given $\Phi(z)$.

11.4.1 Observations of BGK Electrostatic Potentials

The time-stationary BGK electrostatic potentials described above represent a large class of solutions of the time-stationary Vlasov and Poisson equations. Although these potentials are solutions of the time-stationary equations, there is no guarantee that they are either stable or dynamically realizable in any given physical situation. Nevertheless, it appears that at least some of these solutions actually occur in nature.

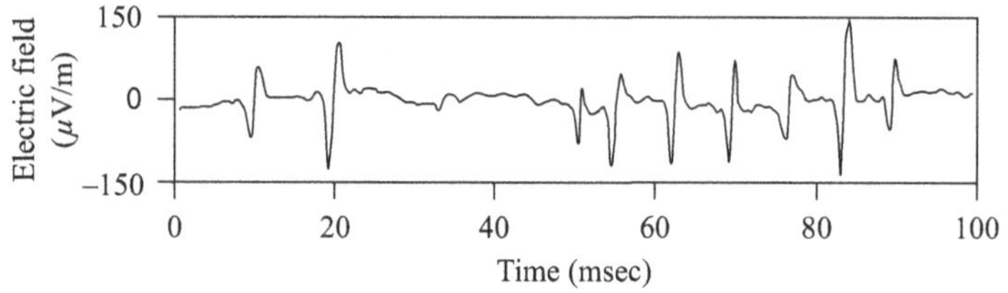

Figure 11.17 A series of bipolar electric field pulses detected in Earth's magnetotail by the *Geotail* spacecraft (from Matsumoto et al., 1994).

For example, both laboratory and spacecraft measurements have shown that large-amplitude time-stationary electrostatic potentials, variously called "electrostatic shocks," "double layers," and "bipolar pulses," can occur. Computer simulations have shown that these electrostatic structures are stable, and that they evolve from electrostatic plasma instabilities, such as the two-stream instability and the current-driven ion acoustic instability.

An example of a series of bipolar (i.e., positive as well as negative) electric field pulses observed in Earth's magnetosphere by the *Geotail* spacecraft is shown in Figure 11.17. Such bipolar pulses are commonly observed in regions where strong field-aligned electron beams exist. Strong evidence now exists from numerical simulations that such bipolar pulses are the final state of an electron-beam driven instability. Figure 11.18 shows the results of a computer simulation that self-consistently solves the nonlinear evolution of an electron-beam driven instability by following the trajectories of individual particles in an electrostatic potential that is self-consistently determined from Poisson's equation. The computer simulation shows that in the final saturated state, the electric field has evolved into a series of nearly time-independent bipolar pulses, shown in the top panel, which very closely resemble the electric field waveforms in Figure 11.17. The electrostatic potentials responsible for these fields appear to be closely related to the time-stationary electrostatic solutions studied by Bernstein et al. (1957). The cause of the bipolar pulses can be traced to the formation of "holes" in the electron phase-space distribution. A phase-space (v_z, z) diagram showing the formation of velocity space holes near the center of the bipolar pulses is shown in the bottom panel of Figure 11.18. In this case, the ions show almost no response, confirming that the bipolar pulses are primarily due to the formation of holes in the electron velocity distribution function.

We conclude by noting another application of the BGK solutions. It has been suggested recently (Hansen et al., 1996) that the resolution of the anti-shielding paradox mentioned in Section 2.2 lies in recognizing that the assumption of a Maxwellian distribution, made in deriving the Debye length formula, is not, strictly speaking, correct. They note that a charged object in a plasma creates in its vicinity

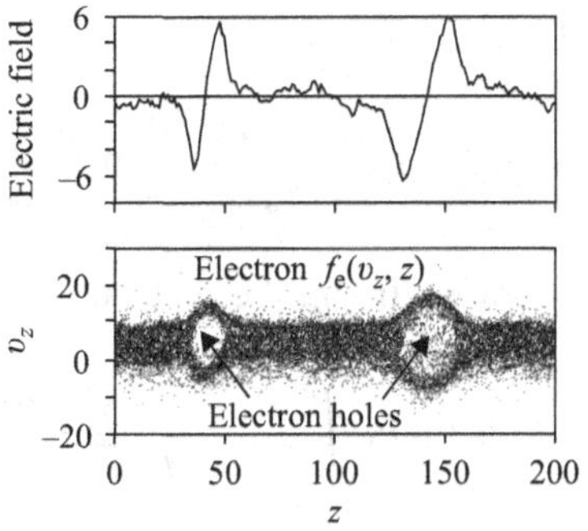

Figure 11.18 The electric field, E, and electron distribution function $f_e(v_z,z)$ obtained from a computer simulation of an electron-beam driven two-stream instability (courtesy of L.-J. Chen).

a potential well that traps particles of opposite charge. The distribution functions thus involve trapped as well as untrapped particles, and are better described as BGK solutions. The particles trapped in the vicinity of the charged object produce enhanced shielding, thus circumventing the anti-shielding mechanism discussed in Section 2.2.

Problems

11.1. Using the identity

$$\lim_{L\to\infty}\int_{-L}^{L} e^{i(k+k')z}\,dz = 2\pi\delta(k+k'),$$

show that

$$\langle W_E\rangle = \frac{\pi\epsilon_0}{2L}\int_{-\infty}^{\infty}\tilde{E}_1(k,t)\tilde{E}_1(-k,t)\,dk = \int_{-\infty}^{\infty}\mathscr{E}(k,t)\,dk.$$

11.2. Show, by taking appropriate moments of the quasi-linear diffusion equation,

$$\frac{\partial}{\partial t}\langle f_s\rangle(v_z,t) = \frac{\partial}{\partial v_z}\left[D_q(v_z,t)\frac{\partial}{\partial v_z}\langle f_s\rangle(v_z,t)\right],$$

that particles, momentum, and energy are conserved.

11.3. Show in the limit of small γ/ω_r that the following equation reduces to a delta function

$$\frac{\gamma}{(\omega_r - kv_z)^2+\gamma^2} \simeq \pi\delta(\omega_r - kv_z).$$

Hint: Consider the left-hand side as a function of ω_r and γ for fixed kv_z. For $\gamma\to 0$, integrate the left-hand side over ω_r to obtain the right-hand side.

11.4. Show that the equation

$$f_s(W_s) = -(2m_s)^{1/2}\frac{1}{e_s\pi}\int_{W_s}^{e_s\Phi_{\max}} dz\frac{d\rho_{s0}(z)}{dz}\frac{1}{\sqrt{z-W_s}}$$

is a solution of the following integral equation:

$$\rho_{s0}(e_s\Phi) = e_s\int_{e_s\Phi}^{e_s\Phi_{\max}} \frac{f_s(W_s)\,dW_s}{\sqrt{2m_s(W_s-e_s\Phi)}}.$$

Hint: After substituting $\rho_{s0}(e_s\Phi)$ into the first equation, show that the resulting double integral involves an integration over the triangular area shown below.

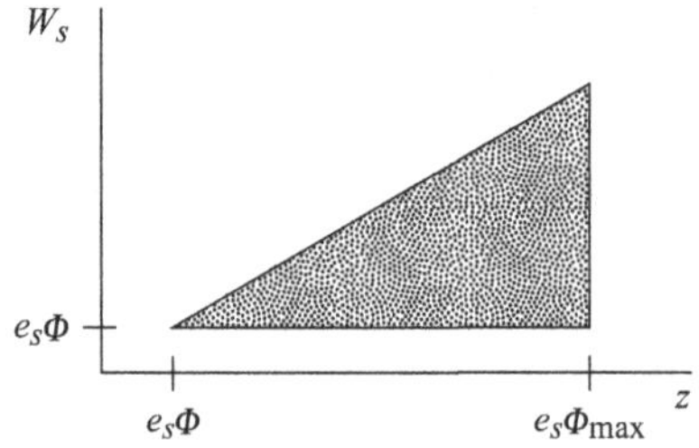

Change the order of integration and make use of the following definite integral:

$$\int_a^b \frac{dU}{\sqrt{(U-a)(b-U)}} = \pi.$$

11.5. Show that if $\mathbf{E}_T$ is parallel to $\mathbf{E}_L$ and $\nabla^2 = -k^2$, then transformations can be found that convert $\mathbf{E}_L$, $\mathbf{E}_T$, and n_A in the equations

$$\left(\frac{\partial^2}{\partial t^2} - c^2\nabla^2 + \omega_{pe}^2\right)\mathbf{E}_T = -\omega_{pe}^2\frac{n_A}{n_0}\mathbf{E}_L \qquad \text{Transverse EM mode,}$$

$$\left(\frac{\partial^2}{\partial t^2} - \frac{3\kappa T_e}{m_e}\nabla^2 + \omega_{pe}^2\right)\mathbf{E}_L = -\omega_{pe}^2\frac{n_A}{n_0}\mathbf{E}_T \qquad \text{Langmuir mode,}$$

$$\left(\frac{\partial^2}{\partial t^2} - C_A^2\nabla^2\right)n_A = \frac{n_0 e^2}{m_e m_i \omega_T \omega_L}\nabla^2(\mathbf{E}_T\cdot\mathbf{E}_L) \qquad \text{Ion acoustic mode,}$$

to x_0, x_1, and x_2 in Eqs. (11.2.7)–(11.2.9) for three coupled harmonic oscillators; see Bellan (2006).

11.6. Show that the following transformations convert the dimensionless forms of the first and second Zakharov equations, given by Eqs. (11.3.12) and

(11.3.13), to the dimensional forms given by Eqs. (11.3.3) and (11.3.9):

$$t \to 2\left(\frac{m_e}{m_i}\right)\omega_{pe}t, x \to \frac{2}{\sqrt{3}}\left(\frac{m_e}{m_i}\right)^{1/2}\frac{x}{\lambda_{De}}.$$

$$E \to \frac{1}{4}\left(\frac{m_i}{m_e}\right)^{1/2}\left(\frac{\epsilon_0}{3n_0\kappa T_e}\right)^{1/2}E_0, \quad \text{and} \quad n \to \frac{1}{4}\left(\frac{m_i}{m_e}\right)\frac{n_A}{n_0}.$$

11.7. Show that the nonlinear Schrödinger equation,

$$\mathrm{i}\frac{\partial E}{\partial t} + \frac{\partial^2 E}{\partial x^2} + |E|^2 E = 0,$$

has a solution

$$E(x,t) = (2\Omega)^{1/2}\frac{\exp(\mathrm{i}\Omega t)}{\cosh(\Omega^{1/2}x)}.$$

References

Atzeni, S., and Meyer-ter-Vehn, J. 2009. *The Physics of Inertial Fusion, International Series of Monographs on Physics*. Oxford: Oxford University Press.

Bellan, P. M. 2006. *Fundamentals of Plasma Physics*. Cambridge: Cambridge University Press, pp. 508–511.

Bernstein, I. B., Greene, J. M., and Kruskal, M. D. 1957. Exact nonlinear plasma oscillations. *Phys. Rev.* **108**, 546.

Cairns, I. H., and Robinson, P. A. 1992. Theory for low-frequency modulated Langmuir wave packets. *Geophys. Res. Lett.* **19**, 2187–2190.

Cairns, I. H., Robinson, P. A., and Smith, N. I. 1998. Arguments against modulational instabilities of Langmuir waves in Earth's foreshock. *J. Geophys. Res.* **103**, 287–299.

Chirikov, B. V. 1969. *Research Concerning the Theory of Nonlinear Resonances and Stochasticity*, Trans. A. T. Sanders, CERN Translation 71-40, Geneva, 1971. Novosibirsk: USSR Academy of Sciences, Report 267.

Dauxois, T., and Peyrard, M. 2006. *Physics of Solitons*. Cambridge: Cambridge University Press, pp. 80–84.

Davidson, R. C. 1972. *Methods in Nonlinear Plasma Theory*. New York: Academic Press.

Filbert, P. C., and Kellogg, P. J. 1979. Electrostatic noise at the plasma frequency beyond the bow shock. *J. Geophys. Res.* **84**, 1369–1381.

Graham, D. B., Cairns, I. H., Prabhakar, D. R., Ergun, R. E., Malaspina, D. M. , Bale, S. D., Goetz, K., and Kellogg, P. J. 2012. Do Langmuir wave packets in the solar wind collapse? *J. Geophys. Res.* **117**, A09107.

Hansen, C., Reimann, A. B., and Fajans, J. 1996. Dynamic and Debye shielding and anti-shielding. *Phys. Plasmas* **3**, 1820.

Kadomtsev, B. B. 1965. *Plasma Turbulence*. New York: Academic Press.

Kim, H. C., Stenzel, R. L., and Wong, A. Y. 1974. Development of cavitons and trapping of rf field. *Phys. Rev. Lett.* **33**, 886–890.

Lin, R. P., Levedahl, W. K., Lotko, W., Gurnett, D. A., and Scarf, F. L. 1986. Evidence for nonlinear wave–wave interactions in solar type III radio bursts. *Astrophys. J.* **308**, 954–965.

Matsumoto, H., Kojima, H., Miyatake, T., Omura, Y., Okada, M., Nagano, I., and Tsutsui, M. 1994. Electrostatic solitary waves (ESW) in the magnetotail: BEN wave forms observed by Geotail. *Geophys. Res. Lett.* **21**, 2915–2918.
Nicholson, D. R. 1983. *Introduction to Plasma Theory*. Malabar, FL: Krieger Publishing, pp. 179–180.
Sagdeev, R. Z., and Galeev, A. A. 1969. *Nonlinear Plasma Theory*. New York: Benjamin.
Sircombe, N. J., Arber, T. D., and Dendy, R. O. 2005. Accelerated electron populations formed by Langmuir wave-caviton interactions. *Phys. Plasmas*, **12**, 012303.
Stenzel, R., and Wong, A. Y. 1972. Threshold and saturation of parametric decay instability. *Phys. Rev. Lett.* **28**, 274–277.
Stix, T. H. 1992. *Waves in Plasmas*. New York: American Institute of Physics, Chapter 16.
Taylor, J. B. 1969. Investigation of charged particle invariants. *Culham Lab. Prog. Report CLM-PR 12*, Th.12.
Zakharov, V. E. 1972. Collapse of Langmuir waves, *Sov. Phys. JETP*, **35**, 908–914.

Further Reading

Bellan, P. M. 2006. *Fundamentals of Plasma Physics*. 2006. Cambridge: Cambridge University Press, Chapter 15, pp. 491–529.
Boyd, T. J. M., and Sanderson, J. J. 2003. *The Physics of Plasmas*. Cambridge: Cambridge University Press, Chapter 11.
Chen, F. F. 1990. *Introduction to Plasma Physics and Controlled Fusion*. New York: Plenum Press, Chapter 8.
Davidson, R. C. 1972. *Methods in Nonlinear Plasma Theory*. New York: Academic Press.
Kadomtsev, B. B. 1965. *Plasma Turbulence*. New York: Academic Press.
Krall, N. A., and Trivelpiece, A. W. 1973. *Principles of Plasma Physics*. New York: McGraw-Hill, Chapter 11.
Nicholson, D. R. 1983. *Introduction to Plasma Theory*. Malabar, FL: Krieger Publishing, Chapter 11.
Sagdeev, R. Z., and Galeev, A. A. 1969. *Nonlinear Plasma Theory*. New York: Benjamin.
Schmidt, G. 1979. *Physics of High Temperature Plasmas*. New York: Academic Press, Chapter 9.
Stix, T. H. 1992. *Waves in Plasmas*. New York: American Institute of Physics, Chapter 16.
Swanson, D. G. 1989. *Plasma Waves*. San Diego, CA: Academic Press, Chapter 7.
Welland, J. 1977. *Coherent Nonlinear Interaction of Waves in Plasmas*. New York: Pergamon Press.

12

Collisional Processes

Collisions between charged particles in a fully ionized plasma are mediated by the Coulomb force, which is a long-range force. By a long-range force, we mean a force that decays no faster than r^{-2}, where r is the distance between two charged particles. In contrast, collisions between charged and neutral particles are similar to those between hard spheres, which interact via a short-range force (Section 2.5). Due to the long-range Coulomb force, collective effects play a significant role in collisional processes. In this chapter, we will focus on collisional processes in a fully ionized plasma.

As discussed in Section 2.6, the number of electrons per Debye cube, $N_\mathrm{D} = n_0 \lambda_\mathrm{D}^3$, characterizes the discreteness of the plasma fluid. When $\mathrm{N_D}$ is much larger than one, the average kinetic energy is much larger than the average potential energy. Under these conditions, the average distance $\langle r \rangle$ between the nearest particles is much larger than the distance of closest approach, r_0. This can be easily seen by rewriting N_D in the form

$$N_\mathrm{D} = \left(\frac{\langle r \rangle}{6\pi r_0} \right)^{3/2}, \qquad (12.0.1)$$

where $\langle r \rangle = n_0^{-1/3}$ and $r_0 = e^2/(6\pi\epsilon_0 \kappa T)$. The average distance of closest approach, r_0, is defined by equating the average kinetic energy, $(3/2)\kappa T$, to the potential energy of a single particle, $e^2/(4\pi\epsilon_0 r_0)$. When N_D is very large, particles rarely come as close as the distance of closest approach and therefore rarely suffer large-angle deflections. However, a large number of small-angle binary collisions can add up to produce a large deflection in the trajectory of the particle. If $N_\mathrm{D} \gg 1$, it is possible to use $1/N_\mathrm{D}$ as an expansion parameter in a statistical description that treats the collision process as the sum of a large number of small-angle collisions. This expansion fails when the condition $N_\mathrm{D} \gg 1$ is no longer satisfied. Such plasmas are sometimes called strongly coupled plasmas (Ichimaru, 1994).

Strongly coupled plasmas occupy the region near and below the line $N_D = 1$ in Figure 2.5.

Before we develop a theory of collisional processes in a fully ionized plasma, it is useful to review the kinematics of binary Coulomb collisions which are the dominant form of collisions in such a plasma.

12.1 Binary Coulomb Collisions

The scattering geometry of a charged particle of type s interacting with a charged particle of type s' is shown in Figure 12.1, as viewed from the center-of-mass frame. The vector $\mathbf{V}$ is the relative velocity before the collision, and the vector $\mathbf{V}'$ is the relative velocity after the collision. Because of conservation of energy, the relative velocities before and after the collision are equal in magnitude, that is, $|\mathbf{V}| = |\mathbf{V}'|$. In the center-of-mass frame, the scattering angle χ and the impact parameter b obey the relation

$$\tan\left(\frac{\chi}{2}\right) = \frac{b_0}{b}, \tag{12.1.1}$$

where

$$b_0 = \frac{e_s e_{s'}}{4\pi\epsilon_0 \mu_{ss'} V^2} \tag{12.1.2}$$

is the distance of closest approach, $V = |\mathbf{v} - \mathbf{v}'|$ is the relative velocity of the two particles at infinity, and

$$\mu_{ss'} = \frac{m_s m_{s'}}{m_s + m_{s'}} \tag{12.1.3}$$

is the reduced mass. In the case of electrons of mass m_e scattering off ions of mass m_i, where $m_i \gg m_e$, the reduced mass becomes $\mu_{ss'} \cong m_e$. The differential cross section $\sigma_{ss'}(\chi)$, where χ is the scattering angle in the center-of-mass frame,

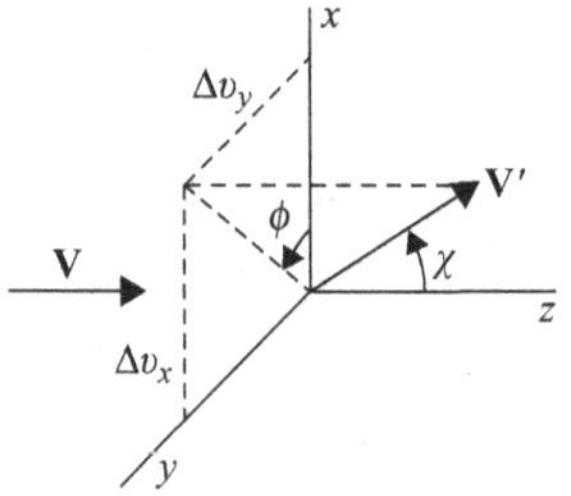

Figure 12.1 The coordinate system used to analyze Coulomb Collisions.

is given by the Rutherford formula

$$\sigma_{ss'}(V,\chi) = \frac{1}{4}\frac{b_0^2}{\sin^4(\chi/2)}, \tag{12.1.4}$$

which generalizes Eq. (2.5.3) of Chapter 2. The total cross section, σ_{T}, can be determined by integrating the differential cross section over all solid angles, i.e.,

$$\begin{aligned}\sigma_{\mathrm{T}} &= 2\pi \int_0^{\pi} \sigma_{ss'}(V,\chi)\sin\chi \mathrm{d}\chi \\ &= \pi b_0^2 \int_0^{\pi} \frac{\sin(\chi/2)\cos(\chi/2)}{\sin^4(\chi/2)}\mathrm{d}\chi \\ &= \pi b_0^2 \left[\frac{1}{\sin^2(\chi/2)}\right]_{\pi}^{0},\end{aligned} \tag{12.1.5}$$

which diverges at the upper limit. It is clear by inspection of the above equation that this divergence arises due to scattering at very small angles (corresponding to very large impact parameters). Due to the long range of the Coulomb force, all particles are scattered, even those at very large distances, thereby causing the divergence.

The cross section for scattering 90° or more in one binary collision is given by

$$\begin{aligned}\sigma_{\mathrm{S}} &= 2\pi \int_{\pi/2}^{\pi} \sin\chi\, \sigma_{ss'}(V,\chi)\,\mathrm{d}\chi = \frac{\pi b_0^2}{2} \int_{\pi/2}^{\pi} \sin\chi \frac{1}{\sin^4(\chi/2)}\mathrm{d}\chi \\ &= \pi b_0^2.\end{aligned} \tag{12.1.6}$$

Unlike the total cross section, this cross section does not diverge because it does not include the contribution of particles with very large impact parameters (or very small scattering angles).

12.2 Importance of Small-Angle Collisions

Next, let us consider an electron moving in a fully ionized plasma. The electron may suffer a large deflection, say 90°, either as a result of a single scattering event, or due to the cumulative effects of a large number of small-angle deflections that add up (in the sense of a random walk). We demonstrate below that the second process is much more important than the first.

Assume, without loss of generality, that the electron is moving initially with speed v in the z direction. As it moves through the plasma, the electron will suffer a sequence of random small-angle collisions. Let us assume that it suffers on average

N_c such collisions per unit length of its path projected along the z axis. For a plasma of average density n_0, N_c is given by

$$N_c = n_0 \int_0^{b_m} 2\pi b \, db, \tag{12.2.1}$$

where b_m is the maximum impact parameter, to be estimated later. The parameter b_m is introduced in order to avoid the divergences associated with the Coulomb force, as discussed in Section 12.1. After N_c collisions, the electron acquires a transverse velocity $\Delta \mathbf{v}_\perp$ with components

$$\Delta v_x = \sum_{i=1}^{N_c} \Delta v_{xi}, \quad \Delta v_y = \sum_{i=1}^{N_c} \Delta v_{yi}, \tag{12.2.2}$$

where, as shown in Figure 12.1, we have

$$\Delta v_{xi} = V \sin\chi_i \cos\phi_i, \quad \Delta v_{yi} = V \sin\chi_i \sin\phi_i. \tag{12.2.3}$$

Here we have assumed, for simplicity, that the relative velocity in each collision is always in the z direction and that $\Delta v_{xi} = \Delta v_{yi} = 0$ before the collision. By inspection of Eq. (12.2.3), it is clear that because the scattering process is isotropic about the z axis, positive values of Δv_x and Δv_y are as likely as negative values. So if we take an average over many collisions per unit length of the particle path (denoted by the symbol $\langle\rangle$), we will obtain $\langle \Delta v_x \rangle = \langle \Delta v_y \rangle = 0$. However, the quantities $\langle (\Delta v_x)^2 \rangle$ and $\langle (\Delta v_y)^2 \rangle$ do not vanish in general. For a single collision, we define $(\Delta v_\perp)^2 = (\Delta v_x)^2 + (\Delta v_y)^2$. Using simple trigonometric identities and Eq. (12.1.1), we then obtain

$$\begin{aligned}(\Delta v_\perp)^2 &= V^2 \sin^2\chi = 4V^2 \sin^2\frac{\chi}{2}\cos^2\frac{\chi}{2} \\ &= \frac{4V^2 (b/b_0)^2}{[1 + (b/b_0)^2]^2}.\end{aligned} \tag{12.2.4}$$

We can calculate $\langle (\Delta v_\perp)^2 \rangle$ per unit length, or $\langle (\Delta v_\perp)^2 \rangle / \Delta\ell$, by taking an average of $(\Delta v_\perp)^2$ over all impact parameters and multiplying the result by N_c, given by Eq. (12.2.1). We obtain

$$\frac{\langle (\Delta v_\perp)^2 \rangle}{\Delta\ell} = N_c \frac{\int_0^{b_m} \Delta v_\perp^2 \, 2\pi b \, db}{\int_0^{b_m} 2\pi b \, db} = n_0 \int_0^{b_m} \Delta v_\perp^2 \, 2\pi b \, db. \tag{12.2.5}$$

Substituting expression (12.2.4) into Eq. (12.2.5) gives

$$\frac{\langle (\Delta v_\perp)^2 \rangle}{\Delta\ell} = 8\pi n_0 V^2 \int_0^{b_m} \frac{(b/b_0)^2}{[1 + (b/b_0)^2]^2} b \, db, \tag{12.2.6}$$

which reduces, for $b_m/b_0 \gg 1$, to

$$\frac{\langle(\Delta v_\perp)^2\rangle}{\Delta \ell} \cong 8\pi n_0 V^2 b_0^2 \ln(b_m/b_0). \tag{12.2.7}$$

Since the long-range Coulomb force becomes exponentially small for impact parameters greater than the Debye length, the maximum impact parameter b_m is usually chosen to be the Debye length λ_D; see Eq. (2.2.8). The parameter b_0, defined by Eq. (12.1.2), can be estimated by replacing $\mu_{ss'} V^2$ by its Maxwellian average, $3\kappa T$; see Eq. (2.1.4). With these substitutions it is then easy to show that the logarithmic term in Eq. (12.2.7) simplifies to

$$\begin{aligned} \ln\frac{b_m}{b_0} &= \ln\frac{\lambda_D}{b_0} \\ &= \ln\left(\frac{12\pi\epsilon_0\kappa T}{e^2}\lambda_D\right) = \ln(12\pi N_D) = \ln\Lambda, \end{aligned} \tag{12.2.8}$$

where the quantity in the parentheses occurs sufficiently often that it is defined to be a new parameter,

$$\Lambda = 12\pi N_D, \tag{12.2.9}$$

called the plasma parameter. For most plasmas, $\ln\Lambda$ ranges from about 10 to 40.

Based on the above analysis, we can now estimate the mean-free path, λ_m, required for multiple small-angle collisions to produce a deflection of the order of 90°. Multiple small-angle collisions produce a change $\langle(\Delta v_\perp)^2\rangle/\Delta\ell$, given by Eq. (12.2.7). To produce a deflection of the order of 90° in a path length, λ_m, we require that

$$\frac{\langle(\Delta v_\perp)^2\rangle}{\Delta \ell}\lambda_m = V^2, \tag{12.2.10}$$

which, after substituting Eq. (12.2.7) for $\langle(\Delta v_\perp)^2\rangle/\Delta\ell$, gives

$$\lambda_m = \frac{1}{8\pi n_0 b_0^2 \ln\Lambda}. \tag{12.2.11}$$

Next, we calculate the mean-free path for a single 90° large-angle collision, λ_S, which, using Eq. (12.1.6), is given by

$$\lambda_S = \frac{1}{n_0\sigma_S} = \frac{1}{n_0\pi b_0^2}. \tag{12.2.12}$$

To compare the relative importance of a 90° deflection caused by a single collision to a 90° deflection produced by multiple small-angle collisions we take the ratio of the above equation to Eq. (12.2.11), which gives

$$\frac{\lambda_{\mathrm{S}}}{\lambda_{\mathrm{m}}} = 8\ln\Lambda. \tag{12.2.13}$$

For a typical plasma, $8\ln\Lambda$ is of the order of 10^2. Thus, the likelihood that an electron will suffer a 90° deflection due to a single collision is much smaller than the likelihood that it will suffer the same deflection due to multiple small-angle collisions.

12.3 The Fokker–Planck Equation

We have demonstrated that in a fully ionized plasma, small-angle collisions are much more probable than large-angle ones. Hence, the trajectory of a particle in a plasma is in general very complicated, composed of a very large number of seemingly random kicks which can eventually add up, in a root–mean–square sense, to a large deflection. To carry out a detailed analysis of such random motions, it is necessary to use a statistical analysis. For a statistical description it is useful to introduce a probability distribution function $P_s(\mathbf{v},\Delta\mathbf{v})$ such that $P_s(\mathbf{v},\Delta\mathbf{v})\,\mathrm{d}(\Delta\mathbf{v})$ is the probability that a test particle of type s with velocity $\mathbf{v}$ at time t will change its velocity by $\Delta\mathbf{v}$ in time Δt as a result of collisions. In writing this distribution function, we have assumed that the probability distribution does not depend on time t explicitly, and furthermore, that it depends only on the particle velocity $\mathbf{v}$ at time t and not on the particle velocity at previous times, that is, on the past history of the particle. In probability theory, such a random process is called a stationary Markov process. We assume that the probability distribution function $P_s(\mathbf{v},\Delta\mathbf{v})$ obeys the normalization condition

$$\int_{-\infty}^{\infty} P_s(\mathbf{v},\,\Delta\mathbf{v})\,\mathrm{d}^3(\Delta v) = 1. \tag{12.3.1}$$

Let $f_s(\mathbf{r},\mathbf{v},t)$ be the single-particle distribution function for particles of type s. If $f_s(\mathbf{r},\mathbf{v}-\Delta\mathbf{v},t-\Delta t)$ is the distribution function which evolves into the distribution function $f_s(\mathbf{r},\mathbf{v},t)$ due to collisions, by using the Markov distribution function $P_s(\mathbf{v},\Delta\mathbf{v})$ defined above, we can write

$$f_s(\mathbf{r},\mathbf{v},t) = \int_{-\infty}^{\infty} f_s(\mathbf{r},\mathbf{v}-\Delta\mathbf{v},t-\Delta t)P_s(\mathbf{v}-\Delta\mathbf{v},\Delta\mathbf{v})\,\mathrm{d}^3(\Delta v). \tag{12.3.2}$$

We note that Eq. (12.3.2) assumes implicitly that the changes in the velocity dependence of the single-particle distribution function occur without a change in $\mathbf{r}$.

In other words, we assume that a typical particle undergoes many collisions in time Δt without changing its spatial location significantly. Assuming that $|\Delta \mathbf{v}|$ is much smaller than $|\mathbf{v}|$ and Δt is much smaller than t, we can expand the integrand on the right-hand side of Eq. (12.3.2) to obtain

$$f_s(\mathbf{r},\mathbf{v},t) = \int_{-\infty}^{\infty} \mathrm{d}^3(\Delta v)$$

$$\times \left[\begin{aligned} & f_s(\mathbf{r},\mathbf{v},t)P_s(\mathbf{v},\Delta\mathbf{v}) - \Delta t \frac{\partial f_s(\mathbf{r},\mathbf{v},t)}{\partial t} P_s(\mathbf{v},\Delta\mathbf{v}) + \cdots \\ & - \Delta\mathbf{v}\cdot\nabla_{\mathrm{v}}(f_s(\mathbf{r},\mathbf{v},t)P_s(\mathbf{v},\Delta\mathbf{v})) + \frac{1}{2}\Delta\mathbf{v}\Delta\mathbf{v} : \nabla_{\mathrm{v}}\nabla_{\mathrm{v}}\{f_s(\mathbf{r},\mathbf{v},t) \\ & \times P_s(\mathbf{v},\Delta\mathbf{v})\} + \cdots \end{aligned} \right], \tag{12.3.3}$$

where a single dot indicates the scalar product of two vectors, and double dots indicate the scalar product of two tensors, which also produces a scalar. Using the normalization condition (12.3.1), we can now cancel the term on the left-hand side of the above equation with the first term on the right-hand side. If we divide the remaining terms in the above equation by Δt, take the limit $\Delta t \to 0$, and neglect the higher-order terms in $\Delta \mathbf{v}$, we obtain

$$\frac{\delta_c f_s}{\delta t} = -\nabla_{\mathrm{v}} \cdot \left[\left\langle \frac{\Delta\mathbf{v}}{\Delta t} \right\rangle f_s \right] + \frac{1}{2}\nabla_{\mathrm{v}}\nabla_{\mathrm{v}} : \left[\left\langle \frac{\Delta\mathbf{v}\Delta\mathbf{v}}{\Delta t} \right\rangle_s f_s \right], \tag{12.3.4}$$

where

$$\left\langle \frac{\Delta\mathbf{v}}{\Delta t} \right\rangle_s = \frac{1}{\Delta t}\int_{-\infty}^{\infty} \Delta\mathbf{v} P_s(\mathbf{v},\Delta\mathbf{v})\,\mathrm{d}^3(\Delta v) \tag{12.3.5}$$

is called the dynamical friction vector, and

$$\left\langle \frac{\Delta\mathbf{v}\Delta\mathbf{v}}{\Delta t} \right\rangle_s = \frac{1}{\Delta t}\int_{-\infty}^{\infty} \Delta\mathbf{v}\,\Delta\mathbf{v}\,P_s(\mathbf{v},\Delta\mathbf{v})\,\mathrm{d}^3(\Delta v) \tag{12.3.6}$$

is called the diffusion tensor. Equation (12.3.4) for $\delta_c f_s/\delta t$ is called the Fokker–Planck collision operator. Substituting $\delta_c f_s/\delta t$ into the Boltzmann equation (5.2.14) gives the Fokker–Planck equation

$$\frac{\partial f_s}{\partial t} + \mathbf{v}\cdot\nabla f_s + \frac{\mathbf{F}}{m_s}\cdot\nabla_{\mathrm{v}} f_s = -\nabla_{\mathrm{v}} \cdot \left[\left\langle \frac{\Delta\mathbf{v}}{\Delta t} \right\rangle_s f_s \right] + \frac{1}{2}\nabla_{\mathrm{v}}\nabla_{\mathrm{v}} : \left[\left\langle \frac{\Delta\mathbf{v}\Delta\mathbf{v}}{\Delta t} \right\rangle_s f_s \right]. \tag{12.3.7}$$

Although the above equation is formally correct, it is incomplete because we have not yet determined the functional forms of the dynamical friction vector and the diffusion tensor. This is done in the next section.

12.3.1 The Dynamical Friction Vector

Next we calculate the dynamical friction vector assuming that the collisional processes are dominated by binary Coulomb collisions, as discussed in Section 12.1. We start by considering collisions between a particle of type s moving with velocity $\mathbf{v}$ and particles of type s' with velocities in d^3v' about $\mathbf{v}'$. If $\sigma_{ss'}$ represents the Rutherford scattering cross section for collisions between a particle of type s and a particle of type s', then $\sigma_{ss'}\,\mathrm{d}\Omega\, f_{s'}(\mathbf{v}')\,\mathrm{d}^3v'$ is the probability per unit length that a particle of type s suffers a collision with particles of type s' with velocities in d^3v' about $\mathbf{v}'$ when the angle of scattering is in $\mathrm{d}\Omega$ about Ω. By Eq. (12.1.4), we note that $\sigma_{ss'} = \sigma_{ss'}(V, \chi)$, where $V = |\mathbf{v} - \mathbf{v}'|$ is the relative velocity of two particles before the collision and χ is the angle of scattering in the center-of-mass frame. Hence, $V\sigma_{ss'}(V, \chi)\,\mathrm{d}\Omega\, f_{s'}(\mathbf{v}')\,\mathrm{d}^3v'$ is the probability per unit time that a collision will occur between a particle of type s moving with velocity $\mathbf{v}$ and particles of type s' with velocities in d^3v' about $\mathbf{v}'$. Since the dynamical friction vector $\langle\Delta\mathbf{v}/\Delta t\rangle_s$ is the average change in the velocity of a particle of type s per unit time, we obtain

$$\left\langle\frac{\Delta\mathbf{v}}{\Delta t}\right\rangle_s = \sum_{s'}\int_{-\infty}^{\infty}\mathrm{d}^3v'\int_0^{4\pi} V\sigma_{ss'}(V,\chi)\, f_s'(\mathbf{v}')\Delta\mathbf{v}\,\mathrm{d}\Omega, \tag{12.3.8}$$

where $\Delta\mathbf{v} = \mathbf{v} - \mathbf{v}_1$, and $\mathbf{v}_1$ is the velocity of the particle of type s after the collision. To proceed further we need to express $\Delta\mathbf{v}$ in terms of V and χ.

Let $\mathbf{v}_1'$ be the velocity of a particle of type s' after a collision. Defining the post-collision relative velocity $\mathbf{V}' = \mathbf{v}_1 - \mathbf{v}_1'$, we obtain the change in the relative velocity

$$\Delta\mathbf{V} = \mathbf{V}' - \mathbf{V} = (\mathbf{v}_1 - \mathbf{v}_1') - (\mathbf{v} - \mathbf{v}') = \Delta\mathbf{v} - \Delta\mathbf{v}', \tag{12.3.9}$$

where $\Delta\mathbf{v} = \mathbf{v}_1 - \mathbf{v}$ and $\Delta\mathbf{v}' = \mathbf{v}_1' - \mathbf{v}'$. By the law of conservation of linear momentum, it follows that

$$m_s\mathbf{v} + m_{s'}\mathbf{v}' = m_s\mathbf{v}_1 + m_{s'}\mathbf{v}_1', \tag{12.3.10}$$

which gives

$$m_s\Delta\mathbf{v} + m_{s'}\Delta\mathbf{v}' = 0. \tag{12.3.11}$$

Eliminating $\Delta\mathbf{v}'$ between Eqs. (12.3.9) and (12.3.11) gives

$$\Delta\mathbf{v} = \frac{\mu_{ss'}}{m_s}\Delta\mathbf{V}. \tag{12.3.12}$$

The vector $\Delta\mathbf{V}$ can be expressed in x, y, and z coordinates as

$$\Delta\mathbf{V} = V\sin\chi\cos\phi\,\hat{\mathbf{x}} + V\sin\chi\sin\phi\,\hat{\mathbf{y}} - V(1-\cos\chi)\hat{\mathbf{z}}, \tag{12.3.13}$$

where we have assumed $\mathbf{V} = V\hat{\mathbf{z}}$, as shown in Figure 12.1. Using Eqs. (12.1.4), (12.3.12), and (12.3.13), we can evaluate the integral

$$\int_0^{4\pi} \sigma_{ss'}(V,\chi)\,\Delta\mathbf{v}\,\mathrm{d}\Omega = \frac{\mu_{ss'}}{m_s}\int_0^{4\pi} \sigma_{ss'}(V,\chi)\Delta\mathbf{V}\,\mathrm{d}\Omega \tag{12.3.14}$$

needed for the dynamical friction vector. We obtain

$$\begin{aligned}\int_0^{4\pi} \sigma_{ss'}(V,\chi)\,\Delta\mathbf{V}\,\mathrm{d}\Omega &= \int_0^{2\pi}\mathrm{d}\phi\int_{\chi_{\min}}^{\chi_{\max}} \sin\chi\,\sigma_{ss'}(V,\chi)\Delta\mathbf{V}\,\mathrm{d}\chi \\ &= -2\pi V\hat{\mathbf{z}}\int_{\chi_{\min}}^{\chi_{\max}} \sin\chi(1-\cos\chi)\sigma_{ss'}(V,\chi)\,\mathrm{d}\chi \\ &= -2\pi\mathbf{V}\left(\frac{e_s e_{s'}}{4\pi\epsilon_0\mu_{ss'}V^2}\right)^2\int_{\chi_{\min}}^{\chi_{\max}}\frac{\sin\chi(1-\cos\chi)}{4\sin^4\chi/2}\mathrm{d}\chi \\ &= -4\pi\mathbf{V}\left(\frac{e_s e_{s'}}{4\pi\epsilon_0\mu_{ss'}V^2}\right)^2\ln\Lambda_s, \end{aligned}\tag{12.3.15}$$

where $\chi_{\min}$ and $\chi_{\max}$ are the lower and upper cutoffs for the scattering angles and $\Lambda_s = \sin(\chi_{\max}/2)/\sin(\chi_{\min}/2)$. We will show below that $\ln\Lambda_s$ is closely related to $\ln\Lambda$, given by Eq. (12.2.9). Without introducing cutoffs, the integral (12.3.14) diverges logarithmically. Substituting Eq. (12.3.15) into Eq. (12.3.8), we obtain the dynamical friction vector

$$\left\langle\frac{\Delta\mathbf{v}}{\Delta t}\right\rangle_s = -\sum_{s'} 4\pi\ln\Lambda_s\left(\frac{e_s e_{s'}}{4\pi\epsilon_0\mu_{ss'}}\right)^2\frac{\mu_{ss'}}{m_s}\int_{-\infty}^{\infty} f_{s'}(\mathbf{v}')\frac{\mathbf{v}-\mathbf{v}'}{|\mathbf{v}-\mathbf{v}'|^3}\mathrm{d}^3v'. \tag{12.3.16}$$

To simplify the notation it is useful to define

$$\Gamma_{ss'} = \frac{e_s^2 e_{s'}^2}{4\pi\epsilon_0^2 m_s\mu_{ss'}}\ln\Lambda_s, \tag{12.3.17}$$

and

$$H_s(\mathbf{v}) = \int_{-\infty}^{\infty}\frac{f_s(\mathbf{v}')}{|\mathbf{v}-\mathbf{v}'|}\mathrm{d}^3v', \tag{12.3.18}$$

and to use the identity

$$\nabla_{\mathbf{v}}H_s(\mathbf{v}) = -\int_{-\infty}^{\infty}\frac{\mathbf{v}-\mathbf{v}'}{|\mathbf{v}-\mathbf{v}'|^3}f_s(\mathbf{v}')\,\mathrm{d}^3v'. \tag{12.3.19}$$

With these definitions the dynamical friction vector can be written in the form

$$\left\langle \frac{\Delta \mathbf{v}}{\Delta t} \right\rangle_s = \sum_{s'} \Gamma_{ss'} \nabla_{\mathbf{v}} H_{s'}(\mathbf{v}). \tag{12.3.20}$$

The function $H_{s'}(\mathbf{v})$, defined by Eq. (12.3.18), is called the first Rosenbluth potential (Rosenbluth et al., 1957). Its mathematical form is analogous to the electrostatic potential $\Phi(\mathbf{r})$ due to a distribution of charge $\rho_q(\mathbf{r})$,

$$\Phi(\mathbf{r}) = \frac{1}{4\pi\epsilon_0} \int_{-\infty}^{\infty} \frac{\rho_q(\mathbf{r}')}{|\mathbf{r} - \mathbf{r}'|} \mathrm{d}^3\chi', \tag{12.3.21}$$

which produces the electric field, $\mathbf{E} = -\nabla\Phi$. This analogy leads to some useful qualitative conclusions. For instance, if a charge distribution is spherically symmetric, then $\mathbf{E} = E(r)\hat{\mathbf{r}}$. Furthermore, $\mathbf{E}$ at any point $\mathbf{r}$ depends only on the charge contained *within* the sphere centered at the origin and of radius r, not on the charge outside the sphere. Analogously, if $f_s(\mathbf{v})$ is a spherically symmetric distribution function in velocity space, that is, $f_s(\mathbf{v}) = f_s(v)$, then the dynamical friction vector on a fast particle of velocity $\mathbf{v}$ must depend only on $\mathbf{v}$ and on particles moving *slower* than the fast particle, not on particles moving faster than the fast particle. By Eq. (12.3.19), the direction of this force is such as to slow down the fast particle, which explains why it is called a frictional force.

12.3.2 The Diffusion Tensor

Using the same approach as used in the last section, the diffusion tensor (12.3.6) can be written

$$\left\langle \frac{\Delta \mathbf{v} \Delta \mathbf{v}}{\Delta t} \right\rangle_s = \sum_{s'} \left(\frac{\mu_{ss'}}{m_s} \right)^2 \int_{-\infty}^{\infty} \mathrm{d}^3 v' \int_0^{4\pi} V \sigma_{ss'}(V,\chi) f_{s'}(\mathbf{v}') \Delta\mathbf{V}\, \Delta\mathbf{V}\, \mathrm{d}\Omega. \tag{12.3.22}$$

In the above equation it is useful to define the following tensor integral, which using Eq. (12.3.13) can be written

$$\begin{aligned} \overleftrightarrow{\mathbf{I}} &= \int_0^{4\pi} \sigma_{ss'}(V,\chi) \Delta\mathbf{V}\, \Delta\mathbf{V}\, \mathrm{d}\Omega \\ &= 2\pi V^2 \int_{\chi_{\min}}^{\chi_{\max}} \sin\chi\, \sigma_{ss'}(V,\chi) \left[(\hat{\mathbf{x}}\hat{\mathbf{x}} + \hat{\mathbf{y}}\hat{\mathbf{y}}) \frac{\sin^2\chi}{2} + \hat{\mathbf{z}}\hat{\mathbf{z}}(1 - \cos\chi)^2 \right] \mathrm{d}\chi. \end{aligned} \tag{12.3.23}$$

It follows from the above equation that $\overleftrightarrow{\mathbf{I}}$ is diagonal with

$$I_{xx} = I_{yy} = \pi V^2 \int_{\chi_{\min}}^{\chi_{\max}} \sin^3\chi\, \sigma_{ss'}(V,\chi)\, \mathrm{d}\chi \tag{12.3.24}$$

and

$$I_{zz} = 2\pi V^2 \int_{\chi_{\min}}^{\chi_{\max}} \sin\chi\,(1-\cos\chi)^2 \sigma_{ss'}(V,\chi)\,\mathrm{d}\chi. \tag{12.3.25}$$

Substituting expression (12.1.4) for $\sigma_{ss'}$ into Eqs. (12.3.24) and (12.3.25) and carrying out the integrals over χ, we obtain

$$I_{xx} = I_{yy} = \frac{1}{4\pi}\left(\frac{e_s e_{s'}}{\epsilon_0 \mu_{ss'} V}\right)^2 \left(\ln\Lambda_s + \frac{\cos\chi_{\max}}{4} - \frac{\cos\chi_{\min}}{4}\right). \tag{12.3.26}$$

As discussed below, $\ln\Lambda_s$ is typically much larger than unity, so we can approximate I_{xx} and I_{yy} by

$$I_{xx} = I_{yy} \simeq \frac{1}{4\pi}\left(\frac{e_s e_{s'}}{\epsilon_0 \mu_{ss'} V}\right)^2 \ln\Lambda_s. \tag{12.3.27}$$

Similarly, Eq. (12.3.25) yields

$$I_{zz} = \frac{1}{8\pi}\left(\frac{e_s e_{s'}}{\epsilon_0 \mu_{ss'} V}\right)^2 (\cos\chi_{\min} - \cos\chi_{\max}). \tag{12.3.28}$$

It is clear by inspection of Eqs. (12.3.27) and (12.3.28) that $I_{xx} = I_{yy} \gg I_{zz}$ because $\ln\Lambda_s \gg 1$. As we have neglected terms of order I_{zz} in reducing Eq. (12.3.26) to (12.3.27), it is consistent to assume that

$$I_{zz} \simeq 0. \tag{12.3.29}$$

Equations (12.3.27) and (12.3.29) have been derived assuming a coordinate system in which $\mathbf{V} = V\hat{\mathbf{z}}$. It is easy to see that the coordinate independent generalization of $\overleftrightarrow{\mathbf{I}}$ is given by

$$\overleftrightarrow{\mathbf{I}} \simeq \frac{1}{4\pi}\left(\frac{e_s e_{s'}}{\epsilon_0 \mu_{ss'} V}\right)^2 \left[\overleftrightarrow{\mathbf{1}} - \frac{\mathbf{V}\mathbf{V}}{V^2}\right] \ln\Lambda_s, \tag{12.3.30}$$

where $\overleftrightarrow{\mathbf{1}}$ is the unit tensor. Substituting the integral (12.3.30) into Eq. (12.3.22), we obtain the diffusion tensor

$$\left\langle \frac{\Delta\mathbf{v}\Delta\mathbf{v}}{\Delta t} \right\rangle_s = \sum_{s'} \Gamma_{ss'} \frac{\mu_{ss'}}{m_s} \int_{-\infty}^{\infty} \frac{f_{s'}(\mathbf{v}')}{|\mathbf{v}-\mathbf{v}'|} \left[\overleftrightarrow{\mathbf{1}} - \frac{(\mathbf{v}-\mathbf{v}')(\mathbf{v}-\mathbf{v}')}{|\mathbf{v}-\mathbf{v}'|^2}\right] \mathrm{d}^3 v', \tag{12.3.31}$$

where we recall that $\mathbf{V} = \mathbf{v} - \mathbf{v}'$. The above equation can be simplified further by using the identity

$$\nabla_{\mathbf{v}}\nabla_{\mathbf{v}} V = \nabla_{\mathbf{v}}\left(\frac{\mathbf{V}}{V}\right) = \frac{1}{V}\left(\overleftrightarrow{\mathbf{1}} - \frac{\mathbf{V}\mathbf{V}}{V^2}\right) \tag{12.3.32}$$

and the definition

$$G_s(\mathbf{v}) = \int_{-\infty}^{\infty} f_s(\mathbf{v}')\,|\mathbf{v}-\mathbf{v}'|\mathrm{d}^3v', \tag{12.3.33}$$

whereupon we obtain

$$\left\langle \frac{\Delta\mathbf{v}\Delta\mathbf{v}}{\Delta t} \right\rangle_s = \sum_{s'} \Gamma_{ss'} \frac{\mu_{ss'}}{m_s} \nabla_{\mathbf{v}}\nabla_{\mathbf{v}} G_{s'}(\mathbf{v}). \tag{12.3.34}$$

The function $G_s(\mathbf{v})$ is known as the second Rosenbluth potential (Rosenbluth et al., 1957). It is related to the first Rosenbluth potential $H_s(\mathbf{v})$ by the identity

$$H_s(\mathbf{v}) = \frac{1}{2}\nabla_{\mathbf{v}} \cdot \nabla_{\mathbf{v}} G_s(\mathbf{v}). \tag{12.3.35}$$

Substituting the dynamical friction vector (12.3.20) and the diffusion tensor (12.3.34) into Eq. (12.3.4) for the Fokker–Planck collision operator gives

$$\frac{\delta_{\mathrm{c}} f_s}{\delta t} = \sum_{s'} \Gamma_{ss'} \left[-\nabla_{\mathbf{v}} \cdot (f_s \nabla_{\mathbf{v}} H_{s'}) + \frac{1}{2}\nabla_{\mathbf{v}}\nabla_{\mathbf{v}} : (f_s \nabla_{\mathbf{v}}\nabla_{\mathbf{v}} G_{s'}) \right]. \tag{12.3.36}$$

It remains to discuss the parameter

$$\ln\Lambda_s = \ln \frac{\sin(\chi_{\max}/2)}{\sin(\chi_{\min}/2)}. \tag{12.3.37}$$

Let us denote the impact parameters that produce the scattering angles $\chi_{\min}$ and $\chi_{\max}$ by b_{m} and $b_{\min}$, respectively. Using relation (12.1.1), we obtain

$$\sin^2\frac{\chi}{2} = \frac{b_0^2}{b_0^2 + b^2}, \tag{12.3.38}$$

which implies that

$$\ln\Lambda_s = \frac{1}{2}\ln\left(\frac{b_0^2 + b_m^2}{b_0^2 + b_{\min}^2}\right). \tag{12.3.39}$$

As discussed in Section 12.2, the parameter b_{m} is usually chosen to be the Debye length λ_{D}, while the parameter b_0 is usually taken to be given by $b_0 = e_s e_{s'}/(12\pi\epsilon_0\kappa T)$, where T is the temperature of the plasma (assuming the species have attained thermal equilibrium). Since there is some freedom in the choice of $b_{\min}$, one may choose either $b_{\min} = b_0$ or $b_{\min} = 0$. Either choice leads to values of $\ln\Lambda_s$ that differ from each other to the extent of $\ln\sqrt{2}$, which is of order unity.

To be consistent with Section 12.2, we choose $b_{\min} = 0$. Since $b_{\rm m} \gg b_0$, we then obtain, by Eq. (12.2.8),

$$\ln \Lambda_s \simeq \ln\left(\frac{b_{\rm m}}{b_0}\right) = \ln(12\pi N_{\rm D}) = \ln \Lambda. \tag{12.3.40}$$

12.4 Conductivity of a Fully Ionized Plasma

As an application of the Fokker–Planck equation, we next compute the conductivity of a plasma using the Lorentz gas approximation. In this approximation, electrons are assumed to collide with fixed immobile ions. Since the ions are immobile (i.e., $m_{\rm i} \gg m_{\rm e}$), the ion distribution function can be represented by a Dirac delta function, that is,

$$f_{\rm i}(\mathbf{v}_{\rm i}) = n_0 \delta(\mathbf{v}_{\rm i}), \tag{12.4.1}$$

where n_0 is the number density of electrons and ions. Substituting the above equation into Eqs. (12.3.18) and (12.3.33), the Rosenbluth potentials reduce to

$$H_{\rm i}(\mathbf{v}) \simeq \frac{n_0}{v} \tag{12.4.2}$$

and

$$G_{\rm i}(\mathbf{v}) \simeq n_0 v. \tag{12.4.3}$$

The Fokker–Planck collision operator then becomes

$$\frac{\delta_{\rm c} f_{\rm e}}{\delta t} \simeq n_0 \Gamma_{\rm ei}\left[-\nabla_{\rm v} \cdot \left(f_{\rm e} \nabla_{\rm v}\left(\frac{1}{v}\right)\right) + \frac{1}{2}\nabla_{\rm v}\nabla_{\rm v} : (f_{\rm e}\nabla_{\rm v}\nabla_{\rm v} v)\right], \tag{12.4.4}$$

where, by Eq. (12.3.17),

$$\Gamma_{\rm ei} = \frac{e^4}{4\pi\epsilon_0^2 m_{\rm e}^2} \ln \Lambda. \tag{12.4.5}$$

We use the identities

$$\nabla_{\rm v}\left(\frac{1}{v}\right) = -\frac{1}{v^2}\nabla_{\rm v} V = -\frac{\mathbf{v}}{v^3} \tag{12.4.6}$$

and

$$\nabla_{\rm v}\nabla_{\rm v} V = \frac{1}{v}\left(\overleftrightarrow{\mathbf{1}} - \frac{\mathbf{vv}}{v^2}\right) \tag{12.4.7}$$

to rewrite Eq. (12.4.4) in the form

$$\frac{\delta_c f_e}{\delta t} = n_0 \Gamma_{ei} \left[-\nabla_v \cdot \left(-\frac{f_e \mathbf{v}}{v^3} \right) + \frac{1}{2} \nabla_v \nabla_v : \left(f_e \frac{v^2 \overleftrightarrow{\mathbf{1}} - \mathbf{vv}}{v^3} \right) \right]. \tag{12.4.8}$$

If we now use the identity

$$\nabla_v \cdot \left(\frac{v^2 \overleftrightarrow{\mathbf{1}} - \mathbf{vv}}{v^3} \right) = -\frac{2\mathbf{v}}{v^3}, \tag{12.4.9}$$

it is easy to see that the first term in the parentheses, which represents the dynamical friction vector, is canceled exactly by an identical term of opposite sign from the diffusion tensor. This cancellation is a consequence of the assumption of infinite mass of the ions, which in turn implies that the electron energy is conserved in a collision with ions, allowing no energy transfer to ions. As a result of this cancellation, we are left with the expression

$$\frac{\delta_c f_e}{\delta t} = \frac{n_0 \Gamma_{ei}}{2} \nabla_v \cdot \left(\frac{v^2 \overleftrightarrow{\mathbf{1}} - \mathbf{vv}}{v^3} \cdot \nabla_v f_e \right). \tag{12.4.10}$$

We note that the right-hand side of the above equation vanishes identically for an isotropic electron distribution function, that is, when $f_e = f_e(v^2)$. This can be seen easily by substituting the identity

$$\nabla_v f_e(v^2) = 2\mathbf{v} \frac{\partial f_e}{\partial v^2} \tag{12.4.11}$$

into Eq. (12.4.10), whereby we obtain

$$(v^2 \overleftrightarrow{\mathbf{1}} - \mathbf{vv}) \cdot \mathbf{v} = v^2 \mathbf{v} - \mathbf{v} v^2 = 0, \tag{12.4.12}$$

thereby annihilating the collision operator. The physical significance of this result is that the collision operator (12.4.10) will tend to relax an anisotropic electron distribution to an isotropic distribution as collisional processes evolve the plasma toward a steady state.

We now proceed with the calculation of the plasma conductivity (see Goldston and Rutherford, 1995). We assume that a small electric field $\mathbf{E} = E\hat{\mathbf{z}}$ is applied to a homogeneous currentless plasma in which the electron distribution function is initially Maxwellian (before the electric field is applied). After the transients have died down, let us assume that the plasma evolves to a spatially homogeneous steady state. From Eqs. (12.3.7) and (12.4.10), we obtain

$$-\frac{e}{m_{\mathrm{e}}}\mathbf{E}\cdot\frac{\partial f_{\mathrm{e}}}{\partial \mathbf{v}} = \frac{n_0\Gamma_{\mathrm{ei}}}{2}\boldsymbol{\nabla}_{\mathbf{v}}\cdot\left(\frac{v^2\overleftrightarrow{\mathbf{1}} - \mathbf{vv}}{v^3}\cdot\boldsymbol{\nabla}_{\mathbf{v}} f_{\mathrm{e}}\right). \tag{12.4.13}$$

It is convenient to write the above equation in spherical polar velocity coordinates (v,θ,ϕ), where $v_x = v\sin\theta\cos\phi, v_y = v\sin\theta\sin\phi$, and $v_z = v\cos\theta$. Since the applied electric field points in the z direction and there is no other preferred direction, we require that the distribution function be azimuthally symmetric, that is, $\partial f_{\mathrm{e}}/\partial\phi = 0$. Under these conditions, Eq. (12.4.13) can be written in the form

$$-\frac{e}{m_{\mathrm{e}}}E\left(\cos\theta\frac{\partial f_{\mathrm{e}}}{\partial v} - \frac{\sin\theta}{v}\frac{\partial f_{\mathrm{e}}}{\partial\theta}\right) = \frac{n_0\Gamma_{\mathrm{ei}}}{2v^3\sin\theta}\frac{\partial}{\partial\theta}\left(\sin\theta\frac{\partial f_{\mathrm{e}}}{\partial\theta}\right). \tag{12.4.14}$$

For a small electric field E, we can linearize the above equation by writing

$$f_{\mathrm{e}}(v,\theta) = f_{\mathrm{e0}}(v) + f_{\mathrm{e1}}(v,\theta), \tag{12.4.15}$$

and neglecting terms proportional to the product Ef_{e1} on the left-hand side. We assume that the zero-order distribution function is a Maxwellian,

$$f_{\mathrm{e0}}(v) = n_0\left(\frac{m_{\mathrm{e}}}{2\pi\kappa T_{\mathrm{e}}}\right)^{3/2}\exp\left(-\frac{m_{\mathrm{e}}v^2}{2\kappa T_{\mathrm{e}}}\right). \tag{12.4.16}$$

The linearized version of the above equation is

$$-\frac{e}{m_{\mathrm{e}}}E\cos\theta\frac{\partial f_{\mathrm{e0}}}{\partial v} = \frac{n_0\Gamma_{\mathrm{ei}}}{2v^3\sin\theta}\frac{\partial}{\partial\theta}\left(\sin\theta\frac{\partial f_{\mathrm{e1}}}{\partial\theta}\right). \tag{12.4.17}$$

Substituting Eq. (12.4.16) into the left-hand side of the above equation, we obtain

$$\frac{eEv\cos\theta}{\kappa T_{\mathrm{e}}}f_{\mathrm{e0}} = \frac{n_0\Gamma_{\mathrm{ei}}}{2v^3\sin\theta}\frac{\partial}{\partial\theta}\left(\sin\theta\frac{\partial f_{\mathrm{e1}}}{\partial\theta}\right). \tag{12.4.18}$$

The above equation can be solved for $f_{\mathrm{e1}}(v,\theta)$ by assuming that

$$f_{\mathrm{e1}}(v,\theta) = Av^4 f_{\mathrm{e0}}(v)\cos\theta, \tag{12.4.19}$$

where A is a constant to be determined by direct substitution in Eq. (12.4.18). After carrying out the necessary calculation, it is easy to show that

$$A = -\frac{eE}{n_0\Gamma_{\mathrm{ei}}\kappa T_{\mathrm{e}}}. \tag{12.4.20}$$

From the expression for $f_{e1}(v,\theta)$ we can calculate the linearized current density along the direction of the electric field:

$$J = -e\int_{-\infty}^{\infty} f_{e1} v \cos\theta \, d^3v$$

$$= \frac{e^2 E}{n_0 \Gamma_{ei} \kappa T_e}\int_0^{2\pi} d\phi \int_0^{\pi} \sin\theta \cos^2\theta \, d\theta \int_0^{\infty} v^7 f_{e0} \, dv. \qquad (12.4.21)$$

Using Eqs. (12.4.5) and (12.4.16) and carrying out the integrals in the above equation, we obtain

$$J = \frac{32\pi^{1/2}\epsilon_0^2(2\kappa T_e)^{3/2}}{m_e^{1/2} e^2 \ln\Lambda} E. \qquad (12.4.22)$$

The plasma conductivity, σ, defined by the relation

$$J = \sigma E, \qquad (12.4.23)$$

is then given, in the Lorentz gas approximation, by the expression

$$\sigma = \frac{32\pi^{1/2}\epsilon_0^2(2\kappa T_e)^{3/2}}{m_e^{1/2} e^2 \ln\Lambda}. \qquad (12.4.24)$$

It turns out that Eq. (12.4.21), which is derived assuming that electron–electron collisions are neglected, overestimates the actual plasma conductivity. This occurs not because electron–electron collisions contribute additively to the effective number of collisions that electron suffer, but more subtly because electron–electron collisions change the electron distribution function and hence, the electron–ion collisional interaction. Note that, except for the weak dependence through $\ln\Lambda$, the conductivity depends only on the electron temperature, and is completely independent of the number density.

The plasma resistivity η is the inverse of the conductivity σ. A more detailed calculation of the resistivity that takes into account the effect of electron–electron collisions, first carried out by Spitzer and Harm (1953), yields the following formula for the resistivity:

$$\eta \simeq 5.2\times 10^{-5}\frac{\ln\Lambda}{(\kappa T_e)^{3/2}} \text{ ohm m}, \qquad (12.4.25)$$

where κT_e is expressed in electron volts. To be precise, the above equation gives the Spitzer–Harm resistivity parallel to an equilibrium magnetic field. The resistivity perpendicular to the magnetic field is approximately twice the parallel resistivity.

Table 12.1 *Comparison of the resistivity of various types of plasmas with some common materials*

Material	Resistivity, η (ohm m)
Copper	2×10^{-8}
Stainless steel	7×10^{-7}
100 eV plasma	5×10^{-7}
5 keV plasma	1×10^{-9}
Interstellar gas (1 eV)	5×10^{-7}
Solar corona (10 eV)	5×10^{-5}
Earth's ionosphere (0.1 eV)	2×10^{-2}

It is interesting to compare the resistivity of fully ionized plasmas at different temperatures to some common materials. For example, in Table 12.1 we compare the resistivities of 100 eV and 5 keV plasmas to those of copper and stainless steel. As one can see, a 100 eV plasma has very low resistivity, comparable to that of stainless steel, but larger than that of copper. However, a 5 keV plasma, typical of large laboratory fusion experiments, is an order of magnitude less resistive than copper.

From Eq. (5.6.8), we can make an estimate of the electron–ion collision frequency, ν_{ei}, given by

$$\nu_{ei} \simeq \frac{n_0 e^2}{m_e \sigma} \simeq \frac{n_0 e^4 \ln \Lambda}{32\pi^{1/2}\epsilon_0^2 m_e^{1/2}(2\kappa T_e)^{3/2}}. \tag{12.4.26}$$

This equation was previously used in Section 2.5.2.

12.5 Collision Operator for Maxwellian Distributions of Electrons and Ions

In the previous section, we calculated the Rosenbluth potentials in the Lorentz gas approximation where the ion distribution function can be approximated by a Dirac delta function, and ions are assumed to be infinitely massive compared with electrons ($m_e/m_i \to 0$). In this approximation, there is no energy transfer to ions. We now generalize this treatment to a case where the electrons and ions both have Maxwellian distributions (with different temperatures) and the ions have a large but finite mass. For both the electrons and ions we assume that the distribution function is given by

$$f_s(\mathbf{v}) = n_0 \left(\frac{m_s}{2\pi\kappa T_s} \right)^{3/2} \exp\left(-\frac{m_s v^2}{2\kappa T_s} \right). \tag{12.5.1}$$

Using the identity

$$\int_{-\infty}^{\infty} \frac{\exp(-x^2)}{|\mathbf{x} - \mathbf{y}|} \mathrm{d}^3 x = \frac{\pi^{3/2}}{y} \operatorname{erf}(y), \tag{12.5.2}$$

where erf(y) is the error function, defined by

$$\operatorname{erf}(y) = \frac{2}{\sqrt{\pi}} \int_0^y \exp(-x^2) \mathrm{d}x, \tag{12.5.3}$$

we obtain the first Rosenbluth potential

$$H_s(\mathbf{v}) = n_0 \left(\frac{m_s}{2\kappa T_s} \right)^{1/2} \frac{\operatorname{erf}(y)}{y}, \tag{12.5.4}$$

where $y = v/(2\kappa T_s/m_s)^{1/2}$. Similarly, the second Rosenbluth potential can be shown to be

$$G_s(\mathbf{v}) = \frac{n_0}{2y} \left(\frac{2\kappa T_s}{m_s} \right)^{1/2} \left[y \frac{\mathrm{d}}{\mathrm{d}y} \operatorname{erf}(y) + (1 + 2y^2) \operatorname{erf}(y) \right]. \tag{12.5.5}$$

The dynamical friction vector, given by Eq. (12.3.20), can now be written

$$\left\langle \frac{\Delta \mathbf{v}}{\Delta t} \right\rangle_s = \sum_{s'} \mathbf{F}_{ss'}, \tag{12.5.6}$$

where the dynamical friction force acting on species s due to species s' is given by

$$\mathbf{F}_{ss'} = -\Gamma_{ss'} n_0 \left[\operatorname{erf}(y) - y \frac{\mathrm{d}}{\mathrm{d}y} \operatorname{erf}(y) \right] \frac{\mathbf{v}}{v^3}. \tag{12.5.7}$$

Similarly, after some algebra, the diffusion tensor can be shown to be

$$\left\langle \frac{\Delta \mathbf{v} \Delta \mathbf{v}}{\Delta t} \right\rangle_s = \sum_{s'} \Gamma_{ss'} \frac{n_0}{2y^3} \left(\frac{m_{s'}}{2\kappa T_{s'}} \right)^{1/2} \frac{\mu_{ss'}}{m_s} \left[\overleftrightarrow{\mathbf{1}} g_1(y) + 3 \frac{\mathbf{v}\mathbf{v}}{v^2} g_2(y) \right], \tag{12.5.8}$$

where

$$g_1(y) = (2y^2 - 1) \operatorname{erf}(y) + y \frac{\mathrm{d}}{\mathrm{d}y} \operatorname{erf}(y), \tag{12.5.9}$$

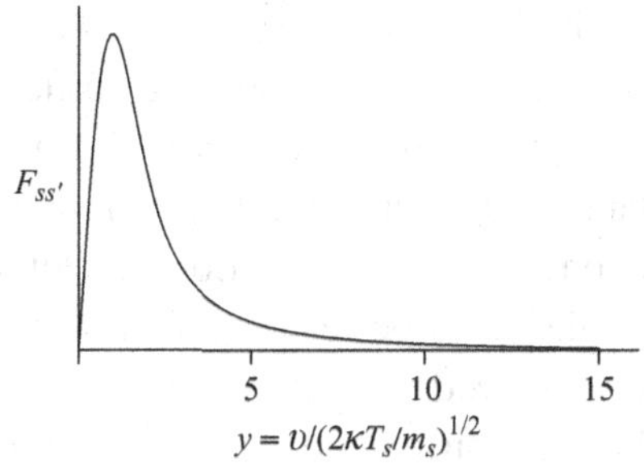

Figure 12.2 A plot of the dynamical friction force as a function of the dimensionless velocity parameter y.

and

$$g_2(y) = \left(1 - \frac{2y^2}{3}\right)\mathrm{erf}(y) - y\frac{\mathrm{d}}{\mathrm{d}y}\,\mathrm{erf}(y). \tag{12.5.10}$$

The magnitude of the dynamical friction force $F_{ss'}$ given by Eq. (12.5.7) is plotted as a function of the dimensionless velocity parameter y in Figure 12.2. This plot shows that the magnitude of the drag force initially increases linearly with velocity. As the velocity approaches the thermal speed (which corresponds to $y = 1$), the drag force reaches a maximum and then starts to decrease, eventually varying as $1/v^2$ at velocities much greater than the thermal velocity. The occurrence of a well-defined maximum in the drag force leads to the phenomenon of runaways, as described below.

When an electric field $\mathbf{E}$ is applied to a plasma, the electrons and ions acquire different flow velocities or drifts parallel to $\mathbf{E}$. Assuming that the plasma is spatially uniform, the momentum equation for the electrons can be written

$$m_e n_0 \frac{\partial \mathbf{U}_e}{\partial t} = -e n_0 \mathbf{E} + m_e n_0 \mathbf{F}_{ei}, \tag{12.5.11}$$

where $m_e n_0 \mathbf{F}_{ei}$ is the drag force exerted on the electrons by the ions and $\mathbf{F}_{ei}$ is given by Eq. (12.5.7). For small electric fields, the drag force increases linearly with v. In this regime it is possible to attain a steady state whereby the left-hand side vanishes and the applied electric field is exactly balanced by the drag force. Indeed, it is this force balance that produces the condition $\mathbf{J} = \sigma\mathbf{E}$ used in Section 12.4 to define the plasma conductivity (Eq. 12.4.24). However, if the electric field is larger than a critical field, E_c, given by

$$E > E_c = m_e (F_{ei})_{max}/e, \tag{12.5.12}$$

where $(F_{ei})_{max}$ is the maximum of the curve in Figure 12.2 (applied to the electrons and ions), then the magnitude of U_e will grow without bound. The critical field

E_c is called the Dreicer field (Dreicer, 1959). For electric field strengths greater than the Dreicer field, electrons are accelerated to sufficiently large speeds that the magnitude of the drag force starts to decrease as $1/v^2$, making it impossible for the drag force to balance the electric field and produce a steady state. Under such conditions, the electron distribution function accelerates without limit. Such runaway acceleration is believed to be responsible for the high velocity electron beams that are frequently observed in both space and laboratory plasmas. When the electron velocities become sufficiently large, the non-relativistic treatment given above breaks down, and needs to be replaced by a relativistic treatment (Connor and Hastie, 1975), which predicts a somewhat different critical field. The maximum velocity attainable by the electrons in a relativistic treatment can approach but never exceed the speed of light.

Problems

12.1. Calculate numerical values of $\ln\Lambda$ for hydrogen plasmas in the range of density 1–10^{20}cm^3 and temperature 10^2–10^8 K. How sensitive is $\ln\Lambda$ to such a wide range of density and temperature?

12.2. Show that the first and second Rosenbluth potentials, given by

$$H_s(\mathbf{v}) = \int_{-\infty}^{\infty} \frac{f_s(\mathbf{v}')}{|\mathbf{v}-\mathbf{v}'|} \mathrm{d}^3 v' \text{ and}$$

$$G_s(\mathbf{v}) = \int_{-\infty}^{\infty} f_s(\mathbf{v}')\,|\mathbf{v}-\mathbf{v}'|\,\mathrm{d}^3 v',$$

are related by

$$H_s(\mathbf{v}) = \frac{1}{2}\nabla_{\mathbf{v}} \cdot \nabla_{\mathbf{v}} G_s(\mathbf{v}).$$

12.3. If the velocity distribution function for the sth species is a Maxwellian

$$f_s(\mathbf{v}) = n_0 \left(\frac{m_s}{2\pi\kappa T_s}\right)^{3/2} \exp\left(-\frac{m_s v^2}{2\kappa T_s}\right),$$

show that the first Rosenbluth potential is given by

$$H_s(\mathbf{v}) = n_0 \left(\frac{m_s}{2\kappa T_s}\right)^{1/2} \frac{\operatorname{erf}(y)}{y},$$

where erf(y) is the error function with $y = v/(2kT_s/m_s)^{1/2}$, and that the second Rosenbluth potential is given by

$$G_s(\mathbf{v}) = \frac{n_0}{2y}\left(\frac{2\kappa T_s}{m_s}\right)^{1/2} \left[y\frac{\mathrm{d}}{\mathrm{d}y}\operatorname{erf}(y) + (1+2y^2)\operatorname{erf}(y)\right].$$

12.4. For the Maxwellian velocity distribution in Problem 12.3, show that the diffusion tensor is given by

$$\left\langle \frac{\Delta\mathbf{v}\Delta\mathbf{v}}{\Delta t} \right\rangle_s = \sum_{s'} \Gamma_{ss'} \frac{n_0}{2y^3} \left(\frac{m_{s'}}{2\kappa T_{s'}} \right)^{1/2} \frac{\mu_{ss'}}{m_s} \left[\overleftrightarrow{\mathbf{1}} g_1(y) + 3\frac{\mathbf{vv}}{v^2} g_2(y) \right],$$

where

$$g_1(y) = (2y^2 - 1)\mathrm{erf}(y) + y\frac{\mathrm{d}}{\mathrm{d}y}\mathrm{erf}(y)$$

and

$$g_2(y) = \left(1 - \frac{2y^2}{3} \right)\mathrm{erf}(y) - y\frac{\mathrm{d}}{\mathrm{d}y}\mathrm{erf}(y).$$

References

Connor, J. W., and Hastie, R. J. 1975. Relativistic limitations on runaway electrons. *Nucl. Fusion* **15**, 415–424.

Dreicer, H. 1959. Electron and ion runaway in a fully ionized gas. *Phys. Rev.* **115**, 242–249.

Goldston, R. J., and Rutherford, P. H. 1995. *Introduction to Plasma Physics*. Bristol: Institute of Physics.

Ichimaru, S. 1994. *Statistical Plasma Physics*, vol. II. Reading, MA: Addison-Wesley.

Rosenbluth, M. N., MacDonald, W. M., and Judd, D. 1957. Fokker–Planck equation for an inverse-square force. *Phys. Rev.* **107**, 1–6.

Spitzer, L., and Harm, R. 1953. Transport phenomena in a completely ionized gas. *Phys. Rev.* **89**, 977–981.

Further Reading

Boyd, T. J. M., and Sanderson, J. J. 2003. *The Physics of Plasmas*. Cambridge: Cambridge University Press, Chapter 8.

Schmidt, G. 1979. *Physics of High Temperature Plasmas*. New York: Academic Press, Chapter 11.

Sturrock, P. A. 1994. *Plasma Physics*. Cambridge: Cambridge University Press, Chapter 10.

Appendix A: Symbols

The following list gives all of the mathematical symbols used in this book and a simple description of the quantity involved. The symbols are organized in alphabetical order, with English symbols first and Greek symbols second. In some cases the same symbol has more than one meaning. For example, the symbol n can mean number density and index of refraction, as well as the integer n. Usually the meaning is obvious from the context. However, to aid in determining the correct meaning, the nearest equation and page numbers in which the symbol first appears are given.

Symbol	Definition	Equation	Page
a	scale length in Harris current sheet	7.6.1	270
a_0	parameter that controls the width of a Gaussian envelope	4.1.3	88
A	area of current loop	3.1.8	24
A	term in equation for index of refraction	4.4.16	109
A	anisotropy factor	10.3.58	418
$A(\mathbf{r} - \mathbf{v}_g t)$	amplitude of wave packet	4.1.25	93
$\mathbf{A}$	vector potential	3.10.1	76
$[A.B]$	Poisson bracket	3.10.8	77
b	impact parameter	3.9.29	70
b_0	impact parameter for 90° scattering	12.1.2	480
B	magnetic field magnitude	2.4.1	11
B	term in equation for index of refraction	4.4.16	109

continued

Symbol	Definition	Equation	Page
B_m	magnetic field magnitude at mirror point	3.4.19	42
B_{max}	maximum magnetic field strength	3.4.23	43
B_t	toroidal magnetic field	3.9.31	72
$\mathbf{B}$	magnetic field vector	3.1.1	23
$\mathbf{B}_0$	externally imposed magnetic field	4.4.1	105
$\hat{\mathbf{B}}$	unit vector in direction of magnetic field	3.3.7	33
c	speed of light	4.1.15	91
c_V	specific heat capacity at constant volume	8.3.56	303
C	constant in energy equation	7.3.36	252
C	integration contour in complex v_z space	9.2.9	330
C_A	ion acoustic speed	9.5.31	370
C_s	thermal speed of sth species	2.1.3	5
C_P	heat capacity at constant pressure	6.1.25	192
C_V	heat capacity at constant volume	6.1.25	192
$d\Omega$	differential solid angle	2.5.2	13
ds	differential distance along magnetic field line	4.4.33	115
D	dielectric tensor element	4.4.8	107
$D_q(v_z,t)$	quasi-linear diffusion coefficient	11.1.44	436
D_W	whistler dispersion	4.4.34	116
$Đ(i\mathbf{k},-i\omega)$	dispersion relation	4.1.13	90
$Đ_\ell(k,\omega)$	dispersion relation for the longitudinal mode	5.5.18	169
$\mathbf{D}$	displacement vector	4.2.1	96
$\overleftrightarrow{\mathbf{Đ}}$	matrix representation of a system of linear equations	4.1.19	91
e	electronic charge	2.2.1	6
e_s	charge for sth species	2.3.6	11
E	electric field magnitude	2.3.1	10
$\mathbf{E}_A$	ion acoustic wave electric field	11.2.47	457
$\mathbf{E}_L$	Langmuir wave electric field	11.2.35	454
$\mathbf{E}_T$	transverse electric field	11.2.34	454
$\mathcal{E}(k,t)$	electric field spectral density	11.1.42	435
$\mathbf{E}$	electric field vector	3.2.2	25
f_{ce}	electron cyclotron frequency, Hz	2.4.2	11
f_{pe}	electron plasma frequency, Hz	2.3.5	10
$f_s(\mathbf{v})$	velocity distribution function for the sth species	2.1.1	4
F	term in the equation for the index of refraction	4.4.18	109

Symbol	Definition	Equation	Page
$\mathbf{F}_\mathrm{P}$	ponderomotive force	3.7.7	51
$F_0(v_z)$	normalized one-dimensional velocity distribution function	9.1.12	321
g	gravitational acceleration	7.2.9	239
$g(\mathbf{r}, \mathbf{v})$	weighting factor in phase space	5.1.3	149
$g(z)$	complex function	9.2.9	330
G	gravitational constant	7.2.9	239
$G(z)$	complex function, $G(z) = g(z)(z - z_0)^n$	9.2.9	330
$G_s(\mathbf{v})$	second Rosenbluth potential	12.3.33	490
$G_{s0}(v_\parallel)$	reduced one-dimensional velocity distribution function	10.3.34	409
h	enthalpy	6.4.16	205
$\hbar$	Planck's constant divided by 2π	2.7.1	17
H	Hamiltonian	3.9.1	64
$H(\omega/k)$	function used to evaluate beam instabilities	9.1.30	324
$H_s(\mathbf{v})$	first Rosenbluth potential	12.3.18	487
H_S	scale height	7.3.53	256
$\mathcal{H}$	magnetic helicity	7.1.20	227
$\mathbf{H}$	magnetic intensity	4.2.1	96
I	current	3.1.7	24
I	integral invariant	3.8.21	56
$I_n(x)$	modified Bessel function of order n	10.2.28	388
J	action integral	3.8.1	52
J	second adiabatic invariant	3.8.18	55
J	Jacobian	5.2.5	153
J_n	Bessel function of order n	10.2.11	383
$\mathbf{J}$	current density	3.2.10	27
$\mathbf{J}_\mathrm{G}$	gradient drift current	3.3.13	37
$\mathbf{J}_\mathrm{m}$	magnetization current	3.3.12	37
$\mathbf{J}_\mathrm{NL}$	nonlinear current	11.2.32	454
$\mathbf{J}_\mathrm{r}$	real current	4.2.1	96
k	wave number	4.1.1	87
k	Hook's law constant	11.2.5	446
k_0	wave number at spectral peak	4.1.3	88
$k_\parallel$	parallel component of wave vector	3.8.28	59
$k_\perp$	perpendicular component of wave vector	10.1.12	381
$\mathbf{k}$	wave vector, also called propagation vector	4.1.8	89

continued

Symbol	Definition	Equation	Page
K	kinetic energy	6.4.20	206
K	dielectric constant	3.6.9	49
$\bar{k}_{\mathrm{a}}$	ambipolar diffusion coefficient	5.7.12	180
$\bar{k}_s$	diffusion coefficient of sth species	5.7.3	178
$\overleftrightarrow{\mathbf{K}}$	dielectric tensor	4.2.6	97
L	Lagrangian	3.9.1	64
L	dielectric tensor element	4.4.11	107
m_s	atomic mass for the sth species	2.1.1	4
M	magnetic moment	3.9.24	70
M_{A}	Alfvén Mach number	8.3.16	291
M_{S}	sonic Mach number	8.3.17	292
$\mathbf{M}$	magnetization (magnetic moment per unit volume)	3.1.10	25
n	index of refraction	4.1.28	95
n	number of roots of $Đ(p) = 0$	9.5.12	360
n_{A}	Alfvén index of refraction	4.4.28	112
n_{A}	electron density of ion acoustic wave	11.2.47	457
n_{L}	index of refraction for left-hand polarized mode	4.4.43	120
n_{n}	number density of neutral gas	2.5.1	12
n_{R}	index of refraction for right-hand polarized mode	4.4.42	120
n_s	number density for the sth species	2.1.2	5
$\mathbf{n}$	index of refraction vector	4.2.9	98
$\hat{\mathbf{n}}$	unit normal vector	8.1.6	283
N	total number of particles in phase space	5.1.2	149
N_{c}	average number of collisions per unit path length	12.2.1	482
N_{D}	number of electrons per Debye cube	2.5.5	14
N_ℓ	line number density	7.1.40	230
N_ℓ	number of turns per unit length	3.3.10	35
N_{w}	winding number	9.5.10	360
p	complex frequency ($p = \gamma - \mathrm{i}\omega$)	9.2.1	329
p	momentum	3.8.4	52
p_i	conjugate momentum coordinate	3.9.3	65
P	dielectric tensor element	4.4.9	107
$\mathbf{P}$	polarization vector	3.6.8	48
$P\int$	principal value integral	9.2.39	338
$P(x)$	integrating factor	10.2.2	382
P_{ij}	pressure tensor	5.1.14	151
$P_s(\mathbf{v}, \Delta\mathbf{v})$	probability distribution function	12.3.1	484
$\overleftrightarrow{\mathbf{P}}_s$	pressure tensor	5.1.7	149

Symbol	Definition	Equation	Page
q	charge	2.2.2	6
q	position coordinate	3.8.1	52
q_i	conjugate position coordinate	3.9.1	64
q_0	safety factor	3.9.40	74
$q(\rho)$	q-factor in screw pinch	7.1.50	232
Q	test charge	2.2.9	7
$Q(x)$	source term in a linear differential equation	10.2.2	382
$\mathbf{Q}_s$	kinetic energy flux	5.4.31	162
r	radius (in spherical coordinates)	2.2.7	7
r	shock strength parameter	8.3.15	291
R	dielectric tensor element	4.4.11	107
$\mathrm{Res}[g(z), z_0]$	residue of $g(z)$ evaluated at z_0	9.2.9	330
R_m	magnetic Reynolds number	6.3.5	197
$\mathbf{R}_\mathrm{C}$	radius of curvature vector	3.3.8	33
$\hat{\mathbf{R}}_\mathrm{C}$	unit vector in direction of instantaneous radius of curvature	3.3.9	33
s	species	2.1.1	4
S	dielectric tensor element	4.4.8	107
S	Lundquist number	7.6.5	271
$\mathbf{S}$	Poynting flux ($\mathbf{S} = (1/\mu_0)\,\mathbf{E} \times \mathbf{B}$)	6.4.14	205
t	time	2.3.2	10
T_{rs}	magnetic pressure tensor element	6.2.3	194
T_s	temperature for the sth species	2.1.1	4
$T_\parallel$	parallel temperature	5.1.16	151
$T_\perp$	perpendicular temperature	5.1.16	151
$\overset{\leftrightarrow}{\mathbf{T}}$	magnetic pressure tensor	6.2.4	195
$\mathbf{U}$	fluid velocity	6.1.2	187
$\mathbf{U}_s$	average velocity of sth species	5.1.5	149
υ	velocity magnitude	2.1.1	4
υ_x	x component of velocity	3.4.4	39
υ_y	y component of velocity	3.4.4	39
υ_z	z component of velocity	3.4.4	39
$\upsilon_\parallel$	velocity component parallel to the magnetic field	3.2.1	25
$\upsilon_\perp$	velocity perpendicular to the magnetic field	3.1.3	23
$\upsilon_{\parallel\mathrm{Res}}$	cyclotron resonance velocity	10.2.35	394
$\mathbf{v}$	velocity	2.1.1	4
$\mathbf{v}_\mathrm{C}$	curvature drift velocity	3.3.8	33

continued

Symbol	Definition	Equation	Page
$\mathbf{v}_{\mathrm{E}}$	$\mathbf{E}\times\mathbf{B}$ drift velocity	3.2.8	26
$\mathbf{v}_{\mathrm{F}}$	drift due to a perpendicular external force	3.2.11	30
$\mathbf{v}_{\mathrm{g}}$	group velocity	4.1.25	93
$\mathbf{v}_{\mathrm{G}}$	gradient drift velocity	3.3.7	33
$\mathbf{v}_{\mathrm{p}}$	phase velocity	4.1.9	89
$\mathbf{v}_{\mathrm{P}}$	polarization drift	3.6.6	48
V	volume	5.1.4	149
V	equilibrium potential of the surface	2.2.12	9
V	velocity space	5.4.1	157
V_0	velocity in ring distribution	10.2.32	392
V_j	beam velocity	9.1.26	323
V_{A}	Alfvén speed	6.5.14	208
V_{S}	speed of sound	6.5.5	207
$w_{\parallel}$	parallel kinetic energy of a particle	3.3.9	33
$w_{\perp}$	perpendicular kinetic energy of a particle	3.1.9	24
w	island width	3.10.23	80
W	total kinetic energy density	5.4.35	163
W	potential energy density	6.4.21	206
W_s	kinetic energy density	5.4.29	162
$W_{\perp}$	perpendicular kinetic energy density	3.3.11	37
$\mathbf{W}_s$	diffusion velocity for species s	6.1.7	188
x	x component of a position vector	3.4.3	39
x	real part of ζ	9.3.7	347
X	dimensionless parameter in CMA diagram	4.4.58	132
X_{WKB}	WKB solution of the time-independent harmonic oscillator equation	3.8.7	53
y	y component of a position vector	3.4.3	39
y	complex part of ζ	9.3.7	347
Y	dimensionless parameter in CMA diagram	4.4.58	132
z	z component of a position vector	3.4.1	38
z	normalized z component of velocity	9.3.2	346
$Z(\zeta)$	plasma dispersion function	9.3.5	347
$\overleftrightarrow{\mathbf{1}}$	unit tensor	4.2.6	97
α	pitch angle	3.4.21	43
α	group velocity angle	4.1.29	95
α	scalar function used in force-free equilibria	7.1.9	224
α_0	loss-cone angle	3.4.23	43
α_s	symbol used for $(k_{\parallel}v_{\parallel}-\omega)/\omega_{\mathrm{c}s}$	10.1.14	381

Symbol	Definition	Equation	Page
β	plasma energy/magnetic energy	6.2.10	196
β_{c}	thermal cyclotron radius times $k_\perp$	10.2.31	389
β_s	symbol used for $k_\perp v_\perp/\omega_{\mathrm{c}s}$	10.1.14	381
γ	power law index, also called the polytrope index	5.4.43	165
γ	ratio of heat capacity at constant pressure to heat capacity at constant volume	8.3.56	303
γ	growth rate	9.1.31	325
γ_0	damping constant for the harmonic oscillator	9.2.5	330
Γ_{ei}	defined parameter in the Fokker–Planck collision operator	12.4.5	491
Γ_n	defined function	10.2.30	389
$\Gamma_{ss'}$	constant factor in dynamical friction vector	12.3.17	487
$\langle\Delta p\rangle$	uncertainty in momentum	2.7.1	17
Δ_b	characteristic width	3.9.41	74
Δs	change in specific entropy	8.3.56	303
Δx	small displacement in the x direction	2.3.1	10
$\langle\Delta x\rangle$	uncertainty in position	2.7.1	17
δ	Dirac delta function	5.1.12	150
$\delta_{\mathrm{c}}f/\delta t$	collision operator	5.2.14	155
δ_{rs}	Kronecker delta	6.2.3	194
ϵ	depth of potential well	11.1.54	438
ϵ	small quantity	9.2.39	338
ϵ_0	permittivity of free space	2.2.1	6
ϵ_t	aspect ratio of tokamak	3.9.35	73
ζ	dimensionless complex frequency	9.3.2	346
η	resistivity	6.1.19	190
θ	polar angle (in spherical coordinates)	3.9.24	70
θ	wave normal angle	4.1.28	95
θ_{Res}	resonance cone angle	4.4.53	127
κ	Boltzmann's constant	2.1.1	4
κ	epicyclic frequence	7.4.7	262
λ	wavelength ($2\pi/k$)	4.1.2	87
λ_{D}	Debye length	2.2.8	7
Λ	plasma parameter	12.2.9	483
μ	magnetic moment	3.1.6	24
μ_0	permeability of free space	3.3.10	35
$\langle\mu_s\rangle$	average magnetic moment for sth species	3.1.10	25
$\bar{\mu}_s$	mobility coefficient for sth species	5.7.3	178
ν_{ei}	electron–ion collision frequency	2.5.5	14

continued

Symbol	Definition	Equation	Page
ν_{ns}	neutral collision frequency	2.5.1	12
ν_s	effective collision frequency for *s*th species	5.6.1	174
ξ	internal energy per unit volume	6.4.15	205
$\boldsymbol{\xi}$	fluid displacement vector	7.3.12	247
ρ	radial distance from the symmetry axis in cylindrical coordinates	3.3.10	35
ρ_c	cyclotron radius	3.1.4	23
ρ_p	polarization charge	4.2.1	96
ρ_q	charge density	2.2.1	6
ρ_{qs}	charge density of *s*th species	11.4.7	472
ρ_r	real charge	4.2.1	96
σ	conductivity	5.6.7	175
σ_n	neutral particle collision cross section	2.5.1	12
σ_s	conductivity of *s*th species	5.6.8	175
σ_C	Coulomb collision cross section	2.5.2	13
σ_H	Hall conductivity	5.6.12	176
σ_T	total scattering cross section	2.5.4	13
$\sigma_{\parallel}$	parallel conductivity	5.6.12	176
$\sigma_{\perp}$	perpendicular conductivity	5.6.12	176
$\overleftrightarrow{\sigma}$	conductivity tensor	4.2.3	97
τ_A	Alfvén time-scale	7.6.3	271
τ_R	resistive diffusion time-scale	7.6.4	271
ϕ	azimuthal angle (in cylindrical coordinates)	3.4.10	40
ϕ	eikonal	4.5.2	138
Φ	electrostatic potential	2.2.1	6
Φ_B	magnetic flux	3.5.8	45
χ	scattering angle relative to incident beam	2.5.2	13
ψ	effective potential	3.9.16	67
ψ	Faraday polarization angle	4.4.41	119
ψ	flux function	7.1.56	234
$\Psi_h(r, \theta_h)$	helical flux function	3.10.19	80
ω	frequency (radian s^{-1})	3.8.2	52
ω_b	bounce frequency	9.2.52	343
ω_c	cyclotron frequency	2.4.1	11
ω_{cs}	cyclotron frequency of *s*th species	2.4.3	12
ω_p	plasma frequency	2.3.8	11
ω_{pe}	electron plasma frequency	2.3.4	10
ω_{pi}	ion plasma frequency	2.3.8	11
ω_{ps}	plasma frequency of *s*th species	2.3.6	11

Symbol	Definition	Equation	Page
ω_{x}	crossover frequency	4.4.35	116
ω_{LH}	lower hybrid frequency	4.4.50	124
ω_{UH}	upper hybrid frequency	4.4.49	123
$\omega_{L=0}$	$L = 0$ cutoff frequency	4.4.30	113
$\omega_{R=0}$	$R = 0$ cutoff frequency	4.4.27	112

Appendix B: Useful Trigonometric Identities

$$\cos A \cos B = (1/2)[\cos(A - B) + \cos(A + B)]$$
$$\sin A \sin B = (1/2)[\cos(A - B) - \cos(A + B)]$$
$$\sin A \cos B = (1/2)[\sin(A - B) + \sin(A + B)]$$

$$\sin^2 A = \frac{1}{2}[1 - \cos(2A)]$$
$$\cos^2 A = \frac{1}{2}[1 + \cos(2A)]$$

$$\sin A + \sin B = 2\cos\frac{1}{2}(A - B)\sin\frac{1}{2}(A + B)$$
$$\sin A - \sin B = 2\sin\frac{1}{2}(A - B)\cos\frac{1}{2}(A + B)$$
$$\cos A + \cos B = 2\cos\frac{1}{2}(A - B)\cos\frac{1}{2}(A + B)$$
$$\cos A - \cos B = 2\sin\frac{1}{2}(B - A)\sin\frac{1}{2}(A + B)$$

$$\cos 2A = 1 - 2\sin^2 A$$
$$\sin 2A = 2\sin A \cos A$$

Appendix C: Vector Differential Operators

Rectangular coordinates (x, y, z):

$$\nabla\Phi = \hat{\mathbf{x}}\,\frac{\partial\Phi}{\partial x} + \hat{\mathbf{y}}\,\frac{\partial\Phi}{\partial y} + \hat{\mathbf{z}}\,\frac{\partial\Phi}{\partial z},$$

$$\nabla\cdot\mathbf{F} = \frac{\partial F_x}{\partial x} + \frac{\partial F_y}{\partial y} + \frac{\partial F_z}{\partial z},$$

$$\nabla\times\mathbf{F} = \hat{\mathbf{x}}\left(\frac{\partial F_z}{\partial y} - \frac{\partial F_y}{\partial z}\right) + \hat{\mathbf{y}}\left(\frac{\partial F_x}{\partial z} - \frac{\partial F_z}{\partial x}\right) + \hat{\mathbf{z}}\left(\frac{\partial F_y}{\partial x} - \frac{\partial F_x}{\partial y}\right),$$

$$\nabla^2\Phi = \frac{\partial^2\Phi}{\partial x^2} + \frac{\partial^2\Phi}{\partial y^2} + \frac{\partial^2\Phi}{\partial z^2}.$$

Cylindrical coordinates (ρ, ϕ, z):

$$\nabla\Phi = \hat{\boldsymbol{\rho}}\frac{\partial\Phi}{\partial\rho} + \hat{\boldsymbol{\phi}}\frac{1}{\rho}\frac{\partial\Phi}{\partial\phi} + \hat{\mathbf{z}}\frac{\partial\Phi}{\partial z},$$

$$\nabla\cdot\mathbf{F} = \frac{1}{\rho}\frac{\partial}{\partial\rho}(\rho\,F_\rho) + \frac{1}{\rho}\frac{\partial F_\phi}{\partial\phi} + \frac{\partial F_z}{\partial z},$$

$$\nabla\times\mathbf{F} = \hat{\boldsymbol{\rho}}\left(\frac{1}{\rho}\,\frac{\partial F_z}{\partial\phi} - \frac{\partial F_\phi}{\partial z}\right) + \hat{\boldsymbol{\phi}}\left(\frac{\partial F_\rho}{\partial z} - \frac{\partial F_z}{\partial\rho}\right) + \hat{\mathbf{z}}\frac{1}{\rho}\left(\frac{\partial}{\partial\rho}(\rho F_\phi) - \frac{\partial F_\rho}{\partial\phi}\right),$$

$$\nabla^2\Phi = \frac{\partial^2\Phi}{\partial\rho^2} + \frac{1}{\rho}\,\frac{\partial\Phi}{\partial\rho} + \frac{1}{\rho^2}\,\frac{\partial^2\Phi}{\partial\phi^2} + \frac{\partial^2\Phi}{\partial z^2}.$$

Spherical coordinates (r,θ,ϕ):

$$\nabla\Phi = \hat{\mathbf{r}}\,\frac{\partial\Phi}{\partial r} + \hat{\boldsymbol{\theta}}\,\frac{1}{r}\,\frac{\partial\Phi}{\partial\theta} + \hat{\boldsymbol{\phi}}\,\frac{1}{r\sin\theta}\,\frac{\partial\Phi}{\partial\phi},$$

$$\nabla\cdot\mathbf{F} = \frac{1}{r^2}\,\frac{\partial}{\partial r}(r^2F_r) + \frac{1}{r\sin\theta}\,\frac{\partial}{\partial\theta}(F_\theta\sin\theta) + \frac{1}{r\sin\theta}\,\frac{\partial F_\phi}{\partial\phi},$$

$$\begin{aligned}\nabla\times\mathbf{F} = {} & \hat{\mathbf{r}}\,\frac{1}{r\sin\theta}\left[\frac{\partial}{\partial\theta}(F_\phi\sin\theta) - \frac{\partial F_\theta}{\partial\phi}\right] \\ & + \hat{\boldsymbol{\theta}}\,\frac{1}{r}\left[\frac{1}{\sin\theta}\,\frac{\partial F_r}{\partial\phi} - \frac{\partial(rF_\phi)}{\partial r}\right] + \hat{\boldsymbol{\phi}}\,\frac{1}{r}\left[\frac{\partial(rF_\theta)}{\partial r} - \frac{\partial F_r}{\partial\theta}\right],\end{aligned}$$

$$\nabla^2\Phi = \frac{1}{r}\,\frac{\partial^2}{\partial r^2}(r\Phi) + \frac{1}{r^2\sin\theta}\,\frac{\partial}{\partial\theta}\left(\sin\theta\frac{\partial\Phi}{\partial\theta}\right) + \frac{1}{r^2\sin^2\theta}\,\frac{\partial^2\Phi}{\partial\phi^2}.$$

Appendix D: Vector Calculus Identities

$$\boldsymbol{\nabla} \cdot \boldsymbol{\nabla}\Phi = \nabla^2\Phi$$

$$\boldsymbol{\nabla} \cdot \boldsymbol{\nabla} \times \mathbf{F} = 0$$

$$\boldsymbol{\nabla} \times \boldsymbol{\nabla}\Phi = 0$$

$$\boldsymbol{\nabla} \times (\boldsymbol{\nabla} \times \mathbf{F}) = \boldsymbol{\nabla}(\boldsymbol{\nabla} \cdot \mathbf{F}) - \nabla^2\mathbf{F}$$

$$\boldsymbol{\nabla}(\Phi\Psi) = (\boldsymbol{\nabla}\Phi)\Psi + \Phi\boldsymbol{\nabla}\Psi$$

$$\boldsymbol{\nabla}(\mathbf{F} \cdot \mathbf{G}) = (\mathbf{F} \cdot \boldsymbol{\nabla})\mathbf{G} + \mathbf{F} \times (\boldsymbol{\nabla} \times \mathbf{G}) + (\mathbf{G} \cdot \boldsymbol{\nabla})\mathbf{F} + \mathbf{G} \times (\boldsymbol{\nabla} \times \mathbf{F})$$

$$\boldsymbol{\nabla} \cdot (\Phi\mathbf{F}) = (\boldsymbol{\nabla}\Phi) \cdot \mathbf{F} + \Phi\boldsymbol{\nabla} \cdot \mathbf{F}$$

$$\boldsymbol{\nabla} \cdot (\mathbf{F} \times \mathbf{G}) = (\boldsymbol{\nabla} \times \mathbf{F}) \cdot \mathbf{G} - (\boldsymbol{\nabla} \times \mathbf{G}) \cdot \mathbf{F}$$

$$\boldsymbol{\nabla} \times (\Phi\mathbf{F}) = (\boldsymbol{\nabla}\Phi) \times \mathbf{F} + \Phi\boldsymbol{\nabla} \times \mathbf{F}$$

$$\boldsymbol{\nabla} \times (\mathbf{F} \times \mathbf{G}) = (\boldsymbol{\nabla} \cdot \mathbf{G})\mathbf{F} - (\boldsymbol{\nabla} \cdot \mathbf{F})\mathbf{G} + (\mathbf{G} \cdot \boldsymbol{\nabla})\mathbf{F} - (\mathbf{F} \cdot \boldsymbol{\nabla})\mathbf{G}$$

$$\int_S \mathbf{F} \cdot \mathrm{d}\mathbf{A} = \int_V \boldsymbol{\nabla} \cdot \mathbf{F}\, \mathrm{d}^3 x$$

$$\int_C \mathbf{F} \cdot \mathrm{d}\boldsymbol{\ell} = \int_S (\boldsymbol{\nabla} \times \mathbf{F}) \cdot \mathrm{d}\mathbf{A}$$

Index

Printed in the United States
By Bookmasters